Empirical Social Research

Michael Häder

Empirical Social Research

An Introduction

 Springer

Michael Häder
Philosophische Fakultät – Institut für Soziologie
Technische Universität Dresden
Dresden, Germany

ISBN 978-3-658-37906-3   ISBN 978-3-658-37907-0   (eBook)
https://doi.org/10.1007/978-3-658-37907-0

This Springer imprint is published by the registered company Springer Fachmedien Wiesbaden GmbH, part of
Springer Nature.
The registered company address is: Abraham-Lincoln-Str. 46, 65189 Wiesbaden, Germany

Contents

The Structure of This Book: An Introduction

1

Abstract

The methods of empirical social research are highly specialized and capable of covering a wide range of tasks. They range from research logic and selection procedures to statistical data analysis. This volume contains the basic knowledge of all relevant methods of empirical social research and lists further publications. Extensive special literature is available on numerous individual problems, to which reference is made at the appropriate points.

In addition, this volume uses various examples to demonstrate how this basic knowledge can be used in researching social facts. After all, a text can convey information on methods and techniques. However, in order to achieve skills in the use of the methods in the reader, they must be applied and practiced by the reader. The volume aims to support this, it has the following structure:

First, the *second chapter* attempts to sensitize the reader to the use of empirical methods. A number of experiments have shown that even seemingly insignificant details of the methods used for researching social facts have a decisive influence on the results of the research. In other words, a research finding depends not only on the facts being researched, but also on the tools used to research them. Thus, it becomes relevant for every user to deal with the prerequisites and the possibilities of the corresponding methods.

The *third chapter* is then initially devoted to the theory of methods. Similar to a cookbook, in which something is also said theoretically in the introduction about the art of correct seasoning and the careful handling of the most important ingredients, aspects of the logic of research should and must also be discussed here. Social problems, which are expressed with the help of certain terms, undergo an initial treatment in the form of hypotheses. They are then further condensed into theories and into laws on the basis of empirical findings obtained. All these steps are to be considered in more detail.

This is followed in the *fourth chapter* by the description of the research and investigation planning. Assuming an interest in working on a particular problem, a suitable design is now needed to bring it closer to a solution. As manifold as the problems are, which are dealt with by empirical social research, as different are its methods and instruments, which it provides for the solution of such problems. In order to be able to deal with a problem using a methodology that is as tailor-made as possible, it is necessary, for example, to develop suitable measuring instruments and to determine their performance. However, it must also be decided, for example, how many investigations are actually required in order to arrive at a satisfactory solution.

Empirical studies usually want to make statements about a whole range of elements with their results. Election forecasts, which are made on the basis of the study of a relatively small number of people in advance of the outcome of the election, have become particularly well known to the public. For studies of this kind, suitable selection strategies for the persons to be interviewed must be found and implemented in practice in order to be able to make reliable statements later and – as has unfortunately happened – to prevent false diagnoses. The *fifth chapter* is devoted to the problems of selection procedures.

Three main strategies are then used in the actual collection of the data: a specific way of interviewing, a form of observation or/and a variant of content analysis. It is both conceivable that one of these methods is used exclusively. However, it is also possible that the survey design provides for a mix of different techniques. These three survey methods are presented in detail in their respective diversity in the *sixth chapter.*

Often it is not sufficient to use only one single method exclusively for the solution of a social science problem. In complex research arrangements, therefore, it is not uncommon for various methods and techniques, with their respective advantages and limitations, to be coordinated and applied as precisely as possible. Examples of such approaches are described in the *seventh chapter.*

Pretests now form a separate group of methods. In pretests, the instruments and strategies intended for an empirical survey must first be tested for their performance before they can be used on a larger scale. Empirical social research has sophisticated procedures for pre-evaluating the instruments intended for a survey. Such pretest designs are presented in *eighth chapter.*

The *ninth chapter* is devoted to the phase that follows data collection. In order to meet scientific criteria, the studies must be prepared in a comprehensible manner, and possible errors must be found and corrected. In quantitative research approaches, this is followed by the mathematical-statistical data evaluation. Here again, an attempt is made to show examples of possible strategies for this phase of a social science investigation.

Finally, the documentation of the study forms the conclusion of an empirical social research project and also of these considerations in the textbook. The *tenth chapter* describes what information is required to describe the findings of a survey in a way that is comprehensible to third parties. Replication studies have their own significance in the arsenal of methods. A prerequisite for such studies is the presentation of all essential steps of a research. After all, science is characterised not least by the fact that it discloses its procedures.

I would like to take this opportunity to thank Ms. Grit Jüttler (M.A.) and Ms. Laura Menze for their persistent and meticulous support in completing the manuscript, as well as my wife, Dr. Sabine Häder, for the many valuable tips and advice, for the useful conversations, and for all other assistance during the work on this book. Meanwhile, my thanks also go to a number of attentive readers who pointed out inconsistencies and problems in the previous editions.

The Importance of Methodological Knowledge for Understanding Empirical Data

2

Abstract

Many people probably like to consume the results of (pseudo-scientific) empirical survey studies in everyday life for entertainment purposes. For example, one could participate in a survey on space tourism on the internet pages of DIE ZEIT and answer the following question: "With SpaceShipOne, the first privately financed space plane has flown into orbit. Now spaceports are planned in the USA or Australia. Off-the-wall tourism plans or realistic market niche?" After casting one's own vote to answer this question, one learns that 27% of visitors to this website who are willing to answer think space tourism is pollution and a waste of money. This should be banned or restricted. Fifty-four percent of the 268 people who have voted so far think: space tourism is a holiday option for wealthy adventurers and not affordable for ordinary people. Finally, 19% of the participants (including the author) opted for the default, "A trip to space is dreamlike and also results in new opportunities for the tourism industry." Slightly fewer people are also professionally involved – and this should be taken considerably more seriously at this point – with results from empirical studies. For the latter group of people in particular, it is important to know exactly how such data has come about and what it is capable of saying. Four seemingly trivial examples will show what a data consumer has to reckon with.

M. Häder, *Empirical Social Research*, https://doi.org/10.1007/978-3-658-37907-0_2

2.1   On the Need to Reflect on Research Methods

2.1.1   First Example: The Evaluation of the CDU

In the context of the first example, a survey asks a number of people a question about their assessment of the CDU. The question text reads: "All in all, what do you think of the CDU in general? Please answer using this scale." The scale consists of an eleven-step ladder, the lower end of which is labeled "nothing at all" and the upper end of which is labeled "very much." In the methodological experiment (compare Schwarz and Bless 1992; Porst 2000), this exact question was asked of three different test groups (A, B, and C). The results, however, differed considerably – despite identical question wording – as Table 2.1 shows.

First of all, it is obvious to attribute the cause of these quite different findings to the respective persons interviewed. Thus, group A may be composed primarily of rejects and group C primarily of supporters of the CDU. However, this assumption is not correct. Numerous other assumptions can also be ruled out, for example that the interviewers had a particular interest in the results of the survey and therefore deliberately exerted a suggestive influence on the respondents, or that the surveys took place at different times. The correct explanation for the different findings can be seen in Table 2.2. According to this, only the order or the wording of the questions was slightly changed in the versions A, B and C. The questions were then asked in a different order.

Further below, we will examine how such a phenomenon can be explained theoretically or how it came about. It should first be noted here that the three sympathy values towards the CDU reported in this test say little about the affection shown towards this party. Thus, it has been shown that such data cannot be readily interpreted. Thus, a prerequisite to properly understand survey results is to study the methodology used in this survey. Thus there are good reasons for every consumer of data not only to look at the figures reported to him, but also to make his own enquiries, for example, about the sequence of questions when he reads or hears about the sympathy ratings of a party or a politician.

2.1.2   Second Example: Television Viewing Time

In the second example (compare Schwarz et al. 1985), a survey asks about the duration of television consumption. The text of the question was: "How many hours do you watch television on a normal working day? Please use the following schedule for your answer." Again, two versions (A and B) were used for this methodological test. Table 2.3 shows the

Table 2.1 Results of the first method test, mean values on an eleven-point scale

Version	A	B	C
Mean value	3.4	5.2	6.5

Table 2.2 Question asked in the questionnaire *before* the question on the evaluation of the CDU in each case

Version	A	B	C
Question text of the preliminary question	Do you happen to know what office Richard von Weizsäcker holds that places him outside the party establishment?	Without preliminary question	Do you happen to know which party Richard von Weizsäcker has belonged to for more than 20 years?

Table 2.3 Response specifications provided for answering the question on television consumption and distributions found (in percent)

Variant A		Variant B	
Up to ½ h	7.4		
½ to 1 h	17.7		
1 to 1½ h	26.5		
1½ to 2 h	14.7		
2 to 2½ h	17.7	Until 2½ h	62.5
More than 2½ h	16.2	2½ to 3 h	23.4
		3 to 3½ h	7.8
		3½ to 4 h	4.7
		4 to 4½ h	1.6
		More than 4½ h	0.0
$N = 65$		$N = 68$	

two different answer schemes presented to the target persons and the empirical findings obtained.

If we now compare the results found in this way, we see once again that they differ greatly from one another – again depending on the method used. According to the results of version A, 16.2% of the respondents watch more than 2½ hours of television, whereas in version B the figure is 37.5%. In this example, effects resulting, for example, from a different composition of the study population can also be excluded.

Here, too, we will initially refrain from asking on the basis of which mechanisms more than twice as many long-term TV viewers were identified in one group as in the other. At this point it should be stated that one should not only ask – as in the first example – about the order in which the questions were asked, but also about the exact answer specifications that were presented to the respondents in order to be able to interpret the results of a survey. Thus, the reported television consumption not only has something to do with the amount of time spent in front of the television set, but also results not least from how one designs the answer schemes for a corresponding question.

2.1.3   Third Example: The Ladders

In the third example, the answer scales presented to the respondents again play an important role. In this experiment (compare Allensbach Institute for Public Opinion Research, IfD study 5007 of July 1988, quoted from Petersen 2002, p. 205) the question was: "How successful have you been in your life so far? Please say it according to this ladder here! It goes like this: 0/−5 means not successful at all and 10/+5 means you have been extraordinarily successful so far. Which number will you use?" Two different ladders have been presented to the target subjects for the experiment on a template sheet. One half of the respondents received version A ($n = 480$) and the other half ($n = 552$) received version B. The scales had the following appearance (compare Fig. 2.1).

In the group presented with version A, the mean value of the responses was 6.4 and in the other group (version B) 7.3. In other words, 34% of the respondents opted for a rating between +4 and 0 in version B and only 13% opted for the (seemingly) analogous specifications −1 to −5.

This third example shows once again how the results of a survey depend on the instruments used. Without a more precise knowledge of the template sheet used with the answer

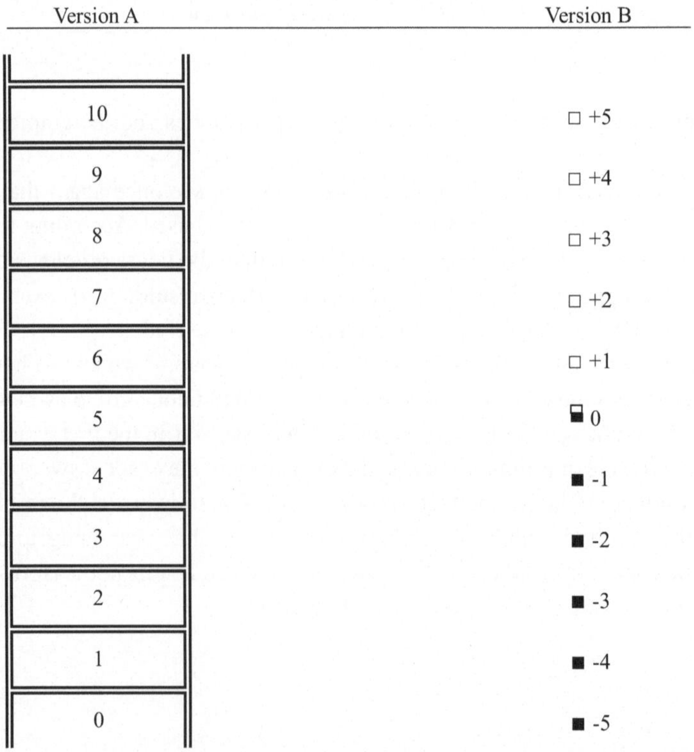

Version A	Version B
10	□ +5
9	□ +4
8	□ +3
7	□ +2
6	□ +1
5	■ 0
4	■ -1
3	■ -2
2	■ -3
1	■ -4
0	■ -5

Fig. 2.1 The response ladders used for the question about one's own success in life

specifications, no appropriate interpretation of the empirical findings can be made. Even the statement that the respondents had rated themselves on an eleven-point scale with regard to their perceived success in life is too general and still not sufficient to be able to adequately understand the results of this survey. Apparently, not only the degree of subjectively perceived success in life determines the answer here, but also the scale presented to the target persons.

This experiment will also be discussed in detail later (see Sect. 6.1.2).

2.1.4 Fourth Example: Prohibited and Not Allowed

Finally, a fourth example: In an American study, two variants used an almost identical question (compare Rugg 1941). In variant A, the question was: "Do you believe that the USA should prohibit public attacks on democracy?" The question text, only slightly changed in Variant B, read, "Do you believe that the U.S. should not permit public attacks on democracy?" The two variants thus differ only in their use of the terms "prohibited" and "not permitted," respectively. If one assumes that these are not infrequently used as synonyms in colloquial speech, then identical answer distributions should also be expected. However, Table 2.4 shows that this is not the case.

While in variant A 54% of the respondents are of the opinion that attacks on democracy should be prohibited, in variant B the proportion of people who think that such attacks on democracy should not be permitted rises to as much as 75%.

This fourth example shows once again the high degree of sensitivity with which respondents react to seemingly simple and insignificant changes in an empirical survey standard. Replacing the term "forbidden" with the term "not allowed" here again leads to clearly different marginal distributions.

In the course of this paper, such surprising findings for the untrained social researcher will be described again and again. These concern not only the design of questionnaires for surveys, but the entire social science survey process. For example, a survey institute wanted to predict the outcome of an election. It interviewed 2,400,000 people for this purpose. Another institute pursued the same goal and interviewed a much smaller number of people. Here, too, the result was surprising: the smaller sample was able to make a significantly better (and correct) statement than the larger one.

Table 2.4 Marginal distributions in response to the question about adequate reactions to public attacks on democracy (data in percent, $n = 1032$)

	Variant A	Variant B
Yes	54	75
No	45	25

From these examples, an attempt will now be made to draw a conclusion. In view of the seemingly uncontrollable findings described above, this conclusion could be that social research should not be conducted empirically, i.e. that the use of empirically obtained data should generally be avoided. After all, the results of surveys appear – according to these examples – to be manipulable to too great an extent to be trusted. However, it is rather obvious to prefer a different conclusion at this point. This must first state the complicated and fragile character of the methods of empirical social research and then, against this background, advocate a thorough study of the functioning of the individual social science instruments. Thus, the examples given were intended to raise sensitivity and awareness of the problem. The use of empirical methods is not justifiable without thorough thought about how they work. On the other hand, the desire for speed, intuition or a certain superficiality are – as these few examples have also already shown – not good advisors in the creation and use of empirical methods. The design of social science surveys thus requires professional skill.

The question whether such logical reflections are not more likely to "cause the somnambulistically safe centipede of our thinking to stumble by unnecessary interposed questions?" (Bayer 1999), can be answered unequivocally in the negative, at least as far as our subject is concerned. Obviously, it is an illusion to assume that social research can be conducted with such somnambulistic certainty and that method-critical questioning of the findings is counterproductive for this reason.

Reference should be made to a further aspect. It is no coincidence that the examples just cited come from tests that were carried out some time ago. These led relatively early on to the initiation of targeted methodological research, especially on the functioning of surveys. This in turn led to the accumulation of a wide range of experience in this field.

A look at various scientific disciplines also shows quite quickly that there is an immense need for empirical knowledge about human behaviour and about social phenomena in numerous fields of knowledge. Here, only a few may be referred to:

1. For example, in the study of social mobility, class structures, occupational prestige and the analysis of changing values, sociology is absolutely dependent on data that are able to provide concrete information about the underlying social relationships. Only with the help of detailed information can theories be derived and tested. Appropriate empirical methods are needed to obtain this information.
2. In political science, no election research, no research on left- or right-wing extremism, on political disenchantment, or even on trust in the institutions of a political system is conceivable without recourse to data and thus to the methods of empirical social research.
3. Especially economics, and here national economics, have contributed to innovations in the field of empirical methods in the past. Thus, knowledge about the development of the cost of living, the business cycle, income distribution and wealth, poverty, territorial mobility and unemployment is unthinkable without the corresponding data.

4. The need of business administration for empirical facts is also manifold. Information on job satisfaction and on turnover intentions within a workforce are worth mentioning. Employee surveys are also being used more and more frequently in companies in order to be able to better control internal processes. Empirical data has gained particular importance in market research.

5. Empirical methods are also indispensable in psychology and social psychology. One only has to think of intelligence tests and the entire personality diagnostics that follow on from them. Clinical psychology, too, cannot perform its tasks without empirical methods.

6. At the latest against the background of the results of the PISA studies on the learning successes of pupils in various countries, which attracted a great deal of attention after their publication, pedagogy and educational science must also admit that topics such as socialisation or the effectiveness of various teaching methods can only be worked out with the aid of appropriate empirical methods and sophisticated sampling designs.

7. In medicine, empirical studies on the health behaviour of the population, on behavioural patterns in connection with the fight against AIDS, on the determinants of dietary behaviour, on the connection between lifestyle and disease, on the motives for or against a visit to the doctor and so on are part of everyday life.

8. In jurisprudence, victim and offender profiles are drawn up by means of criminological studies. Surveys on self-defence legislation (cf. Sect. 4.1) reveal the agreement or rather the disagreement between the jurisprudence practised and existing public opinion.

9. The science of history analyses the contents of historical texts as well as numerous other sources and also uses empirical methods for this purpose.

10. In geography, studies are conducted on urban development, spatial planning, and migration movements that would be inconceivable without the tools of empirical social research.

11. In demography, issues such as the desire to have children and preferred partnership lifestyles play a central role. In turn, the methods of empirical social research, such as population surveys, provide access to such facts.

12. Theology not only approaches the problem of distanced churchiness with empirical means (compare Kretzschmar 2001), but also the development of structures for community education work is supported, for example, with the help of empirically collected information (compare Steinhäuser 2002). Practical theology makes use of empirical studies on conflicts at the middle church leadership level in the Protestant Church in Germany (compare Häder and Hermelink 2017).

13. In linguistics, conversation analysis is used to investigate aspects of our everyday language, such as the nature of argumentation and the verbal expression of conflicts (cf. Deppermann 1999).

14. Last but not least, the technical sciences are dependent on the methods of the empirical social sciences, for example in the context of acceptance studies of modern technologies or products (cf. Rammert 1994).

This list could be continued. All of the disciplines mentioned have in common that purely theoretical thinking – knowledge produced exclusively at a desk, as it were – is not sufficient to solve the problems at hand in each case. A profound knowledge of the functioning of the methods of empirical social research is necessary both for the various consumers of empirical data and for the producers of such data. Intuitive work, on the other hand, merely leads to assumptions that do not adequately meet the quality requirements.

2.2   Some Basic Concepts

Up to now, terms such as empirical social research, methods and also the concept of theory have been used as a matter of course. In the following, these will be briefly explained. Later, the concrete rules for defining terms will be discussed.

2.2.1   Empirical Social Research

Empirical social research is understood as a set of methods, techniques and instruments for the scientifically correct conduct of studies of human behaviour and other social phenomena. As shown in the previous section, Empirical Social Research is conducted in the context of a whole range of different fields of knowledge. Thus, Empirical Social Research is a cross-cutting discipline. This quickly becomes apparent when one imagines, for example, that certain basic rules according to which a content analysis is conducted are the same for all objects of study, regardless of whether we are dealing with medieval church books, party programs from the Weimar Republic, or the EBay offers on the Internet of the twenty-first century.

The common concern of empirical social research is the collection of knowledge about social reality.

2.2.2   Methods

Methods are a component of empirical social research. Methods represent systems of instructions and rules in order to be able to realise certain findings, or to achieve certain results, or to collect specific information. Methods thus always serve to achieve a certain goal, such as the acquisition of social information. Since – as shown in the definition of the term empirical social research – such methods are not a priori bound to very specific contents, they are mostly formal rules.

This formal character can be well illustrated, for example, by the selection methods. It is relatively unimportant whether a random selection is to be made from a population of persons in the context of an election study or from a population of end products of an industrial enterprise in the context of a quality control. If the aim is to obtain information on the nature of the study population, it is almost irrelevant what type of elements are involved. In both approaches, the same formal considerations have to be made, for example, when it comes to the probability of error which has to be assumed for the statements made on this basis.

Another thought is important: Empirical social research uses numerous methods originally derived from everyday life. This is not surprising, since it is also about collecting information about everyday life. Examples are individual communication – in social research we speak of personal oral interviews – observation techniques – for example during a holiday in a foreign environment – and content analyses – also known in everyday life from studying advertisements in a newspaper. Interviews, observations and content analyses are also the three basic methods for data collection in empirical social research (cf. Chap. 6).

It has now become established practice to distinguish between qualitative and quantitative methods (cf. Endruweit et al. 2014, p. 298 ff.). While the first aim to understand individual actions and analyse them in detail, the second type of method seeks generalisable statements and uses standardised data collection for this purpose.

2.2.3 Techniques

The concrete forms of the methods mentioned are referred to as *techniques*. Thus, there are numerous variants of content analyses, each of which focuses on certain aspects, such as the examination of a text or an image. The same applies to observations and interviews. Especially in the latter, numerous different techniques can be used. For example, in the context of personal oral interviews for market research, samples of goods or template cards are used. In other survey techniques, for example, a specific approach is taken to the object of investigation. Thus, one can speak of a postal, a web-based and a telephone survey technique.

2.2.4 Methodology

The term *methodology*, synonymously also used the terms *research logic* and *philosophy of science*, includes the meta scientific discussions about science. Within the framework of methodology, it is examined, for example, whether the chosen methods are appropriate to the presupposed purpose and why this is so. Methodology seeks suggestions for improved social science practice to promote the advancement of knowledge; it also asks, for example, about the presuppositionlessness of research.

In order to better understand the terms methods, techniques and methodology, an analogy to a craftsman's toolbox can be helpful. All the instruments in the toolbox serve a specific purpose or enable a specific result to be achieved. An inadequate handling of the tools would be of little help. Therefore, there are appropriate instructions and hints – quasi a methodology – for what the individual methods (such as an access panel, a personal oral or a postal survey) can be used for. The correct use of the various tools also needs to be learned and practiced. In some cases, there are also different techniques that can be used to achieve a certain goal.

Another analogy to the toolbox could consist of a reference to a danger of trivialisation. Since the methods originate from everyday life, the false assumption could arise that anyone could master them at first go without having read the instructions for use or having dealt with the methodology beforehand.

2.2.5   Theory

Theory is understood to be a system or a network of consistent statements for ordering knowledge about a range of facts, for explaining facts and finally for predicting them. When speaking of a system of statements, it is meant that the individual statements are logically consistent with each other. Theories exist, for example, about the different determinants of the voting behaviour of certain social groups. For example, the expected benefit resulting from a decision in favour of a certain party or the long-term commitment of a person to a certain party are taken into account. Learning theories contain statements about the fact that behaviours occur more frequently when they have been rewarded in a certain way in the past. Thus, a key difference between method and theory is that the latter do not contain instructions for action. Methods contain instructions for action to realize a certain goal, but theories do not.

Theories help to explain reality, to uncover the causes of certain phenomena, such as the decision to vote for a particular political party. Finally, theories are also expected to allow predicting future behavior. Thus, according to learning theory, if a certain behavior is rewarded, it is expected that this behavior will also occur more often in the future than a behavior that was not rewarded. Furthermore, theories are characterized by the fact that they have already been proven in practice. They thus have a certain empirical truth content, but this does not necessarily mean that they are also completely free of errors. The task of research and science is now to elaborate such theories, to test them and finally to improve them.

2.2.6   Empiricism

Empiricism refers to knowledge based on systematic experience as well as theoretical models. Empiricism is a specific form of statements describing reality. In contrast to theory, however, these have not yet been proven (sufficiently and comprehensively) in practice. The transition from empirical or experiential knowledge to theoretical knowledge is fluid. Theory and empiricism are – as will be shown later (compare Chap. 3) – in a dialectical relationship.

2.2.7 Qualitative and Quantitative Data

Data is all information that has been obtained with the help of social science methods. A distinction can be made between two types: One can describe an issue, such as satisfaction with one's life, with the help of numbers. For example, a value of one means that someone is very dissatisfied with their life and a value of seven means that someone is very satisfied with their life. However, it is also possible to obtain more detailed verbal information from a person about their satisfaction with their life, for example by asking a person to simply tell you how they are doing. In this case, too, we should then speak of data.

"The set of all characteristic measurements is called (quantitative) *data of* an investigation. If characteristics or characteristic values are described verbally, one speaks of *qualitative* data" (Bortz and Döring 2002, p. 6, emphasis as in the original).

2.2.8 Variable

Variables are certain characteristics of objects. These characteristics can, in turn, each take on certain expressions. For example, a variable would be the gender of a person. This variable can – if one follows the newer considerations – take on the characteristics male, female and diverse. First of all, it is important that all expressions of a variable are named exhaustively. This requirement is fulfilled if it is assumed that there are no other expressions besides "male", "female" and "diverse" to designate the gender of a person. Also, the expressions of a variable must be specified without overlap. This is the case if it can be precisely stated for each person whether he or she is male, female or diverse; an in-between is thus excluded.

Further, a distinction is made between dependent and independent variables. The dependent variable is the one for which an explanation is to be sought within the framework of a study. The independent variables, on the other hand, serve to provide such explanations.

Binary variables are characteristics that can only take two forms. The traditional distinction of gender into male and female was a typical binary variable. But an electoral decision can also be transformed into such a binary variable. Thus one distinguishes the voters (expression one) and the non-voters (expression zero) of a certain party.

For more information on the general terms of empirical social research, compare Reichertz (2019, p. 31 ff.).

Philosophy of Science

3

Abstract

In order to find their way in the world, people need explanations and instructions for their actions. Such instructions and explanations are provided by implicit and explicit theories. As already shown (compare Chap. 2), theories are systems of statements, with the help of which knowledge about a range of facts can be ordered, facts can be explained and predicted. Now these theories are to be further divided into explicit and implicit ones. The housewife, who has invested much love and effort in the preparation of a feast, announces insights (derived from implicit theories) when presenting her cooking skills, such as: "Dear guests, eat quite slowly, for then you will manage more of this delicious food!" And when asked, "That's what my mother used to say." So this theory has been passed down from generation to generation – after all, eating slowly is all about enjoyment, and it also represents praise for the cook when everything is eaten. Also, slow eating hardly seems to have harmed anyone so far. So this implicit theory may even be self-contradictory, it is helpful in this case anyway. Furthermore it explains certain facts resp. predicts which (wanted) consequences slow eating has. So, why not continue to follow this implicit theory and enjoy the food in peace?

3.1 The Concern of the Philosophy of Science

The conductor of an ICE train, who considers it his duty to explain to fellow passengers the delay of his train, which can no longer be overlooked after departure from Frankfurt am Main main station, calls out over the train radio: "Our train is 13 minutes late because of the late provision in Frankfurt am Main." Here, too, there is quite obviously a theory to explain a fact (the delay). The cause of why the train is currently late is established. The previous delay is stated as the cause of the current delay. Again, this is ultimately an every-day interpretation of a phenomenon using an implicit theory.[1] So it will have to be asked what expectations can now be attached to scientific explanations, however.

The explanations and instructions used in everyday life also include, not least, the "journalistic abridgements of reality" in the media (Atteslander 1984, p. 14). These repre-sentations are also not subject to the strict requirements that are attached to explicit theo-ries. It will thus have to be shown what structure such explicit, instructions should have if they want to earn the predicate scientific.

Science, too, is concerned to explain facts and, derived from them, to provide instruc-tions for action. However, science uses a very specific approach which distinguishes it from the housewife or the train conductor. Before looking at these in more detail, however, some parallels should first be discussed. The philosophy of science thus has a certain prox-imity to both epistemology and the methods of research.

Scientific and everyday interpretations of reality are together concerned with interpret-ing processes in the complex world and then representing them as simply as possible. These representations are ultimately used to obtain instructions for the (better) design of life. This concern of scientific and everyday explanations sometimes causes irritation, especially when no clear distinction is made between explicit and implicit theories and when the character or the history of the origin of a theory is not made clear. It can also happen that scientific explanations have an erroneous, for example, everyday structure. This points once again to the necessity of specifying rules according to which scientific theories are elaborated and of developing criteria for describing scientific theories. This is precisely the task to which the philosophy of science is dedicated (compare also Chalmers 2001).

First, scientific representations of reality and thus explicit theories appear with the claim to be comprehensible and criticizable – of course, the housewife and the train con-ductor do not expect any criticism. Thus, it is sufficient for implicit theories to be helpful in some way, seemingly free of contradiction, and plausible. Scientific theories, on the other hand, must be documented so as to always remain verifiable. As has already been shown, the findings of social science surveys are also influenced, among other things, by the method used for them. In order to be comprehensible, the method used to develop the theory must also be disclosed.

Secondly, the procedure of scientific work itself is again subject to scientific consider-ations. In contrast, we assume that the implicit theories described are rather spontaneous, unsystematic and only little empirically verified. Philosophy of science, or methodology, is concerned with the question of how scientific work is to be carried out in order to pro-duce findings that are consistent with reality. The transitions between explicit and implicit

---

[1] Apparently, the Bahn AG has now noticed certain deficits with regard to the announcements of their train conductors, so that a separate training programme has been set up to improve the quality of train announcements. This is reported by the Frankfurter Allgemeine Sonntagszeitung of 29 August 2004 on page 35.

theories are not particularly clear at first glance because of the common concern of both types of theory. A certain danger may result from this. However, only those who follow the methodological rules can claim to arrive at a correct representation of reality.

The philosophy of science is thus concerned with the logic of research, it elaborates the rules of the game by which persons engaged in scientific activity must orient themselves. The philosophy of science thus goes beyond simply laying down recipes for action: it also justifies these rules. It says not only in what way scientific research should be done, but also why it should best be done that way. In this way, philosophy of science is in part comparable to a good cookbook. It not only tells you how to prepare a meal, but also why certain procedures should be followed as described there, for example to preserve vitamins or to bring out the natural flavor of all the ingredients.

Science should always take action when the implicit everyday theories are no longer sufficient to achieve a satisfactory result in solving a problem, when this everyday knowledge is overstretched to answer certain questions, and finally when the objects on which information is needed are too complicated.

Three aspects are particularly relevant when it comes to finding such statements (compare Patzelt 1986, p. 59 ff.). These are *perspectivity*, *selectivity* and *normativity*. Perspectivity determines the way in which a particular object is to be viewed. When a carpenter looks at a table, the quality of the wood processed there may play a role above all. If a physicist looks at the same table, he might see the atoms and their movements. This perspective might be even different for a hungry eater who takes an expectant seat at this table (compare Dahrendorf 1971). Statements are not only perspective, but furthermore selective. Thus, only excerpts from the complex reality or from the object under consideration are ever perceived and expressed. Similar to a photograph, only certain details are ever included in a picture. Finally, statements are connected with a certain normativity. Here we are reminded once again of the housewife who, with her remark, recommended a certain eating speed to her guests.

Statements from everyday life and those from science now differ in that the latter *consciously* carry out these three aspects. Statements from everyday life are also connected with a certain perspective, are selective and contain – as seen – norms. However, these norms flow unconsciously and unquestioned into such statements. In science, however, there is conscious reflection on these aspects: For example, which perspective is the most appropriate to answer a question or to solve a problem. For this purpose, the decisions made in science are consciously designed and controlled.

Science is further characterized by the endeavor to find true statements. This also means that this striving for truth must be systematic and rule-governed. In scientific work, it is assumed that statements, for example in hypotheses, may well not be true. However, it is important that there are concrete rules for the verification of the truth content, which determine scientific work.

The need for methodological rules for proper scientific work is thus manifold. They will be dealt with in detail in this section, following the logic of the research process:

1. First of all, the problem or the knowledge objective of a scientific paper must be stated (Sect. 3.2).
2. In order to communicate this knowledge goal and to solve the problem at hand, terms are needed. Defining them is an important stage of research (Sect. 3.3). In order to gain new knowledge, the formulation of corresponding assumptions in the form of hypotheses is a further important step. Such hypotheses can be structured in many ways (Sect. 3.4).
3. In empirical research, it is important to confront the assumptions contained in the hypotheses with reality. To do this, it is necessary to break down or operationalise the complex question accordingly. This makes the question empirically workable (Sect. 3.5).
4. For the description of reality and thus also for the design of scientific work, the empirically based elaboration of models and the formulation of laws and explicit theories is a further essential step. The individual empirical information collected beforehand is synthesised again for this purpose (Sect. 3.6). The goal of the scientific work is then achieved.
5. After all, these methodological rules are not entirely uncontroversial. Rather, discussions are taking place that question their meaning and address their scope. A brief insight into these discussions is given in Sects. 3.7 and 3.8.

3.2   Problems and Their Formulation

3.2.1   Problems as Conflicts Between Goals and Means

A research process usually begins with the perception of a problem, in other words: problems act as triggers for a research process. It will therefore be discussed what is meant by a problem, what a problem must be like, and what other conditions must be fulfilled in order to actually trigger a research process, because obviously not all problems also entail scientific processing. Furthermore, it will be discussed which types of problems can be distinguished.

The starting point is the observation that problems always occur when conflicts arise between the intended realization of a goal and the possibilities available for realizing this goal – here, above all, the available knowledge is to be mentioned.

Problems thus result from conflicts between goals and means. As will be shown, conflicts between goals and means can arise due to different constellations. It is conceivable that such conflicts affect very different areas. Three examples will illustrate this:

1. In a company, management positions were newly filled. Now the effect of this new appointment on the job satisfaction of the employees is to be determined. For this purpose, information about the current level of satisfaction of all employees in this company is needed. In order to be able to realize the goal – to determine the effect of a change within the company – certain means are missing – current information about the level of job satisfaction of the employees. Thus, there is a problem here.

2. The German right of self-defence is regarded by legal scholars as particularly dashing (see Wittemann 1997 for comparative law). This judgement is based on the fact that the right is granted in Germany independently of a weighing up of interests, which relates the defended good to the good injured in the attacker's case. A duty to evade an unlawful attack is also denied. The basis of these features of German self-defence law is the idea that right need not give way to wrong, and therefore the right of self-defence serves not only to protect the property attacked, but also at the same time to "preserve the law". Information is now to be obtained as to whether and to what extent this practised jurisprudence on self-defence is shared by the population. The aim here is to establish whether case law guides people's behaviour or merely evaluates it.

3. There seem to exist – if one believes certain media – various strategies, for example, to lead a comfortable life at the expense of others. However, many people are not familiar with such strategies. Here, too, we are dealing with an end-means conflict. For the achievement of a certain goal – the comfortable life – many people lack the appropriate means. This is thus also a problem.

It can be assumed that not all of these goal-means conflicts will result in a scientific investigation. This is to distinguish between problems that can initiate scientific research and issues for which this is not the case. In order to be able to make such a distinction, certain criteria are needed that allow the problems to be differentiated accordingly. One such criterion is whether sufficient algorithms are already available for solving the goal-means conflict. In the case of 1. it can be assumed that the existing methodological and company-sociological routines are sufficient to empirically determine the current level of job satisfaction of a workforce and to compare this, if necessary, with a measurement collected at an earlier point in time. This includes, for example, the knowledge about how to draw a sample in a company, how a questionnaire has to be structured and in which way the findings have to be interpreted. Thus, no scientific research would be necessary to solve this conflict of goals and means, but an empirical investigation would. Also, the processing of the question does not primarily contribute to the further development of a theory and/or the methods of science.

Thus, we should only speak of a problem that needs to be addressed scientifically if the knowledge or methods required to solve it are not (yet) available. For such goal-means conflicts, the term "question" is appropriate to describe this situation. Consequently, we should always speak of a problem when a solution to the goal-means conflict is possible with the help of the existing methodological and/or theoretical knowledge. However, the existence of solution routines is only a necessary criterion, but not a sufficient one.

In addition, it must be asked whether there is also a need for scientific research into the problem in question. This need must be of such a nature that it is also able to secure the resources for the solution of the problem. Now here at 3. serious doubts can be raised as to whether the interest in living a comfortable life at the expense of others can be a sufficient trigger. Here, at least at the level of society, there should be no need to deal with such an issue.

A need for the resolution of a goal-means conflict may exist in society, in an institution or, in some circumstances, in an individual, if the necessary resources are provided. Such resources may come from a variety of sources. Parties and unions, for example, have established endowments that budget for research. Universities and other research institutions meet their resource needs primarily through research funding agencies such as the German Research Foundation (DFG), the Volkswagen Foundation, or the Thyssen Foundation. Numerous research projects are also funded internally by companies. One example of this is market research studies that look for ways to determine and increase the sales opportunities of a product. Last but not least, there may also be private researchers – admittedly less so in the social sciences – who put their lives at the service of a scientific project and devote themselves, for example, to finding a new species of butterfly.

Depending on the scope of the planned project, there should be a more or less detailed description of the need to solve the problem, highlighting the necessity of working on it and, under certain circumstances, enabling an external review. When looking at the three examples from above, it thus turns out that only in the case of 2. is there a situation that could trigger a scientific investigation and which – as in the present case – has been funded by the Volkswagen Foundation. Some steps of the scientific research process will now be described and discussed on the basis of this project.

In order to become the trigger for a scientific paper, the problems must be narrowed down and specified. For example, in the context of a project on the acceptance of jurisdiction in the case of self-defence, it may be a matter of evaluating different situations. In doing so, it could be asked whether and to what extent the good attacked in each case (such as honour, property, health and life) is decisive for the evaluation of the situation. It is also conceivable to consider the degree of superiority of an attacker. Finally, it is likely to be decisive for the decision as to whether a conduct is justified by the population as self-defence, what damage the attacker suffers as a result of the counter-defence by the attacked person.

In this way, the problem is broken down into several sub-problems, narrowed down and thus made more precise. Main and secondary goals can thus be identified and formulated (compare Vetter 1975, p. 164 ff.).

3.2.2 Types of Problems

Inspired by Opp (2004) and Vetter (1975), different types of problems are distinguished. It is always taken into account that problem solving is intended to resolve the underlying goal-means conflict. The most common types of problems are:

Problems Taking Action or Value Problems
Problems taking action are "when the question is what can be done to achieve certain goals … rather than what should be done." Value problems, on the other hand, are about "what the case should or must be. This question arises, first, with respect to the goals that

a practitioner wants to achieve. … Further, value problems refer to the question whether one should or must take a certain action" (Opp 2004, p. 1 f.).

Another example can be used to demonstrate this distinction. For example, it has been shown that the recent increase in the price of cigarettes has led to a decline in tobacco consumption, especially among young people. From this, both a problem taking action and a value problem can now be inferred. In the case of a taking action problem, the question would be what can be done in principle to reduce cigarette smoking among young people even further. Here, the answer to be further supported by scientific studies might be: Raise the price of cigarettes even further in order to make it more difficult for young people to start smoking. A higher price could be cited as the reason for the reduction in sales.

The question of whether tobacco tax should actually be increased or whether it should be refrained from, on the other hand, is a value problem. The decline in tobacco consumption has also led to a decline in tax revenue, which is urgently needed by the state. This makes it clear that an efficient measure (in this case the price increase) is by no means necessarily also an acceptable measure for the state. Last but not least, reference must also be made at this point to moral-ethical questions that must always be taken into account when solving a problem of values.

What should one do now? The answer to this question is again by definition a value problem. While it has been empirically proven that higher costs (of cigarette prices) lead to lower consumption among young people – this was a taking action problem – the value problem, on the other hand, cannot be solved empirically. Rather, one will have to resort to other value judgments (compare Opp 2004).

Theoretical and Practical Problems
Depending on the existing deficit that has led to the problem, a distinction can be made between a theoretical and a practical problem. Whereas in the case of the former it is necessary to remedy a theoretical deficit in order to achieve the desired goal, the latter results from a lack of methods or routines to solve the goal-means conflict. A synthesis of both problem types is also conceivable.

The problems to be solved in the context of the self-defence project mentioned above represent such a synthesis. First of all, there is a lack of theoretical knowledge that could be used to explain people's attitudes to these issues. At the same time, however, there is also a practical problem, since there are as yet no suitable instruments for querying evaluations of acts of self-defence within the framework of an empirical study.

Description Problems
When solving this type of problem, the aim is to represent the external appearance of an issue exactly, to describe it precisely. Such a problem exists, for example, if the methods of empirical social research are to be used to find out how widespread self-defence experiences are in Germany in general, whether there are regional differences here, whether certain social groups are confronted with such events more frequently than others, what types of attacks have already been experienced and so on.

Explication Problems

Problems of explication arise when theoretical statements have initially been formulated only with the help of fuzzy or unclear terms. Such problems of explication can arise especially with new theoretical approaches and with the use of terms from colloquial language. For example, the distinction made above between the terms "question" (for the designation of goal-means conflicts that can be resolved with the help of existing routines) and "problem" (if this is not the case) represents the solution to such a problem of explication.

Definition Problems

Closely related to problems of explication are problems of definition. If it is a question of explaining the essence of a phenomenon or a state of affairs with the help of an appropriate description (compare Sects. 3.2 and 3.4), such a problem of definition exists. For example, the question of what is the essence of a healthy diet or of democracy would be such definitional problems.

Explanation Problems

This type of problem occurs when both the nature of an issue and the conclusions to be drawn from it are unclear, and therefore both are to be explored with the help of appropriate studies. Hypotheses are used to solve problems of explanation. These then describe, for example, possible causes of a phenomenon to be explained. An example might be the PISA study on student achievement. Although it was perhaps initially expected that Germany had a good education system, it became clear that this was not (any longer) the case in the results of the first study. Now a discussion started in Germany to explain these disadvantageous findings in order to find measures to remedy the deficiency that had occurred.

Forecasting Problems

These aim to predict, based on theories, the form, course and/or outcome of certain developments. The prediction of certain behaviour is a particularly demanding goal of social science research (see Sect. 3.6). If the effect of an election campaign strategy is to be predicted, if the sales prospects of a new product on the market are of interest or if it is to be forecast how the numerical ratio of fixed network connections to mobile telephones will develop in the future, then we are dealing with typical forecasting problems.

Problems to Be Solved in the Context of Contract Research as Well as Problems in Self-Initiated Research

A further distinction becomes possible if the initiator of a research project is taken into account. According to this, a distinction can be made between problems that are worked on within the framework of contract research and those that are solved through self-initiated research. In the case of the former, the problem in question is, as it were, given to the researcher from outside. The facts to be explained or the dependent variable are thus known. It is important that the resources required for the project are already available and are also provided by the client. Such contract research is usually announced by means of calls for tenders. Initiators can be, for example, ministries, federal institutes and welfare associations.

In self-initiated projects, there is initially greater scope for fixing and concretising the problem. Taking into account the respective possibilities and using inspiration, research projects can, for example, address controversies in the literature, test (new) theories or replicate existing studies. Furthermore, it is a particularly worthwhile attempt to compare the explanatory power of different theories for a very specific issue on the basis of empirical data. One restriction here, however, is the procurement of the resources required for a research; above all, it is a question of the necessary competence, sufficient time and financial resources. Examples of self-initiated research can be found in numerous diploma theses, master theses and dissertations. So-called third-party funding applications are also often required to support such projects.

Vetter (1975) names some more problem forms, such as demonstration problems, proof problems, diagnosis and statement problems, relation problems, and finally optimization problems.

3.3   Terms and Their Definitions

Words such as role, norm, action, working climate, satisfaction, power, healthy food or freedom but also representative, significance and explanation are terms from everyday language that are also not infrequently used by the social sciences. The fact that there are, for example, different expectations when speaking of explanation is not surprising and, as long as it is a colloquial use of the term, does not usually pose any particular difficulty. However, a need for action arises when there is no common understanding of the term among scholars. After all, communication about problems presupposes a common understanding of the communicated content. Only in this way can the findings obtained be understood and criticised, i.e. scientifically processed. The introduction of definitions thus serves to ensure a common understanding of terms among specialist scientists and is thus an essential basic prerequisite for scientific work.

Terms have at least four functions (compare Patzelt 1986, p. 113 f.): First, an ordering function. Terms such as human, animal or city serve to perceive a particular segment of reality. Secondly, terms have a communication function. Persons can only exchange their thoughts with the help of concepts, which are therefore irreplaceable. Thirdly, there is an evaluation function, so in communication not only the naming of segments of reality takes place, but also their desirability is expressed. Thus the terms revolution and overthrow may name at least a very similar section of reality, but its evaluation is carried out very differently with these two terms. Fourthly, following Patzelt, an appeal function of terms can also be identified; terms are able to give instructions for action.

Such terms initially consist of character strings . Such a character string could consist of the letters (= characters) T, I, G, E and R for example. The term "tiger" resulting from this is now in a certain relation to real facts or things, which are called designata.

The designata are the real facts that are named by a string, a term. Also the expression referent is used for this. All persons who are sufficiently familiar with the German language will now have no problem to relate the term "tiger" with the designata and to imagine these in the further.

However, certain real facts can also be designated with different terms, an identical idea can be triggered with the help of different words. For example, the words butcher and-slaughterer denote an identical profession. The term designans is now used for these triggers.

With terms, it is further about sign(−chains) which are connected with certain phenomena. On closer examination, however, it turns out that even the term "tiger" can be associated with different designata in the German language. For example, a tiger can be, first, a cat of prey, second, mean a certain pop singer, or, third, denote a breed of horse with large roundish or elongated spots. (As the author has discovered, the term tiger is also used by a company that purports to manufacture innovative door hardware.)[2] This variety usually does not lead to any problems in the everyday linguistic usage, since mostly already exactly becomes clear due to the context, which designata is meant with the term tiger.

Definitions are those conventions or agreements which serve to show the meaning of the terms used and thus guarantee that the terms are uniformly understood by all users. In scientific language use, semantic rules, also known as correspondence rules, are needed to bring about the unambiguous assignment of the signs (of the term) to the designates (the real facts to be designated). These rules govern the necessary unambiguous attribution of meanings. Figure 3.1 illustrates this relationship.

In science, definitions serve to assign certain meanings to signs or words. However, even in science not every term needs to be redefined. Therefore, it is useful to distinguish between terms such as "explanation", "working atmosphere" or "healthy diet" on the one hand, and such as "and", "not", or "never" on the other. The former are empirical terms that refer to objects of reality and to their characteristics. They are therefore spoken of as descriptive. The other expressions, on the other hand, are logical terms (compare Bergmann 1966, p. 12). While logical terms do not require definitions, empirical terms do.

Following Opp (2004, p. 102 ff.) and many other authors (Patzelt 1986, p. 147 ff.; Babbie 2002, p. 124 f.; Bortz and Döring 2002, p. 63 ff.; Diekmann 2001, p. 160 ff.), different types of definitions will now be distinguished. Here, above all, nominal and real definitions will be considered in more detail.

---

[2] Compare: http://www.tiger.de/unternehmen.html, last accessed Jan 24, 2019.

Fig. 3.1 The relationship
between signs, reality and
semantic rules (cf. Opp
1999, p. 103)

3.3.1   Nominal Definitions

In a sense, nominal definitions have the same structure as a mathematical equation: the expression on the right-hand side of the equation has to correspond exactly to that on the left-hand side. In the simplest case, therefore, one plus one equals two (compare Suppes 1957, p. 154). Thus it is ultimately irrelevant whether one uses the expression "one plus one" in mathematics or whether one says, somewhat more simply, "two". The same is true of nominal definitions. The term to be defined (it is to be called A1), is equated with other terms (here denoted A2) which are presupposed to be known. The term to be defined (A1) is named as Definiendum and the terms presupposed as known (A2) are named as Definiens. The terms A1 and A2 are – as shown – arbitrarily interchangeable. So in the end we are dealing with tautologies.

An entertaining and illustrative literary example of the substitutability of one term for another is provided by Heinrich Böll in his story Doktor Murkes gesammeltes Schweigen. Here, the Definiendum to be replaced by another term is "God" (A1). God is now to be replaced by the Definines "that higher being whom we revere" (A2). Specifically, what happens in Böll is this:

Herr Doktor Murke, editor of the Cultural Word department, is plagued with a less than attractive task: the eloquent essayist Professor Bur-Malottke regrets having so clearly called God by his name in his post-war lectures. Therefore Murke has to cut the word God out of the old broadcast tapes and paste in the formula "Jenes höhere Wesen, das wir verehren" ("That higher being we worship") in its place.

Assuming that A1 is equal to A2, this should not actually lead to any problem from the point of view of scientific theory. As already shown, nominal definitions are tautologies. They are also characterized by several other properties:

- First of all, they have no empirical reference. Thus, it cannot be empirically verified whether the definiendum (in Böll's case "God") is actually equal to the defienienien. The question whether such a positing is true could not be answered. This comes about because of the fact that nominal definitions are not assertions about reality. Such an assertion would, of course, be empirically testable – against that very reality. Thus, it cannot be discussed whether the presented definition is justified.
- Another argument for the use of nominal definitions concerns the economy of language. For instance, in the lectures of Professor Bur-Malottke quoted above, the word

"God" occurs 27 times. From this arises for the editor, Doctor Murke, the problem that one broadcast minute and 20 s are required for speaking "that higher being whom we revere," whereas only 20 s were required for speaking "God" 27 times. From which it is immediately clear that nominal definitions also serve an important function in language economy: A1 is shorter than A2.

- Finally, definitions must satisfy the criterion of expediency. It would therefore be purely theoretically conceivable to replace the word God also by a completely different word, for example by tiger. Since – as already shown – however, already very different conventions exist for the use of the term tiger, it would be highly inexpedient and confusing to make such a designation. It is thus necessary in the context of nominal definitions to ask in each case first whether the term to be defined has not already been determined in some way.

Sometimes it may be useful and necessary that the terms used in the definiens to describe the definitive must also be defined again. For example, it might be asked what is meant by a higher being. In this way, whole chains of definitions would be generated, whereby one must assume that at some point all terms used in the process can be assumed to be known.

Two approaches can be distinguished in the creation of nominal definitions: the intensional and the exensional. In the intensional approach, a set of features is invoked, all of which must be present in order to speak of the term denoted. Obviously, in designating the term "God" (A1) as "that higher being whom we worship" (A2), we are dealing with such an intensional definition. Thus, we can only ever speak of "God" if (first) it is a higher being and if (second) the condition is also fulfilled that this is worshipped by human beings.

In the case of extensional definitions, on the other hand, the set of objects to be designated by the respective term is enumerated. Thus, the definition of the term "social sciences" as (= A1) science dealing with the problems of human society (= A2) would be an intensional definition. All the characteristics are enumerated which must be fulfilled in order to speak of "social sciences". The definition of social sciences (= A1) as sociology, social psychology and political science (A2) would be an extensional definition. Here, all three elements are enumerated for which the designation "social sciences" is to apply.

For example, as part of the study of the public's attitudes to self-defence, the term 'clandestine self-defence' had to be defined. By means of a nominal definition, it was established that secret self-defence is an act of self-defence that is designed in such a way that its reason remains obscure. This too represents an intensional approach to the nominal definition of a term.

Diekmann gives two more interesting examples of nominal definitions: The term "xenophobia", the definiendum, is determined as the common presence of discrimination at work (DA) and in the neighbourhood (DW), of legal discrimination (RS) and of violence (GE) between natives and foreigners (definiens).

The term "political party" (PA) is defined as an organization (O) with registered members (M) and a democratic internal structure (dB), which participates in election campaigns (W) and applies for government participation (R) (compare Diekmann 2001, p. 139 f.).

This makes another important property of nominal definitions clear: the definition of a term, such as "political party" as O, M, dB, W and R, has important consequences for further work. If, for example, one defines a certain fact to be researched – as happened here – with the help of an extensional definition, then only certain elements (here O, M, dB, W and R) are elevated to constituents of the respective term, precisely those that are named in the definition. All others, for example in this case such organisations with a totalitarian internal power structure, have been defined out and are no longer considered a political party in the context of a possible research project.

In summary, nominal definitions play an important function in social science research. They regulate the necessary common understanding of terms, are important for the economy of language and thus exert influence on the further design of the research work.

3.3.2 Real Definitions

In addition to nominal definitions, there is also a certain need for real definitions in the social sciences. These are characterized by the attempt – in contrast to the nominal definitions – to express the essence of an issue or to describe it. The question, for example, what is meant by "healthy nutrition" can be understood in such a way that information is desired about what the essence of such a diet consists of, or to state what "healthy nutrition" actually is. It quickly becomes clear that the relative arbitrariness of nominal definitions is of no use here.

> A 'real' definition, according to traditional logic, is not a stipulation determining the meaning of some expression but a statement of the 'essential nature' or the 'essential attributes of some entity'. The notion of essential nature, however, is so vague as to render this characterization useless for the purposes of rigorous inquiry. (Hempel 1952, p. 6)

Thus, real definitions have an empirical reference. Such a real definition, if it wanted to define the term "healthy diet", for example, would have to name all those dimensions that constitute such a diet. If it fulfils this requirement, the definition is correct; if it does not, then this definition would be wrong. A real definition can thus be right or wrong, for in the end it states something about reality. Such real definitions are also called definitions of essence. They are expected to show the meaning of a certain fact. For further aspects and the problems of such determinations of essence with the help of real definitions, compare Opp (2004, p. 109 f.).

One problem with real definitions results from the fact that there are no fixed criteria for determining whether the essence of an issue has actually been correctly captured in the definition. If, for example, we take the term "healthy nutrition" and imagine that we want to determine it by means of a real definition, it should quickly become clear that this is a problem that can hardly be solved, not least because of the difficulties that would be involved in checking this definition against reality.

3.3.3 Operational Definitions

Operational definitions determine the way in which a theoretical concept is to be linked to those concepts that can be observed empirically. The problem of operationalisation is then dealt with in detail in Sect. 3.5.

Other types of definitions can also be distinguished. Opp (2004, p. 115 ff.), for example, mentions the complex definitions.

It is important at this point to summarize by pointing out that it is necessary to indicate which type of definition is involved in each case, after all, as has been shown, there exist some essential differences between nominal and real definitions. It must therefore be made clear which of these two types of definition is meant. For this purpose, it has proved useful to emphasize nominal definitions by marking them "= df.". Accordingly, one would write, for example:

Or in general:

God	= df. that higher being we worship,
Secret self-defence	= df. Self-defence act designed so that its reason remains in the dark

$$A1 = df.\ A2.$$

In contrast, in the case of real definitions, it must be explicitly pointed out that the sentence that now follows is such a definition.

3.4 Hypotheses and Their Treatment

It was shown in Sect. 3.1 that problems can act as triggers for a research process. Before using the example of a specific social science project to demonstrate the role of hypotheses in the research process, their function should be described in general terms.

A problem occurs when a certain goal cannot be achieved with the available means. For example, a problem exists when there is insufficient knowledge (the means) to explain a phenomenon (here, the goal). This can now be used to answer the question of what it means to solve a problem: To solve a problem means to uncover the causes which bring about the state of affairs to be investigated. Thus, a problem (E) is solved when the causes (C) that can be held responsible for the occurrence of the problem have been determined. Or, to put it more briefly, when it is possible to name the relation C → E. In this context, the letter E stands for effect.

Problem solving means looking for relationships between phenomena. An important step is to determine the dependent and the independent variable of a research process. The dependent variable in our case would be E. It is that state of affairs for which an explanation is to be sought. The independent variables are then those that can be held responsible as causes for the occurrence of the dependent variable, in our case C. It is of central importance to accurately identify the dependent and independent variables of a research process in each case. It is not uncommon for different problems to be dealt with at the same time, i.e. in parallel, within the framework of a social science research process. In this case, several independent variables can be named. As already discussed in Sect. 3.3, it is also necessary to define the terms used adequately in order to communicate the problem.

At this point, the function of hypotheses can be explained. First of all, hypotheses = df. scientifically founded assumptions about a fact or about a connection between at least two characteristics. They contain both confirmed knowledge and as yet unconfirmed assumptions. Hypotheses are to be tested with the help of empiricism. They thus serve the further development of knowledge. Confirmed knowledge provides both a contradiction-free formulation of the conjectures contained in the hypotheses and a classification of these conjectures in a theoretical edifice. The testing of the hypothesis thus already begins with its formulation.

One could also say that hypotheses are presumed solutions to problems or attempts to explain them. Furthermore, it must be stated that hypotheses must be of such a nature that they can, in principle, be empirically processed. As a consequence, the logical structure of hypotheses is identical to that of a problem solution: $C \rightarrow E$. The difference, however, is that this relation has not (yet) been empirically confirmed.

It should be pointed out that hypotheses have a research-guiding function. Such a research-guiding function results from the fact that the researcher's entire epistemic interest is encoded in the hypotheses. In principle, it would be conceivable to search for a finite number of causes for E. However, this totality of possible explanations for E is restricted in the hypotheses to one or a few possible ones. From the broad spectrum of possible explanations for E, the decision is made to work on very specific ones. Only on these can statements be made on the basis of the research results. Thus, there is a scientific reduction of reality.

In the further course of the research process, a number of decisions are made and determinations are made that result from the concrete assumptions contained in the hypotheses, for example, a survey of a specific group of people on their views on self-defence is designed and implemented. The assumptions contained in the hypotheses – if they are later confirmed – finally make it possible to make concrete changes in reality.

To illustrate this, a few court decisions will be presented in an exemplary and highly abbreviated form. These form the background for the hypotheses developed within the framework of the Dresden self-defence project, which are then presented and discussed. Afterwards, various requirements for the formulation of such hypotheses will be described, as well as ways of distinguishing between different types of hypotheses.

The right to self-defence is granted in Germany – as already briefly mentioned – independently of a weighing up of goods. A relationship between the good to be defended – for example a material value – and the good injured in the attacker as a result of self-defence – for example the attacker's health – does not need to be established. See also the brief descriptions of some cases and the judgments handed down by the courts in Table 3.1.

Table 3.1 Abstracts of some judicial decisions on self-defence cases

Description of the case	Court decision
Case 1: N guards his fruit trees against night-time theft during the period of shortage after the First World War. Early in the morning he notices two men stealing fruit from his trees. When the men do not stop when he calls them, N shoots at them with buckshot, thereby injuring one of them considerably	The Reichsgericht justifies the shooting of N by self-defence, because in the application of the right of self-defence the idea of proportionality does not play a role in principle. (RGSt 55/82)
Case 2: Innkeeper N is at odds with a group of "rockers". One evening the "rockers" enter N's pub with the intention of "taking revenge" and bring him down backwards. N fears that he will be beaten and that his pub will be destroyed. He therefore grabs an iron bar and hits two of the attackers with it, which would have been fatal if the injured had not been taken to hospital at the last minute	The Federal Supreme Court considers a justification of N's blows by the right of self-defence in defence of property and the right of domicile to be possible. (BGH StV 1982/219)
Case 3: For unknown reasons, O tries to break into N's flat. In doing so, he delivers boxing blows, which, however, do not hit N. N first tries to ward off the attack by hitting him with a walking stick, but gets caught in the door frame with the stick. He therefore grabs a knife and inflicts a fatal injury on O with it	The Federal Court of Justice considers a justification of self-defence possible because N was allowed to defend his domestic right even with a fatal stab. (BGH GA 1956/49)
Case 4: N's girlfriend owes O DM 6000.00. O therefore tries to forcibly prepare a seizure of the girlfriend's car. He first tears open the car door and tries to take the car keys away from N, who is sitting at the wheel. When this fails and N drives off, O pursues him together with his son. At a traffic stop, O's son places his moped half across in front of the car driven by N, and O drives up so close from behind that N cannot simply back up either. Then O goes to N and asks him to hand over his girlfriend's car. N refuses to do so and, after being threatened to do so, drives off in front. In doing so, he seizes the moped and pulls it to the ground together with O's son. The latter suffers abrasions and bruises	The Higher Regional Court justifies N because of the defence of his freedom of will and movement against an unlawful coercion. A careful shunting out of the car had been possible, but not necessary, because the principle of the defence of rights legitimised N's driving towards the coercer. (OLG Karlsruhe NJW 1986/1358)
Case 5: N gets into an argument with the pub guest O. The latter ambushes N in front of the pub. The latter ambushes N in front of the pub. Therefore, N hesitates to leave the pub for an hour. When he finally steps out of the door, O lunges at N with the words "I'll kill you". N, fearing that O is carrying a knife, first fires a warning shot from his pistol. As this does not stop O, N strikes down the attacker, who has come to within 2 m, with a fatal shot	The Federal Court of Justice considers it possible to justify the fatal shot by self-defence, because N had not been obliged to "get out of the way of injustice." (BGH GA 1965/147)

(continued)

Table 3.1 (continued)

Description of the case	Court decision
Case 6: N and a companion are physically harassed by the heavily intoxicated O when returning to their parked car. However, they manage to get into the car. On approaching, O first jumps in front of the car, then back onto the pavement and twists the right side mirror. N gets out to straighten the mirror. As O approaches N with his hands up, N hits O on the head with a fire extinguisher. O suffers a fractured skull	The Frankfurt Higher Regional Court denied N the plea of self-defence because it had been possible to evade a drunkard without fleeing in disgrace. (OLG Frankfurt VRS 40/424)
Case 7: N is travelling home from work by train in a first class compartment. He feels disturbed by O, who takes a seat in the compartment with an open beer can without first class authorisation because the train is overcrowded. Therefore, N opens the train window to drive O out of the compartment by the winter cold draught. O closes the window again. When N opens it a second and third time, a verbal fight ensues in which O threatens N with a beating if he opens it again. When N opens the window a fourth time, O grabs N's face with both hands. N fears that O will "go for his throat" and defends himself by stabbing the attacker in the stomach. The latter dies	The Federal Court of Justice denied N the plea of self-defence. N had provoked O's attack by his socio-ethically objectionable preliminary behaviour and was therefore obliged to first ask the fellow passengers standing in the aisle for help. (BGH NStZ 1996/380)
Case 8: Mr. O and Mrs. N have been married for a long time. Between N and O, who is not significantly physically superior to his wife, there are frequent physical confrontations. However, there had been no serious physical injuries. When O again assaults N, Mrs. N stabs O in the left side of the chest with a knife. The knife penetrates both chambers of the heart, so that O dies immediately	The Federal Court of Justice confirmed the jury court which sentenced Mrs. N for bodily harm resulting in death. The obligation of spouses to sympathetically respond to their partner and to be considerate required that the attacked spouse – provided that the other spouse was not threatening his or her life – had to be content with a milder form of defence. This applied even if this form of defence did not end the attack with complete certainty. (BGH NJW 1975/62)
Case 9: N, who was a prisoner of war in Russia for a long time, and O argue about the invasion of the Soviet Union by the German Wehrmacht. When O reproaches N in an excited suada that the imprisonment was "N's fault, he just shouldn't have gone to Russia", N lunges at O. The latter falls onto the pavement and dies	The Federal Court of Justice denied N the right to invoke the defence of honour. Although O's assertion was an insult to N, the defamation was minor. Under these circumstances, a physical attack by N on O could not be justified; N should have limited himself to a retort with words. (BGHSt 3/217)

(continued)

Table 3.1 (continued)

Description of the case	Court decision
Case 10: N has committed a robbery. O, N's uncle, learns of the robbery from N's diary. He therefore takes the diary and demands DM 60,000.00 from N for its return; otherwise he will hand the diary over to the police. N agrees to the blackmail on the pretence, but takes O's emissary hostage during the agreed handover of money in order to wrest the diary back from O in this way	The Supreme Court did not recognise a right of self-defence of N against the extortion by the uncle and his assistants. It therefore convicted N of hostage-taking. (Supreme Court of the Republic of Croatia after Novoselec NStZ 1997/218 ff.)

At this point, there is a *problem,* or rather a *conflict between the goal and the means,* since it is not clear whether the type of jurisdiction shown above is shared by the population. The phenomenon that is to be *explained with the* help of research is thus the approval or disapproval of behaviour in which an attacked person defends himself and in doing so causes harm to the attacker. The *dependent variable represents the* approval or disapproval of a particular behavior. What is sought are the *reasons* why certain behaviors are evaluated (as permissible or as impermissible) in the case of an assault on a certain good. The formulation of hypotheses now forms an important step towards the solution of the research problem. While the problem resulted from a lack of knowledge or of methods, the hypotheses for it now contain a solution approach. In the hypotheses, the totality of possible determinants is reduced to a few selected ones. Specifically, the following *hypotheses* should be addressed (compare Amelung and Häder 1999; Amelung and Kilian 2003; Kilian 2011):

H1	The strident views of the case law (compare cases one to five) and also of the literature on the scope of the right of self-defence do not resonate with the general population. While in case law the conditions that must be present in order to affirm a right of self-defence are relatively broad, the general population does not follow such a line of thinking
H2	Self-defence is rather justified in the population primarily by a weighing of goods. The greater the importance attributed to the good to be protected, the more likely it is that self-defence will be agreed to in order to protect these goods. The smaller the damage to the attacker, the more likely it is to be regarded as legitimised by self-defence
H3	The idea, originating from the jurists, that the right of self-defence serves the purpose of proving the law (compare case four), plays only a minor role among the population in the legitimisation of acts of self-defence
H4	The same applies to the legal dogma that one does not need to evade a present unlawful attack in principle
H5	Insofar as legal doctrine denies an obligation to flee, this is based less often on the idea of legal rights than on the "disgracefulness" of the flight. This applies at least to men, especially those of young age and from lower social strata
H6	For this reason, the popular view of limiting self-defense against the mentally disturbed, drunks, and children (compare case six) is more consistent with legal doctrine than it is with the exclusion of the balancing of interests idea
H7	The restrictions on the right of self-defence made by case law and literature in the case of so-called self-defence provocation (Roxin 1997, p. 550, compare case seven) resonate with the population

(continued)

(continued)

H8	The restrictions on self-defence between spouses, some of which are postulated in case law and literature (compare case eight), are generally approved of by the population. However, if – As is usually the case in the cases decided in real life – The right of self-defence of an abused woman is restricted, women and younger people are more likely to decide against such restrictions
H9	A violent defence of an immaterial legal right such as honour (compare case nine) is not regarded as justified by the population. The case law here tends to follow the views of the population
H10	The clandestine self-defense against blackmail through the threat of compromising disclosures (compare case ten) is not considered legitimate by the public
H11	Views on self-defence assistance are strongly influenced by the degree to which the victim is in need of help. In particular, the defense of the elderly and children is associated with corresponding positive attitudes
H12	The population's perceptions of the right to self-defence, in particular the subjective norms, are strongly differentiated by social-demographic factors
H13	More important than knowledge of legal regulations for the evaluation of a concrete self-defence situation is the extent to which a person affected by an unlawful attack sees himself physically able to defend himself

When formulating hypotheses, a number of requirements had to be observed. In general, hypotheses should be logically formulated, comprehensible, testable, criticisable and true. Following Opp (2004) and Bortz and Döring (2002, p. 10), hypotheses have to meet the criteria listed here:

- Hypotheses are intended to represent statements, but they are not intended to be questions or commands. For example, one such statement is that the general population does not follow the views of the judiciary (H1).
- Hypotheses must contain at least two semantically meaningful terms. In H1, these would be, firstly, the views of the jurisprudence on self-defence and, secondly, the views of the population on this issue.
- These semantically rich terms are to be connected in the hypotheses by logical operators. Hypotheses contain conditional sentences. In H1, the hypothesis is that views on self-defense are not congruent. In H2, the assumption is that if the good to be protected is considered highly important, self-defense is more likely to be considered justified. In this case, one also speaks of "if – then" parts of a hypothesis. Such statements can:
  - (a) be necessary. This means that it must be a matter of goods to which a high degree of importance is attached so that self-defence is regarded as justified. However, other conditions must also be met in order for the expected effect to occur, for example, the damage caused to the attacker by the act of self-defence must not be too great.
  - (b) sufficient and
  - (c) be necessary and sufficient.

- The statement must not be tautological, that is, the two concepts must not completely cover each other. If one were to define "justification of self-defence" in H2 as an "attitude that comes about as a function of the importance of the good to be protected", one would be dealing with such a tautology. The explanation of a train delay in everyday language given above is also such a tautology – which is inadmissible in science.
- The statements contained in the hypotheses must be free of contradictions. On the other hand, the statement that "if a good that is recognised as important is protected with the help of self-defence, this self-defence behaviour may be regarded as justified, but it may also be that this is not the case" would not be free of contradictions.
- The conditions of validity are to be enumerated under which the statement of a hypothesis is to be true. For example, the conjecture formulated in H5 about the nefariousness of flight is said to be valid primarily for young men. In H1, on the other hand, an assertion is made that claims validity for the general population. No such restriction is found in a number of other hypotheses. This implies that they should be valid for all persons subject to German jurisdiction. In the event that such correlations are also formulated independently of time, they are so-called all-statements.
- The concepts of a hypothesis must be operationalisable, it must be possible to find rules to determine whether and to what extent the described facts exist in reality (compare Sect. 3.5). For example, it must in principle be possible to determine empirically how highly the importance of a good under attack is rated by people, and whether they regard the behaviour described as justified.
- Finally, there is the requirement that the statements contained in the hypotheses must be falsifiable. This premise can also be satisfied relatively easily in the case of H1, for example. Finally, one could look for significant goods whose defense with the help of self-defense is not considered justified. In such a case, the hypothesis would be disproved. In order to satisfy this requirement, the facts addressed in the hypotheses must be real, existing in reality.
- Last but not least, the formulation of hypotheses should also include ideas about the theoretical context in which the respective phenomenon is to be investigated. Hypotheses thus contain both unconfirmed knowledge in the form of a presumed connection and already confirmed, theoretical knowledge. If one assumes that there should be freedom from contradiction between the two forms of knowledge, one can conclude that hypothesis testing already begins with their formation. The demand for freedom from contradictions in the formulation of the hypothesis also represents the first hurdle to be overcome in the verification of a hypothesis.
- This also addresses the fact that hypotheses must have a certain degree of general validity. It would therefore be insufficient for a hypothesis to merely state: Respondent X regards the restrictions on the right of self-defence in the case of a self-defence provocation as justified.
- One danger in hypothesis generation should be pointed out. Thus, it is not sufficient for a hypothesis to take the form that (presumably) the phenomenon E occurs when C is present, but it must also be stated that a certain theoretical background exists that

supports precisely this assumption. Thus, when formulating a hypothesis, it is always necessary to name the theoretical context that is to be used to clarify the problem. For example, it has been shown that specific behavioral intentions can be explained on the basis of norms, values, and control beliefs. Now one might suppose that a theory that has already proved itself in explaining various other behavioural intentions is now also suitable for explaining behavioural intentions in connection with self-defence. Later (compare Sect. 3.6). it will be necessary to show what such a theoretical model might look like in concrete terms. In the case where there is no relationship between a theory and the hypotheses in the statements, we speak of empiricism.

Not suitable for inclusion in scientific hypotheses are the so-called there-is propositions. If a sentence were to read, for example, "There are views on self-defence assistance that are not influenced by the degree of the victim's need for help," then it could not be falsified. If one wanted to make a corresponding attempt, one would have to examine all people to see whether there is not at least one person for whom this statement is not true. Also the premise, according to which hypotheses must have a generally valid character, is violated with such a statement, because finally only the one certain existence is asserted.

The so-called may-sentences are also unsuitable for hypotheses. They are ultimately tautological and thus also not falsifiable. If one were to say, "A violent defense of an intangible legal good such as honor *cannot be* regarded as justified in the population," this would mean that both the justification and the rejection of this assertion constitute a confirmation of this assertion. This is inadmissible.

Different types of hypotheses can now be distinguished. In the case of the self-defence justification hypothesis (H2) shown above, this is a so-called Je-Desto hypothesis. Also, hypotheses that have already undergone empirical testing can be distinguished from those for which this is not yet the case. Finally, the level of reference represents a criterion for distinguishing hypotheses. First, however, the if-then hypotheses will be presented.

3.4.1 If-Then Hypotheses

An if-then hypothesis requires that both variables included in the hypothesis are dichotomously expressed. In the social sciences, such cases are, for example, pensioners and non-pensioners, car drivers and non-car drivers, men and women, people who live alone or those who live with a partner, people from East or West Germany, and so on. This would mean for our H2 developed above that there is only approval or disapproval for a given action. Also, only the "significant good" and "insignificant good" states would be allowed. However, the latter criterion is not satisfied in H2. Thus, H2 is not an if-then hypothesis.

In H9, on the other hand, such an if-then hypothesis is present. If it is about the violent defence of honour, the behaviour described there is disapproved of. As already mentioned, these hypotheses can again be subdivided into (a) those in which the dependent character-istic only occurs if the independent characteristic mentioned in the hypothesis is present. But it can also be (b) that the dependent characteristic occurs without the independent characteristic contained in the hypothesis to be registered. In our example, we are dealing with type (b), because an act of self-defence can presumably also be considered unjustified for other reasons. This may be the case, for example, if the attacker is a small child or an intoxicated person.

In this type (b) only a statement is made about what happens when honour is defended by violent means. However, nothing is said about what happens when a significant good (such as the life or health of a person) is defended by violent means.

This is different for type (a). Here, the independent characteristic is both necessary and sufficient. That is, only if one characteristic occurs, does the other also occur. Or in other words: For the justification of self-defence, only the significance of the attacked property plays a role, but nothing else.

If we again use the letters C and E instead of concrete events, these two types of if-then hypotheses can be well illustrated with the help of a four-field table. We are either dealing with the presence of a characteristic, for example C, or the case occurs that. C is not pres-ent, symbolized by ~C. The same is true for E. This is done in Fig. 3.2.

In the four (I to IV) cells of the two crosstabs it is indicated in each case whether this speaks in favour of the hypothesis being true, i.e. we are dealing with confirmers (K), or whether this speaks against the assertion contained in the assumption, i.e. we are dealing with a falsifier (F). Thus, in the case of deterministic implication, there is only one falsifier in the third quadrant if C occurs but E fails to occur. Thus, our hypothesis H9 would only be falsified if it involved the violent defense of honor (C is present) and this was consid-ered justified. Finally, no assumption was formulated in H9 about other determinants of the non-justification of self-defense (~C). The situation would be different if there were deterministic equivalence. If a hypothesis were to read, only in the case of a violent defense of honor is self-defense considered unjustified, this would be true. Here the number of falsifiers then increases. Thus the constellation ~C and E would also disprove the assertion contained in the hypothesis.

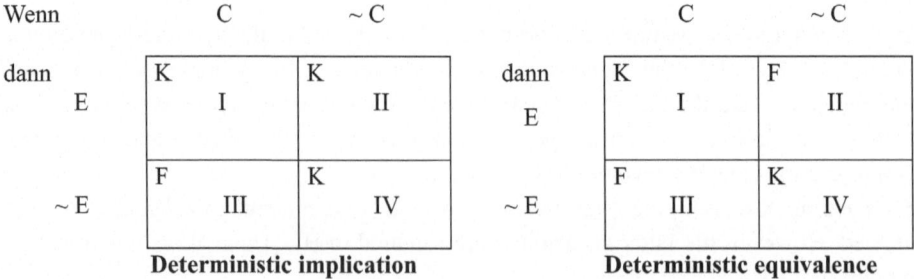

Fig. 3.2 Representation of the if-then hypotheses as well as confirmers (*K*) and falsifiers (*F*) in a cross table (compare Diekmann 2001, p. 126)

3.4.2 Je-desto Hypotheses

In addition to if-then hypotheses, the social sciences also work relatively frequently with je-desto hypotheses. Je-desto hypotheses contain facts or variables whose categories can (at least) be interpreted as rankings, i.e. they have an ordinal scale level (cf. Sect. 4.3). H11 can be regarded as an example of such a je-desto hypothesis. This hypothesis is slightly reformulated here in order to better demonstrate its je-desto character:

H11a:	The more needy a victim is, the more self-defense assistance is considered justified

It is irrelevant whether the claimed relationship is positive or negative. It would also be conceivable to formulate the hypothesis: The less a victim is able to defend himself against an attack, the more self-defence assistance is considered justified.

3.4.3 Deterministic (or Nomological) and Probabilistic Hypotheses

Nomological or deterministic laws or hypotheses possess by definition (df.) a validity or a claim to validity, which is independent of space and time, i.e. they are valid everywhere in the world and also at all times. As indicated, one speaks here also of all-statements. Of course, the elaboration of such statements is a particularly demanding task. The uncovering of all-statements is a goal to be striven for by every empirical science. In the social sciences, however, compared with the natural sciences, this goal is still some way off (cf. Sect. 3.6). In the meantime, there are two possible solutions:

1. Instead, one looks for statements of medium scope. Such statements do not apply everywhere in the world, but for example only to developed Western industrial societies. This reduces the information content of the hypotheses and at the same time limits the number of falsifiers. A similar approach has been taken with hypotheses H5 and H8. Here it is assumed that the presumed relationship can only be proven for a certain group of people. In the case of H5, these were men, especially those of young age and from lower social strata. In the case of H8, it is assumed that if the right of self-defence of an abused woman were to be restricted, it would be women and younger people who would be more likely to decide against such restrictions.
2. One searches for statistical laws with the help of corresponding hypotheses. According to this, one would assume that a statement can only be made with a certain probability. An example is provided by H12, which has been reformulated for this purpose: "The higher a person's level of education, the more likely his or her attitude is to correspond to the attitude advocated by the courts." The assumption here is that not all persons who have a high level of education will also vote in accordance with the law when questioned. However, it is assumed that the probability of such a decision increases with the level of

education. In other words, the number of legally compliant judgments is higher for 100 people with high education than for the same number of people with low education. Even with a ratio of 51 to 49, the aforementioned hypothesis would be sustainable.

It may be left open whether such probabilistic statements always represent a reference to (still) insufficient knowledge about reality. If one knew all the determinants of the facts to be explained, it would then also be possible to describe them with the help of all statements. Or whether social reality – in contrast to technology, for example – is not of such a nature that it *always takes the* form described and it therefore makes no sense at all to search for "complete" explanations. In part, this insight even leads to disappointment (compare this discussion with, for example, Halfmann 2001, p. 78 ff.).

3.4.4 Individual, Collective and Contextual Hypotheses

Another criterion according to which hypotheses can be distinguished is their reference level. As already described, hypotheses contain statements about a relationship between (at least) two variables. For the distinction, it is important at this point what kind of characteristics these are. Thus, individual, collective and context hypotheses can be distinguished. *Individual hypotheses are* those where the dependent and independent variables are both individual characteristics. For example, the hypothesis that there is a relationship between the duration of education and the level of income is such an individual hypothesis. Both components of the hypothesis – the duration of education and the level of income – relate to the characteristics of an individual.

We speak of *collective hypotheses* when the dependent and independent variables are collective characteristics. For example, Marx's well-known philosophical thesis that social being determines consciousness could be called such a collective hypothesis. "It is not the consciousness of men that determines their being, but conversely their social being that determines their consciousness" (Marx 1971, p. 9). Marx is clearly not concerned with making statements about single individuals, but rather understands both being and consciousness as characteristics that can be attributed to collectives. Another example would be the already well-known hypothesis from the self-defence project: (H1) The strident views of case law and literature on the scope of the right of self-defence do not resonate with the general population.

It is important to note in each case that conclusions from the collective level to the individual level are not necessarily logical. Thus, Marx's thesis cannot be refuted by finding individuals whose individual being is not shaped by their social being.

In the case that the dependent variable is an individual characteristic and the independent variable is a collective characteristic, one speaks of a *context hypothesis*. Such context hypotheses have a special significance in sociology. For example, a conjecture about the influence of social structure on the conditions of individual action is such a hypothesis. But also many of the hypotheses mentioned above are context hypotheses.

3.4.5 Differentiation of Hypotheses According to Their Stage of Development

Hypotheses make claims about relationships, whereby the truth of such claims has not yet been (fully) established. Since it is usually not sufficient to carry out a one-off empirical confrontation of the hypothesis with reality (cf. Sect. 3.6), a hypothesis will have to undergo a verification process over a longer period of time before it can be given the status of a law. Patzelt (1986, p. 180 ff.) suggests distinguishing hypotheses according to how far such a review has progressed. The following classification is adopted for this purpose:

Firstly, there are hypotheses which have not yet been empirically tested; these should be called *conjectures*. They are at the beginning of the scientific work process. Second: There are hypotheses that are difficult to test or (as yet) untestable. These pose a particular challenge to science and the creativity of scientists, and they thus drive the process of knowledge. Third, there are hypotheses that have already been tested; they will be called *test hypotheses*. Fourth, there also exist hypotheses that have already been subjected to *empirical* testing. This testing may in turn have led to different results. Accordingly, confirmed hypotheses are to be distinguished from those that have been refuted. Quite often, however, the findings are not so clear that this clear distinction can be made. In this case, the hypotheses are divided into those that have been (a) absolutely refuted or absolutely confirmed, (b) well refuted or well confirmed, (c) sufficiently refuted or confirmed, and (d) possibly refuted or confirmed.

3.5 Operationalisation

The first step was to name the problem to be worked on with the help of clear terms. In the next step, hypotheses were derived that contained theoretical knowledge and an as yet unconfirmed assumption regarding the solution to the problem. Now the next task is to check the truth of the hypotheses against reality. To be able to do this, the facts contained in the hypothesis must now be made measurable. In the context of the self-defence project, for example, it must be determined whether or to what extent a certain behaviour is regarded by people as justified, how important a good under attack is considered to be, and so on. Operationalization is concerned with this concern (compare Bridgman 1927). The aim of operationalisation is thus to make it possible to measure or to create the preconditions for the empirical investigation of complex and/or latent facts.

An important methodological principle of Critical Rationalism states that all statements of an empirical science – insofar as they are inaccurate – must in principle be capable of failing on the basis of experience (Popper 1971, p. 15). If one follows this premise, the following three conclusions can be derived for the statements contained in the hypotheses:

- The claimed relationships must have an empirical reference. That is, the corresponding designata must in principle be perceptible or observable. Concepts for which this does not apply, such as "that higher being we worship", would thus be useless. In contrast, it can be assumed that the "justification of a behaviour" and the "significance of a good to be protected" have such an empirical reference.
- The terms used must be precise, i.e. it must be clear what they refer to. To this end, the relevant definitions (see Sect. 3.3) are introduced.
- The terms must be operationalizable. In other words, it must be possible in principle to empirically ascertain the designata denoted by the terms with the aid of appropriate operations. Operationalisation thus means specifying operations or actions that are suitable for empirically depicting a state of affairs. In the case of a hypothesis that says: If Peter is bowling in heaven, then it is thundering on earth (the structure $C \to E$ required for hypotheses is given, after all), we would have to look for ways to determine: (a) whether bowling is currently taking place in heaven, and that by Peter, or whether this is not the case, and (b) whether it is thundering on earth or not. The latter should be possible, while the first claim cannot be met with the instruments currently available. Thus, this hypothesis does not meet the requirements.

In order to work on a social science hypothesis, the terms used in the hypothesis must be defined operationally. This can be done by specifying indicators or observation operations for a term – for example, gender. In the context of a postal survey, for example, gender could be operationalised by the instruction: "Please tick the appropriate box to indicate whether you are male or female!"

In this way, it becomes clear to the social scientist whether or to what extent the fact meant by the term exists in reality. The following illustration shows how objects of a concept are determined by research operations (compare Fig. 3.3).

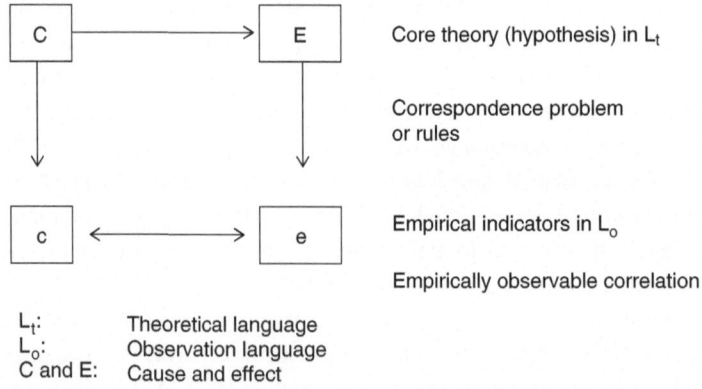

Fig. 3.3 Theoretical language, observation language and correspondence rules, cf. Schnell et al. (2013, p. 69)

Figure 3.3 is to be interpreted as follows: Between the states of affairs denoted by the capital letters C and E there exists in reality a cause-effect relationship, shown by the symbolismC → E. This means that whenever C exists E occurs. Both facts (C and E) must now be operationalized for verification in the context of a social science investigation, that is, rules must be found that specify how the respective asserted facts can be empirically measured. This is done by the rules of correspondence. The result is the development of the indicators c and e, denoted by small letters. They allow empirical observation of the facts. On the basis of this empirical observation, it is then established that there is a correlation between u and w. Thus, it can be seen that whenever one of the two quantities occurs, the other quantity can also be empirically demonstrated. Furthermore, it can be seen from Fig. 3.3 that the correspondence problem must be solved as part of the operationalization. This is done by a whole series of steps. First, the subject matter has to be defined and decomposed into dimensions. Secondly, a decision has to be made on the data collection method to be used (whether, for example, a content analysis, an Internet survey or participant observation is suitable). Thirdly, it must be clarified how the individual dimensions collected are to be summarised again at the end.

Suppose C is the gender and there is a presumption that if a particular gender is present, a statement of fact E can be made. Such a presumption was contained in H8. Thereafter, it was assumed that if an abused woman's right of self-defense is restricted, women are more likely to opt out of such restrictions.

The correspondence rule would now state that if, for example, in the context of a postal survey, a person is presented with the instruction (c, or the empirical indicator): "Please tick whether you are male or female", this person will actually state which gender (C) applies to him or her. One would now have to proceed accordingly with E and e respectively.

As already indicated, correspondence rules are used in empirical social research. As will be shown later (see Sect. 6.1.2). these rules cannot, however, guarantee 100% that a particular empirical indicator (for example, the questionnaire question c about gender C) actually reflects the facts of interest (for example, the gender of a person). In the case of complex facts, this problem is correspondingly more difficult than in the case of the question about gender.

The operationalisation process will be illustrated using the example of "welfare". Welfare is a central category of social reporting. Its aim is to gain information about the development of welfare in a country with the help of empirical analyses. It quickly becomes clear that it is not straightforward to ask people about the degree of welfare they have achieved. Rather, the complex term "welfare" needs to be operationalized step by step. Following the Data Report (2002, p. 425 ff.), this is summarized in Fig. 3.4.

In a first step, four dimensions of welfare are distinguished: a more intellectual dimension, denoted by the term "satisfaction with one's own life", a strongly emotional dimension, denoted by the term "happiness in life", a dimension containing mental stresses, denoted by "concern", and an "anomie" dimension, which encompasses people's possible fears about the future. On the way from the abstract category (welfare) to observable indicators, facets have thus been named and defined by which the term sought can be described in more detail. Now, for example, questionnaire questions can already be formulated on

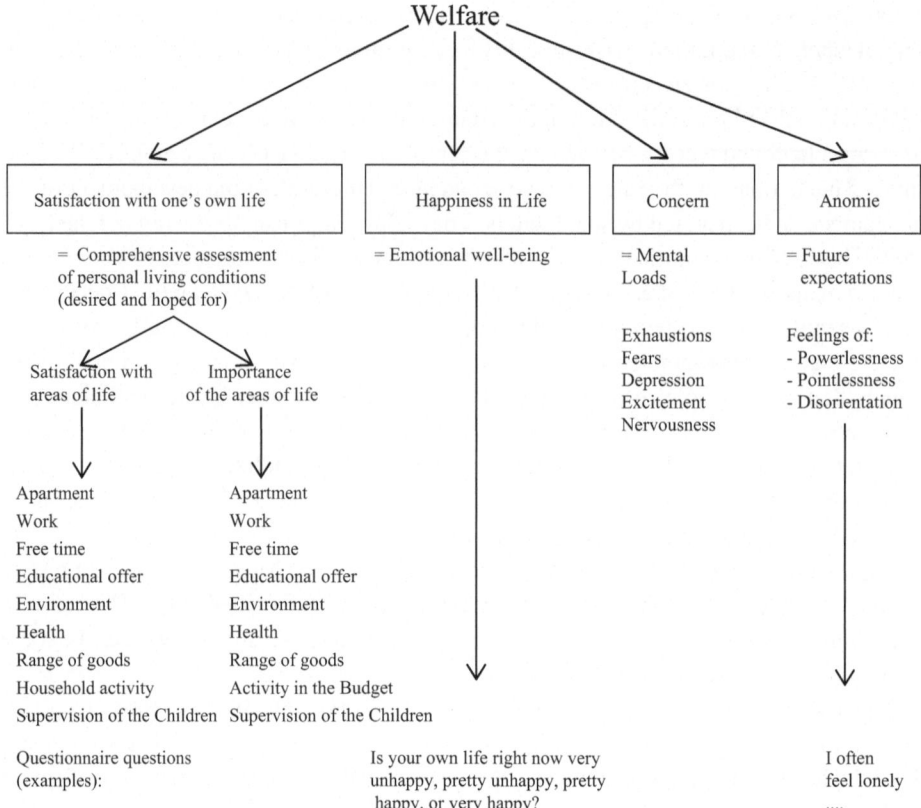

Fig. 3.4 Steps in the operationalisation of welfare

the dimension of "happiness in life", such as: "Is your own life at the moment very unhappy, fairly unhappy, fairly happy or very happy?" (Data Report 2002, p. 435).

It is thus assumed that people are in a position to provide information on these matters within the framework of a survey. It is also assumed that the answers obtained in this way have an indicator function. They can indicate certain sensitivities of the interviewees in which the social researcher is interested.

For other dimensions of this model, the operationalisation must be continued. In the case of the anomie dimension, three sub-dimensions are distinguished: firstly, feelings of powerlessness, secondly, senselessness and thirdly, disorientation. A corresponding indicator to be answered by the respondents is then: "I often feel lonely." Finally, the general satisfaction dimension was divided into two further sub-dimensions. These concern (a) subjective satisfaction and (b) the importance attached to specific areas. The idea is that there may well be a difference for the level of well-being if one is particularly (un)satisfied with an area rated as important or with an area rated as unimportant. Here, too, it is now relatively easy to formulate the corresponding indicators for an empirical survey.

The basic consideration was first of all that it would not make sense and probably not be possible to ask directly about such a complex and at the same time fuzzy issue as welfare. Presumably, the interviewees' understanding of this term would be rather diffuse, and thus their answers would be difficult to compare, making them unusable. If, on the other hand, one is dealing with simpler facts, where it can be assumed that they are understood in the same way by all persons, such an operationalisation is not necessary. Presumably, it is not difficult for a person to give information about his year of birth. Thus, at this point, it is unnecessary to ask, for example, about the fullness and colour of the hair, about the voice, about walking and visual aids, the completeness of the dentition or similar things. Here, the indication of the year of birth usually already provides sufficiently reliable information.

Later (cf. Chap. 9) it will be shown how the individual pieces of information obtained in this way can be brought together again. Ultimately, the aim is to use a number of pieces of information to draw conclusions about the actual issue of interest, in this case welfare. The question of perceived loneliness, for example, is merely a tool for this purpose.

The operationalisation process will be illustrated by a second example. In the context of the self-defence project, the focus was on the extent to which the principles of case law are followed by the population (compare, for example, hypotheses H1 and H2). The design of the survey assumes that the persons to be interviewed are presented with various cases for evaluation. The behaviour of a person under attack is then to be evaluated by the respondents as justified or as not justified. The individual cases are now arranged in such a way that three aspects vary systematically: firstly, the good attacked, secondly, the degree of superiority of the attacker, and thirdly, the good damaged in the attacker's defence. This is also the first step of operationalization. At the same time, it is also fixed that further facts such as the gender of the attacker, his country of origin, a possible alcoholization and so on, will not be the subject of consideration.

In the following, it has been determined which goods are to be attacked. This concerns (1.1) immaterial goods such as honour, (1.2) material goods such as property and (1.3) life and health. The superiority of the aggressor was distinguished according to whether (2.1) such a superiority can be seen or, (2.2) whether such a superiority obviously does not exist. Finally, the property damaged in the defense is also further specified. Here we are concerned with (3.1) a slight injury, (3.2) a serious injury and finally (3.3) a fatal injury to the attacker. Now the respondents are to evaluate certain cases. For this purpose, these were previously formulated accordingly at a further stage of operationalization. In total, an evaluation is obtained for $(3 \times 2 \times 3 =)$ 18 cases.

For example, a case involving an attack on health in which the attacker is fatally injured and in which the attacker is initially superior is: "An 18-year-old vocational student has been violently bullied and teased by a fellow student for some time. One day he is brutally attacked again. This time, however, he pulls a knife in his defense and fatally wounds his tormentor. In your opinion, is the vocational student justified in fatally injuring his tormentor or not?"

Provided with a corresponding question text, the complex issue of evaluating self-defence cases can now be surveyed empirically. This has required a number of work steps. Operationalisation can be considered complete when empirically observable indicators have been derived for all the terms contained in the hypotheses.

3.6   Theories and Laws

After a problem had been named and an understanding had been reached on the terminology in which it could be described, it was a matter of creating hypotheses, which represented an important function in and a first step towards solving the problem. It was emphasized that hypotheses are characterized above all by the fact that they contain – in addition to confirmed findings – as yet unconfirmed conjectures about relationships. In order to verify these assumptions, there is a need for suitable instruments. With this objective in mind, the next step was to operationalise the complex and/or latent issues, i.e. to make the phenomena that were initially not directly observable measurable. After the subsequent confrontation of the hypotheses with reality by means of empirical instruments, it is now a matter of condensing the findings obtained into theories and laws. Thus, in this section it will be shown what distinguishes theories and laws, what structures they have and what differences they possess.

Here, a specific, rather narrow understanding of theories, laws and hypotheses is represented. According to this, they have certain common features in their structure: They assert connections between at least two states of affairs, and they further contain logically contradiction-free statements about phenomena that can in principle be confronted with reality. Finally, they presuppose unambiguous definitions of the terms they use.

In this sense, theories and laws have the same structure as hypotheses: $C \rightarrow E$. However, there are important differences between these forms of knowledge. Laws, for example, are additionally characterized by the fact that the statements they contain have already been frequently empirically proven.

A theory is said to exist when there is a whole system of interrelated statements comprising several hypotheses and laws. A subset of these statements refers to empirically testable relationships between variables. Furthermore, one often speaks of a theory only if at least one other law can be derived from these laws (compare Opp 2004, p. 38; Albert 1964, p. 27).

Examples of such theories in the social sciences include learning theory, rational choice theory, and socialization theory. Thus, learning theory asserts that behaviors that are rewarded occur more frequently than those for which this is not true. Rational choice theory assumes that behavior is based on a cost-benefit trade-off and that those behaviors that produce higher benefits and lower costs, respectively, are to be expected. In contrast, socialisation theory maintains that behavioural patterns are established in people, particularly in the formative years, and that these then remain more or less constant over a lifetime.

Thus, one could start from a hierarchy. At the bottom of this hierarchy is a mere assumption about a relationship between two variables. This conjecture undergoes a first test in the formulation of the hypothesis. Here it must be possible to classify it without contradiction alongside knowledge that has already been confirmed. After a series of empirical tests have confirmed the validity of the assumption or led to modifications, one can speak of a law. The term theory – the upper end of our hierarchy – is used when a whole series of similar facts can be described with the help of a set of laws.

Now the question has to be discussed, in which way a hypothesis can be confirmed and thus transformed into a law or into a theory. In order to verify an assumption (if C, then E), one could take different paths. It would be conceivable to look for cases in which exactly the connection contained in the assumption has turned out to be true. One observes, for example, that the violent defence of particularly highly valued goods, such as one's own health, is regarded as more justified than the violent defence of goods regarded as less important, such as honour, and thus declares the corresponding hypothesis to be confirmed. Such an approach is practised by qualitatively oriented research and is generally rejected in quantitative social science (cf. Sect. 3.8). The alternative approach here is to first formulate a counter-hypothesis containing an alternative assumption. Now it is empirically checked whether cases can be found in which – as in our example – the defence of less important goods is considered justified. As long as this is not successful, the original assumption is retained. Thus, one looks to see if the alternative hypothesis can be rejected. In that case, there is reason to believe that the initial hypothesis can be maintained. If, however, the initial assumption proves to be untenable, it would have to be rejected or modified. This procedure is also called the falsification principle. According to this principle, theories do not represent a systematic accumulation of true knowledge, but are rather characterized by the gradual elimination of false knowledge.

To understand the problem, different types of laws and theories are now presented and discussed.

3.6.1 Theories with Different Scope

According to König (1973, p. 4), theories can be ordered according to the respective degree of abstraction of their statements. Following this idea, empirical regularities are initially (firstly) observed, and on this basis (secondly) so-called ad hoc theories are developed. Such theories are (initially) developed without any reference to existing theories. An example that is used again and again would be the observation that there is a correlation between the birth rate in a region and the number of storks in that region. An ad hoc theory could now formulate a corresponding dependency. In the third phase, which then follows, the aim is to work out theories of medium range (Merton 1957). Such theories establish connections to other theories. Very probably, the end of the stork hypothesis for the explanation of the birth rate would already be reached here. Further, intermediate-range theories specify the conditions under which the claimed relationships operate. For example, Inglehart (1977) explicitly limited his theory of value change to modern industrialized societies.

Finally, in rare cases (fourthly) it is possible to establish theories of higher complexity. Here, a relativization, such as the restriction of validity to a certain temporal period, would then be omitted.

3.6.2   Nomological and Probabilistic Laws

In the case of nomological or deterministic laws – similar to nomological and deterministic hypotheses (cf. Sect. 3.4) – statements are made which are always true and therefore also true without temporal restrictions, i.e. also in the past and in the future. This means that they can never be confirmed. It is only possible to falsify them or to confirm them provisionally. As is well known, it is not possible to make definite statements about future processes. Such all-statements are called deductive-nomological explanations, or in short D-N-models.

Further, this means that the statements made are not only valid indefinitely, but also for all elements. The sentence: "If an attacked good is given high importance, then its violent defence is also agreed to" is such a deterministic statement. It is not limited in space or time. Thus, it also claims validity in the future. Further, it claims to be applicable to all assailed property to which high importance is attached. If one were to find only one case in which such a type of defense is deemed inappropriate, the statement would be refuted.

In the social sciences, one mostly encounters non-deterministic or probabilistic laws. Such laws firstly restrict their scope of validity and secondly claim to be valid only with a certain probability. Here there are further analogies to the remarks on hypotheses already presented. In H12 it was famously assumed: The higher a person's level of education, the more likely his or her attitude is to be the same as that held by the judiciary. The restriction here is that the statement is not true for all persons with higher education. Only with a certain probability can a statement be made about what attitude a person with higher education will have.

3.6.3   The Role of Theories and Laws in Explanations

In the context of scientific explanations, theories and laws play a crucial role. The prerequisite for scientific explanations is the existence of laws. In laws the causes for a phenomenon to be explained are named. Let us assume that we are interested in a fact E, and that this fact is to be explained. If we had a law containing a statement of the form "if C exists, then E occurs" (C → E), we would be able to explain the appearance of the state of affairs E scientifically. C is then the cause of E. According to Hempel and Oppenheim (1948), an explanation is thus present if it is possible to name the cause of an occurrence as well as to determine the presence of these causes (also called boundary conditions). The causes of certain phenomena are fixed in laws. Thus scientific explanations have the following general form:

(1)	Nomological law (If C, than E)	}	Explanans
(2)	Boundary conditions (C is available)		
(3)	Theorem describing the result to be explained (W is available)		Explanandum

Specifically, it is to be assumed that the aim is to explain why the general population considers a certain type of behaviour to be self-defence and thus justified. According to the notation used so far, this would be E. Thus E would be the explanandum. It is assumed that E exists and at the same time it is asked why this is so. Now we have to look for a law which states the reasons why a conduct is justified by self-defence. Let us further suppose that such a law existed. It states that whenever a good considered to be of special importance is attacked (C), such an act is considered justified. So the cause of E would be C. Now, the next step is to ask what the actual boundary conditions are. In the case that in the behaviour to be explained a good considered to be particularly valuable was attacked, the boundary condition would then be fulfilled (C is present) and thus the explanation to be provided for E is given.

However, Hempel and Oppenheim's scheme can also be used to illustrate the procedure for a forecast. In forecasting, the typical question is: What will happen if? Specifically, it might be, "What response can be expected (E) if a good that is considered valuable is attacked (C)?" In other words: What event will occur if a particular cause is present? A theory of the form C → E is also needed to answer this question. Here the explanandum is known and the explanandum is searched for. Forecasts are thus always possible if one has a certain theory and at the same time knows all the boundary conditions.

Finally, a question motivated by social science can also be in what way an intervention has to take place (C) so that a certain goal (E) can be achieved. Or also, what must happen so that an action is justified by the general population as self-defence. In social planning, such a question could become relevant – admittedly with a different content orientation. A theory with the known structure is also needed for such a problem. It is known what must happen for E to occur. In order to justify self-defence, the answer could be, it is necessary that the property attacked and violently defended is regarded as a particularly valuable one.

In addition to these scientific explanations, we also speak of hypothetical explanations or potential explanations. These are not nomological laws that are used for explanation, but conjectures whose empirical truth is still largely unclear, or explanations that are not based on nomological laws.

3.6.4   Presentation of Theories

In practice, far more complicated theoretical models are adopted than have been shown here in a highly simplified form (U → W). Thus, a large number of independent variables are usually used to determine a state of affairs. Path diagrams or models are often used as tools to represent such a complex theory structure. In principle, models are characterized by the fact that they possess certain essential features of a different set of facts. For example, scaled-down models of a motor vehicle are not suitable for moving from one place to another. Nevertheless, they can be useful in determining the wind resistance of a vehicle. A map is also a type of model; although it does not photographically depict the landscape, it highlights key features and facilitates orientation. A theoretical approach frequently used for social science purposes is that of Ajzen and Fishbein, it is the Theory of Planned Behaviour, (compare Ajzen 1985, 1988, 1991; Ajzen and Fishbein 1980).

Figure 3.5 shows an example of the application of this theoretical approach to explain the ideas about, or potential behavioural intentions in self-defence (see also Häder and Klein 2002).

3.7   The Value Judgement Problem

The previous sections have discussed general rules to be followed in scientific work. Some of these rules are followed by consensus in the scientific community, while others are the subject of controversy. One such discussion that has been going on for a long time in empirical social research concerns the value judgment problem.

The value judgement controversy emerged in Germany at the beginning of the twentieth century from a debate in the Verein für Sozialpolitik. The subject of this discussion was the influence of values, i.e. personal opinions, political views and so on, on scientific work. Max Weber demanded a clear separation of scientifically justified statements from statements which, on the other hand, cannot be justified scientifically. The latter are value judgements. Weber was thus primarily concerned with the renunciation of evaluative statements in scientific work. The dispute between Max Weber (1951) and the chairman of the standing committee, Gustav Schmoller, thus concerned primarily the role of value judgments in scientific work.

The basis for this dispute was probably a different basic understanding of the role of science in each case. While Weber held the view that value-free research should be pursued, the ultimate goal of which is knowledge, Schmoller assumed the inevitability of value judgments and saw the goal of science as changing the world. Both points of view have far-reaching implications, as will be shown. First, it will be clarified more precisely what value judgments actually are. Then the subject of the value judgement controversy will be examined in more detail.

Value judgments are nominal sentences such as the following:

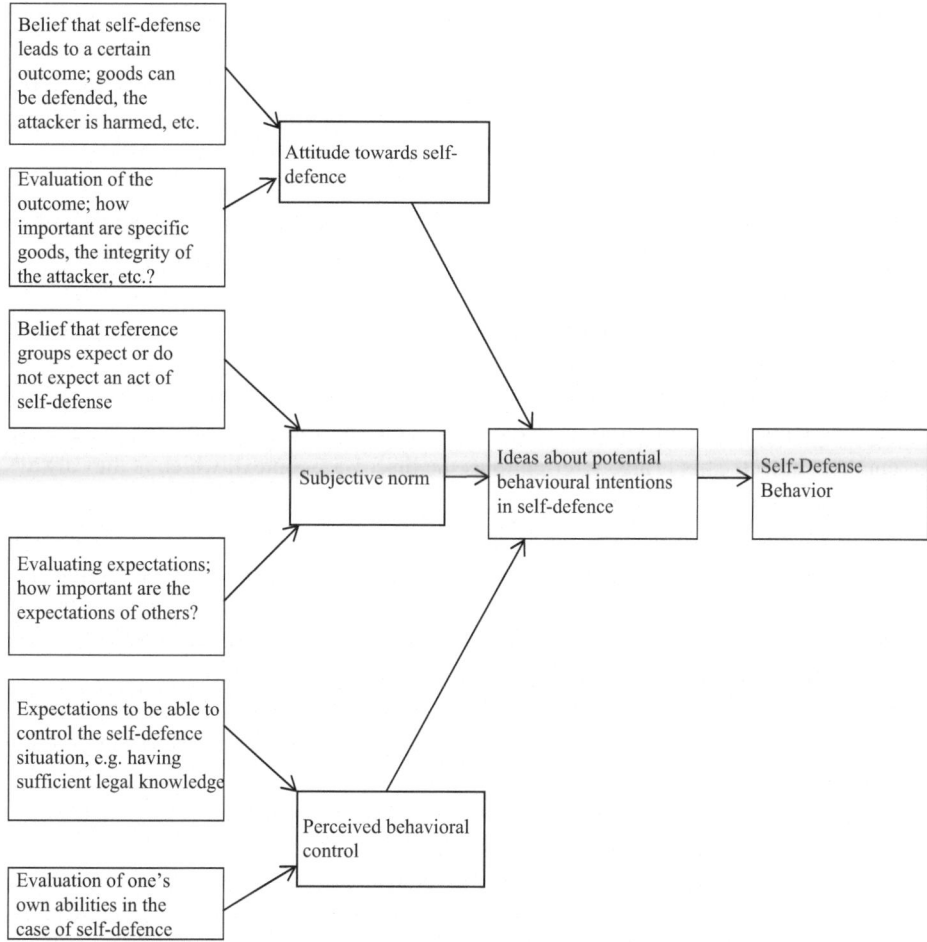

Fig. 3.5 Model for the explanation of ideas about or potential behavioural intentions in self-defence

- The distribution of income in Europe is more equitable than that in America.
- Unemployment in Germany is too high.
- Telephone interviews may be conducted unannounced between the hours of 9:00 am and 9:00 pm only.
- The death penalty should be reintroduced in Germany.
- Also, the sentence that a social scientist should not engage in speculation is a value judgment.

In such prescriptive[3] sentences, either facts are evaluated (as here, for example, the distribution of income or unemployment) or certain actions (such as non-participation in

---

[3] The word "prescribe" comes from Latin and means to prescribe.

speculation or the reintroduction of the death penalty) are demanded. The actual decisions for or against a certain alternative action are thus again not value judgements.

According to Opp, the literature on this dispute is now impossible to survey (Opp 2004, p. 215). First, however, some aspects concerning value judgments are undisputed. These will be presented first. Following this, the disputed issues will then be discussed.

1. For example, it is undisputed that the choice of a research question to be addressed by a particular research project involves a certain value judgement. Within the framework of the so-called relevance problem (Albert 1956, 1960, 1965), there is the question of what is important enough to be researched. The answer to this question at the same time evaluates as less important the topics that are not mentioned or those that are not considered worth working on. The time and resources spent by the researcher on researching one subject cannot be used at the same time or simultaneously for researching another subject. This is also referred to as the discovery context, which is thus always based on certain evaluations.

2. There are also values that relate to scientific work. A ban on falsification seems self-evident, but respect for the personal rights of test subjects, for example in covert social experiments in the field, documentation of the procedure in order to make a project comprehensible to others, adherence to the principles of data protection, and so on, are also values that must be observed in (social) scientific research (cf. Sect. 4.9).

3. Furthermore, evaluations can become the subject of scientific work. Here we are thinking of studies in which people are asked to give evaluations within empirical investigations. On a meta-level, empirical statements about such evaluations are created. For example, it is a value judgement if 85% of the people surveyed from the general population believe that a certain type of behaviour is not covered by self-defence – while the Federal Court of Justice, in a judgement in the name of the people, has found exactly the opposite to be true (cf. Sect. 9.3). The same applies to statements such as that a majority of those questioned were of the opinion that the economy should do more for the environment or also the finding that in the Federal Republic of Germany a certain percentage of the population is of the opinion that a government in office is incompetent and should be replaced. Again, these are not the value judgements that the value judgement controversy has been led to.

4. The results of a social science study are used to achieve certain goals. They are thus exploited. However, while the empirical statements should not yet logically contain any indication of a certain action, such are then derived in practice. If, for example, a study has been made on the structure of the supporters of a political party, the corresponding results can be used both by the supporters of this party, for example to recruit new members, and by the opponents of the party, for example to develop a certain election campaign counter-strategy. This is the so-called exploitation context, where the responsible use of research results must be demanded. It is also indisputable that valuations are used at this point.

The argumentation presented by Weber, which triggered the discussion on value judgements, primarily concerned another aspect of social science research: It concerned the maxim that the scientific description of facts should be objective, that is, value-free. Weber's position on this was that the subjective wishes of the scientist must not have any influence on the design of the context of reasoning.

Up to now, we have mainly been talking about the discovery context and the exploitation context. Now the concept of the context of justification is also introduced. Within the framework of all these contexts, decisions are made in each case that concern the social scientific research process, that are intended to move it forward. Since reality can never be researched as a whole, it is in the context of the context of *discovery that decisions* must be made as to which section research should be devoted to. As shown, value concepts play a certain role in this.

This is followed by decisions within the framework of the reasoning context. In the reasoning *context, the* most appropriate approaches and methods for dealing with a particular problem must be determined. Again, there are numerous decisions to be made. In making these decisions, however, it is assumed that – given a particular problem – the course of an investigation must be independent of the basic political and ideological convictions of the persons conducting the research, of their wishes and (methodological) preferences. Thus, to use a common metaphor in contemporary politics, there are only suitable and unsuitable methods for addressing an issue. It is precisely this context of justification that is meant when we speak of the postulate of freedom from value judgement. Assuming a certain topic as given, it should be irrelevant what gender the researcher is, what religion he belongs to, how old he is, and so on.

Furthermore, when dealing with the context of justification, it should be pointed out that the possibility for an intersubjective verifiability of the methodological procedure should always be given. This applies to the disclosure of all steps, starting with the definition of the research questions, through the principles of the selection of the objects of investigation and the instruments used (such as questionnaires, observation protocols, etc.), to the error correction of the raw data matrix and the statistical procedures used (cf. Chap. 9). In some cases, different strategies are possible for dealing with a topic. For example, it would be conceivable to use both a face-to-face interview and a telephone interview to investigate a question. However, reasons must be given as to why the decision was ultimately made in favour of a particular variant.

The decisions in the context of discovery are followed by those in the *context of exploitation*. At this point, subjective evaluations again play a role – as has also already been shown. It depends, for example, on the political orientations of a person how the results of a social science investigation, for example the prediction of a certain election outcome, are dealt with. The researcher is thus free – without violating the postulate of freedom of value judgement – to make such subjective and evaluative use of his research results (once they are available).

In this context, reference must be made to another important aspect. For example, the principle applies that subjective value judgements within a social science study must be made clearly recognisable when documenting it and that these value judgements must be

clearly distinguished from the empirical findings. For example, the following sentence constitutes an impermissible value judgment: "The study found that self-defense law in the
Federal Republic must be changed." On the basis of the demand to do something, to make a
change in the law here, a value judgement is made. However, such a judgement cannot be
empirically proven. The following sentences, on the other hand, would be correct: (1) "The
study showed that 85% of the respondents evaluate certain behaviors in the case of self-
defense differently than the judiciary." (2) "This should lead to the conclusion to discuss
changing the law." Here, the first sentence makes an empirically verifiable and therefore
admissible claim. The second sentence, on the other hand, represents a clear value judgment.

The sentence: "The study shows that in order to increase the age at which people start
smoking cigarettes, tobacco taxes must be increased," is also not admissible. It is true that
a scientific study can be used to clarify which causes can be held responsible for a phenomenon. Here, for example, it has been found that the independent variable of price or
cost influences the dependent variable of consumer behaviour. Thus, the statement that the
age at which people start smoking cigarettes can be increased by raising tobacco taxes is
formally correct. After all, such a relationship is in principle empirically verifiable.
However, it would be inadmissible to deduce from this that tobacco taxes *must be increased
in order* to achieve a certain goal. A political decision (such as raising a tax) always has
other consequences (in this case, the lack of tax revenue from the sale of cigarettes).

With this, it would be more appropriate to state as follows, "A scientific study has
shown that increasing the cost leads to behavioral changes. The conclusion should be to
raise taxes to reduce cigarette consumption." In this way, the first sentence quotes the scientific evidence without any value judgement, and the second then makes a value judgement on this basis, while at the same time clearly stating it as such.

Now the problem arises in this context whether there can be other reasons for the provisional acceptance of a hypothesis or a theory than a correspondence between theory and
reality. The researcher's understanding of science is central to answering this question.
The position that holds the postulate of value freedom assumes that science has to provide
true knowledge about the world. The opposing position points out that in social science
studies – in contrast to natural science studies – the individual scientist is at the same time
himself a member of the society he is researching. He has certain interests which he pursues as a member of a class or stratum within this society. Thus, his work can never be
value-free. The understanding of science would then be that it is not only a matter of gaining knowledge about social reality, but also of criticizing it and, if necessary, changing it.
Thus, critical theory demands the scientist's interference in politics.

The value judgement dispute found a certain continuation in the methodological dispute between the representatives of critical theory (to be mentioned are the Frankfurt
School, Max Horkheimer, Theodor Adorno) and the representatives of critical rationalism,
above all associated with the name Karl Popper. In the following Sect. 3.8 we will return
to this problem.

Opp takes an interesting standpoint on this. He first specifies what the actual value judgment postulate is about: "A scientist should make it clear which of his statements are

evaluations and which of his statements are object-language, i.e. factual statements". (Opp 1999, p. 218, compare also Opp 2004, p. 218). Now, according to Opp, it is important, for example in the scientific preparation of a decision, not to mix factual statements and value statements. While factual statements can be checked against reality, this is not the case for value statements. An empirically verifiable factual statement would be that 85% of the population values a particular self-defense case differently than the case law. A value statement that is not verifiable against reality would be that legislation must therefore be changed.

In the discussion about the freedom of value judgement, the argument is further put forward that a separation between factual and value statements is not possible at all, since all terms are in themselves already value-laden. Thus, expressions such as freedom, crime, victims, capitalism or half-full and half-empty are always associated with certain values. This would be different for a natural scientist, who can work on his subject with a completely different distance. Opp now emphasizes that this is correct, but that a distinction must always be made between the connotative (evaluative) and the denotative (descriptive) meaning of such terms. Science, however, uses only the descriptive meaning of the terms. Thus, Opp (1999, p. 221) concludes, the above argument about the impossibility of separating factual and value statements must be regarded as refuted.

Reference should also be made to the following arguments:

- The postulate of freedom of value judgement, i.e. the demand to distinguish clearly between evaluative statements and factual statements, is not identical with the demand for political abstinence on the part of a scientist. The latter is free to be politically active as well. In his scientific work, however, he has to follow the above-mentioned premise within the framework of the context of reasoning.
- Another argument in the value judgement controversy is the prevention of misuse of scientific research results by anchoring value judgements in research. However, as the past has shown, the misuse of a theory developed by science cannot be prevented by, for example, incorporating moral judgements into this theory. Rather, what is needed here are socially responsible individuals who stand up against such misuse.
- The reference that a value-free science has a conservative function, since it only depicts the existing without questioning it, is also untenable in this way. After all, the value judgement postulate merely amounts to clearly separating the personal opinions of a scientist and the results of science. With the help of a scientific theory, social changes can very well be justified and promoted.

3.8 Qualitative and Quantitative Research Approaches

In addition to the value judgement dispute, there has also been – at least in the past – a sometimes bitter dispute about the relationship between primarily qualitative and primarily quantitative empirical research. Both approaches will now be examined in more detail.

By way of introduction, the terminology should be discussed. It should be noted that, strictly speaking, there can be no such thing as purely qualitative research, just as there can be no such thing as a purely quantitative approach. Rather, quality and quantity form a unity. If, in a quantifying study, the characteristics of an issue are to be determined, the quality sought, i.e. the issue to be measured, must be specified. Thus, a quantitative study presupposes the precise determination of the quality to be counted. On the other hand, for a qualitative study, at least quantity one must be present in order to be studied at all. It is therefore not necessary to assume that there is a fundamental contradiction between the two research approaches.

The specifics of primarily qualitative approaches as well as those of primarily quantitative research can be vividly illustrated by a comparison with a criminal case. The detective who sets out to find the perpetrator is engaged in what can be understood as qualitative work. He pursues various traces, without knowing in advance which of them will be successful. At the beginning of his investigations, it is not clear to him what exact result his work will lead to. He cannot determine his (investigative) methods exactly in advance; they arise successively from the preceding findings. He also merely sets out to solve a particular case. The detective's goal is achieved when he has identified a person sufficiently suspected of the crime and has a well-founded assumption about his guilt. To do this, he may have gathered an immense amount of material.

Now begins the work of the judge, which, on the other hand, strongly resembles quantitative research. While the detective has been in the field, the judge's work takes place in his office. This is a neutral or artificial environment. Here, very concrete arguments for and against the defendant are compiled. Even before the trial begins, it is clear that there are really – we simplify a little – only two possible outcomes: The charges are tenable or they are dismissed. All erroneously pursued leads and suspicions are no longer of interest at this point. The task to be solved by the judge is precisely pre-structured. He takes note of all the arguments at hand and then decides – with a certain probability of error as small as possible – whether the arguments (or the hypotheses) are in favour of the charge or whether the alternative hypothesis is correct and the accused must be acquitted. For everything else, he should be blind. A criterion of quality for a good judge might be whether he has succeeded in being as objective as possible, that is, in judging as any other judge would have done at that point. The detective's work, on the other hand, will have to be measured by whether the case, whether the object of interest, has been properly investigated, whether he has meticulously followed up all possible leads, whether he has properly documented them and, finally, whether he has drawn the right conclusions from the wealth of his investigative work.

Such a comparison of the two approaches shows quite well the different (methodological) approaches and objectives in social science research. However, it unfortunately only insufficiently clarifies the theoretical premises underlying the qualitative and quantitative approaches. Primarily quantitative social research is based on an analytical-nomological understanding of science. This is based on the following assumptions:

- Society is – analogous to nature – an ordered, structurally regular world.
- If one follows this, then the cause-effect principle, of the kind: if C exists, E occurs (C → E), can thus also be applied to the understanding of society.
- The task of science is to uncover these principles and rules and thus ultimately to make the world controllable.
- The sciences of nature and society differ only on the basis of their respective subject matter, but not on the basis of the principled nature of the methodological approach to the study of their respective subject matter.
- This makes it obvious to work with hypotheses in the social sciences as well. As shown in Sect. 3.4 above, hypotheses make assumptions about reality (C → E).

This analytical-nomological understanding of science is contrasted with the interpretative approach. According to this approach, qualitatively oriented social science is based on the following partly contradictory assumptions:

- There are no predetermined structures in society. Rather, people create these structures themselves through their actions. There are therefore no analogies to natural science.
- Each person interprets the world according to his or her experiences. Consequently, reality is the result of complex sequences of interactions.
- A search for causal laws (C → E) operating independently of human beings would thus be pointless in the social sciences.
- The aim of social science is rather to gain experience in the respective field of investigation. The principle of openness prevails. Understanding thereby becomes a central category.
- One task of qualitative interviewing approaches is, for example, the perception of the actors' definitions of the situation. In this way, interpretative assignments of meaning are made.

The theoretical premises only briefly outlined result in different methodological strategies for the two approaches. Thus, quantifying research will strive for highly standardized instruments in order to keep the subjective influence of the investigators to a minimum and thus to secure findings that are as objective as possible, i.e. independent of the respective researcher. In contrast, qualitative research will devote itself in detail to individual cases and investigate them as intensively as possible. A previously standardized research program can thus not be used.

If one compares the different dictions of qualitative and quantitative research, the system shown in Table 3.2 emerges (compare also Brüsemeister 2000, p. 21 ff.).

In order to illustrate the specificity of both approaches even more clearly, the different methodological procedures for the evaluation of the research results in two contrasting studies are described below as examples. As an example of a quantitative study, a project on the determinants of people's enjoyment of life (compare Clark and Oswald 2002;

Table 3.2 Comparison of qualitative and quantitative research approaches

Quantitative	Qualitative
A nomothetic self-understanding, it is then about the search for natural laws, these are treated in a generalizing way	An idiographic self-conception is present, the facts are described, the procedure is individualizing (compare Windelbrand 1894)
The model is the scientific approach	A typical humanities approach is present
The examinations take place under standardized (also laboratory) conditions	The tests shall be carried out as far as possible under unadulterated field conditions
A deductive procedure (general → particular) is used, this is considered truth-preserving	An inductive procedure (particular → general) takes place, it is considered truth-expanding
The testing of established hypotheses is the goal	It is about discoveries, the principle of openness applies (compare Hoffmann-Riem 1980)
The facts are to be explained, for this the external causes (C) must be consulted	The facts are to be understood, for this the inner reasons are to be considered
Sampling is used to obtain information, a probability of error can be specified	Individual cases are analysed, a deliberate theoretical sampling takes place
The work happens with large numbers of cases	Relatively few cases are investigated
Hard, i.e. fully standardized methods are used	Soft, hardly standardized methods are used
The principle of measurement and operationalization is practiced	It is about describing cases and raising awareness (compare Strauss and Corbin 1996, p. 25 ff.)
Knowledge goals are statements about aggregates and statistical correlations	The statements are made on a case-by-case basis, these are reconstructed (compare Kelle and Kluge 1999, p. 14 f.)
Generalisations are sought (frequencies)	Type formations (inner logic) are made
Quality criteria are objectivity, reliability and validity	The quality criterion is the object-relatedness of the methods (cf. Lamnek 1995, p. 152 ff.)

Blanchflower and Oswald 2002) will be presented; the qualitative study was concerned with the living conditions of homeless people.

The first approach uses – and this is quite common in quantitative research – a regression model for the evaluation (compare Sect. 9.2 as well as Backhaus et al. 2000, p. 1 ff.; Andreß et al. 1997, p. 261 ff.; Ghanbari 2002, p. 215 ff.). The project was guided by the following research question: Different events such as illness, marriage and unemployment, have an important impact on people's lives. In this context, a way is sought to measure the different strengths of these influences on human happiness and on mental health in a standardized way in a large sample. Regression equations represent the solution in data analysis. They have the following appearance:

$$y = b_0 + b_1 x_1 + b_2 x_2 + \ldots + b_j x_j + e_i \tag{3.1}$$

Here, the expression of happiness, as the dependent variable (y), is estimated based on a set of determinants (x_i). Where y is also referred to by the term regressand. In the present case, it is the level of enjoyment of life, this was standardized with the help of various indicators. b_0 is a constant element, b_j is the regression coefficient of the jth regressor and x_j is the jth regressor. Regressors that influence happiness include, for example, level of income, living with a partner (as a binary variable), health status, and similar variables. These are then the independent variables that together are supposed to explain the level of happiness. They are also referred to as predictor variables. Finally, an error variable, e, was included in the equation (cf. Kühnel and Krebs 2001, p. 388 ff.).

The above project has now empirically collected the relevant variables on a random sample (this was the British Household Panel)[4] and can thus estimate the regression model shown. This estimation yields, first, a coefficient of determination. This measure provides information about the proportion of variance explained by the independent variables. In other words, it gives an indication of how well or poorly happiness (y) can be explained by the independent variables included in the equation. Second, the estimation determines the respective regression coefficients. These in turn indicate how strong the influence of each variable (for example, income) is on the dependent variable. It is thus obvious that not every determinant has an equally strong effect on the degree of enjoyment of life.

The regression model thus offers the possibility of predicting what change will occur in the dependent variable (How does enjoyment of life change?) when there is a change in each of the predictor variables (for example, higher income, poorer health, and similar events). The coefficients for income and for the various life events further allow one to calculate the amount of money that each life event compensates for. Thus, one can say by what amount income must increase to compensate for a change in the other predictor variables. For example, one result is that a wedding yields the same amount of happiness as an extra annual income of £70,000. Losing a partner yields a level of unhappiness that could be offset by an amount of £170,000. The corresponding results are shown in Table 3.3. The question of which is worse for a person's happiness, divorce or unemployment, can now be easily answered using the table.

These results were presented to demonstrate typical findings of quantitative research. As shown, the views of individual respondents no longer play any role here in fact. Due to a far-reaching standardization of the method, the answers of the target persons are

Table 3.3 Value of life events (expressed in £) when changing status from … to …

Employed to unemployed	−15,000
Single to married	Not significant
Married to separated	−8000
Married to divorced	−1000
Married to widowed	−7000
Very good health good health	−10,000
Very good health to something healthy	−32,000

Source: British Household Panel data, from: Clark and Oswald (2002)

---

[4] Compare to more information: http://www.iser.essex.ac.uk/survey/bhps, last accessed 12/16/2013.

summarized and statistically processed. The statements made also do not refer to very specific individual people.

The strategy is completely different with a qualitative approach. Here, the procedure is to be demonstrated by means of a survey of homeless people (cf. Krieg 2004) (cf. Brinker 1996). In Dresden, eight homeless people were interviewed using an open-ended survey standard (see Sect. 6.1.3). were interviewed. The main focus was on the biographical trajectories that led to homelessness, the everyday organisation of homeless people and their reception of the social assistance systems. The interviews were recorded and transcribed (see Sect. 9.1). Table 3.4 shows a small excerpt.

On the basis of such texts, it is now possible to look for certain (typical) cases, to work out conspicuous features and so on. For example, it turned out that among the homeless people interviewed, the types help-seeker, help-refuser, newcomer and resigned (Krieg 2004, p. 117) can be distinguished, whereby certain types are at the beginning of a homeless career and others already have great experience in the homeless milieu.

In this way, some essential differences between the two approaches should have become clear. In summary, some problems in the use of qualitative or quantitative methods of

Table 3.4 Extract from a transcript of a qualitative interview with a homeless person in Dresden (Krieg 2004, Annex, p. 23)

249	A	Yes … private tutor … = although I lay
250		no value to it but he makes a point of being a private tutor, + well is
251		such a distinguished man there and after A. back to the Maxim gorgie again
252		Street … Beautiful … very beautiful … I mean here it also goes:: but it is, ä
253		difference is that already … well that's the way it is … better than nothing there
254		I tried times times tried = = here in so'n I had though ne
255		apartment … = = how it is when you sleep here as a homeless person outside was
256		I have been in such a demolition house I spent the night there, ;; there came the rats
257		on, the ten = but I had boots on there nothing can happen and there hab
258		I said, + I won't do it again (laughs), = there I have made out of mill barrels
259		eaten but + they have sometimes extra already the (…) sausages so put down
260		that this is already packed that you can take:: and so everything … sometimes
261		a whole pack, coffee = what do I want with it (laughs). That was just a
262		a bit broken and so what you then from … (…) which was then just not
263		been more vacuum, yeah … and then oh no and then eggs sometimes eggs
264		and butter, so I'm not starving, as you can see, = I'm so not
265		a little bit fatter, + so anyway there,,, from there … I am then then then
266		to A. then back again, I told them yes, back to the Maxim
267		gorgie after Maxim Gorgie Street, + where did I go? (laughs). =
268		I must have gone somewhere … well … + where could that have been
269		be?
270	Int.	I have unfortunately no idea
271	A.	… did I get a place there? (pause)
272		I can't tell him that now (pause)

social research should be pointed out at this point. These are still the subject of sometimes highly controversial discussions.

1. No doubts should be raised about the raison d'être of either approach. Both projects, presented only briefly here, should provide interesting information for social scientists.
2. The selection of samples in qualitative studies is usually done deliberately (theoretical sampling), often in an attempt to identify typical cases. This leaves open the generalizability of the results found on such a basis.
3. In order to be able to draw conclusions from quantitative surveys – as presented here – criteria such as objectivity, reliability and validity must be guaranteed in the surveys (cf. Sect. 4.4). The extent to which this is actually the case would have to be examined in each case.
4. Qualitative researchers need to be aware that the perceptions of the target interviewees can be highly selective. This can become even more of a problem because, although in-depth interviews are conducted, these are only with relatively few interviewees.
5. A self-fulfilling prophecy effect in qualitative research approaches needs to be discussed. While an attempt can be made to minimise the influence of the interviewer in quantitative studies through a high degree of standardisation, this is in fact futile in qualitative approaches. This raises the question of whether or how intensively the target persons in a qualitative interview could be influenced (even unconsciously) by the interviewer's preconceived opinions (cf. Schnepper 2004).
6. The result of qualitative studies is usually a considerable amount of material. The eight interviews in the project described above lasted between half an hour and two hours. The transcribed text is approximately 220 pages long (cf. Krieg 2004, Appendix).
7. While in the case of quantitative interviews it is usually possible to employ non-specialist interviewers, high demands must be placed on the interviewer in the case of qualitative surveys. In most cases, the interviewer and the researcher are one and the same person. Problems resulting from this have already been addressed (cf. point 4).
8. A high level of linguistic competence on the part of the target person is a prerequisite for qualitative interviews. As the transcript above also shows, it is not always easy for the respondent to articulate himself. Quantitative studies, on the other hand, strive to do without this to a large extent, for example with the help of predefined answer categories.
9. Both qualitative and quantitative studies sometimes require a considerable amount of time.

The sometimes heated debate about the relationship and the justification of qualitative and quantitative methods in empirical social research has probably entered a new stage in the meantime; it has calmed down considerably. The juxtaposition and confrontation of qualitative versus quantitative approaches has rather developed into a common coexistence. In the practice of empirical social research, the possibilities of a mutual supplementation or a combination of both approaches are now increasingly practiced.

Research and Investigation Planning

4

Abstract

The concrete course of a social science investigation depends on various things. Above all, the respective research interest, the available resources and the complexity of the research question exert a decisive influence on the design of a research project. But also a possible client or the regulations of a doctoral thesis can significantly influence the course of the investigation. Thus, in the practice of empirical social research, one has to deal with very different and diverse designs. In addition to permanently funded projects for the long-term observation of society, for which a relatively large staff is responsible, such as the European Social Survey (ESS) or the General Population Survey of the Social Sciences (ALLBUS – compare Sect. 6.1.4 on the two surveys), there are numerous smaller studies, which are carried out, for example, for graduate theses and which essentially have to manage without external funding. Last but not least, there are a number of studies that serve the purpose of market research and are financed by companies. While survey series such as the ALLBUS and the ESS have meanwhile developed certain methodological routines according to which the individual surveys are carried out, the development of the design of a one-off survey always presents a particular challenge to the researcher. Between these two poles, various other research projects can be identified, whose approach can always be very different. Despite this diversity in approaches, an attempt is made here to demonstrate a general phase model for the course of a social science investigation. A division of social science investigations into phases offers various advantages. For example, it supports time and resource planning and, similar to the checklist to be worked through by a pilot before take-off, it points out all the essential work steps, such as the absolute necessity of a preliminary investigation. In this way, despite the differentiated nature of the individual research projects, it can help to prevent duplication of work or to structure the entire process optimally according to a certain logic. In the following Sect. 4.1, such a phase model is first presented and then applied to the Dresden self-defence survey (Sect. 4.2).

M. Häder, *Empirical Social Research*, https://doi.org/10.1007/978-3-658-37907-0_4

4.1   Phases of an Empirical Project

The starting point for an empirical study is a specific problem or question (cf. Sect. 3.2). As an example, reference has already been made to a study which ascertained the views of the population on self-defence legislation in the Federal Republic of Germany. Other projects may be concerned with finding out the determinants of environmentally friendly behaviour, exploring the causes of declining job satisfaction among a company's employees in a sociological survey, making an election forecast, analysing which media are used by which people, who uses which means of transport to travel to work and why, and so on.

These projects will develop a specific design to solve a problem or clarify an issue and then implement it step by step. According to Diekmann (2012, p. 186 ff.), five phases can be distinguished. The authors of other textbooks, such as Kromrey (2002), make a similar division to Diekmann: First, the formulation and specification of the research problem; second, the planning and preparation of the survey; third, the data collection; fourth, the data analysis; and finally, fifth, the final reporting.

By their very nature, such classifications have a primarily heuristic character and are therefore relatively arbitrary. In further development of the systematization proposed by Diekmann, the phases of an investigation are to be described here in a result-oriented manner. This means that a phase is always considered to be completed when a certain result – economists would possibly speak of intermediate products or milestones – can be presented. Depending on this result, the next phase is then tackled. The following phase model is assumed:

First phase	The preparation of the project plan. Depending on the type of project, this may result in a proposal to a research funding institution or an application for a call for tenders, or even a concept for a graduate thesis as an intermediate product
Second phase	The elaboration of the design of the study including the necessary survey instruments. For this purpose, the ideas previously developed in the project plan are to be specified and the corresponding instruments are to be elaborated
Third phase	The collection of data in the field. In quantitative studies, for example, this phase ends with the creation of a machine-readable data set
Fourth phase	The evaluation of the study, including the preparation of tables, overviews, statistical calculations, graphical representations and similar materials
Fifth phase	The documentation of the methodology used and the publication of the findings, for example in the form of articles in journals and anthologies, form the conclusion of the project

All steps are linked to a series of decisions that the researcher must make. Each decision specifies the research problem and helps to make it empirically workable. The first stage involves formulating and specifying the research problem. It is completed when an appropriate project description is available. The second phase of design development and preparation for the survey aims to develop the appropriate instruments, for example, an observation protocol, a questionnaire for an interview, or a coding scheme for a content analysis. The third phase of data collection ends with the production of a machine-readable data file for quantifying studies, and the fourth phase of data analysis then produces the necessary tables and graphs. Finally, reporting involves documenting the project experience and publishing the substantive findings obtained to make them accessible for practical use. The phases will now be presented in detail.

First Phase: The Creation of the Project Plan
In order to be able to describe an envisaged study more precisely in the context of the preparation of the project plan, it is useful to distinguish between certain types of such studies. The following criteria are used for this typification: First, it is asked whether it is basic research or an applied project. Secondly, it is important to know whether the project is externally or self-initiated by the researcher. Thirdly, it deserves attention whether an exploratory or a confirmatory concern is to be pursued with the research. Thus, the following pairs of types exist.

(a) *Basic* social science *research* projects and *applied research* projects
In the context of basic research, methodologically demanding new developments of instruments and designs are sometimes required. Examples include theory comparisons in which the aim is to empirically determine the performance or explanatory power of different theoretical models for uncovering a concrete problem. Such complex projects sometimes also require considerable innovations in terms of the methodology of the study.

Applied research projects, on the other hand, can be processed with the help of more or less proven methodological routines. For example, existing methodological standards can be used in regular employee surveys to collect information about the working atmosphere, turnover intentions and information about internal company matters. However, election research equipped with appropriate routines can also be counted among this category of projects. Such research usually provides comparable data over longer periods of time.

(b) *Contract research* projects have different characteristics than *self-initiated studies*
Commissioned research projects are publicly announced by public institutions such as federal ministries with the help of calls for proposals or are addressed to a specific group of potential applicants (cf. Figure 4.1). However, projects that can be worked on as part of a graduate thesis and are sponsored by a company or public institution also belong to this category. In addition, there are the commissions taken on by survey institutes in the private sector. Here, the problem to be researched is largely predetermined. Social scientists can apply for the direction and must, for example, calculate the resources they require. The methodological scope for working on the project is thus usually relatively clearly fixed. A certain difficulty can arise from the fact that the clients of a study do not themselves have sufficient experience to be able to realistically assess the capabilities of empirical social research. In this case, social scientists are obliged to provide clarification, if necessary. After all, not all expectations placed

---

Conducting a survey on the topic "Young people in new learning environments"

The project "Young People in New Learning Worlds" of the German Youth Institute, which is funded by the Federal Ministry of Education and Research, is concerned with extracurricular activities and learning processes of young people. In order to obtain more detailed information on this topic, a nationwide survey is to be carried out among pupils from different schools

The planned survey period is March – April 2002. It is intended to conduct a written class survey focusing on grades 9 to 12 and thus on the age group 15 to 19. In addition, different school types (Main school, Secondary school, Gymnasium and Vocational school) are to be taken into account in the design of the survey, as well as a regional differentiation according to the criteria East/West and urban/rural

The size of the sample should be at least 1000 respondents net, for the questionnaire about 60 questions, 10 of them open, are planned. Detailed information on the design and the required leads can be found on the DJI website at: www.dji.de/Forschung/AbteilungFSP1/, current-projects/youth-in-new-core-worlds

Offers should be sent in a sealed envelope marked "Do not open. Survey offer for pupil study" to be sent by 6 Jan. 2002 to: Deutsches Jugendinstitut, for the attention of Mr W. xxxx, Nockherstr. 2, 81541 Munich. For further information please contact Mrs. xxxx, Tel.: 089/62306-145 or Dr. xxxx, Tel.: 089/62306-167

The opening of tenders will take place on 7 January 2002 and the deadline for the award of contracts will be 8 February 2002

D J I

Fig. 4.1 Example of a tender text of the German Youth Institute (DJI) for conducting a survey

on social research can be fulfilled by it. This is especially true when one takes into account the resources available for the project work

Self-initiated projects, on the other hand, are characterized by greater latitude in their design. Here, the researcher's imagination can unfold more freely. Of course, it should be noted that the resources required for the work must be available or must be acquired. In addition to financial resources, one must also think of material resources and a corresponding time fund. The project plan to be drawn up plays an important role here.

(c) *Exploratory* and *confirmatory* research also have their own characteristics.

Explorative studies aim to obtain empirical basic knowledge for dealing with a problem. Thus, the procedure here is only slightly structured and subject to a relatively large subjective influence. The result of such projects is, for example, initial hypotheses for the solution of a problem. It is very difficult to formalise the process of such projects (cf. Sect. 3.8). In this case, the individual steps result from the knowledge gained in the previous step. If necessary, individual steps may have to be repeated, others may be omitted, and so on.

In *confirmatory research* projects, concrete hypotheses are tested according to precise methodological rules. The subjective influence of the respective social scientists is thus to be kept correspondingly low in the solution of the problem. The result is to be expected to be as objective as possible in terms of whether the tested hypotheses have proved themselves empirically or not.

In the further presentation of the research process, the focus will be primarily on these types of investigations

For every (social) scientific study, it is important to face up to professional criticism and, if possible, to stand up to this criticism. Only in this way will the results presented find recognition and attention and thus be able to contribute to practical decisions. In order to be able to face criticism successfully, a number of rules should already be followed when drawing up the project plan.

The decision of wanting to work on a particular problem will come about due to various aspects. First of all, the creativity of the researcher in question can play a major role in raising an original and exciting question. If the researcher succeeds in convincing others of the necessity and usefulness of the research idea, the project has a good chance of gaining the support of colleagues in the field – and thus, if necessary, a basis for funding. In addition to the creativity of the researcher, the circumstances prevailing at a particular point in time also play a role. Sometimes it is particularly attractive to use opportunities such as the collapse of the political system of the GDR, the spontaneous relief actions on the occasion of flood catastrophes on the Elbe or the Oder, the unification of Europe or a recent change in legislation as an occasion for a research project. Controversies in the literature or a replication of existing studies or studies from other regions can also lead to an interesting problem. Last but not least, political and other public debates can also be taken up when it comes to finding ideas for the selection of a research problem. Depending on the taste of the times, topics such as the integration of foreigners, the revision of Paragraph 218, new regulations in the granting of unemployment benefits, the treatment of terrorists, or the electoral successes of politically extreme parties are particularly discussed. This results in a more or less urgent need to deal with such issues in a sociological way.

In order to be meaningful, a project plan should outline certain key points of the envisaged research. It is advantageous if the project plan to be developed enables outside specialists to critically assess the research project. Such assessments or appraisals are based, among other things, on the theoretical foundation of the problem, on the chances of success, on the professionalism of the approach, on the competence of the investigators, on the work results already presented by them, on the usefulness of the targeted problem solution, on the realistic assessment of the resources required for the work, and on similar criteria. Depending on the complexity of the investigation, the scope of such a project plan will vary. For example, a relatively short synopsis of three to four pages may be sufficient for the assignment of a topic for a Master's thesis. In contrast, the application for a Collaborative Research Centre at the German Research Foundation will be correspondingly more extensive.

There are sometimes certain formal requirements for the preparation of a project plan. For example, it is usually expected that the proposal is based on a detailed study of the literature, that the ideas underlying the project are presented comprehensively, and that existing theoretical approaches are reviewed for their viability in solving the problem. Two aspects in particular should be pointed out:

It is important to consider whether the problem at the centre of interest can be dealt with at all using the methods of empirical social research. The assessment of this question requires a relatively high level of methodological competence. For example, it may be of the utmost importance for an institution such as the Federal Criminal Police Office to use an empirical study to gain a more detailed knowledge of the practice and exact prevalence of protection racketeering in a large city and to tender for a corresponding empirical study for this purpose. However, the practical and/or theoretical need for such a study does not guarantee its methodological feasibility. Thus, it would be necessary to conduct a survey of potentially affected persons, preferably on the basis of a random sample. Both the drawing of a random sample and the generation of motivation to participate among the persons concerned bring empirical social research to or beyond the limits of its own capabilities. Thus, the presentation of the basic workability of a problem is an important part of the project plan.

As part of the development of the project plan, it is also necessary to determine which will be the dependent and which will be the independent variables. While the dependent variable represents the very facts for which an explanation is to be sought, the independent variables are expected to contribute to such an explanation. For example, it could be of scientific interest how it comes about that people are prepared to help a person who has been attacked. The question would then be: What determinants must be present for individuals to express a willingness to come to the aid of others when assaulted. The dependent variable to be determined here would be the willingness to help in self-defence reported by the persons interviewed.

The definition of the dependent variable further specifies the objective of the study. In the example given, it would be conceivable to determine as independent variables such variables as the number of attacks on persons that have already taken place in a particular place of residence, the age of a person, their physical condition, the existing control beliefs, past experience with self-defence assistance and the like. It is important that the independent variables are determined on the basis of the assumptions contained in the hypotheses (cf. Sect. 3.4).

Second Phase: The Elaboration of the Research Design and Survey Instruments
After the first step, project planning, has been completed, for example with a project proposal or with an application to conduct contract research, the second step is to work out the research design and, above all, to create the necessary survey instruments. Various decisions also have to be made in this context: It must be clarified which form of survey (for example, a technique of observation, a form of interview or a content analysis) is to be used. The survey instrument (the questionnaire, the observation protocol, and so on) must be designed and pretested before use. Finally, the question must be clarified on the basis of which selection procedure (a simple random selection, a lumped selection or also a quota sample can be considered, compare Sect. 5.2) the survey units are to be determined. Once all these steps have been completed, the research design is available.

The elaboration of the research design takes place in the context of the justification context. While subjective evaluations are certainly incorporated into the discovery context and also into the exploitation context – to be described in more detail later – the justification context is to be kept free of preferences and other personal preferences, for example, for individual methods. It is important to select the most suitable methods for solving the problem from the multitude of methods available in the arsenal of empirical social research.

Ideally, researchers acting independently of each other would propose the use of the same design to address a particular problem.

With each decision made in favour of a methodological variant, the universe of possibilities is narrowed down and at the same time the empirical solution is advanced. Comparable to a funnel, at the beginning a wide range of methodological variants is available, these are then worked out more and more concretely, narrowed down and finally lead to the design of the study.

The choice of the respective form of investigation is of course made depending on the problem previously fixed in the project plan. This must be optimised, taking into account, firstly, the resources that are usually limited in any investigation and, secondly, the interest in knowledge. Each method not only offers certain possibilities for solving the problem (and is at the same time equipped with certain limitations), it also causes certain costs when applied. In addition to the financial and personnel costs, the time possibilities of an investigation also play a significant role. It may be best for the solution of a research question if a mix of methods were used, for example, social experiments scheduled over a longer period of time and various observations in parallel. However, if in practice only a limited period of time is available to develop the findings, appropriate compromises will have to be made.

In detail, decisions have to be made in the preparation of each of the following documents or instruments:

- The original concept of the study, the project plan, must be specified. In this context, working terms are to be defined, the hypotheses to be worked on are to be drawn up and, if necessary, concretised, and the research question is to be operationalised (compare Sects. 3.3, 3.4 and 3.5).
- For the quantitative study, the corresponding measurement instruments must be newly developed or already proven instruments must be selected. When developing such instruments, care must be taken to ensure that the relevant quality requirements are met. Both problems are discussed in Sects. 4.4 and 4.5.
- The design of the study must be defined in detail. For example, it must be decided how many investigation times are appropriate and which survey methods are suitable for dealing with the problem, whether, for example, a laboratory or a field experiment is to be carried out. It must be determined which level of investigation is required to solve the problem. In principle, studies at the individual or collective level are possible. Multilevel studies are also available to empirical social research and can be used (cf. Chap. 7). At the individual level, information is usually collected from an individual, for example about his or her intended voting decision(s). At the collective level, it may be of interest to know how large the proportion of voters for a particular party is in the respondent's place of residence, how high unemployment is there, and what proportion of the total population foreigners make up in that place. The latter are the so-called collective characteristics. Sections 4.5, 4.6 and 4.7 present the techniques in more detail.

- The units of study must be selected. In rare cases, all the elements of interest in a population may be studied, or a sample may be drawn from this universe. The methods available for this purpose are again extremely diverse and will be discussed in Chap. 5.
- Finally, the actual data collection instruments must be developed. In addition to the questionnaires and observation protocols already mentioned, coding schemes for a content analysis or time-use protocols, for example, may also come into question (compare Chaps. 6 and 7).
- Before the actual survey begins, the instruments or the entire design must be reviewed as part of a pretest. If necessary, these should be revised on the basis of the pretest findings. Numerous methodological instruments are also available for this purpose (cf. Chap. 8).

Third Phase: The Survey in the Field and the Creation of the Data Set
The development of the research design and the survey instruments can be followed by the actual data collection – or fieldwork. Its aim is to create a raw data file with the help of the information obtained. For data collection, the aforementioned face-to-face oral interviews, standardized observations, content analyses, non-reactive methods, and others can be used. Where the opportunity exists, commercial institutes are contracted to undertake this work. Such institutes have the necessary infrastructure, such as a staff of interviewers that can be deployed throughout the country, laboratories for group discussions, access panels, and central telephone studios. They then provide the social scientists with the finished data file, including a field report. Data collection (see Chap. 9) is mostly part of such an assignment. In the case of various data collection methods, such as telephone interviews or face-to-face interviews conducted with the aid of a laptop, this is now done automatically by computer. Initial data cleaning and error checking can also be agreed.

However, data collection can also be carried out by means of personal contributions, for example in the context of research seminars, master's or diploma theses. If this method is chosen, the printing of questionnaires, interviewer training, interviewer controls and the like may have to be planned and carried out during this phase of the project. Particularly often, qualitative surveys are carried out by the researchers themselves.

Fourth Phase: The Data Evaluation
During data evaluation, different steps can again be distinguished. First, the raw data file supplied by the survey institute should be checked. These checks include logical checks and the systematic treatment of missing values (cf. Sect. 9.1). In cases where the data collection was the responsibility of the social scientist, these are necessary tasks anyway. To facilitate the analysis, a code plan should be drawn up. This contains the names of the variables and the respective response options. Finally, the actual data analysis can be done afterwards. Statistical programs such as SPSS, STATA and R are usually to be used for this purpose. The results of the statistical data analysis phase are mainly tables and graphs. These are processed further in the next step.

Since the data evaluation will be strongly influenced by the content and the specifics of the respective survey, this phase could only be described in very general terms. What is certain is that here, too, the hypotheses have a research-guiding function. Their processing is ultimately the goal of the data evaluation.

Fifth Phase: Reporting and Documentation
The concrete type of reporting is also strongly dependent on the character of the respective project. Publications in specialist journals, graduate theses, research reports to the respective client and expert reports for the practical implementation of the findings are probably the most common forms of reporting on the results of a research project.

Empirical projects sometimes require considerable resources. For example, in the ALLBUS survey in 2000, over 3000 people were interviewed. This required two years of preparation and the survey costs alone amounted to over DM 1,000,000. These were financed by the public sector (Koch 2002). Now it stands to reason that such studies should be made available for secondary analysis to as wide a circle of social scientists as possible. But even for purely scientific reasons, outsiders should be given the opportunity to precisely reconstruct the findings of a study. Thus, a precise documentation of all methodologically relevant aspects of the study is indicated. Last but not least, it should be examined whether the respective data set should be made available to the Data Archive and Data Analysis Department of GESIS, Leibniz Institute for the Social Sciences.

In the following, the steps discussed will now be demonstrated on the basis of a social science project. Again, this is the Dresden self-defence survey s.

4.2   The Example of the Dresden Self-Defence Study 2001/2002

In order to illustrate the planning of an investigation, it will be demonstrated by means of a concrete example. Firstly, this is suitable for illustrating those aspects that are also valid for various other social science investigations. Secondly, such examples always have facets that make them unique and unrepeatable. The Dresden Self-Defence Project (compare Amelung and Häder 1999; Amelung and Kilian 2003; Häder and Klein 2002; Kilian 2011), which will be further presented here – in a highly abbreviated and simplified form – is intended as a contribution to basic research. It is also situated at an intersection between legal and sociological interests. In the context of this self-initiated project, the aim was to test a number of hypotheses.

Phase 1: Preparation of the Project Plan
The project plan presupposes a thorough study of the literature. The study of literature in preparation for the Dresden self-defence project focused primarily on three main areas: firstly, on relevant court records, secondly, on theories of action to explain certain types of behaviour, and thirdly, on that literature which contains information on the development of instruments. Initially, various pieces of information drawn from court records were

particularly stimulating. For example, some cases (compare again Table 3.1) particularly stimulated the interest of the researchers.

Now the idea that grew out of the literature study was that people were unaware of this legislation and or held different views on self-defence than the case law. This would then have the consequence that the law does not regulate people's behaviour, but at best only evaluates it.

Furthermore, lawyers associate the right of self-defence with the idea of legal protection. Jurisprudence gives an attacked person great freedom in the choice of his means of defence. At the same time, this places an attacker largely outside of legal protection. In this way, the legal system would receive some support from responsible citizens. The presumption, however, is that even this doctrine will not be embraced by those subject to the law. In order to investigate such presumptions, it became necessary to design and implement a General Population Survey.

Since empirical social research has sufficient experience with the survey of opinions and attitudes – which is what the project was about – it seemed that the question could be dealt with in principle.

On the basis of this project idea, the next steps had to be planned. For example, the necessary research funds had to be procured, extensive instrument development was necessary, a sample had to be drawn, hypotheses had to be derived from existing social science theoretical approaches, and so on.

As a dependent variable, the approval or disapproval of an action in which an attacked person defends himself and thereby causes more or less serious harm to the attacker was determined.

Section 3.4 already presented a number of hypotheses on this subject, from which the independent variables can also be seen. After the Volkswagen Foundation approved the requested funds, the project could finally enter the next phase.

Phase 2: The Elaboration of the Research Design and the Survey Instruments
The first step was to tackle the construction of the survey instrument. For this purpose, the working terms had to be defined. For example, Section 32 of the Criminal Code defines self-defence as follows: "(1) Anyone who commits an act that is required by self-defence is not acting unlawfully. (2) Self-defence is the defence necessary to avert a present unlawful attack against oneself or another." These provisions were followed in the processing of the project. Further examples of definitions from this project have already been presented in Sect. 3.3.

The next decisions had to be made in the context of the concept specification. Thus, the population was defined as the adult population in the Federal Republic of Germany with German language skills living in private households with a telephone connection.

As a dependent variable, the interpretation of a defensive action as justified (or as not justified) was determined – as also already indicated. The following were counted among the independent variables: firstly, the good attacked, here a distinction was made as to whether it was an immaterial good, such as honour, a material good (a person's property) or whether the attack was directed against health or life. Secondly, the property damaged by the attacker during the attack was considered as an independent variable. This could be either a light, a medium or a heavy damage that the attacker had to accept. Finally, thirdly, the superiority of the attacker was also included as an independent variable. Here the two expressions a superiority is recognizable and a superiority is not recognizable were provided. Possible further explanatory variables were left out.

A telephone survey was to be used as the survey instrument. For this survey technique, questions had to be formulated and indicators developed. For operationalisation, the cases known from the study of the court files (see above) were prepared in such a way that (a) they depicted the independent variable and (b) the cases were structured in such a way that they could also be presented to the target persons on the telephone for a decision. Here is an example as well. The following question was developed from the case description cited above (compare Table 3.1, Case 3):

> A stranger intrudes into an apartment. The apartment owner first defends his premises in vain with a walking stick, after which he fends off the intruder by fatally stabbing him. Do you consider the behaviour of the apartment owner to be justified or not?

The next step was the construction of the entire instrument. For this purpose, a draft questionnaire was compiled from a number of such case evaluations, the demographic questions about the person were supplemented, and additional time measurements were taken at various points in the questionnaire in order to be able to estimate the duration of the entire survey or individual parts. A professional speaker was asked to record the cases on a tape (saved as *.wav files) in order to be able to play them in a standardized way during the interview in the interest of the highest possible comparability. A rotation of the cases had to be carried out in order to avoid possible sequence effects (compare in detail Sect. 6.1.2) and various other things.

Again, a number of decisions had to be made and justified when determining the specific form of the research. The individual level was determined as the level of investigation, i.e. the target persons were to be questioned about their personal views. Further, a cross-sectional design was decided upon, that is, the hypotheses were to be tested in a single study. Finally, a non-experimental research design was envisaged. Artificial interventions – as is usual, for example, in social experiments – were thus not called for in dealing with these problems.

Based on the hypotheses and the definition of the population (see above), the procedure for selecting the target persons then had to be determined. A total survey was ruled out, so a suitable sampling procedure had to be found. The choice initially fell on a version of random selection. Since the survey was to take place as a telephone survey, an appropriate sampling procedure (cf. Sect. 5.2.3) had to be used. The size of the sample was to allow for analysis possibilities even for small subpopulations and was thus to be correspondingly large. It was planned to collect n = 3500 cases.

The newly developed instrument was tested in a three-step pretest (see also Chap. 8). In the first step, the cases to be assessed were read out to the target persons, followed by paraphrasing, in which the participants were asked to repeat the text of the question in their own words. The second step involved expert ratings on the (probably) most difficult indicators. Once these were identified, they were also questioned using a cognitive instrument. Only in the third step did this pretest involve the use of the questions in a telephone interview. The main purpose here was to test the questionnaire programmed for a computer-assisted telephone interview and to obtain information about the length of the entire instrument. This then completed the survey instrument.

Phase 3: The Creation of the Data Set
For the telephone survey, a contract was awarded to a commercial institute (USUMA Berlin) on the basis of the results of a tender procedure. This institute took over all steps in the survey, ranging from programming the CATI questionnaire, including the definition of logical controls, to drawing the sample, controlling the interviewers and preparing a field report and the data set.

Phase 4: Data Evaluation
The next step is to create a data file that can be analysed. Due to the CATI technology used, the error checks, error correction and data collection are not too time-consuming. These were also carried out by the USUMA Institute in consultation with the organisers of the study (cf. Sect. 6.1.3). The overall responsibility for the content-related data evaluation then lay once again with the social researchers who had initiated the project.

The statistical data analysis, which included such operations as the formation of indices, item analysis, univariate statistics and the creation of correlation analyses (compare Sect. 9.2), followed at this point. Due to the specific nature of the individual analysis steps, it is not helpful to give a standard procedure here. In the end, all projects will follow their own paths when it comes to data evaluation.

Phase 5: Reporting
A report was initially expected from the Volkswagen Foundation, which funded the project. Here an interim report and a final report were required. In addition, a contribution to the practical implementation of the results was made in publications (compare Amelung 2002, 2003).

4.3   Measurement and Index Formation

4.3.1   Problem Definition

As shown, quantifying empirical social research is concerned with uncovering interrelationships in the social world. To this end, hypotheses are formulated using well-defined terms (see Sects. 3.3 and 3.4). Hypotheses could look like this in terms of their logical form:

$$C_1 \ldots C_n \rightarrow E$$

This means that the state of affairs E, or the dependent variable, occurs when the states of affairs C_1 to C_n, the independent variables, are present. One would be dealing here with a cause-effect relationship. For example, a defensive action is evaluated as justified (E) whenever the attacked good is valuable (C_1), the attacker is strongly superior (C_2), and the good damaged in the defense (C_3) is considered less valuable. It is also possible, in the context of a hypothesis, to merely assume a relationship between two quantities. Whenever a certain state of affairs occurs, another state of affairs also occurs.

In addition, these hypotheses must be subjected to empirical testing to determine whether the relationship assumed in the hypothesis actually exists in reality. In order to be able to make a corresponding statement, it is necessary to quantify all variables contained in the hypothesis (C_1, … C_n, E) or to measure them. For this purpose, complex and/or latent facts must first be operationalized – made measurable – (cf. Sect. 3.5) before they can be subjected to measurement. This is usually done by breaking down complex issues into different, simple dimensions. In the following, we will discuss what it means to measure an issue in the social sciences. After that, it is a matter of recombining the individual results of such measurements and then expressing the original complex dimension to be measured as a numerical value (index).

4.3.2   The Principles of Measurement

In order to demonstrate the principles of measurement in the social sciences, a simple comparison will first be used again. If one were to measure the size of a credit card with the help of a ruler, one would find that this piece of plastic is about 8.5 cm wide and 5.4 cm high. All you have to do is place the ruler, on which a certain set of numbers is arranged according to a certain principle, on the check card to be measured and then read off the corresponding result. Later still, by multiplication, one can find that the size of the cheque card is thus 45.9 cm^2. This is a trivial process. However, it already contains certain characteristic steps which must also be present in social science in order to be able to speak of a measurement.

Measurement is defined in the literature either as the comparison of something unknown with a normalized known. In this process, numbers are given the task of signifying properties. Similarly, one could also say that measurement is the "assignment of numbers to objects or events according to rules" (Stevens 1959, p. 18). For example, assigning numbers to members of a football team would already be measurement. Other views, on the other hand, assume that measurement is always tied to some metric (Schnell et al. 2013, p. 135 ff.). In this case, the players' shirt numbers would then have to be assigned according to the number of goals scored by each individual, for example, in order to meet the criterion of measurement.

An important element in the discussion of measurement concerns the relationship between the object to be measured and the way in which the measurement can be made. For example, measuring size using a ruler was trivial: the larger the object to be measured, the higher the number read off the ruler. It is also no problem to measure the age or income of a person: The older a person is or the higher his or her income, the higher the corresponding numerical value will be. In the social sciences, however, the researcher is sometimes interested in latent and complex facts such as the degree of social integration of a person, their marital status or the working atmosphere in a company. Now, it is above all the peculiarities of these objects of measurement that determine the procedure for their measurement. Some special features of social science measurements will be presented in more detail:

Structural Measurements
Different relationships can exist between the object to be measured and the numbers assigned to this object during the measurement. Under certain conditions, it can be a mapping that is true to the structure. Such mappings presuppose that the object to be measured can be ordered according to a system. While it is in principle conceivable to classify the degree of social integration of a person into a system from very high to non-existent, this would not be feasible for a measurement object marital status. This means that only in the first case do we have to deal with a true-to-structure mapping. On the other hand, a structurally faithful mapping of marital status is not possible.

It is conceivable to assign corresponding numerical values to the various forms of marital status, e.g. for single the value one, for married and living together the value two, for married, living separately the value three, and so on. However, there is no order behind the numbers one and following, which could result from the object to be measured. The situation is different if a low degree of social integration of a person into a system is assigned a small value and a high degree of social integration of a person is assigned a correspondingly large value. A structurally faithful measurement thus presupposes that the measured objects can be ordered. The numbers that are assigned to the meanings are then also ordered accordingly. This is not the case with the measured object marital status.

Also, the demand "to make as many of the phenomena to be investigated as possible accessible to metric recording" (Schreiber 1975, p. 279) is too sweeping and should be relativized to the effect that it is better to first examine the character of the object to be measured.

Empirical and Numerical Relative
A further distinction is now to be made between an empirical relative and a numerical relative. An *empirical* relative represents a set of objects over which a relation is defined. Thus, a set of persons (here the objects to be ordered) can be ranked according to various criteria, for example, according to their height, according to their income, or according to their degree of satisfaction with life. A *numerical* relative, for example a scale – represents a set of numbers by which a relation is defined. Here, for example, we may be dealing with numerical values expressing the degree of satisfaction with life, where a lower degree of satisfaction is expressed by a smaller number and a higher degree of satisfaction by a

correspondingly larger number. To return to the trivial example from above: The empirical relative here would be the check card and the numerical relative would be the ruler containing a certain set of numbers. Later (see Sect. 4.4), we will see how such instruments, which are similar to rulers, can be constructed in social science research.

When measuring, it is important to express the order present in the empirical relative in the numerical relative. It would therefore be completely nonsensical, for example, to designate a low degree of integration of a person in a social system with the value 15, a medium degree of integration with the value zero and a correspondingly high degree with the value eight. In a true-to-structure mapping, small values would therefore mean a correspondingly lower degree of integration – and vice versa.

Test Theory and Measurement Error
Another assumption that is made in measurements concerns the measurement errors that occur in the process. Test theory describes the conditions under which it can be assumed that measurement errors are minimized. The basic assumption is that a true value of the unobservable theoretical construct exists at all. It may be, for example, a certain degree of acceptance towards a defensive action that a person represents. This is then the so-called true value or the true score generally abbreviated to T. It is further assumed that a value is determined with the help of an appropriate measuring instrument. The observed value is to be denoted by X. However, it only indicates the true value approximately. Therefore, test theory assumes that the observed value is subject to a certain measurement error. This is abbreviated with E (Error). Thus, the observed value is composed additively of the true value and the measurement error. Thus, the following applies:

$$X = T + E.$$

Measurement theory is based on the idea that measurement errors (E) scatter around the true value for many measurements. In other words, it happens that the measurements deviate randomly in both directions from the true value. The expected value of the measurement error – to be called $\mu(E)$ – is thus zero, provided that there are enough measurements and that the deviations are always random. It thus holds:

$$\mu(E) = 0.$$

If we imagine an interviewer who enters the answers of his target person as numerical values in a list, it may happen that he notes down – admittedly without intention – once a value that is too large and once one that is too small. These errors are not systematic, but balance each other out. It is therefore valid:

$$T = \mu(X).$$

This means that the true value is identical to the expected value of the measurements for a sufficiently large number of interviews. Again, it is assumed that the sum of the measurement errors is zero. However, as soon as, for example, an interviewer systematically influences the interviewee and in this way provokes socially desirable responses from him or as soon as the measurement instrument does not function properly, this assumption would be violated. Some further assumptions have to be made in addition:

- The true value and the measurement error are not correlated. If, for example, the degree of integration of a person into a social system is to be measured, the same measurement error must occur for each degree of integration. The above basic assumption would therefore be violated if, for example, the measuring instrument tended to produce different errors at an extremely low level of integration than at a different constellation. Take as another example the question about a person's income. Here, this assumption would be violated if different measurement errors occurred in the case of persons with a higher income than in the case of persons with a lower income.
- It must also be assumed that the measurement errors in repeated measurements do not correlate with each other. A high measurement error at one point in time may not yet say anything about the error to be expected at another point in time. To stay with the interrogation of income: Suppose an individual's first query reports – inadvertently – an income that is too high. If, in a repeated query – but now on purpose – exactly this information were to be replicated, for example because the person is trying to answer consistently and remembers his or her last answer, this assumption would also be violated.
- The last assumption is: There must be no correlation between the measurement error of a first measurement and the true value of a second measurement with the same measuring instrument.

4.3.3   The Formation of the Index

An index can be defined – quasi evaluatively – as a variable that results from certain arithmetic operations with several other variables.

Following the measurement of a fact, it is often a matter of relating the individual measured values to each other. For example, the size of the credit card already measured above in its length and width could now be given with the product of the two sizes length and width (= 45.9 cm^2).

Such summaries of individual measurement results into indices are particularly useful when dealing with complex, multidimensional issues. As a rule, such complex facts must be collected in a first step with the aid of a large number of indicators. The information thus obtained is correspondingly extensive. When the index is formed, this information is then reduced again or the individual pieces of information obtained about the complex issue are summarized.

Economics has provided various examples of how the problem of measuring complex issues and then condensing them into an index can be solved. One example is the purchasing power index, which can be used to compare different countries or to show trends in purchasing power within a country. The procedure is as follows:

To measure price trends, around 600 price collectors in 188 municipalities record the prices of the same products in the same shops month after month. In addition, a central price survey is carried out for many types of goods, for example on the Internet or in mail-order catalogues. In total, over 300,000 individual prices are recorded each month. Once an article has been selected for price monitoring, it is then exchanged for another if it is no longer sold or only sold in small quantities. Changes in quantity are also included in the price comparison. If, for example, a supplier reduces the packaging size of a product while the price remains the same, this is recorded in the price statistics as a price increase. Furthermore, quality changes are also taken into account – for example, in the case of goods with technical progress. For the price measurement, the purchase prices including value added tax (VAT) and excise duties are observed[1]

As a result of this relatively complicated procedure, it may now be possible to speak of a price change amounting to a certain percentage.

Various steps of the procedure are typical and transferable to social science problems. For example, it is convenient if the prices of as many goods as possible are registered in the course of this census in order to prevent random special offers from providing a distorted picture. Correspondingly, one would also use a whole series of questions for the survey of the degree of integration of a person into a social system in order to reduce the proportion of random measurement errors. The survey in the middle of each month is also an interesting methodological aspect, which guarantees that the index values determined are comparable over time. The fact that the survey is carried out in different regions prevents the specific features of individual areas from having too great an impact. The selection of very specific goods and services helps to ensure that the findings can be generalised.

The following steps can thus generally be distinguished in index formation:

1. First of all, it is a question of determining the complex problem which is to be represented with the aid of an index, in this case the development of purchasing power.
2. This is followed by the operationalisation of this issue and the derivation of a (larger) number of indicators – 750 goods and services are asked for. The actual measurement is then carried out with the help of these indicators.
3. The summary of the results in the form of an index and thus again the reduction of the information variety, about the numerous individual goods to a common value.

Similar examples of the use of indices are:

---

[1] Source: https://www.google.com/url?sa=t&rct=j&q=&esrc=s&source=web&cd=1&ved=2ahUKE wiDx-Xz0KHgAhXBJFAKHT2ABzkQFjAAegQICRAC & https://www.itb.en/business-economist/ add-ons-to-bw-hwo/pt-i-vwl2.html?file=files/content/itb/downloads/Add-Ons%20fur%20Scripts/ VWL2/AO%20VWL2%20103%20ConsumerPriceIndex.pdf&usg=AOvVaw3bCip6 LQw8T9McRcH-UmWJ, last accessed 04 Feb 2019.

- The *Human Development Index*: This aims to capture the general living conditions in a country. For this purpose, the life expectancy at birth of a person, the literacy rate, the per capita social product of a country and the enrolment rate of the population in primary, secondary and higher education are correlated.[2]
- The *marks of* a multiple-choice exam: The complex issue to be covered is the performance level of a person. For this purpose, a relatively large number of questions is asked, again in order to keep the random error as low as possible. The weighting of the individual questions or the points to be achieved in each case can vary here. A correct answer to a difficult question should yield more points than a correct answer to an easy question. The overall score, the index, is then calculated by adding the points achieved and transferring them according to a certain scheme.
- The *stratification index* (cf. Scheuch and Daheim 1970, p. 102 f.) combines the dimensions of education, income and occupational position and aims to measure social prestige (for further information on the formation of this index, cf. Schnell et al. 2013, p. 157 ff.).
- In the social sciences, moreover, the *Inglehart Index* has now achieved a certain notoriety. This index aims to provide information on how the change in values is proceeding in developed Western industrialised countries. For this purpose, the numerical ratio of people with materialistic and those with post-materialistic value attitudes is shown. In the short form of the instrument, the operationalization provides for the evaluation of four statements. The proportion of materialists and post-materialists is determined on the basis of the answers. Finally, the actual index results from the subtraction of both values.

Table 4.1[3] shows the responses to the Inglehart[4] values questions, Table 4.2[5] shows the proportions of materialists and post-materialists, and Table 4.3 shows the resulting index. The basis is the results of the ALLBUS studies (cf. Sect. 6.1.4) 1980 to 2006 for German citizens living in West Germany and for German citizens living in the old federal states. The procedure is based on the pattern presented above.

---

[2] Source: http://www.bpb.de/themen/26G2CN,0,0,Human_Development_Index_ (HDI).html, last accessed on February 04, 2019.

[3] The ALLBUS data collected according to the ADM design were transformation weighted (see Sect. 5.8).

[4] The question text was: Even in politics you can't have everything at once. On this list you will find some goals that can be pursued in politics. If you had to choose between these different goals, which goal seemed the most important to you personally? A Maintaining peace and order in this country, B Giving citizens more influence on government decisions, C Fighting rising prices, D Protecting the right to free speech.

[5] A materialistic value attitude exists if the specifications A and C were named on the ranks one and/ or two. A post-materialistic value attitude is present if the specifications B and D are mentioned on the ranks one and/or two. The 100% missing values are the so-called mixed types.

Table 4.1 The response to the Inglehart questions by the West German population in selected ALLBUS studies since 1977

Rank	1980	'84	'88	'92	'96	2000	'04	'08	'12
Importance of A, peace and order									
1	48	39	41	36	40	38	32	27	30
2	22	23	20	23	23	26	28	27	27
3	15	18	21	24	23	23	26	25	28
4	15	20	18	17	14	13	15	21	15
Total	100	100	100	100	100	100	100	100	100
Importance of B, citizen influence									
1	16	24	25	32	31	36	37	30	34
2	25	22	25	26	30	28	24	26	25
3	27	27	28	23	26	23	23	25	27
4	32	27	22	19	13	13	17	20	14
Total	100	100	100	100	100	100	100	100	100
Importance of C, inflation control									
1	21	18	9	13	6	7	15	24	9
2	33	26	21	27	18	15	22	25	15
3	27	29	29	28	28	25	25	22	21
4	19	27	41	32	48	53	38	29	55
Total	100	100	100	100	100	100	100	100	100
Importance of D, freedom of expression									
1	16	20	26	20	23	19	17	21	27
2	21	29	34	25	29	31	27	23	33
3	31	25	22	24	23	29	26	27	23
4	32	26	18	31	25	21	30	29	16
Total	100	100	100	100	100	100	100	100	100

Table 4.2 Proportions of materialists and post-materialists (according to Inglehart) in the West German population since 1980

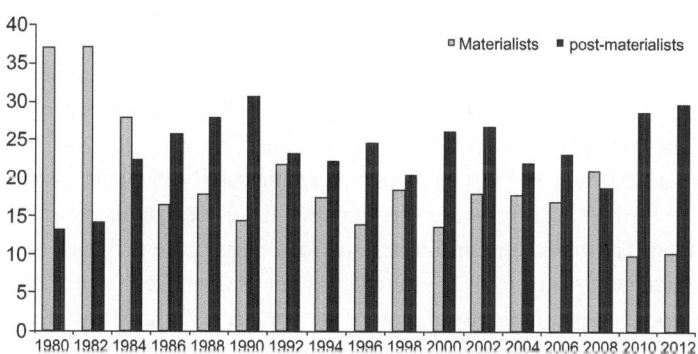

Table 4.3 Difference between the proportion of materialists and post-materialists (according to Inglehart) in the West German population since 1980

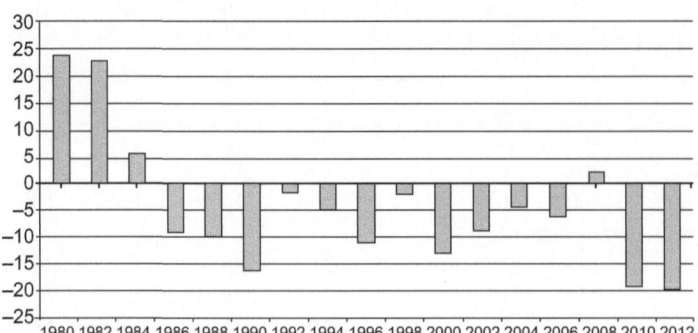

1. Step, determination of the facts to be depicted: state of development of a western industrial society on the basis of the value attitudes of the citizens.
2. Step, operationalization and development of appropriate question texts.
3. Step, summary of the results: (a) Determination of the proportions of materialists and post-materialists, (b) Subtraction and presentation of the indices.

Several problems related to the formation of indices should be pointed out: A first prerequisite is the most complete and successful operationalization of the complex issue. Thus, in the formation of the purchasing power index, it was important to select specific – as typical as possible – goods and services. The success of the creation of an index on value attitudes also depends on whether the derived indicators fully cover the issue sought. Secondly, it must be ensured that all indicators cover only this one issue. In the case of the Inglehart Index indicators, for example, the question arises as to the extent to which stable value attitudes are actually being surveyed over a longer period of time and whether they are not also being used to record attitudes that are current on a daily basis (cf. Hollerbach 1998). Thirdly, it is necessary to decide with what weight the results of the individual variables measured are to be included in the formation of the index. In the case of the purchasing power index, for example, the question arises as to how much weight should be given to the prices of chocolate bars and housing rents. In the case of the Inglehart index, all four specifications are included in equal weight in the formation of the index.

Schnell et al. (2013, p. 157 ff.) show a total of four possibilities for index construction: Indices can be formed additively. For this purpose, the corresponding variable values are simply summed. They can be multiplicative indices. These take the value zero as soon as a single characteristic – a factor – is evaluated as not present. Such a procedure would be conceivable, for example, in the formation of an index for life satisfaction. In this way an area of life could be treated which is completely unimportant for the respondent (value zero). It is also possible to form weighted additive indices. Thus, with the help of expert assessments or factor analyses, the individual values surveyed can be incorporated into the index to different degrees.

The question that may arise in the end, to what extent the index formed is also correct, cannot be answered in this form. Various steps of the index construction have been determined by the underlying theory. External criteria, which allow an objective examination of the procedure, are usually not available. Thus, the practical use of the indices must determine whether they can contribute to an increase in knowledge.

Summary
Operationalisation (cf. Sect. 3.5) initially involved breaking down complex and/or latent issues according to specific instructions and thus making them empirically measurable in the first place. The result was a certain set of indicators. This section now described what it means to measure these facts. In addition, index formation was used to deal with how the path back from the indicator battery to the complex facts can be followed.

In the following, the scaling procedures are discussed, with the help of which such measurements can be realized in empirical investigations.

4.4   Scales and Scaling Methods

The term scale is used in different ways in empirical social research. In a broader sense, the instruments for the empirical investigation of a specific issue are referred to as scales. In a survey, this then includes both (1) a specific question text, (2) the answer specifications presented to the persons to be interviewed, and (3) finally, any instructions that the interviewers have to follow when asking the questions.[6] Such use of the term scale was probably supported in no small part by the ZUMA Scale Manual. This contains and documents instruments for surveying particularly relevant social science questions, such as Inglehart's Materialism-Postmaterialism Scale (compare Sect. 4.3).[7]

In the narrower sense, however, scales are the numerical relatives used to map an empirical relative. In the narrower sense, the term scale is also used in this passage. The point is to distinguish between different types of scales. Then the interest will be directed to those procedures with the help of which scales can be developed. See also Blasius (2019, p. 1437 ff.) and Latcheva and Davidov (2019, p. 893 ff.).

4.4.1   The Different Measurement (Scale) Levels

It has already been shown that a measurement should be a structurally faithful representation of the empirical relative. Thus, the object of measurement essentially determines the

---

[6] For the definition of the term, compare also Blasius (2019, p. 1437 ff.).

[7] The former ZUMA Scale Handbook has since been (more appropriately) renamed the Compilation of Social Science Items and Scales (ZIS), expanded and/or extended, and is available at the following URL https://zis.gesis.org/ last accessed on 2019-02-04.

character of the numerical relative. To illustrate this point, some questions used in the social sciences will be presented, and in the process the various types of scales will be discussed. In Figs. 4.2,[8] 4.3, 4.4 and 4.5, four different questions are shown. All of them come from the ALLBUS survey series.

The question in Fig. 4.2 contains a nominal scale. The digits of this scale serve to distinguish the individual types of housing from one another. Apart from this distinction, it is

The next question is about the apartment in which you or your family live here
Please tell me which of this list applies to you or your family.

(Interviewer: submit list. Only one entry possible)
-- -

01. For subletting	4
02. In a company flat	1
03. In a rented flat of social housing	16
04. In a rented flat (not social housing), in rented owner- occupied flat	31
05. In a rented house	3
06. In a condominium (own or family property)	7
07. In your own home (or the family home)	37
08. Other form of housing, which one?	1
97. Refused	1 (M^1)
99. Not specified	1 (M)

Fig. 4.2 Question about the target's home and the corresponding marginal distribution of responses. (Source: ALLBUS 1980, n = 2953) Figures in percent

There is a lot of talk these days about the different strata of the population. Which class do you consider yourself to belong to: the lower class, the working class, the middle class, the upper middle class or the upper class?

1. Under class	1
2. Working class	29
3. Middle class	59
4. Upper middle class	9
5. Upper class	1
6. None of these layers (Int.: Do not read out)	1
7. Classification refused (Int.: Do not read out)	54 (M)
8. Don't know	59 (M)
9. Not specified	1 (M)

Fig. 4.3 Question on subjective stratification and the corresponding marginal distribution of responses. (Source ALLBUS 1980, n = 2841) Figures in percent

---

[8] The data set of the ALLBUS 1980 is available from GESIS Leibniz Institute for the Social Sciences. The study no. is 1000.

On this list are some sentences that you have heard at some time or another when it came to guest workers. For each sentence, please tell me how much you agree with it. Using the scale at the bottom of the list, you can again grade your opinion.

(Interviewer: hand over list – please note scale value)

Guest workers should adapt their lifestyle a little better to that of the Germans.

01. Do not agree at all	8
02.	5
03.	8
04.	13
05.	21
06.	15
07. Agree wholeheartedly	30
97. Resused	4 (M)
98. I don't know	2 (M)
99. Not specified.	6 (M)

Fig. 4.4 Question on attitudes towards foreign workers (guest workers) and the corresponding marginal distribution of responses. (Source: ALLBUS 1980, n = 2943) Figures in per cent

(If respondent has not lived there since birth) Approximately how many kilometers do you live from your previous residence?

(Interviewer: explain in case of further questions: that is, the place where you lived before you moved to here)

	Frequency	Percent
No removal	1.139	38
Less than 25 km	732	25
Between 26 and 50 km	222	8
Between 51 and 75 km	110	4
Between 76 and 100 km	85	3
Between 101 and 200 km	151	5
More than 200 km	497	17
Refused	1	-
Don't know	1	-
No answer	17	1

Fig. 4.5 Question about the distance to the last place of residence and extract from the corresponding marginal distribution of responses. (Source: ALLBUS 1980, n = 2955)

not possible to map other relations of the housing types. Only equality and inequality can be interpreted. Thus, there is a simple classification that only distinguishes between equal and not equal. It is important that all housing types can be classified and that each housing type can only be assigned to one category. In accordance with the character of the characteristic to be depicted, nominal scales are also used, for example, in circumstances such as the classification of marital status, leisure activities and the like.

The formation of the arithmetic mean is not permitted. The modal value can be indicated. It indicates the category in which the highest frequency can be found. According to this, most respondents in West Germany lived in their own house in 1980.

In Fig. 4.3[9] an ordinal scale is used. These are distinguished from nominal scales by the fact that, in the case of the object to be measured, an additional ranking can be interpreted empirically. This rank order must be expressed accordingly in the scale by numerical values. While it would be possible to swap the numbers used in the nominal scale without further ado, this possibility does not exist here.

The median is considered the acceptable mean. It separates the lower and the upper 50% of the answers and lies in the 1980 study with the category "middle class". However, the distances between the scale values 1–2 – 3 – … are not interpretable. According to the object to be measured, only a hierarchy is assumed. It cannot be assumed that, for example, the same distance exists between the working class and the middle class as between the upper upper class and the upper class.

It should be noted that again all possible answers can be classified. For this purpose, it is also necessary to provide for various special levels, such as "classification refused", if necessary.

Figure 4.4 shows an interval scale. With interval scales – in contrast to ordinal scales – the distances between the measured values can also be interpreted. It is therefore assumed that the distances between the seven scale points are equal. This assumption then allows the arithmetic mean to be calculated.

It should be pointed out that it is actually necessary to empirically test the assumption that there are actually equal distances between the scale points in each case. For example, anyone who calculates an average value from different grades at school, which is common practice, makes such an assumption. In social science surveys, a scale sheet is often used to support this assumption. On this scale sheet, which is handed over to the target person, a series of numbers is then shown at equal intervals. In this way, it is also hoped to give the interviewees the impression that the values are equally spaced.

However, the data of the people who moved were combined into categories, so that in the end it is only an ordinal scale level. Thus, the distances between the individual response levels differ significantly.

Figure 4.5 contains an example of a ratio scale, which is rare in social science research. This is characterised by the existence of a natural zero point. Here it would mean that the place of residence of the target person has not changed since his/her birth and that the surveyed distance therefore takes on the value zero. This is true for 38% of the respondents.

With ratio scales, it is permissible to make percentage comparisons, such as person A lives twice as far away from their place of birth as person B. As indicated, ratio scales are relatively rare in the social sciences. Other examples would be questions about the duration of marriage or the duration of schooling in years.

---

[9] The letter M here indicates that these values are treated as missings (see also Sect. 9.1). They have not been taken into account in the addition of the valid cases.

4.4.2 Scaling Procedure

Empirical social research has at its disposal a series of procedures in the form of scaling methods that make it possible to measure latent facts – such as attitudes and opinions. Scaling procedures are thus special cases or continuations of operationalizations (cf. Sect. 3.5). The aim is to create a suitable instrument, a scale, here in the broader sense of the term. At this point, only an overview of the most common techniques can be given. For details, please refer to the special literature, for example by Borg and Staufenbiel (1997) or Häder (2008b).

First, one can again define a sequence of steps to be completed in the development of scales.

First Step
The problem is operationalised. As shown (cf. Sect. 3.3), a complex issue must be broken down into more concrete dimensions in various steps, as far as possible on theoretical grounds. For example, one could assume that a certain attitude towards an object has a cognitive, an evaluative and a behavioural dimension.

Second Step
A collection of statements (sometimes up to 100 or more) is then made on the dimensions. This can be based on various sources. Possible sources would be a literature review, own constructions, preliminary investigations, diaries and the like. It should be noted that such statements are needed in about four times the amount compared to the final scale. Because of the expected weeding out of unsuitable statements, such a corresponding set must be calculated.

An example could be the instruction for an empirical collection of statements: "In public there is a lot of talk about the problem of integration of foreigners. Please write down three possible opinions about it!"

Answers: 1. foreigners enrich our lives. 2. foreigners threaten our jobs. 3. foreigners do not interest me.

Third Step
In addition, duplicate statements are eliminated and the language is revised. It should already be taken into account that the statements are to be processed into questionnaire questions one day.

Fourth Step
This is followed by the actual calibration of the scale. For this purpose, different scaling methods such as the Likert technique, the scalogram analysis according to Guttman or the method of equal appearing distances according to Thureston can be used. These techniques will be discussed in more detail below.

Fifth Step
Finally, the final scale is constructed from the remaining statements.

The Likert Scaling
Likert scaling is also called the technique of summated assessments (Likert 1932). It is a particularly popular technique because it is relatively easy to implement. A set of statements is required (compare steps one to three). These are presented to a study population with a request to rate each one. Five-point intensity scales are usually used for this purpose (see also Faulbaum et al. 2009, p. 24 ff.). They have the following appearance:

To what extent do you agree or disagree with the statement [...]? Do you ...

1. agree totally,
2. agree more likely,
3. agree partially,
4. agree more likely not,
5. agree not.

A score is assigned based on the evaluation of each item, for example, a three for the answer "partially agree". These values are added up for each person to a so-called sum score. Now those statements have to be identified that *do not* (sufficiently) express the issue sought. After all, it is conceivable that statements have also been included in the list that measure a dimension other than the one sought. For this purpose, one needs a reference value for each person under investigation that expresses the sought attitude of this person as accurately as possible. All individual assessments would then have to be compared with this value. Only those statements can then be regarded as suitable which are closely related to the reference value. Since such a reference value does not exist, all items are correlated with each other. It is assumed that in the sum of the answers given the attitude of the person sought is at least approximately well represented. Therefore, such an item intercorrelation is calculated. A high correlation indicates a good quality of the statement. In this way, the question of whether all items on the list measure the same dimension becomes answerable.

The analysis also determines how well each statement discriminates. The attitude to be measured is likely to be present to varying degrees among the subjects. The measuring instrument is now expected to be sensitive enough to pick out these differences or to discriminate. To this end, people with a particularly high score (the top 25%) are compared with those with a particularly low score (the bottom 25%). It is obvious that such statements, which were answered in the same way in both groups, have only a small or even no discriminating effect. Such specifications should also be separated out from the pool of statements. In this way, it is possible to gradually separate suitable and unsuitable items from one another.

In the following, an example of a scale development according to the Likert method is shown. It is about the determination of the attitude towards surveys. In the following list, the statements to be evaluated are shown. These have been rated by 64 people according to the pattern shown above.

Collection of Statements for a Scale Development

- Surveys are a nice change of pace and fun.
- Surveys reliably depict the climate of opinion.
- Polls are often rigged.
- Polls are basically irrelevant.
- The money used for surveys should be available for more important things.
- I have confidence in the accuracy of poll results.
- Surveys usually make a serious impression.
- Surveys have no benefit to the individual.
- Politicians need survey results for their decisions.
- Surveys contain awkward questions.
- I find polls boring.
- Interviewers who conduct surveys are pushy.
- Surveys are indispensable for scientists to verify their assumptions.
- People very often lie when answering surveys.
- Surveys are an excuse to solicit customers.
- Surveys are developed by experienced specialists.
- By participating in surveys, I feel like I am being helpful.
- How survey results are arrived at is concealed from outsiders.
- Surveys are important to our society.
- Surveys are often abused.
- Survey results are interesting.
- It annoys you when you're asked to participate in surveys too often.
- Surveys are used to sound out the population.
- Survey results help me form my own opinion.
- Surveys provide interesting information to the general public.
- Surveys ask indiscreet questions.
- I think it's a waste of time to participate in polls.
- Surveys are used by the advertising industry to manipulate consumers.
- Surveys encourage people to think about unusual things.
- Surveys are very important for research.

Due to a tendency towards affirmative answers (see Sect. 6.1.3 for more details), some specifications have been formulated negatively. This applies, for example, to "Surveys are often falsified" and "Surveys are basically unimportant". The corresponding answers had to be recoded in a first evaluation step, so that the extent of agreement with the individual specifications is now also a measure of the positive attitude towards surveys. The sum score formed for each respondent makes it possible to calculate the discriminatory power (compare Brosius 2002, p. 768 f.; Wittenberg and Cramer 2000, p. 139 ff.; Wittenberg 1998, p. 95 ff.). This is the correlation of the respective item with the overall result of the

Table 4.4 Selectivity of the examined items and information on α, if the respective item would be deleted

	Item description	Discriminatory power	α if item deleted
V1	Nice change and bring fun	0.54	0.913
V2	The climate of opinion is mapped	0.46	0.914
V3	Often falsified	0.58	0.912
V4	Basically unimportant	0.61	0.912
V5	Money for more important things	0.60	0.912
V6	Confidence in survey results	0.72	0.910
V7	Make a serious impression	0.63	0.911
V8	No benefit for the individual	0.31	0.916
V9	Politicians need survey results	0.27	0.918
V10	Contain unpleasant questions	0.42	0.915
V11	Are boring	0.78	0.909
V12	Interviewers are intrusive	0.42	0.915
V13	Indispensable for scientists	0.38	0.916
V14	Lies are told very often	0.33	0.916
V15	Pretext to attract customers	0.48	0.913
V16	Developed by experienced specialists	0.33	0.916
V17	Participation gives feeling of being helpful	0.63	0.912
V18	Results are concealed	0.42	0.915
V19	Important for society	0.67	0.911
V20	Frequent abuse	0.30	0.916
V21	Results are interesting	0.75	0.910
V22	It inconveniences to participate	0.26	0.917
V23	Sound out the population	0.59	0.912
V24	Help to form own opinion	0.39	0.916
V25	Interesting information	0.67	0.911
V26	Indiscreet questions	0.54	0.913
V27	Waste of time	0.66	0.911
V28	Consumer manipulation	0.58	0.912
V29	Encourage reflection	0.29	0.917
V30	Very important for research	0.36	0.915

test. This is shown in Table 4.4.[10] In the SPSS program, the discriminatory power is shown as Corrected Item-Total Correlation.

The discriminatory power is a measure of the contribution and suitability of the individual items for the overall scale. If the discriminatory power is high, there is a particularly close correlation between the items concerned. Another important clue for the construction of the final scale comes from the column "α if item deleted". As can be

---

[10] The answers were coded in such a way that for "it applies completely", a 1, for "rather applies" a 2, for "partly partly" a 3, for "rather does not apply" a 4 and finally for "does not apply at all" a 5 was coded. A total of 64 people were interviewed.

seen, item V22 correlates only relatively weakly with the summerscore. Thus, it is expected that the quality of the overall scale will improve if this item is excluded. This quality is indicated by Cronbach's α. The value is an expression of the reliability of the scale.

According to a rule of thumb, Cronbach's α should be at least 0.8. In the example shown above, there was even a value of 0.92, which can thus be interpreted as completely sufficient. It is therefore not necessary to delete individual items.

It is important to note that Cronbach's α is not yet a sufficient measure of the dimensionality of the scale (compare Brosius 2002, p. 767). The value is higher the more strongly the individual items correlate with each other. It would then be reasonable to assume that this is indeed a one-dimensional scale. It is conceivable, however, that two or more dimensions are being measured, but that these are in turn highly correlated with one another. In order to determine such correlations, a factor analysis must be calculated. Table 4.5 shows the rotated component matrix (Varimax rotation).

This seems to confirm the above assumption: although initially a markedly high value of Cronbach's α indicated a homogeneous scale, the factor analysis produced a different finding. Reducing to those ten items that load most strongly on factor 1 (*in italics*) and repeating the reliability analysis now yields a Cronbach's α value of 0.89. This still meets the required benchmark. The repetition of the factor analysis now yields a one-component solution. Thus the items V2, V6, V7, V13, V17, V19, V21, V24, V25 as well as V30 now form the final scale and express the sought dimension "attitude towards surveys".

The Skalogram Analysis

The initial idea of the scalogram analysis is the following: All statements used in a scale can be arranged according to their difficulty. This can be well imagined for an exam, for example. It should contain questions with different degrees of difficulty. Then one can assume that persons with high knowledge can answer all questions. People with medium knowledge will only be able to answer the easy questions and the moderately difficult questions. Finally, people with little knowledge will only be able to answer the easy questions correctly, and those who know nothing will not be able to answer any question correctly. One could now – mentally – arrange all exam questions in such a way that they become more and more difficult. Up to a transition point then all questions are answered correctly by a person and after that all questions are answered incorrectly. The location of this point would then give the grade. The assumption of such a transition point is also called deterministic jump function in scalogram analysis (compare Guttman 1944, 1947).

If one substitutes the expression of a certain attitude characteristic for knowledge in an examination, the logic of the scalogram analysis becomes even clearer. It assumes that all items can be arranged one-dimensionally. It is also assumed that the items are answered

Table 4.5 Rotated component matrix with factor loadings 100 and greater

	1	2	3	4	5	6	7	8
V1	0.18	0.27	0.12	*0.65*	0.16	0.11	0.16	−0.13
V2	*0.59*		0.16			0.36		*−0.45*
V3		*0.62*		0.23	*0.44*	0.26		
V4	0.39	*0.70*	0.29					−0.21
V5	0.25	*0.66*	0.23		−0.10	0.33	0.20	0.15
V6	*0.67*	0.36	0.15	0.11	0.24	0.17		−0.19
V7	*0.69*	0.21		0.30	0.17	0.13	−0.16	−0.12
V8	0.19	0.12	−0.14			*0.89*		
V9	0.15	0.14			0.26		*0.80*	
V10		*0.47*		0.33		0.45	−0.27	−0.29
V11	0.34	*0.47*	0.26	*0.47*	0.15	0.24		0.17
V12	0.17	*0.50*	0.12		0.21		*−0.51*	
V13	*0.72*	0.11	−0.14			0.12	0.23	0.26
V14	0.13	0.32	0.17	−0.35	*0.52*	0.16	−0.29	−0.19
V15		0.34	*0.62*	*0.42*				−0.15
V16	0.19	0.11			*0.64*	−0.14		0.24
V17	*0.44*	0.21	0.18	*0.45*	0.38	−0.12	−0.25	0.11
V18	0.11			0.17	*0.76*	0.27	0.26	
V19	*0.76*		0.34	0.14	0.16		0.14	
V20		0.16	*0.67*		0.13	−0.16	0.14	*0.47*
V21	*0.52*	0.22	0.32	0.39	0.22	0.23		−0.11
V22			0.14	0.17	0.23	0.17	−0.17	*0.78*
V23	*0.18*	0.30	*0.50*		0.13	*0.57*		
V24	*0.44*	−0.22	0.12	0.37	*0.45*	−0.11	0.15	0.15
V25	*0.61*	0.24	0.17	*0.46*			−0.18	
V26		0.37	*0.41*		*0.44*	0.26		
V27	0.24	*0.73*	0.33	0.14				0.16
V28	0.22	0.21	*0.79*	0.26			−0.16	
V29	0.15			*0.80*				0.19
V30	*0.54*	0.28	−0.33		0.21		*0.41*	

dichotomously, for example with "correct" or with "incorrect" or with "applies" or with "does not apply".

In our example, however, the variables were provided with a five-point answer scale, so the answers had to be recoded accordingly: The default "strongly agree" is now contrasted with the others. In this case, variable V30 (surveys are very important for research) would contain the lowest level of agreement or the easiest answer. This was the view that most of the target respondents agreed with. In other words, the view that surveys are important for research does not yet indicate a particularly strong positive attitude towards surveys. The opposite pole is V24 (Surveys help me form my own opinion). Respondents who also hold

this view now have a decidedly positive attitude towards surveys. The remaining specifications lie accordingly between these two poles, whereby the following ranking was
determined:

V30	(Surveys are) important for research
V21	(Survey) results are interesting (see above)
V25	Offer interesting information to the public
V19	Are important for our society
V13	Are indispensable for scientists
V17	Participation gives feeling of being helpful
V7	(Surveys) make a serious impression
V2	The climate of opinion is mapped
V6	Trust in the accuracy of survey results
V24	(Surveys) help in forming one's own opinion

Table 4.6 Order of six respondents and four questions by number of agreements

Befragten ID	V30	V19	V7	V24
3	1	1	1	1
2	1	1	1	2
10	1	1	2	2
8	1	2	2	2
25	2	2	2	3
39	2	2	2	1

Table 4.6 now shows an arbitrarily selected excerpt from the data matrix of the above-
mentioned survey for illustrative purposes.

While the first four questions and the first five persons (with the ID 3, 2, 10, 8 and 25)
can be arranged without error in Table 4.6, this is not possible for person 39. Similar to a
football match, in which the last team in the table sometimes wins against the top team, the
prerequisite of one-dimensionality is sometimes not fulfilled in the construction of scales.
To determine the quality of such a scale, the reproducibility coefficient (Rep) can be determined. It provides information on the extent to which the assumption of one-dimensionality
is fulfilled.

The Thurestone Scaling
In Thureston's scaling, steps one to three are again completed and in this way a pool of
statements on the relevant issue is created. Then a group of heterogeneously composed
calibration persons is asked to classify the statements (compare Thurestone and Chave
1929). The task for them is to place the individual statements into predefined categories.
For example, an eleven-point intensity scale with verbalizations at the poles is used:
11 = positive statement and 1 = negative statement. The respondents are asked not to

express their own views on the subject matter (as was the case with Likert scaling), but rather to evaluate the statements presented to them independently. The task can be made easier with the aid of a template sheet on which the eleven-point scale is visualised, and by means of cards on which the individual statements are printed. The assessors thus exercise a certain expert function.

The result shows firstly whether statements are available in the item pool for all eleven scale values or whether, for example, extreme expressions of the attitude continuum are not yet sufficiently covered by statements. Secondly, it becomes clear which scale items are covered by several statements. Here, a corresponding reduction of the relevant specifications can then take place. Thirdly, the measure of dispersion says something about the quality of the individual specifications. Statements which, for example, were evenly distributed by the calibration persons over all eleven scale values are of little use in depicting this attitude. In contrast, the unanimous allocation of a specification to only one scale point would be interpreted as an indication of a very clear formulation.

The final scale then contains the remaining statements. The respondents are then asked in the main study to agree or disagree with the respective statements. In the case of agreement, the target persons are credited with the scale value of the item in question and this is then summed up accordingly for all items.

4.5   The Quality Criteria of Objectivity, Reliability and Validity

An essential aspect of research planning is the question of the quality of the survey or measurement instruments intended for use. The general goal of a quantifying empirical survey is to provide data that can be used to make reliable and intersubjectively comprehensible statements about reality. The quality of survey instruments depends on a number of factors. Here we will describe the criteria that can be used to determine the quality of a measurement. It will be discussed what expectations the user of a survey method or the user of data may have of it. The focus will now be on the objectivity of the measurement, its reliability and the validity of the data obtained.

4.5.1   Objectivity

The results of an investigation should reflect certain dimensions of the object under investigation, for example the expression of an opinion in a target person, their intelligence quotient or the intensity of an intention to act. Of no interest, on the other hand, are the characteristics or the influence of the person carrying out the measurement, for example the respective interviewer or the test user. The objectivity of a measurement means the independence of the findings from the persons carrying out the investigation.

Objectivity thus exists when different people achieve the same results in an examination. Objectivity would not exist if everyone obtained a completely different result from the same subject using the same instrument, for example an intelligence test. The requirement to obtain the same result applies to all stages of a social science survey. An objective survey must, for example, be independent of the interviewer or evaluator, both when conducting a survey and when evaluating a survey.

The maxim that quantitative survey methods should exhibit maximum objectivity applies. However, it would not be correct to interpret the criterion of objectivity as meaning a complete absence of any subjective influence. After all, it is always people – subjects – who carry out social science surveys.

Experience with the creation and testing of objectivity is available primarily from psychological testing practice. Ways of increasing objectivity are then above all the elaboration of precise instructions to the interviewers for their behaviour as well as the compilation of glossaries, for example in order to guarantee objectivity in the context of observations (compare Sect. 6.2). The simple idea behind this is that such instructions reduce the subjective influence of the person being interviewed as much as possible.

One criterion that distinguishes scientific work from everyday experience is the intersubjective verifiability of the results obtained. This requires that the procedure in scientific work is standardised with the help of rules and finally documented.

4.5.2 Reliability

Reliability (the terms accuracy and precision are also used) is a measure of the reproducibility of measurement results. The same result should always be achieved when the instrument is used repeatedly.

Reliability is indicated by means of a correlation coefficient r (a correlation measure between 0 and + 1). 0 means that the measured value consists of only one error and, correspondingly, 1 means that the measured value is error-free. Various possibilities are available to determine the reliability of an instrument. The parallel test, test-retest design, test halving and consistency analysis will be presented.

Parallel Test
In this method of quality determination, the measurement is carried out – in parallel – with two different measuring instruments. It is assumed that two completely equivalent instruments are available for this purpose. In medicine, for example, two different instruments for measuring blood pressure would be conceivable. The agreement between the two forms of measurement is then an indication of their reliability. In the social sciences, however, parallel tests are likely to be used only rarely because of this prerequisite.

Test Retest Design
Measurement stability can also be determined by repeatedly measuring the same object with the same instrument after a certain time interval. This is a special form of panel design (see Sect. 4.6). The minimum value required for the correlation between the two

measurement results according to a rule of thumb is r = 0.85. If this value is multiplied by 100, the proportion of the variance that can be attributed to the true differences is obtained. In other words, the acceptable influence of error is 15%. Various prerequisites must be observed in the test-retest design.

- It only makes sense to use this design if the facts to be measured can be assumed to be stable. In the case of rapidly changing circumstances, it would not be possible to say to what differences determined between two surveys can be attributed, to faulty survey instruments, to changes in the object of investigation, or to both.
- All changes in contextual conditions must be avoided in the test-retest design. This applies both to the instrument – for example, the succession of questions in a question-naire should be maintained – and to the interviewer.
- When choosing the time interval between the two surveys, it is important to bear in mind that learning or memory effects must be avoided. Thus, a relatively long interval between the surveys should be chosen. However, the avoidance of changes in the object of investigation argues in favour of the shortest possible succession. It is thus necessary to make an optimization between both aspects.
- Finally, possible influences by the first measurement must be taken into account. These could, for example, have triggered a mental examination of the object of the survey and now also have changed the object to be measured. This issue was relevant, for example, in a study that sought to determine the desire for children among young adults. Here, the question partially triggered a cognitive engagement with the topic among the participants in the first study, so that changes already occurred after a relatively short time interval.

Using the ALLBUS 1984 as an example, the results of such a test-retest study will be presented (compare Zeifang 1987a, b). The design of the survey provided for:

- The target persons were interviewed on three survey dates, with the main survey of the ALLBUS 1984 as the first wave, followed by two further waves. It is thus a test-posttest design.
- The time interval between the waves was supposed to be a uniform four weeks. Due to problems with the field work, however, it was then three to five weeks.
- For reasons of research economy, a concentration on only one sample network was carried out. This means that surveys were conducted at 210 sample points (see Sect. 5.2).
- A gross approach of 420 interviews was aimed for. Initially, aresponse rate of 70% was achieved. This corresponds to 294 addresses. Of these, 210 persons were prepared to participate again in the second wave of the survey. Of these, 181 were then interviewed, 154 of whom were also involved in the third wave. Compared to the main sample (n = 3004) of the ALLBUS, the test sample (n = 154) shows no significant differences

in the demographic variables and in most of the attitudinal variables (compare Zeifang 1987a, b).
- Only a questionnaire reduced by half was used, which contained the most important questions of the main survey.
- It was possible to use the same interviewers in 84% of the cases.
- The original goal of checking all interviews by telephone after the test-retest was only achieved in 88% of the cases.

The results are shown in an overview (compare Table 4.7).[11] This table shows for the demographic variables in what percentage of cases a match was encountered in all three waves. Such demographic items often serve important functions as independent variables.

An interpretation of the findings is relatively difficult. For example, the reason for the not completely identical answers to the question about one's own gender remains in the dark. The less than optimal design of the test-retest study can also be blamed for unstable responses (see above). However, there was no actual alternative to this during the field-work. However, reference should be made to some conspicuous features:

Table 4.7 Results of the test-retest study in the ALLBUS 1984, number of questions answered identically by the 154 respondents in all three survey waves (in percent)

Subject of the question	Constant response behaviour
Number of children	100
Age	100
Gender	99
Marital status	98
Graduation of the father	90
Highest school degree	89
Denomination	89
Current employment	81
Predominant livelihood	78
Occupational status	73
Training certificate	72
Last professional position	56
Occupational status (father)	52
First professional position	42
Age at first marriage	69
Weekly working time	84
Income	27

---

[11] After all, income turned out to have a correlation of r = 0.90 (compare Zeifang 1987b), indicating a close relationship.

- As the number of answers increases, the stability of the answers decreases significantly. This was observed, for example, in the question about the respondents' own professional position and the professional position of their father. There were 32 and 37 different possible answers to this question respectively.
- Open questions are sometimes answered more stably than closed ones.
- The stability between the second and third waves is greater than that between the first and second waves. According to one assumption, the seriousness in answering the questions and the preoccupation with the topic increased from wave to wave (compare Zeifang 1987a, b).

In addition to these results related to the demographic variables, interesting findings were also obtained for the attitudinal variables. Here, generally lower values were obtained than for the questions shown so far. For the indicators of subjective stratification, voting intention, and retrospective voting decision, which are used relatively frequently in survey research, the stability of responses was 80%. Here, too, some general tendencies are to be reported:

- Those issues that the respondents have presumably already dealt with to a greater extent are assessed as more stable than others. This applies, for example, to some questions on the welfare state.
- In ranking questions such as the Inglehart Index (see also Sect. 4.3), the most important and least important items were answered most stably.
- Likewise, stable responses were found for the evaluation of extreme parties such as the NPD and the DKP.
- The results are worse when ambiguous stimuli are used. This could indicate an intellectual excessive demand on the part of the target persons, especially in such cases where there is actually no opinion (cf. also Sect. 6.1.2). Older persons in particular were also overtaxed in constantly stating their (highest) school-leaving qualification.

Test Halving

Test halving in addition to the test-retest design and the parallel test, test bisection (also called split-half) offers a further possibility for reliability testing. Only indicator batteries that consist of a larger number of items and that are all directed at the same subject matter are suitable for test halving. The demographic questions discussed above, for example, would therefore not be suitable for test halving.

Test halving involves dividing the test into two halves. Here, the first and the second half of a test can be selected, but a random division into the test halves can also take place. The advantage of this is that the test instrument only needs to be used once. Then it is calculated how strongly the answers of both halves correlate with each other. Due to the halving and the associated reduction of the items, the reliability is underestimated. Thus, the actual reliability is likely to be higher. This makes sense, since one can assume that a higher number of items and thus of measurements ensures a reduction in random errors. Reliability thus increases with the length of the test.

Consistency Analysis

Finally, consistency analysis lends itself to reliability testing. The procedure is comparable to that for the test halving technique. This also applies to the prerequisites to be applied. In the case of consistency analysis, the instrument is not divided into two halves as in the case of test halving, but broken down into as many parts as there are items (for an example of consistency analysis, see Sect. 4.4).

In test psychology, it is common to also determine the degree of difficulty of a scale. For dichotomous questions, the degree of difficulty indicates how many of them were answered correctly and, accordingly, how many were answered incorrectly. In survey research, the degree of difficulty is used to determine the proportion of yes answers in dichotomous questions.

In order to achieve a high discriminatory power, a medium level of difficulty is desirable. Applied to a performance test, neither too high nor too low a level of difficulty yields satisfactory results. In the former case, questions would be so difficult that only very few participants would be able to answer them correctly; in the latter case, they would be so easy that (almost) everyone would answer them correctly. In both cases, however, it would not be possible to adequately assess the performance of all test takers. Reliability can thus be increased by items with a medium degree of difficulty.

4.5.3 Validity

Besides objectivity and reliability, validity is the third important quality criterion. Validity is the main objective in the development of survey instruments. Objectivity and reliability are to be regarded as necessary prerequisites. If an instrument were to measure only objectively and reliably, one could say: I know that I (or any other person) always achieve the same result with the instrument, but I do not know *what* I have actually measured with it.

Validity is also referred to as content functionality or validity. According to Lienert (1969, p. 16), validity is the degree of accuracy with which a test measures what it is supposed to measure. Validity also has a relation to the theoretical concept, because this ultimately determines what is actually to be measured. Thus, there can be no objective criteria for testing validity. When empirically testing validity, it is important to take this theoretical reference into account. It has thus proved useful to distinguish between different forms of validity depending on this. In addition to content validity and criterion validity, construct validity should also be mentioned. These forms each require a different strategy of validity testing. Finally, triangulations and the Multitrait-Multimethod-Matrix (MTMM) are used to validate measurement instruments.

Content Validity

Content validity (also face validity) assumes that the individual questions contained in the instrument fully represent the property to be measured. The subject matter to be measured and the questions used for it are virtually identical. For example, a vocabulary test would have very high content validity if the test contained about 50 words drawn at random from the 2000 most used words in a language. However, examples of content validity in the social sciences are scarce. In part, therefore, expert ratings replace this type of validation.

Criterion Validity

This form of validity assumes that the social scientist has access to a criterion that also describes the issue of interest independently of the measurement instrument being tested. The correlation between the results obtained with the measurement instrument and the external criterion then provides information about the validity of the instrument. For example, the (non-)membership in a home association (provided that this could be determined, for example, with the help of appropriate reliable lists) and the results of a scale to survey the attachment to the region could be confronted with each other. Another example would be the results of a professional aptitude test on the one hand and the – later determined – actual professional success on the other hand.

A major problem in the use of criterion validity lies in the determination of a suitable external criterion. In such a case, the question arises as to whether one should not dispense with the measurement altogether and use the safe criterion instead. Thus, the use of criterion validity in social science practice is probably also subject to relatively narrow limits.

Construct Validity

A construct is a bundle of questions or variables whose answers are related because of a common origin. Again, a robust theory is required to establish this common origin. For example, a person could be assumed to have a certain basic political attitude, they could possibly be conservative. This would then be the origin that controls the answering of a whole series of further questions. Thus, on the basis of theoretical considerations, it would be expected that such a person would prefer a certain political party, would place himself in a certain position on the right-left scale, would have corresponding ideas about the upbringing of children (educational values), and so on.

Construct validity is based on the assumption that various constructs can be measured with the help of a measurement instrument, such as a questionnaire. The validation of these constructs checks whether all expected theoretical correlations can actually be proven empirically. Appropriate prior knowledge is required as a basis. Statistical test procedures are not usually used in this process. The following steps are taken:

- A construct is determined on the basis of theory, and it is likely to be certain behavioural dispositions that are related to other characteristics.
- Afterwards, hypotheses are to be derived about which correlations can be expected within the construct.
- Finally, the hypotheses are tested. This conclusiveness can then be determined with the help of factor analyses.

The Multitrait-Multimethod-Matrix (MTMM-Matrix)

The idea of Campbell and Fiske (1959) was to empirically collect (at least) two theoretical constructs (U and K) with (at least) two different procedures (for example, by means of a survey and an observation). This is a special variant of construct validation. With this strategy, also known as triangulation, both convergent and discriminate validity can be determined. Convergent validity highlights the correlation between the measurement results of different methods for capturing the same construct, and discriminant validity highlights the correlation between the measurement results of different constructs using the same method. Now, according to Camphell and Fiske, the postulate is that convergent validity should be greater than discriminant validity. This means that the measurement results should firstly be method-independent and secondly that it should be possible to discriminate between different constructs.

In his example (see Fig. 4.6), Diekmann presents the measurement of the constructs environmental action (U) and cooperation behaviour (K). Both were measured firstly by indicators in an interview and secondly with the aid of behavioural observation in the same sample. Figure 4.6 now contains the (fictitious) data obtained in the process. The values 0.59 and 0.50 represent the validity data. They show that the same construct (U: 0.59 and K: 0.50 respectively) was determined in a very similar way by different methods. In contrast, the other coefficients are clearly lower. This indicates that the method effects are significantly lower and that the measurement was valid.

The interested reader can find more information on the quality criteria for quantitative social research in Krebs and Menold (2019, p. 489 ff.).

4.6   Cross-Sectional and Longitudinal Studies

In the context of research planning, attention has so far been paid primarily to problems that need to be taken into account when devising the research instruments. Attention has been paid, for example, to the creation of scales in order to be able to survey facts such as latent value orientations, opinions and intentions to act. The scales then become

Method		Interview		Observation	
	Construct	U	K	U	K
Interview	U				
	K	.40			
Observation	U	(.59)	.25		
	K	.28	(.50)	.34	

Fig. 4.6 Example of a multitrait-multimethod matrix (cf. Diekmann 2012, p. 262)

components of questionnaires, coding schemes or of observation protocols, i.e. the actual survey instruments. The question of the quality of the scales or instruments also had to be asked in this context.

The following section focuses on the way in which these survey instruments are finally used.

Regardless of whether it is intended to collect the data by means of an interview or an observation, it must be decided whether a survey at only one point in time is sufficient, or whether it is necessary to provide for several study points in time. In the latter case, it is also necessary to answer the question of whether the information needs to be collected again from the same research units – for example, from identical respondents – or whether it is necessary to carry out the study using the elements of a different, new sample. One-time social science studies are called cross-sectional studies. Studies in which the same respondents are interviewed again are called panel studies. Finally, a trend study is one in which the study is repeated at different points in time on independently drawn samples. Trend studies and panel studies are considered longitudinal studies.

Thus, the question arises for which knowledge objectives cross-sectional and longitudinal studies are suitable in each case. It is obvious to elucidate correlations (A ↔ B) by means of one-off studies. This is also an important step in the knowledge of social reality. The expression of a certain state of affairs A is related to the expression of another state of affairs B. However, a methodologically more demanding question would now be to ask about the cause-effect relationship. The typical instruments for investigating cause-effect relationships are experiments (cf. Sect. 7.1) or panel studies. The extent to which an effect (E) is produced as a result of a (targeted) intervention (C) is observed. The cause precedes the effect, so that not only can a relationship between two facts be established, but the cause can also be identified. This is only possible to a limited extent in cross-sectional studies. However, with the help of logic and comparisons with other studies, it should be possible to approach causal statements by means of cross-sectional studies.

Cross-sectional and longitudinal studies each offer the social researcher their own opportunities for gaining knowledge and at the same time place specific demands on the design of the study. These will be discussed below by way of example.

For the study of the expected social change in East Germany after the collapse of the GDR, it was appropriate to design *longitudinal studies*. Researching the restructuring of people's attitudes that was foreseeable as part of the transformation process posed a particular challenge for sociologists. To this end, for example, in the studies in the series "Life in the GDR/East Germany" (cf. Häder and Häder 1995a, 1998), following the model of social reporting, questions were asked within the framework of a population survey about subjectively perceived life satisfaction as well as satisfaction with different areas of life. The series of investigations began with a cross-sectional survey as early as spring 1990. In order to be able to follow the course of social change, including changes in people's life satisfaction, this instrument had to be used again at different points in time with random samples drawn from almost the same population (compare Götze 1992; Häder et al. 1996). It was found that, if the beginning of 1990 is taken as the starting point, there was a marked increase in general life satisfaction in the general population of East Germany in the following period. Other studies, such as the Welfare Survey, also come to this conclusion for the period mentioned (cf. Datenreport 2002, p. 431 ff.). Specifically, the questions of the spring 1990 study were replicated[12] in 1993 and 1996. At all three study time points, as is common in longitudinal studies, the definition of the population was maintained. Above all, the use of identical questions made it possible to interpret the results obtained in each case as a trend.

In order to demonstrate the different possibilities of panel and trend studies, another substantive problem shall serve as an example. One question in the above-mentioned series of studies was concerned with retrospective satisfaction. Thus, the base year 1990 or, more specifically, the economic, social and monetary union as a particularly memorable event was chosen as the reference point. In the studies of 1993 and 1996 the target persons were asked to indicate whether they were more satisfied or less satisfied in 1990 than in the present situation or whether they could not detect any difference in this respect. The background to the question was the assumption that memories of the situation at the time of monetary union can be transfigured due to the complete change in the overall living conditions of people in East Germany (cf. Häder 1998). In Table 4.8[13,14,15] the results of the mentioned surveys are shown.

An initial examination shows that satisfaction in eastern Germany has risen in a number of areas of life since 1990. This statement is based on a direct comparison (mean comparison) of the three points in time examined. Thus for example the satisfaction with the environment in East Germany rose particularly clearly since 1990. This increase was already noticeable in 1993 (+1.00 on a five-point scale) and continued in 1996 (+1.26). This statement is based on data obtained in longitudinal studies.

Whereas in 1993 42% of respondents indicated an increase in their own satisfaction with the environment, this proportion *fell* to 39% in the 1996 survey (base: cross-sectional data). Although satisfaction with the environment has risen dramatically (basis: longitudinal data), fewer respondents say they are currently more satisfied with the environment than in 1990. The situation is similar for satisfaction with wages. This first increased slightly in 1993 (+0.05) and then more significantly in 1996 (+0.64). Retrospectively, however, in the individual cross-sectional surveys the share of those persons who state that

---

[12] The cumulative data set as well as a corresponding codebook are available from GESIS Leibniz Institute for the Social Sciences in the Data Archive and Data Analysis Department and are listed here under study number 6666.

[13] Mean values on a five-point scale with 1 = very satisfied and 5 = very dissatisfied. Higher values thus indicate higher dissatisfaction.

[14] Positive values indicate an increase and negative values indicate a decrease in satisfaction compared to 1990.

[15] Answers to the question "Please compare the time before the monetary union with today". How has your satisfaction with the following things or sides of life changed? Where are you more satisfied than you were then, where has nothing changed, and where were you more satisfied then than you are now? Choice of 'more satisfied today than then', figures in percent.

Table 4.8 Satisfaction with various areas of life in East Germany in 1990, 1993 and 1996 (mean values and mean value differences) as well as positive retrospective satisfaction in 1993 and 1996 compared with 1990 in each case

	1990	Difference 1990/1993		Difference 1990/1996	
Habitat	Satisfaction	Satisfaction	Comparison	Satisfaction	Comparison
Work	3.60	+0.06	17%	−0.22	15%
Free time	3.12	+0.56	25%	+0.45	25%
Social security	3.57	−0.76	11%	−0.84	10%
Environment	1.70	+1.00	42%	+1.26	39%
Reside	3.51	+0.30	27%	−0.38	36%
Education	3.35	+0.07	24%	+0.07	17%
Partnership	4.35	+0.07	10%	+0.08	9%
Health	2.79	+0.76	29%	+0.64	27%
Living with children	3.69	−0.15	8%	−0.27	7%
Wage	1.83	+0.05	24%	+0.64	22%

they are currently more satisfied with their wages than in 1990 *decreases*. The situation is again different with social security. In 1993 a clear (−0.76), and in 1996 even more pronounced (−0.84) decline in satisfaction can be observed. Accordingly, the share of respondents who retrospectively assume an increase in satisfaction with social security decreases slightly in 1996 compared to 1993.

These results raise several questions: Can it actually be compellingly concluded from the findings shown that people in East Germany are gradually coming to view their own past differently. Various things can be said about this:

First, the individual cross-sectional surveys each assess the current situation in relation to the situation at the time of monetary union. The two retrospective questions involve statements by people who come from changed populations. For example, the 1996 survey involved six age cohorts that were not counted in the population in 1990. It can also be assumed that, due to mortality, certain persons are missing from the 1996 survey who were, however, still part of the population in 1990.

Secondly, these data cannot be used to make a statement about the extent to which *individual* satisfaction has changed since 1990. It is theoretically conceivable, for example, that satisfaction with a particular area has not changed for the majority of people or may even have deteriorated slightly, while for a few people it has risen significantly. In such a case, the mean increases, that is, the overall level of satisfaction increases. At the same time, however, fewer people are currently more satisfied with this area than in the past. This could also explain the seemingly paradoxical findings shown above.

In summary, it is therefore only possible to make corresponding statements at the aggregate level, i.e. for the entire adult population in eastern Germany at the time of the study. On the other hand, it is not possible to track the individual course of satisfaction development with these data. If one wanted to answer the question of whether in retrospect there is a change in the assessment of the situation in spring 1990, *panel data* would be required for this. (On the problem of retrospective assessments of the situation in East Germany, see also Häder 1998, p. 7 ff.). This design will now be presented in more detail.

In a panel, an identical group of people is surveyed repeatedly. In Germany, the Socio-Economic Panel (SOEP) of the German Institute for Economic Research in Berlin (DIW) is particularly well known and of great scientific importance. Access panels are another type of panel.

The SOEP has been surveyed since 1984. Surveys are conducted in the old and, since 1990, also in the new federal states. In addition to Germans, foreigners and immigrants are also included. For this purpose, the survey questionnaires are translated into the corresponding languages. In 2000, the sample comprised about 12,000 households with more than 20,000 persons. A special design was used in each case, which will be presented below. The SOEP focuses on household composition, employment and family biographies, labour force participation and occupational mobility, as well as income patterns, health and life satisfaction. At this point, we would like to point out some general methodological problems with panel studies:

- In panel surveys, it must be expected that a certain percentage of participants will drop out when the survey is repeated. One reason for this, apart from refusals, is for example the departure of participants. For example, more highly educated and younger panel members are more likely to leave than others. Thus, these dropouts are systematically caused. Substitutes may have to be recruited. Good address maintenance is also necessary to keep dropouts as low as possible. Thus, panel studies are associated with a considerable research organisational effort. They are therefore usually significantly more expensive than trend studies.
- To enable comparisons between points in time, the constancy of the measurement instrument must be ensured. Of course, this applies to longitudinal studies in general. As already shown at the beginning (cf. Chap. 2), respondents sometimes react extremely sensitively if even slight changes are made to the text of the question.
- Furthermore, it is possible that a panel effect occurs. This means that changes can occur in the participants of the survey simply due to the repeated measurements. For example, they may remember their answers from the last survey or become more familiar with the questions over time, have dealt with the questions cognitively, and so on.
- As a rule, the results of panel surveys cannot be regarded as representative population surveys. Over time, the population changes, for example, there may be a gradual increase in the level of education in the population. In panel surveys, however, the level of education remains more or less constant.
- Finally, in panel studies, participants' addresses must be stored until the end of the last wave. (In contrast, for one-time surveys, these are destroyed at the next possible date, which is usually after all interviewer checks have been completed). This means that special data protection regulations must be observed in panel studies. Certain restrictions therefore apply to the handling of panel data.

Due to the special features described above, different methodological variants of the panel design are used. These attempt to compensate for certain disadvantageous aspects of the panel design. Some of them will be shown (compare also Zimmermann 1972; Schnell et al. 2013, p. 228 ff.).

In the case of an alternating panel (cf. Table 4.9), two groups are interviewed alternately. Thus, fewer panel effects are to be expected than in a more frequently conducted study. However, a correspondingly large sample is a prerequisite.

The rotating panel (see Table 4.10) uses an even larger number of groups. One group is then exchanged per survey wave. Assuming five waves, five groups would have to be formed. One new group is included in the panel per wave, while the oldest group leaves the panel. The Microcensus of the Federal Statistical Office uses this design, which of course involves a particularly large amount of work. However, the advantages are also obvious. Firstly, a new sample is drawn at each point in time, so that social change can be reflected. With a conventional panel, on the other hand, it is hardly possible to make statements about social change. In a simple panel, for example, certain people drop out of the panel, while the next generation has no access to the panel. Secondly, the panel effect is kept to a minimum, since participation is terminated after five interviews.

Finally, the split panel (see Table 4.11) should be mentioned. This is characterized by the respective combination of the survey of a panel group (G1) with a cross-sectional survey. These then also make it possible to capture the change in the population. Compared to the rotating panel, however, the panel effect is not reduced. As with the other designs, the split panel requires a high research organizational effort.

Table 4.9 Procedure for an alternating panel

	t_0	t_1	t_2	t_3	t_4
G1	M		M		M
G2		M		M	

G respondent groups, M measurement

Table 4.10 Procedure for a rotating panel

	t_0	t_1	t_2	t_3	t_4
G1	M	M	M	M	M
G2	M	M	M		
G3	M	M	M		
G4	M	M			
G5	M				
G6		M	M	M	M
G7			M	M	M
G8				M	M
G9					M

G respondent groups, M measurement

Table 4.11 Procedure for a split panel

	t_0	t_1	t_2	t_3	t_4
G1	M	M	M	M	M
Q1	M				
Q2		M			
Q3			M		
Q4				M	
Q5					M

G respondent groups, M measurement, Q persons interviewed in a cross-sectional survey

Table 4.12 Length of stay in and probabilities of leaving poverty

Number of years in poverty since poverty began	East Germany		West Germany	
	Staying in poverty	Probability of exit	Staying in poverty	Probability of exit
1	100	–	100	–
2	47.9	52.2	54.3	45.7
3	31.6	34.0	38.7	28.6
4	15.1	52.1	31.8	18.0
5	9.4	37.4	25.5	19.6
6	7.8	17.8	24.1	5.6
7	7.8	0.0	22.0	8.6
8	7.8	0.0	18.3	16.8

Table 4.12[16] shows an example of the use of panel data from the SOEP. This involves the length of time spent in poverty and the likelihood of leaving poverty. Poverty, as defined, is that part of the population that has less than 50% of the average income of the population as a whole. In the following, only those persons from the longitudinal population of the SOEP are considered for whom a period of poverty began in 1993 or later.

The figures in the table show that in eastern Germany about half of the people affected by poverty were in this status for a period of two years, and that after three years just under a third of these people were still affected by poverty. This picture is somewhat different in West Germany: "In West Germany, more than half of all poor persons had to experience a poverty phase of at least two years; for almost two fifths it was a poverty phase of at least three years and for a close third a poverty phase of at least four years." (Otto and Siedler 2003, p. 1 ff.).

When looking at the exit probabilities from poverty, one can see that in West Germany this decreases with the duration of poverty. The longer someone is poor here, the less likely it is that this status will be left again. This relatively entrenched status, on the other hand, does not exist to the same extent in eastern Germany. While, for example, after four years of poverty the chance of leaving poverty in eastern Germany is a good 50%, in the

---

[16] Source: https://www.diw.de/sixcms/detail.php?id=284881#HDR0, last accessed February 04, 2019.

west it is only just under 20%. It should have become clear at this point that such panel data are indispensable for interpretations of such changes at the individual level.

In this context, it will also be shown which social science instruments are available to determine life-cycle, age and cohort effects. A cohort is a (def.) = population group that is defined by a common starting event that has a longer-term impact. Examples of cohorts include age or birth cohorts, marriage cohorts, and occupational entry cohorts. All were exposed to a particular event. *Cohort effects* now arise because the individuals in question (the cohort) are collectively exposed to certain events. *Period effects,* on the other hand, are those effects that are generated due to historically unique events – such as an economic crisis, the fall of the Berlin Wall, or the moon landing – among the people who experienced them. Finally, *life cycle effects* are correlations between a certain characteristic and the time that has elapsed since then. This will be illustrated by an example.

It has been shown that the exclusive ownership of a mobile phone (i.e. doing without a landline) is found above all among younger people – aged between 16 and 25 (cf. Häder and Häder 2009). If this were a cohort effect, in ten years these mobile-only people would have to be 26 to 35 years old. If we are dealing with a life cycle effect, the group of people in question would still be as old in ten years as they are today. In order to separate out such effects, longitudinal data, and in particular panal data, are required.

The question now to be discussed is: Is it not also possible to obtain panel data with the help of (relatively simpler) cross-sectional surveys? The answer is: in principle, yes, but. The rest of this section is devoted to a more detailed discussion of this problem.

Obviously, a distinction must be made between, first, the designs discussed so far and, second, the structure of the data obtained with these arrangements. Thus, we speak of *cross-sectional data* when they refer to only one point in time. For example, Table 4.3 showed the data on retrospective life satisfaction in various areas obtained in the 1993 survey in eastern Germany. In principle, such cross-sectional data can also be obtained in longitudinal studies. This is done with the help of questions that refer only to the point in time of interest.

In addition, there is *time series data*. This is a sequence of values of a variable at different points in time. In economics, for example, the unemployment rate or the German stock index (DAX) are known as time series data. Time series data are characterized by many measurement points. In addition, a constant collection methodology is required. Time series data are usually generated with the help of trend studies. In order to obtain time series data, it is necessary that the respective population remains constant, which usually results in the need to draw new samples. In the social sciences, compared with economics, only data obtained at relatively few points in time are currently available (cf. Sect. 6.1.4).

With *panel data,* statements can be made about the same units of investigation (these can be persons or households, for example). Table 4.12 showed the length of stay in poverty of individual units of investigation. Such statements can be made at different points in time by identical persons. It is important to note at this point that, theoretically, cross-sectional studies are also suitable for collecting panel data. In principle, it would be possible to retrospectively ask target persons about the vocational training they had completed at certain points in the past or about the qualifications they had obtained in the process. It should be noted, however, that the quality of such data is likely to be low. For example, it may still be possible to conduct a study on individual career paths using a cross-sectional design. However, many other questions, such as the duration of unemployment in a given period or even satisfaction before a given time limit, are not suitable for this purpose. Even the retrospective question about how well someone has slept in each of the last nights yields different results than if one inquires about this very fact after each individual night (compare Grau et al. 2000, p. 20 ff.). In this case, the respondents are often cognitively overwhelmed in order to provide reliable information.

Finally, *data on* the *course of life* or *events* will be presented. For example, these make statements about the duration of marriage, the duration of unemployment, about certain occupational histories, the duration of residence, and so on. Thus, historical data are time intervals between certain events. In some cases, they have a particularly high information content for social research. In principle, historical data can also be obtained in different ways. The most reliable way seems to be to use panel studies for this purpose, as already shown above on the basis of the length of time spent in poverty. Certain trajectories, such as periods of education, can probably be obtained directly from the persons concerned, even in a cross-sectional study.

In summary, the most important characteristics of cross-sectional, trend and panel studies will be presented once again.

Cross-Sectional Studies

The study takes place at a point in time or during a relatively short period of time, i.e. it is designed as a one-off quasi-photographic survey recording the current state of affairs. The aim of cross-sectional studies can be, for example, to determine the attitudes of the population to the right of self-defence.

Trend Study

Here, several cross-sectional studies are organised on the same topic. As a rule, the same indicators are used for this purpose, in order to be collected at several points in time from different samples (persons) and the same population. In this way, changes at the aggregate level can be identified, for example changes in satisfaction with certain aspects of life among certain groups of the population. Subject to appropriate selection, these are generalisable results for individual points in time.

Panel Studies

Similar to trend studies, these are multiple surveys, i.e. a specific form of a longitudinal study. The main difference is that the same research subjects (the so-called panel) are interviewed several times. Panel studies can identify changes between the study time points at the individual level. However, they are not necessarily also representative population surveys, as the population (the general population) may have changed between surveys but the subjects remain the same, or due to panel mortality there may be systematic omissions.[17]

4.7   Non-Reactive Approaches

Surveys, such as face-to-face interviews, are highly complex situations in which the behaviour of all participants is controlled by numerous facets (this problem is discussed in more detail in Sect. 6.2). In the control of human behaviour during an interview, the effect of very subtle mechanisms must also be assumed. This is true even in the case of fully standardized surveys, which are characterized by detailed specifications of all essential components of the survey situation – especially with regard to interviewer behavior.

When we speak of behavioural control, we may think of a person who is weak-willed, uncertain about his intentions, who prefers to seek advice from others rather than to give information about himself, and who, because of his character, can easily be made the object of influence, for example by the advertising industry or even by an interviewer. The opposite could also be imagined. For example, there are people who willingly and knowingly influence others through their behavior, such as the door-to-door salesman type. The term behavioural control can also be associated with people who consciously orient themselves to majority opinions, for example in order to avoid being socially marginalised. From the point of view of an empirical social researcher, it is interesting to note that behaviour control also occurs unconsciously and that quite everyday behaviour can be affected by it. Various experiments provide evidence of such influence.

Rosenthal and Jacobson (1968) describe an experiment involving students and teachers. First, the intelligence of students was tested. Then their teachers were given manipulated feedback concerning the results of the intelligence tests. Then, after a period of time, intelligence tests were administered to the students again. Now it turned out that those children whose test results had been falsely signalled to the teachers as being above average, in the meantime actually achieved better results than in the first attempt. The organizers of this experiment interpreted the result to mean that the teachers had turned more toward the supposedly more intelligent students because of the information they had been given, thus eventually producing the effect that the experiment uncovered. This experiment has become known as the Pygmalion experiment, and the effect thus obtained as the Pygmalion effect.

Diekmann (2012, p. 623 ff. following Watzlawick 1976) presents an example from the circus where even a horse (clever Hans) was able to solve apparently numerical tasks. The results were communicated to the spectators by a corresponding number of hoof movements. The secret of this animal's ability to calculate, however, was that the horse always

---

[17] For more information on cross-sectional and longitudinal studies, compare Mochmann (2019, p. 259 ff.) and on panel studies Schupp (2019, p. 1265 ff).

kept an eye on the questioner and, when the questioner leaned forward to observe whether the horse was pausing at the right place with its hoof movements, interpreted this leaning forward – usually an almost imperceptible movement of the head – as a signal. Clever Hans then adjusted the hoof movements and had thus apparently solved the task correctly on his own.

This is another example of subtle behavioral control. It should be emphasized that it does not have to be a conscious manipulation of behavior, but that the animal directs its movements based on an inconspicuous detail.

If one transfers these two findings to interviewer-assisted survey research, the fear is obvious that the target person can also be influenced by the behaviour of their counterpart – imperceptible movements of the head cannot be ruled out here either. In empirical social research, the term reactivity is used in this context. Reactivity means that the result of a measurement is influenced by the measurement process itself. Especially in laboratory experiments (but of course not only here), which take place in an artificial environment, reactivity is to be expected.

One way of dealing with this phenomenon in face-to-face interviews is to add certain characteristics of the interviewer, such as age, gender and level of education, to the survey data set. This makes it possible for a later evaluator to check whether and to what extent, for example, the answer to a particular question came about as a result of certain characteristics of the interviewer or is related to them. Reactivity cannot be neutralized in this way, but its occurrence can at least be estimated. It should also be noted that in this way it is only possible to identify reactivity that occurs as a result of the characteristics that were collected about the interviewer. Head movements and similarly subtle reactions cannot, of course, be registered. Especially in the case of face-to-face interviews, a high degree of reactivity cannot be ruled out under certain circumstances.

Non-reactive methods represent a solution here. Empirical social research distinguishes between various forms of these approaches. These include above all content analyses, which are dealt with in Sect. 6.3. The following techniques should also be mentioned (for more detailed suggestions, see Webb et al. 1966): non-reactive field experiments, the analysis of behavioural traces and the use of process-produced data.

4.7.1 Non-Reactive Field Experiments

In the case of field experiments, non-reactive experiments (for the treatment of social experiments, see Sect. 7.1) initially involve a targeted manipulation of social reality. In other words, the independent variables are changed. Then – as is usual in experiments – the effects produced in this way are observed. The non-reactive character of this type of experiment is due to the fact that the consequences of the manipulation are observed in the actual environment and that the persons involved in the experiment are not informed about their role in the experiment. This may result in research ethics issues.

Non-reactive experiments take place covertly, and this is also important, in the natural environment of the test subjects, in order to exclude any possible influences of an artificial atmosphere here as well. Concrete examples of non-reactive field experiments are presented in more detail by Diekmann (2012, p. 623 ff.); Friedrichs (1990, p. 309 ff.).

In order to work on the hypothesis that people who are in a good mood are more likely to help than others, a telephone booth was prepared. Some of the test subjects were put in a good mood – one hopes – by finding a coin in the phone booth in a seemingly random but actually manipulated manner. In the comparison group, such an intervention was dispensed with. On leaving the telephone booth, both groups were now confronted with a person who, apparently by accident, had a stack of documents fall out of his hand. It can now be observed non-reactively which of the two groups offers the stranger more frequent help in picking up the documents. The prerequisite is, of course, that the subjects are not made aware that they are part of an experiment. This also applies to the following designs.

The aim of the study was to investigate aggressive behaviour in road traffic and the social status of the vehicle driver. For this purpose, a driver did not move his car at an intersection after the traffic lights had changed to green. The time that elapsed before other road users began to honk was then recorded. The indicator function for social status was the type of car of the driver who was the first to start honking. The honking of the horn served as an indicator of aggressive behaviour in road traffic.

The *Lost Letter Technique* (cf. Milgram et al. 1965) has become known to a wide audience. This technique can be used to determine the prestige of a sender or recipient of a letter. For this purpose, stamped and addressed envelopes are apparently lost, but in reality they are deliberately deposited in certain places. It is observed how large the proportion of those items is that finally reaches the recipient. The hypothesis is that the prestige of the recipient or the sender influences the return rate. The higher the prestige of an institution, the more letters will be dropped into a mailbox by – as the organizers hope – unsuspecting test subjects. One can work with various modifications of this technique. For example, uncertainty about who the finder of the letter is can be reduced by targeting mail to be placed in front of certain apartment doors or tucked behind the windshield wipers of certain types of vehicles. Unfortunately, this technique has now become so well known that some finders of the letters are well aware that they are part of an experiment.

Under the *Misdirected Letter Technique*, letters are again delivered seemingly unintentionally to wrong addresses. Here, too, the handling of these mail items is observed. With the help of the variation in the addressee, differences in the image of the recipient in question can be identified.

The *misdial-up technique* involves the following procedure. An apparently misdirected caller asks a person by telephone for a little help, for example a call to another person. Now it is seen how high the proportion of people willing to help is who comply with such a request. Non-reactive observations can also be made with the misdialling technique by means of variations in the design, such as a foreign-sounding voice.

Another method is the use of fictitious *newspaper advertisements*. These can be, for example, personal ads, job advertisements or job offers. One varies a detail in the texts concerned and observes the different effects produced by this. Finally, it is also conceivable to use the reactions to certain advertisements for non-reactive experiments. In this case, different variants of response letters would also be used and the effects achieved with the advertiser in each case would be observed.

Reference has already been made to the ethical problems associated with non-reactive methods. In the case of individual designs, for example, it is necessary to assess the extent to which the manipulations carried out and the behaviour thus produced in the test subjects and associated with certain costs for them are justified on the basis of the purpose of the research. Presumably, it can be assumed that the insertion of a found letter into a letterbox causes only very low costs or only a low effort for the participating persons and therefore represents an unobjectionable strategy. It is debatable to what extent one is justified in exploiting the bona fide helpfulness of other people for a social science experiment. However, experiments in which the subjects incur major costs are problematic. This could include fictitious job advertisements or manipulated personal ads. The expenses incurred by the test subjects in the case of fictitious job advertisements are difficult to justify.

A balance must be struck between the benefits of the scientific work and the costs incurred by the persons concerned as a result of the experiment. Reference should be made to various guidelines for dealing with test subjects drawn up by the Association of German Market and Social Research Institutes (ADM, see also Sect. 4.9). These can also serve as a good guide for social researchers outside the ADM.

4.7.2   Behavioural Traces

In addition to non-reactive experiments, there is the possibility of passive observation of behavioural traces. Diekmann (2012, p. 644 ff.) has also compiled some interesting variants on this. One of them concerns the relation between International Politics and pizza consumption at the White House in Washington DC. Thus, there was the possibility of reading the seriousness of the world situation on the basis of behavioral traces or, more specifically, on pizza consumption in the White House. The more tense the work here and the more overtime had to be worked, the more pizza was consumed. However, this indicator is obviously a historically transient indicator. The relationship between the seriousness of the world situation and pizza consumption is also linked to the respective president or his work style.

The indicator function of pizza consumption – or so it seems – has meanwhile lost its reliability. In the course of the ABC policy (Anythink But Clinton) of the president who followed Clinton, George W. Bush, other framework conditions arose for the work of the White House staff. So now the meetings start on time. "Gone are the days of the sacrilegious who ordered pizza service to meetings in the Roosvelt Room, the sanctuary of the forefathers. No one curses in the hallways anymore; dozens of new rules of conduct govern the White House." (Kleine-Brockhoff 2003, p. 14) Thus, under President Bush Jr. it was customary to provide one's work at set hours, at the expense of spontaneous consumption of that Italian national dish. Unfortunately, the seriousness of the world situation shows itself unimpressed by this.

However, the physical wear of book pages and similar imprints can still be used to determine reading, viewing, and listening habits. The degree of soiling of the carpet in front of certain pictures in an exhibition, the number of soiled pages in a book or the number of dog-ears to be found there, and the like are non-reactive indications of the causation of these marks. The situation is similar with the number of loans in libraries. These too represent a non-reactive indication of certain underlying behaviours. What all indicators have in common is that they are generated without any presumption on the part of those involved that this will form the basis of a later analysis.

Another strategy is to determine the frequencies set on the car radios during a workshop visit (Babbie 2002, p. 304). In this way, one would also obtain non-reactive information about listening habits while driving and could additionally tie these to specific vehicle types.

Waste research, i.e. the analysis of things thrown into the garbage, also provides a non-reactive insight into consumer habits. Sexual behaviour can be determined non-reactively with the help of condom sales figures. The reception of television programmes could be estimated using the water consumption during the commercial breaks and at the end of the respective programme. Warning signs and signposts can be observed. Thus, the prohibition of children's games in the yard can be seen as an indicator of the hostility towards children in a residential area.

Obviously, there is plenty of room for creativity on the part of social researchers at this point. However, one problem that must always be taken into account is the validity and reliability of the information collected in this way. The necessary renunciation of communication with the research subjects, on the other hand, opens up room for misinterpretation of behaviour on the part of the social researcher. For example, market researchers once predicted that a product would have few chances on the market. However, this prediction turned out to be wrong. Although there was in fact hardly any demand for the product on the market, its packaging (a nice tin can) was very attractive for numerous other purposes and so the contents of the tin can sold well.

4.7.3   Use of Process-Produced Data

In our society, information is collected and stored about numerous processes. This applies to the listing of bicycle thefts as well as to information on the frequency of births. Newspaper articles can contain sociologically relevant information, as can, for example, sales offers in Internet auctions (compare Diekmann and Wyder 2002, p. 674 ff.). What all this information has in common is that it has not been registered for the purpose of research and is therefore also non-reactive. This can make such information attractive for secondary analysis. Even the analysis of graffiti and grave inscriptions can provide information about social behaviour.

An approach presented by Jenkel and Lippert (1998) can serve as an example for the discussion of possibilities and limitations in the use of process-produced data. This proves that different sources, which have not been created for the purpose of scientific analysis and which apply to an identical set of facts, can certainly also arrive at different findings. The project in question is devoted to participation in political protest actions in Leipzig in the period from 1992 to 1996. In the comparison, firstly the issues of the Leipziger Volkszeitung (LVZ) and secondly the records of the Ordnungsamt (public order office) of the city of Leipzig were used. It was checked which protest actions were documented in which medium in each case. The evaluation resulted in 484 actions (= 100%). The following overview (cf. Table 4.13) now shows how these were documented in the respective media.

Here it becomes clear that there are substantial differences between the two non-reactive data sources – for various reasons. For example, one of these analyses showed that the specific purpose for which this information was originally generated must always be taken into account.

The main problems arising from the use of non-reactive methods can be summarised as follows:

- The reliability and validity of the analysis must be questioned. Here, hardly any standardizations are possible due to the required naturalness of the situation.
- First of all, it must be possible to find a non-reactive design for a relevant social science question.
- The selection of the sample, or rather the definition of the relevant population beforehand, can become a problem. It must be clarified, for example, which group of people can be referred to in the evaluation when organising non-reactive experiments.
- Supplementing the non-reactive design with, for example, a survey is unlikely to be avoidable in many cases.
- Finally, ethical issues should always be considered when dealing with non-reactive field experiments.

4.8   Secondary Analyses of Data Sets

When planning a study, it should be examined at the earliest possible stage whether an empirical survey is necessary or whether a satisfactory solution to the problem could also be found with the help of secondary analyses of existing data sets.

Table 4.13 Documentation of protest actions in the LVZ and the public order office of the City of Leipzig

Only mentioned in the LVZ	48.6%
Recorded only in the public order office	32.2%
Mentioned in LVZ *and* public order office	19.2%

The maxim should be to only collect data from individuals if this is absolutely necessary for solving the problem. Otherwise, it is advisable to resort to secondary analyses. Firstly, in accordance with the provisions of data protection (cf. Sect. 4.9), there must be no interference with people's personal privacy without sufficient reason. Also, misuse of data can best be prevented by not collecting it in the first place. Secondly, overburdening the population with surveys should also be avoided. In the context of the discussion on the non-response problem (cf. Sect. 5.8), we will discuss the phenomenon of survey fatigue (also known as oversurveying) later on.

There are no hierarchies in the methods of empirical social research with regard to their value. In each case, a method is more or less suitable for clarifying a particular issue. Thus it is also not possible to deny the importance of secondary analyses of external data sets. An argumentation, for example, that it is fundamentally a flaw not to use self-collected surveys for an analysis is thus completely misguided.

It can be assumed that numerous data sets generated in the past still contain sufficient information material of interest to the social sciences. A complete and exhaustive analysis of a particular empirical survey, it can still be assumed, is unlikely to exist.

The main *advantage of* secondary analyses is that they are available to interested parties at low cost and relatively quickly. If one takes as a comparison an exclusive study, the funds for which must first be applied for – for the preparation and processing of an application for material funds by the research funding institution, more than half a year must usually be expected in each case – it is not unrealistic to assume that the desired data may only be available for evaluation after two to three years.

Another argument for the use of secondary analyses is their suitability for re-analyses. Re-analyses make it possible to check primary analyses. In order to detect deliberate falsifications and unconscious errors, this strategy is of great importance in science.

The decisive *disadvantage of* secondary analytical data analyses is the necessary search for and procurement of adequate data sources that are able to sufficiently cover one's own intended research goals. At the same time, it must be ensured that these data sets are exhaustively documented and are also available to social research for secondary analyses.

In the Federal Republic of Germany and in many other democratically constituted societies, various data sets from surveys can be used for such analyses for scientific work and in some cases also for the entire interested public. In the Federal Republic of Germany, the institutional prerequisites for this have been created by the GESIS Leibniz Institute for the Social Sciences and, in particular, by the Data Archive and Data Analysis Department in Cologne. This department is concerned with the documentation, archiving and dissemination of social science survey data (see also Mochmann 2019, p. 263).

In addition, it is also possible to use data generated by other (non-scientific) institutions, for example – as already shown above – from court records or from other authorities for secondary analytical research. These are usually process-produced data, the use of which has already been reported on in another section (cf. Sect. 4.7). and will be discussed later in Sect. 6.3.

The European Social Survey (ESS),[18] the General Population Survey of the Social Sciences (ALLBUS),[19] the Microcensus of the Federal Statistical Office (MZ)[20] and the Socio-Economic Panel (SOEP)[21] are also likely to be of particular interest in the Federal Republic. Detailed documentation on all survey series can be found on the Internet. This also applies to the microcensus. If you are interested in evaluating other studies, you can ask the GESIS Leibniz Institute for the Social Sciences in Cologne for help in obtaining the data set (cf. Sect. 6.1.4).

A prerequisite for the secondary analytical evaluation is comprehensive *documentation of* the present study (cf. also Chap. 10). It is important to obtain information on all steps of the survey in order to be able to follow it up and thus understand it correctly.

Big Data

In modern everyday life, digital data is generated almost constantly. No matter whether when making a phone call or using the Internet, when making purchases with a credit card, when doing banking transactions or visiting the doctor, in surveillance cameras, when monitoring the weather, while on social networks or when booking a hotel, such data is generated everywhere. These are process-produced and are largely machine-generated. At the same time, ever greater capacities are available to store this data electronically for longer periods of time. The volume of these sometimes very heterogeneous data sets runs into terabytes, petabytes and exabytes. Finally, it is particularly attractive for social researchers to include complex causes in the explanation of social facts and to make systematic use of the most extensive data pools possible in order to increase the proportion of explained variance.

Such large amounts of data are referred to by the term Big Data. Three criteria are central here: Big Data are available in digital form and, firstly, they have a large volume, secondly, they arise quickly and, thirdly, their origin is diverse (compare Trübner and Mühlichen 2019, p. 143). Empirical social research does not need its own methodology to use them. The same principles apply to their use as they do, for example, in survey research

---

[18] The ESS is described at URL http://www.europeansocialsurvey.org/ (last visited Feb. 05, 2019) the study data can be obtained at URL http://www.europeansocialsurvey.org/user/login?from=/download.html?file=ESS6e01_2&c=&y=2012 (last accessed also Feb. 05, 2019).

[19] Information about the ALLBUS can be found at URL http://www.gesis.org/allbus/ (last visited February 05, 2019). Some of the ALLBUS data must be requested from GESIS in the Data Archive and Data Analysis work area, but some can also be downloaded directly from the web, https://dbk.gesis.org/dbksearch/SDesc.asp?nf=1&search=ALLBUS&field=TI&DB=D&sort=MA+DESC&maxRec=100&product=on&ll=10&tab=0& (last accessed on 05/02/2019).

[20] Details about the Microcensus are available at URL https://www.destatis.de/DE/ZahlenFakten/GesellschaftStaat/Bevoelkerung/MikrozensusInfo.pdf?__blob=publicationFile (last accessed February 05, 2019).

[21] The SOEP presents itself at the URL http://www.diw.de/de/diw_02.c.222725.de/soepinfo.html (last accessed on 04 February 2019). Due to data protection regulations, special conditions apply for obtaining the data.

with comparatively smaller amounts of data. This Big Data, i.e. the concatenation of numerous observations, therefore does not replace theories.

The application of Big Data opens up numerous new possibilities, not only in the social sciences. Sometimes there is even talk of a quantum leap in the availability of information. It promises a high added value, but at the same time is associated with various problems. These relate above all to access to such data, data protection, and the technical requirements for research and calculations using this data, to name but a few.

One example of the use of Big Data is the election campaign in the 2012 presidential election in the USA. This is a Big Data problem in its purest form. The individual voters are first described with as large a number of characteristics as possible, and on this basis their concrete voting behavior is then predicted, i.e. with what probability they go to the polls and which candidate they support. What is now particularly interesting is the question of whether and how they can be swayed in their decision. A sophisticated algorithm will provide information on the appropriate medium for a change of mind, whether in such a case a home visit by an election worker, a phone call or even just an e-mail would be helpful. But conclusions can also be drawn about the respective contents of such change of vote campaigns. For example, young, white, male voters in Florida are best persuaded by the protection of the Everglades. Because such a method attempts to influence individual voters in a targeted and tailored manner, it is also referred to as microtargeting.[22] In part, this gives the impression that the handling of such Big Data lacks a theoretical embedding (compare also Mahrt 2015; Heise 2015, and for some examples Trübner and Mühlichen 2019, p. 143 ff.).

4.9   Data Protection, Anonymity and Confidentiality

In empirical social research, data protection, the anonymity of the target person and the assurance of confidentiality to the persons concerned play a major role. It seems self-evident, for example, that people who voluntarily take part in a survey and provide information about private matters must not suffer any disadvantage as a result. Thus the social researcher assumes no small responsibility. As will be shown, there are various rules that must be observed.

The term data protection could initially be misunderstood, namely if it is assumed that data protection, in the literal sense of the word, is merely about protecting data from loss, regularly creating updates and backup copies, ensuring safe storage, and so on. In this context, however, data protection means above all the protection of the personality of the person concerned. In general, data protection regulations are concerní with the handling of data stored about a person and the protection of the personality that this data relates to.

---

[22] Compare: http://www.silicon.de/41575201/obama-wahlsieg-dank-big-data-und-analytics/ (last accessed February 06, 2019).

In the Federal Republic of Germany, the Federal Data Protection Act (BDSG)[23] and the data protection laws of the federal states in particular regulate the handling of personal data. These laws were considered strict in Europe in the sense that a high priority is given to the protection of personality. It has since been amended in accordance with the new EU standards and thus brought into line with the other EU states at a level that remains strict. Section 4 of the BDSG is particularly relevant. It regulates the permissibility of data processing and use. It states:

"(1) Processing of personal data and the use thereof shall be permitted only if this Act or another provision of law permits or directs it or to the extent that the data subject has consented thereto."

The collection of personal data is thus initially not permitted in principle. Only if a corresponding passage of the law provides for this can a corresponding authorisation be derived from it. This is a prohibition with the reservation of permission. This means that the collection of personal data is structured in a different way than, for example, participation in road traffic. Here, all persons can participate and must then observe the corresponding rules and exceptions.

The regulations on data protection are based on the idea of very extensive protection of personal rights. Exceptions relate, for example, to the security interests of the state, investigations into criminal offences and – following a ruling by the Federal Court of Justice (BGH) – also the microcensus (cf. Sect. 6.1.4). However, the legislation also takes into account the interests of science and market research in the collection of data.

A number of institutions such as registration offices, insurance companies, health insurance companies, mail order companies, banks, personnel departments and, last but not least, furniture stores have to create extensive files with personal data in the course of their activities. The specific regulations that apply cannot be dealt with here.

The provisions of the Works Constitution Act apply to the organisation of internal employee surveys. This replaces or supplements the BDSG in this case: According to this, employee representation must be involved in the preparation and implementation of such a survey in the case of employee surveys. The collection of data and the handling of this data (for example: who from the workforce has access to which data) must be regulated here.

According to the BDSG, the collection of personal data from the general population in the context of scientific research is possible if the consent of the target person has been obtained. The requirement of informed consent by the data subject applies. This means that a person who is asked to provide information must first be informed by whom and for what purpose this information is required. Then he or she must give his or her consent (or, as the case may be, his or her non-consent) to the collection of this data. The passing on of personal data, for example by the institute collecting the data to the client of a study, is

---

[23] On 25 May 2018, the EU Data Protection Amendment and Implementation Act as amended on 30 June 2017 (BGBL 1 No. 44) shall enter into force; at the same time, the Federal Data Protection Act as published on 14 January 2003 shall cease to apply.

prohibited without such information and without the consent of the person concerned (cf. also Metschke and Wellbrock 2002).

In the practice of empirical social research, various aspects of data protection are now particularly relevant. The following will be referred to in more detail:

- In the case of face-to-face interviews, contact records are drawn up by the interviewers (cf. Sect. 10.1) and forwarded to the survey institute. These contain the full address of the target person interviewed and can be related to the specific information provided by this person in the interview. They are necessary to control the work of the interviewers. When handling such sensitive data, the survey institutes have to follow certain rules.

Keeping such records is considered essential for monitoring the work of interviewers, most of whom are freelancers, and thus for survey research. These records serve, among other things, to verify the selection made by the interviewer of the target person in the target household. For this purpose, it must also contain, for example, the number of all persons living in a household who belong to the population (cf. Sect. 5.2.2). If possible, the telephone number of the respective target households should also be given and stored for telephone enquiries. It goes without saying that any interviews identified as falsified by the interviewer must be removed from the data set. For this purpose, a principled link between the contact records and the completed questionnaires is required. This is done by means of an identification number (ID). The survey institutes must keep the contact logs and the completed questionnaires separately. Both contain the ID, which makes it possible to allocate the data.

Firstly, this separation should be made spatially, i.e. the documents should not be kept together in one room. Secondly, there should also be a personnel separation of the two tasks. Employees of the survey institute should not have anything to do with both files. Further, the ID, i.e. the actual link between the contact logs and the original question-naires, should be deleted immediately when the checks of the interviews are completed. The data are anonymised in this way.

Finally, all employees who handle personal data in the data collection institute must be instructed in a manner that is documented in the files. The principle that each employee only has access to the data that he or she has to deal with due to his or her function in the respective institute has proven itself.

According to the BDSG, a company data protection officer must be appointed at the survey institute to ensure compliance with these regulations. The data protection officer represents the interests of the respondents, performs his tasks on his own responsibility and is therefore not subject to any obligation to give instructions.

- Population survey data files are often made available for use by a wider clientele for secondary analyses, for example by the GESIS Leibniz Institute for the Social Sciences (cf. Sect. 4.8). They are also available to others for their analyses. There is thus a certain risk of being able to re-identify the target persons.

As already described, the destruction of the transfer code (ID) is intended to ensure that the data is made anonymous and that it is therefore no longer possible to subsequently identify the persons surveyed. In this case, the data is no longer personal data. This enables the now necessary archiving and also transfer of the files, for example for re-analyses and replications. Such analyses are completely indispensable for compliance with scientific standards and thus for professional practice in general.

A problem can arise if certain information on the person of the respondent, which can be taken from the anonymised data set, such as the profession (district chimney sweep, mayor, state bishop), age and gender as well as information on the place of residence (federal state and size of town) can ultimately lead to the identification of at least individual target persons.

This problem is exacerbated if data are to be added to the survey data set in order to enable multi-level analyses, for example. In this case, information such as the unemployment rate in the respective residential area of the interviewee, the election results in the last federal election or the proportion of foreigners in the place of residence of the target person are of interest for the sociological evaluation.

The solution of the problem or the avoidance of a re-identification of target persons has to be done in a particularly responsible way, after all, the respondents were assured of anonymity during the survey. First of all, the various risks of such a re-identification have to be weighed up. First and foremost, one must consider the possible harm to a respondent as a result of the re-identification. It must also be considered how likely and how complicated such a re-identification is for a data attacker. Secondly, the scientific interest and the benefit that is to be pursued with the investigation and that makes the addition of further information necessary must be set against this.

One solution may be to specify the sensitive variables of a data set, which could be used by a data attacker for re-identification, such as the unemployment rate and the election result at the place of residence, only in coarsened categories (compare Wirth 1992). In this way, the percentage of votes obtained by all parties would not be fed into the data set, but only coarsened levels, e.g. one party obtained between 20 and 30% of the votes. Another or even additional solution would be to link the disclosure of the data set with a special obligation to the users. They must be informed about the consequences of misuse, including possible penalties. A limitation of the disclosure to a certain group of persons or the disclosure of only an incomplete data set could also be considered.

- Telephone surveys are conducted using generated telephone numbers. In this way, households and target persons not listed in publicly accessible directories can and should be reached (by chance). In addition, a supervisor must be able to listen in on the interviews to ensure quality control of the interviewers' work.

Here, too, patent solutions are difficult to find. Once again, a balance must be struck between, firstly, the interests of the target persons concerned in the protection of their privacy and the effort they may have to incur in notifying a refusal and, secondly, the justified scientific research interest. Here, the ADM, the Arbeitsgemeinschaft Sozialwissenschaftlicher Institutes e. V. (ASI) and the professional association of sociologists have agreed on concrete regulations.[24]

According to this, unannounced telephone contacts for the purpose of an interview should only take place between the hours of 9.00 am and 9.00 pm. It is reasonable to expect a contacted target person to take a call during this period and, if necessary, to signal non-participation to the caller. It is also permissible thereafter to listen in on interviews in the interests of quality control and compliance with data protection. The problems that nevertheless arise in practice are dealt with, for example, by Wiegand (1998, p. 19 ff.).

The misuse of scientific surveys for commercial advertising is strictly prohibited. This also applies to a combination of market surveys with sales efforts.

A judicial decision[25] has largely confirmed the ADM's view. Background: The complaint was filed by a target person who had been repeatedly called by a survey institute and asked to participate in a survey for the purpose of market research. The target's telephone number was not contained in a directory, but was generated by a random process. The target person, a lawyer, requested the survey institute to refrain from such calls in the future by means of a cease-and-desist declaration with a penalty clause and to pay an "enclosed bill of costs." However, the AG Hamburg – St. Georg rejected this request in its ruling of 27 October 2005. The reasons for this rejection[26] are interesting.

Thus, although unsolicited telephone calls for advertising purposes to private individuals generally violate the prohibition of unfair competition (Section 7 (2) No. 3 UWG), this does not apply to market research. The judgment goes on to say: "A weighing of the conflicting interests shows that the plaintiff in the injunction had to accept the defendant's calls as lawful."

The telephone is comparable to a mailbox. If this is used, there is no reason to fear that one cannot be reached by others. Whoever – according to the court's reasoning – acquires a telephone connection expresses the wish to communicate via this medium.

Even the fact that the caller pursues commercial purposes with his call is not objectionable; on the contrary, the activity of a survey institute is also protected by the constitution within the framework of professional freedom.

• The BGH requires informed consent of the persons concerned when participating in interviews and experiments.

---

[24] Compare   https://www.adm-ev.de/wp-content/uploads/2018/07/RL-Telefonbefragung.pdf   (last accessed Feb 06, 2019).

[25] Source: http://www.kanzlei-prof-schweizer.de/bibliothek/urteile/index.html?id=12951&suchworte=Umfragen, last accessed February 06, 2019.

[26] Friedrichs 2019 describes other examples from survey research practice in which similar conflicts have arisen.

In Germany, the collection of data or even the respective questionnaire used for this purpose does not need to be approved by a (central) institution. This would contradict the principle of freedom of research.

However, the BDSG stipulates that the written consent of the person concerned must be obtained for the collection of personal data. Such consent makes the interviewer's work more difficult, it may arouse mistrust in the persons approached, increase the non-response rate and thus worsen the data quality. An exception has therefore been made for social science research. According to this, in such a case – and only here – the written consent of the target person is *not* required (compare Kaase 1986, p. 3 ff.).

However, panel studies are an exception (cf. Sect. 4.5), where the IDs and thus the addresses of the target persons have to be stored over a longer period of time. In addition, the risk of possible re-identification increases with the number of survey waves and thus justifies a higher effort to protect the data or the personality of the survey participants.

The persons to be interviewed must be informed about the purpose of the interview and about compliance with data protection regulations. For example, a data protection statement is given to the target persons by the institutes associated in the ADM, as well as a text that provides information on what happens with the information given. Both documents are available on the Internet. It goes without saying that these statements must also be complied with. Figure 4.7 shows the declaration on data protection.

Another sensitive aspect is refusal to participate by the contact or the target person. If a target person clearly states that he or she is not willing to participate in the survey, the survey institute has to accept this without further ado. A second attempt at contact, for example by a different type of interviewer (a man instead of a woman; a young interviewer instead of an old interviewer and so on), is not allowed. Also, a target person has the right to request deletion of his or her information even after the interview has been completed. This should prompt the collecting institute to anonymise the personal data as soon as possible, i.e. to destroy the ID.

In general, it is important to implement the principle of informed consent of the persons concerned. This is particularly important in the case of group discussions for which audio and/or video recordings are to be made or in the case of experiments. Various other guidelines exist for this purpose, which are also made available by ADM via the Internet.[27]

- For social scientists, it is also possible to use data from the civil registers kept in the municipalities for research. Here, too, certain rules apply.

Population registers contain various details on the people living in the respective municipality and therefore represent an interesting source of information for empirical social research. In addition to postal surveys (cf. Sect. 6.1.3), register samples (cf. Sect. 5.3) in particular require this data source. The legal provisions stipulate that population registers may only make their data available for scientific purposes – not for market

---

[27] Source: https://www.adm-ev.de/standards-richtlinien/#anker2, last accessed February 06, 2019.

Declaration on data protection and the
absolute confidentiality of your information during
oral or written interviews

(Name of the institute; supplement for ADM institutes:
– Member of the Arbeitskreis Deutscher Markt- und
Sozialforschungsinstitute e.V. (Working Group of
German Market and Social Research Institutes).
(ADM)) operates in accordance with the provisions of
the Federal Data Protection Act (BDSG) and all other
data protection regulations

This also applies to a repeat or follow-up interview,
where it is important to conduct another interview
with the same person after a certain period of time and
to carry out the statistical analysis in such a way that the
information from several interviews is linked together
by a code number.

Again, there is **no disclosure of data that
identifies you personally.**

The results are -just as with one-off surveys – presented
exclusively in anonymous form. This means that no one
can tell from the results which person provided the
information. On the back of this statement, we show
you an example of the path of your data from collection
to the completely anonymous results table.

If the person asked to participate is not yet 18
years old and there is no adult present at the
time: Please also show this leaflet to your parents
with the request that they take note of it

Responsible for compliance with data
protection regulations:

(Name and full address of the institute)

(Name of the head of the institute)

(Name of the company data protection officer)

Fig. 4.7 ADM documents on data protection

research. For this purpose, the collecting institution must, if necessary, submit a declaration of no objection issued by the respective competent authority for data protection (this is usually the Ministry of the Interior of the federal state in which the collecting institution is located). This certifies the scientific character of the institution.

The residents' registration offices then compile the required information themselves. However, they are not obliged to provide information. There is also no time frame within which such information must be provided. Nor is there any entitlement to receive the data in a specific form, for example as a machine-readable data file. Finally, there are no uniform rules on the fee payable for this information. These facts complicate the work with the registers. Albers (1997, p. 117 ff.) gives a very clear account of the problems involved.

Finally, it is important for the institute to note that the data received may not be used for purposes other than those known. For example, information is usually only provided for a specific survey that deals with a specific topic. It is also not permitted to store large quantities of addresses for possible future surveys. This means that addresses not used in the study in question must be destroyed.

For more information on data protection in the social sciences, compare Häder (2009) and on research ethics Friedrichs (2019, p. 67 ff.).

Sampling Procedure

5

Abstract

In the context of research planning for a social science study, the question of the selection procedure to be used in the process plays an important role. First, however, a general decision must be made as to whether it is necessary to draw a sample or whether a total survey, i.e. the investigation of all elements of interest, should be carried out. The arguments in favour of a total survey are that certain errors that can occur in sampling are excluded and the findings obtained may be more robust and accurate. The main argument against a total survey is the usually immense effort involved. By way of introduction, the decision in favour of a sample or a total survey will be considered in more detail. Afterwards, different sampling strategies will be discussed. There are problems for which it is in principle not possible to carry out a total survey. For example, medical blood tests will always have to make do with random samples, since otherwise the organism to be examined would be destroyed. This also applies to the tasting of a new vintage of wine; here, a total survey would consume the entirety of the new wine. Thus it is to be stated: In cases where a total survey would destroy or damage the population, only sampling can be used.

5.1 Basic Concepts and Classification

In the context of research planning for a social science study, the question of the selection procedure to be used in the process plays an important role. First of all, however, it must be decided in general whether it is necessary to draw a sample or whether a total survey, i.e. the investigation of all elements of interest, should be carried out.

It has to be taken into account that censuses can be quite expensive. The 1987 census in the Federal Republic of Germany, for example, caused financial costs of the equivalent of more than €500,000,000. In addition, there are further costs arising from political debates about the necessity of such surveys (the microcensus and the census are sometimes regarded by their opponents as "elements of the surveillance state" and should therefore be boycotted by the population; cf. Rottmann and Strohm 1987, p. 7). Therefore, total surveys are often not feasible and the use of random samples is unavoidable.

Another aspect of the decision-making process between sampling or total survey will be demonstrated by an example. The mayor of Neustadt, a municipality with about 50,000 inhabitants, had decided to ask all adult citizens about the investments to be planned for the next budget year. He wanted to try this with the help of a total survey, after all, all citizens should be allowed to give their opinion on this important problem. But the financial means of the mayor of Neustadt were limited. He was not able to employ trained interviewers, nor was he in a position to launch the reminder campaigns required for such surveys (cf. Sect. 6.1.3). He therefore distributed the questionnaires to each of the approximately 25,000 households with the help of a town gazette published in his town and asked his citizens to return the completed questionnaires to the town hall or another office on their own initiative. The response rate of only about 5% was completely insufficient from a sociological point of view. It remained largely unclear who had taken part in the aforementioned survey and for what motivation. Initially, only one questionnaire was sent out per household. But whoever liked could download further copies from the internet.[1] The result obtained in this way would therefore have to be described as unusable in the context of a serious social science study. From the point of view of social research, it would have been more promising and the results would have been meaningful if only a randomly selected sample – for example, of 1000 citizens – had been surveyed, and if the focus had been on using all available resources to survey precisely this group of people. Scheuch (1974) already advocated such a strategy some years ago. Unfortunately, the available resources in Neustadt were not used in a target-oriented way.

It may be appropriate and sensible to design employee surveys as total surveys (cf. Borg and Treder 2003, p. 81). If the population is not too large, if it is also possible to motivate all elements of the population to participate, and if the necessary resources for such a survey are available, it is even possible to obtain particularly robust findings with the help of a total survey. Total surveys would also be conceivable in associations, church congregations and similar institutions.

If, however, a decision has been made in favour of a sample survey, various variants are again available. Their basics and some variants are discussed in the following. Sampling does not only play a role in surveys. As will be shown, selection procedures are also used for observations (see Sect. 6.2) and content analyses (see Sect. 6.3).

---

[1] Compare: http://www.neustadt-weinstrasse.de/burgerhaushalt/Download/download.html,   last accessed 02/02/2006, unfortunately now no longer available.

The importance of the correct selection of the target persons to be interviewed is usually illustrated in methodological textbooks by an election survey that was conducted in the USA as early as 1936. The purpose of the survey was to predict the likely winner of the presidential election. If one assumes that there was already a similar interest in forecasting among the public at that time as there is today, then the explosive nature of the prediction enterprise becomes conceivable. The magazine Literary Digest dared to make such a prediction. To this end, it sent out sample ballots to members of a sample of n = 10,000,000. The addresses were selected from directories in which telephone owners and car owners were registered. Of the ballots sent out, about 2,400,000 were completed and returned to the sender. The result of the poll was then clear: Candidate Roosevelt would clearly lose the election, his opponent Landon would win.

An institute called Gallup chose a different approach to sampling. A much smaller sample was recruited on the basis of certain quota specifications. However, the composition of this sample corresponded fairly closely to the social structure of the American voting population. On this basis, Gallup was able to predict the correct election result, Roosevelt would be the winner of the election. Thus, instead of the 43% predicted by Literary Digest, Roosevelt received 62% of the votes (cf. Freedman et al. 1978, p. 302 ff.). This correct prediction and, in contrast, the relatively large error in the forecast made by Literary Digest – it was off by a whole 19 percentage points – gave the Gallup Institute long-lasting world fame.

Thus, an important finding can be made: The size of a sample alone does not guarantee a good representation of the properties or parameters of the population. Or, correspondingly: Small samples can be better than large ones under certain circumstances.

Two problems with the Literary Digest study deserve special mention: First, the low response rate – more than three-quarters of those contacted had not cast a vote – must be blamed for the error that occurred. It can thus be assumed that it was not a matter of chance whether someone took part in this survey, but rather that this motivation was related to the issue being surveyed (cf. Sect. 5.8).

A second reason for the failure of the forecast concerns the sampling frame used from which the addresses were drawn. It should have been clear to the organizers that a list of telephone and motor vehicle owners would not include all eligible voters. Again, whether or not someone is on one of these lists is not random. Rather, it is to be expected that higher earners, for example, are disproportionately represented here. From these again a different voting behaviour is to be expected than from the rest of the eligible voters. This is how it finally came about.

Two important concepts can now be introduced: If a list is used for selection that does not contain all elements of the population, it is called undercoverage. If one had a list in which elements are contained more often than only once, one speaks of overcoverage. Both phenomena occur in the practice of survey research.

Before Sects. 5.2 and 5.3 present selection strategies that are of particular importance in the practice of empirical social research, some other basic terms should be explained, some of which have already been used. The most important terms are "population" (or simply abbreviated to "N") and "sample", which is also used in English.

The *population is* a number of elements that are of interest to the researcher because of a particular property. For example, in an electoral study, the population can be defined as the set of people (these are the elements) who are eligible to vote on a given date (this is then the common property).

If all elements of the population are now included in a survey, as was the case for the 1987 census, for example, this is referred to as a *total survey*. This is another relevant term.

A *sample* is a selection of elements from the population. There is no complete agreement in the literature as to whether one can only speak of a sample if such a selection has been made according to certain mathematically-statistically justified rules, or whether it is always a sample if not all elements of the population are considered. Here, the second view is taken and it is assumed that any selection of elements from a population is also a sample. Another problem arises, of course, with the question of what information can be drawn from such samples in each case.

Survey units is another important term in this context. These are those elements of the population to which the selection can in principle refer. To stay with the example of a survey on the intended election decision, it quickly becomes clear that it is almost impossible to make people who, for example, live abroad at a certain time or who currently have to stay in hospital into elements of the sample to be surveyed.

Sample size (also referred to as just "n") refers to the number of selected items to be included in the study.

On the basis of what considerations can we now derive the justification for making statements (also referred to as forecasts) about the population on the basis of a selection of only relatively few elements? This question will be examined in the following. For this purpose, a small thought experiment is presented first.

Let us imagine an urn containing all 100 elements of the population in the form of spheres. Let us further imagine that ten of these spheres are white and 90 are black. Finally, let us imagine that we draw a sample of n = 10 elements (spheres) from this urn. Now, in principle, it is conceivable that all ten balls drawn are white. It is also conceivable that nine of the balls are white and one is black, or even that all ten balls are black and numerous other possibilities. From this an important conclusion can be drawn: That is, *random sampling yields uncertain results.* In other words, until you know the outcome of the draw (the colours of the ten balls), you cannot say exactly what the outcome will be. This is true even if – as in our case – we know the structure of the population, i.e. how many balls of each colour are contained in the urn.

Even in the event that one repeated this little experiment, one still could not completely rule out the possibility that, for example, all ten drawn balls are white again. This is very unlikely, but it cannot be ruled out. This leads to another important conclusion: drawing samples *always* gives uncertain results. To provide certainty here, we would have to draw all the balls from the urn. However, the fact that it is then no longer a random sample has already been shown above.

Now one can look for possibilities to make the findings in sampling a little more satisfactory. Since we are always dealing with uncertain statements here, it makes sense to use probability theory. Let us simplify our thought experiment a little further. Let's assume we have a coin and we would toss it four times and observe whether after the toss we see the side with the head or the side with the number. Again, we cannot initially rule out the possibility of arriving at the same finding, for example "number", four times in a row. Before we do this, however, let us consider what results can be expected in principle from these four throws. A total of 16 different constellations can occur. Table 5.1 shows these.

As can be seen in column 16, it can happen that the result of a coin toss four times is "number". Column 1 shows the case where the result is "heads" four times. The probability for the occurrence of "heads" or "number" can now be calculated relatively easily. To do this, the number of possible variants (for four rolls there are 16) is put in relation to the result to be considered in each case. Since in one of 16 cases the result "number" never occurs (compare again variant 1 in Table 5.1), this means here 1/16. Furthermore, in four of 16 cases the result "number" occurs once (compare variants 2, 3, 5 and 9) or three times (compare variants 8, 12, 14 and 15). The entire result is shown in Table 5.2.

As stated at the beginning, it is not possible to make reliable statements on the basis of random samples. It is therefore not possible to know exactly what the result will be if the coin is tossed four times. However, with the help of probability calculation, it was possible to determine how likely it is that a certain result will be achieved in each case. The most likely outcome (37.5%) is that "number" and "heads" occur the same number of times. This assumes that the coin tosses are not manipulated, the results are properly recorded,

Table 5.1   16 possible result variants for a fourfold toss of a coin

Variant																
Litter	1	2	3	4	5	6	7	8	9	10	11	12	13	14	15	16
1	H	H	H	H	H	H	H	H	T	T	T	T	T	T	T	T
2	H	H	H	H	T	T	T	T	H	H	H	H	T	T	T	T
3	H	H	T	T	H	H	T	T	H	H	T	T	H	H	T	T
4	H	T	H	T	H	T	H	T	H	T	H	T	H	T	H	T

K head, Z number

Table 5.2   Probability of no, one, two, three and four occurrences of the event "number" (Z) for a coin tossed four times

Event	Frequency	Quotient	Percentage
N (0)	1×	1/16	6.25
N (1)	4×	4/16	25.00
N (2)	6×	6/16	37.50
N (3)	4×	4/16	25.00
N (4)	1×	1/16	6.25

Reading note: Z (0), the event "number" does not occur, it occurs in one out of 16 cases, which again corresponds to 6.25%

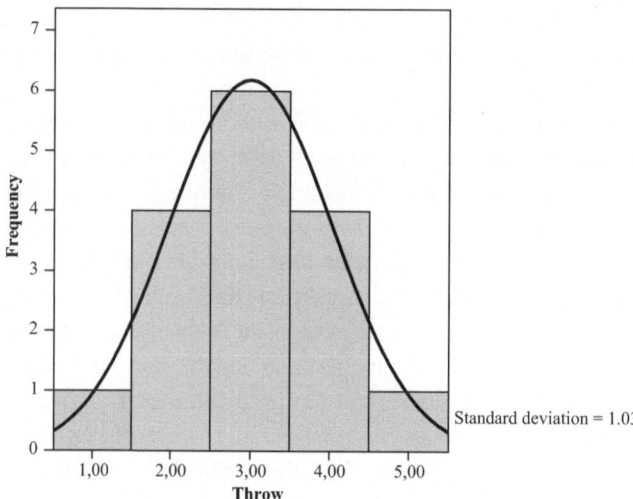

Fig. 5.1 Results of the four coin toss and normal distribution curve

the coin is never left on the edge, and so on. In such a case – i.e. when random selections are used – the results of these experiments can be mapped in the form of a normal distribution and thus become calculable (compare Fig. 5.1).

A more detailed account of such experiments can be found in Kühnel and Krebs (2001, p. 107 ff.). This assumption of normal distribution, which arises due to the law of large numbers in the results, now provides the basis for answering various further questions.

If one assumes that in a population (for example, 100 balls) a certain characteristic (for example, the colour of a ball is white) is present with a certain frequency (for example, 10% of all balls), then one cannot expect that in any random sampling exactly 10% of all balls drawn will also be white. Rather, at one time exactly 10%, at another time more, and at another time fewer balls of that color will be drawn. The difference between the expression of a parameter in the population and its expression in the sample is called *sampling error*. The distribution of these sampling errors is – if the sample is drawn randomly and if there are enough repetitions – also normally distributed. This now means that 95% of the values lie in the range of ±1.96 standard deviations (compare Borg 2003, p. 183; Bortz and Döring 2002, p. 510 f.). This range is referred to as the *95% confidence interval*.

In other words, for 100 samples drawn, in 95 cases the value found is within the above-mentioned range of ±1.96 standard deviations. Accordingly, it is outside this range in five cases. Since in social research one is usually not in a position to repeat the drawing of a sample as often as desired in order to be able to secure the result accordingly, this knowledge is now used for the one-time sampling and the interpretation of the results.

The starting point is again the already well-known observation that samples always provide uncertain information. Now, however, as just indicated, specific calculations can be made to assess more precisely the degree of certainty with which, firstly, the parameters determined with the aid of a sample can be transferred to the population. Secondly, the question can also be answered as to how many persons must be selected from a population in order to be able to draw sufficiently reliable conclusions about this population. Both problems are addressed below.

5.1.1   Sample Size

Another example is used to demonstrate the procedure for determining the sample size. For this purpose, it is assumed that statements about the population of all approximately 35,000 students[2] at the TU Dresden are to be made with the help of a random sample survey. The aim could be to determine the satisfaction with the canteen food. One will now ask how many students must be included in the sample in order to be able to make a statement about satisfaction with the canteen food among all students with a certain degree of certainty. Borg has published a table for determining the required sample size, which is helpful here. This is shown in Table 5.3.

The procedure for using this table is as follows: First, one must decide with what degree of certainty one wants a result – or, rather, what magnitude of error e one is willing to accept. (If one wants to be sure that one does not make this mistake, a total survey must be carried out. However, for reasons of cost in particular, this will usually have to be dispensed with, as has already been shown several times above). It should next be assumed that the estimate should be accurate to ±3%.

Table 5.3 Minimum sample size n for given absolute sampling error e with probability of error $\alpha = 0.05$ for proportions p = 0.05 and p = 0.08 (or p = 0.02)

| p = 50% | | | p = N | 80 % (or 20%) e= 0.03 | |
N	e = 0.03	e = 0.05			e= 0.05
200	168	132	200	155	110
300	234	168	300	208	135
400	291	196	400	252	152
500	340	217	500	289	165
750	440	254	750	357	185
1000	516	278	1000	406	197
3000	787	341	3000	556	227
7500	934	365	7500	626	238
10,000	964	370	10,000	639	240
50,000	1045	381	50,000	674	245
100,000	1056	383	100,000	678	245

According to Borg (2003, p. 188)

---

[2] Compare: https://tu-dresden.de/tu-dresden/profil/zahlen-und-fakten#ck_stud,   last   accessed February 08, 2019.

Since we assume (for example, as a result of a previous survey) that approximately $p = 50\%$ of the students are satisfied with the canteen food, we have to resort to a conservative estimate. Conservative estimates are the safest type of estimate. Thus, it is easy to see that an error of $e \pm 3\%$ at 50% is larger than the same error size at only 20%. If you choose the 50% mark, you will always be in the safe range.

In the next step, the size of the population must be taken into account. It is $N = 35{,}000$ students. In the event that the true proportion value or the proportion of students satisfied with the canteen food is 50%, the sample size must be approximately $n = 1000$. In the table, it is therefore between 964 and 1045.

This means that if 1000 students are selected purely at random, for example from a complete enrolment list, if all those selected take part in the survey that then follows, if all give a valid answer and if these answers signal that 50% of the respondents are satisfied with the canteen food, the actual value (the proportion of satisfied people among all students) will lie between 47 and 53% with a 5% probability of error.

5.1.2   Confidence Interval

Let us now turn to another question that is essential in this context and concerns a very similar problem. Here we again assume that an intended random sample can also be studied completely, i.e. that all selected target persons participate in the survey, that all of them state their actual opinion, that no errors occur in the data transmission, and so on. Let's assume that a certain percentage of the eligible voters surveyed indicated that they were satisfied with the government. The question that now arises is what level of confidence can be placed in the determined value. After all, we know that samples always yield only uncertain statements and that one must therefore expect a sampling error e. In order to specify this certainty, an interval is to be determined in which – with a certain probability – the real value lies in the population. The initial considerations are therefore again very similar to those above.

Answering would not be a problem if we could repeat the sampling sufficiently often, say 10,000 times. However, as already shown, this is not feasible in practice. Therefore, a different approach must be taken.

In a first step, we again specify that the probability of error should be 5% – a value that is common in the social sciences. It means that with a theoretically assumed 100 different samples, an error occurs in only five samples. We then determine an interval in which the true value lies in the population and accept that we are mistaken about this with the given probability. To completely eliminate this error, as has also been repeatedly pointed out, all eligible voters would have to be surveyed. Further important to answering the question is the sample size. Based on the assumption of the normal distribution of the sampling error, one can now calculate again in which range the actual value in the population will lie. This is called the confidence interval. Again, a helpful table will be printed for this purpose (compare Table 5.4).

Table 5.4 Error tolerances (in percent) for different proportion values in a sample

n = ...	1/99	5/95	10/ 90	15/8	20/ 80	25/ 75	30/ 70	35/ 65	40/60	50
100	–	–	–	7.3	8.2	8.8	9.4	9.7	10.0	10.2
300	–	–	3.5	4.2	4.7	5.0	5.3	5.6	5.7	5.8
500	–	2.0	2.7	3.2	3.6	3.9	4.1	4.3	4.4	4.5
700	–	1.7	2.3	2.7	3.0	3.3	3.5	3.6	3.7	3.8
1000	–	1.4	1.9	2.3	2.5	2.7	2.9	3.0	3.1	3.2
1300	–	1.2	1.7	2.0	2.2	2.4	2.6	2.7	2.7	2.8
1500	–	1.1	1.6	1.9	2.1	2.2	2.4	2.5	2.5	2.6
1700	0.5	1.1	1.5	1.7	1.9	2.1	2.2	2.3	2.4	2.4
2000	0.4	1.0	1.3	1.6	1.8	1.9	2.1	2.1	2.2	2.2
2500	0.4	0.9	1.2	1.4	1.6	1.7	1.8	1.9	2.0	2.0
3000	0.4	0.8	1.1	1.3	1.5	1.6	1.7	1.7	1.8	1.8
5000	0.3	0.6	0.9	1.0	1.1	1.2	1.3	1.4	1.4	1.4
10,000	0.2	0.4	0.6	0.7	0.8	0.9	0.9	1.0	1.0	1.0

From ADM (1999, p. 150)

An example should clarify how to deal with the table. The size of the population should be N = 69,157,300. These are all Germans eligible to vote. What is sought is the limits within which the confidence interval lies. In the study n = 1000 randomly selected persons were interviewed. In the table the number 3.2 is found at the corresponding place. If 50% of the eligible voters interviewed in the sample are satisfied with the government, the real value thus lies between (50–3.2 =) 46.8% and (50 + 3.2 =) 53.2%. It is accepted that for five out of 100 samples the real value will lie outside the confidence interval given (compare Diekmann 2004, p. 401 ff.; ADM 1999; Bortz and Döring 2002, p. 414 ff.). The following formula applies to the determination of the confidence interval in the case of unrestricted random selection:

$$I_{1,2} = p \pm z_{\alpha/2} \cdot \hat{\sigma}_p,$$

with

$I_{1,2}$:	The confidence interval to be determined,
p:	Proportion of the feature of interest in the sample (in our case p = 0.5),
$\hat{\sigma}_p$:	$$\hat{\sigma}_p = \sqrt{\frac{p(1-p)}{n-1}},$$ Estimated standard error of the proportion,
α:	Probability of error,
$z_{\alpha/2}$:	Tabulated value from the standard normal distribution, for probability of error is α = 0.05 $z_{\alpha/2}$ = 1.96,
n.	Sample size (it should be 1000)

Substituting the appropriate values from our example yields the values for z_w. In detail:

$$I_{1,2} = 0.5 \pm 1.96 \cdot \sqrt{0.5(1-0.5)/1000},$$
$$= 0.5 \pm 1.96 \cdot 0.0158,$$
$$= 0.5 \pm 0.031.$$

In the considerations so far, the concrete implementation possibilities for the drawing of random samples have not yet been taken into account. The main question is whether there is a register that contains all the elements of the population, comparable to the urn mentioned at the beginning – for example, a central register of residents. At the same time, such a register would also have to be accessible for social research. This question must be answered in the negative for the Federal Republic of Germany. Thus, it is not possible to draw unrestricted random samples for population surveys according to the model described above.

Thus, in Germany and also in many other countries, different strategies for drawing samples for social science surveys had to be developed and implemented. Basically, these strategies have to be distinguished between, first, random selections, second, deliberate selections, and third, arbitrary selections.

Only in *random selections,* or probability selections (the term random sampling is also commonly used), does each element have a specifyable non-zero probability of being included in the sample. This is the criterion for a random selection, similar to a lottery drum, a coin toss, or the urn model. As will be shown later, there are again different variants of how such a selection can be made in practice (compare Sects. 5.2, 5.3, 5.4, 5.5, 5.6 and 5.7). All the considerations made in this section up to this point assume that such a random sample has been drawn. Only in the case of such samples can it be assumed that the sampling errors are normally distributed. Random selections are therefore not those that are – to use colloquial terms – arbitrary and indiscriminate. In order to obtain the predicate random selection, the condition mentioned above – each element of the population has a specifyable, non-zero probability of being included in the sample – must be fulfilled.

In the case of *deliberate selection,* criteria, e.g. quotas, are set according to which the elements of the sample are specifically determined – usually by the interviewers. Such quota characteristics can be, for example, age, gender, place of residence and the like (cf. Sect. 5.8). The specification of particularly typical cases – in the sense of a theoretical concept – would also be a form of deliberate selection. In the case of deliberate selections, it is thus completely open as to the probability with which a particular element will end up in the sample. The Gallup survey cited above, for example, was based on a quota sample.

Finally, samples can also be compiled completely without a predefined plan. In such cases, the term *arbitrary selection is used.* The characteristic of arbitrary selections is that the actual sampling is not controlled. Such procedures are implemented, for example, in street-corner surveys or river sampling. For certain purposes, for example for psychological experiments or in general, for testing correlations, such strategies are quite suitable. It will therefore not always be necessary in empirical social research to work with random selection.

However, the aim of surveys is often to estimate parameters of a population, for example the proportion of students who are satisfied with the canteen food. In such cases, random sampling – with the exception of total surveys – is currently the only way to obtain information.

5.2   Random Sampling in Survey Practice

5.2.1   The ADM Design

Random selections according to the classical lottery drum or urn method (see Sect. 5.1) can only be implemented in practice under certain conditions. First of all, a complete list with the elements of the population must be available; this is also referred to as a frame. This can be the case if, for example, a company is planning a customer survey and has previously compiled a list of all relevant customers. Such lists may also exist for the membership survey of a leisure club or for interviewing all students at a university. Some countries have central registers in which all citizens of the respective country are registered; Germany, as already explained, is not one of these countries. Even if such lists do exist, it is still not certain that social research will have access to these lists for its own purposes. Moreover, it is questionable how reliable (up-to-date) they are.

Furthermore, it must be ensured that all selected units (persons) can be reached. In the case of a face-to-face survey – the abbreviation F2F for face-to-face surveys has also become common, especially in ADM – (cf. Sect. 6.1.3), interviewers would have to be available throughout the country and could work in any region. However, this is not always possible.

If these conditions are met, simple (also called unrestricted) random selection can be used. This is done in a single stage; all elements of the population have an *equal* chance of being selected. To do this, the elements can be numbered and then a certain number of random numbers are generated. The elements determined by the random number generator become constituents of the sample.

In the practice of survey research, however, other procedures must be practiced in Germany. Multi-stage random selections are used – some exceptions have already been indicated above. This includes the design of the Association of German Market and Social Research Institutes (ADM design), which will be discussed here because of its great importance (compare, for example, ADM 1999; Hoffmeyer-Zlotnik 1997, p. 33 ff.). This design is constantly being further developed at considerable expense. According to its own information, ADM spends about 300,000 € per update.[3]

---

[3] See at https://www.adm-ev.de/leistungen/arbeitsgemeinschaft-adm-stichproben/ last accessed on February 08, 2019.

In this ADM design, the selection is done in several steps, specifically it has been decided to proceed in three stages: First, areas are randomly selected in the first stage, followed by the selection of households in the second stage and finally the identification of the actual target persons in the third stage.

In the sampling according to the ADM model, a so-called stratification is carried out. Transferred to the model of the lottery drum, one would not only have one drum containing all the elements, but several, for example one for each federal state, from which the elements are then drawn. Now it would be possible to draw as many elements from each drum as correspond to the proportion of the population in Germany – assuming the corresponding knowledge. This means, for example, that fewer elements are drawn from the "Saarland" lottery drum than from the "North Rhine-Westphalia" lottery drum. This guarantees that all federal states are actually represented in the sample according to their share of the population. This is important because, as has already been shown (cf. Sect. 5.1), only uncertain results are obtained with random sampling. However, the result of stratification at least ensures the regional representativeness of the sample. At the same time, stratification ensures that the selection is random, since the selection chance of each element can be specified.

Stratification in sampling presupposes not only prior knowledge about the characteristics – for example, about the population shares in the individual federal states in the total population – but also that the technical possibilities exist to carry out an appropriate drawing. Stratification is particularly important when the composition of the population is highly heterogeneous. This is the case, for example, with the federal states in Germany, which differ greatly in terms of population size. If the sample size is not very large, it would not be possible to exclude the possibility that, for example, a small federal state is not represented in the sample at all or only with a very small number of elements.

Specifically, the ADM design for in-person oral interviews provides for three steps, which have the following characteristics.

First Step: Selection of the Areas

The entire inhabited area of the Federal Republic can be divided into about 53,000 areas on the basis of official statistics. These contain at least 350, on average around 700 private households. This is the finest division supported by data from official statistics. In the case of municipalities with more than 10,000 inhabitants, these areas can also be delimited with the aid of digitised street maps (cf. Fig. 5.2). A random sample is drawn from these areas by the ADM, and the elements identified in the process are referred to as sample points. While from 1997 to 2003 the approximately 80,000 voting districts of the Federal Republic of Germany formed the basis for determining the sample points, since 2003 a more refined division of the sample points into street sections has been carried out.

For the formation of the sample points, the smallest available administrative area units are used down to building block level. In the meantime, official inner-city breakdowns and

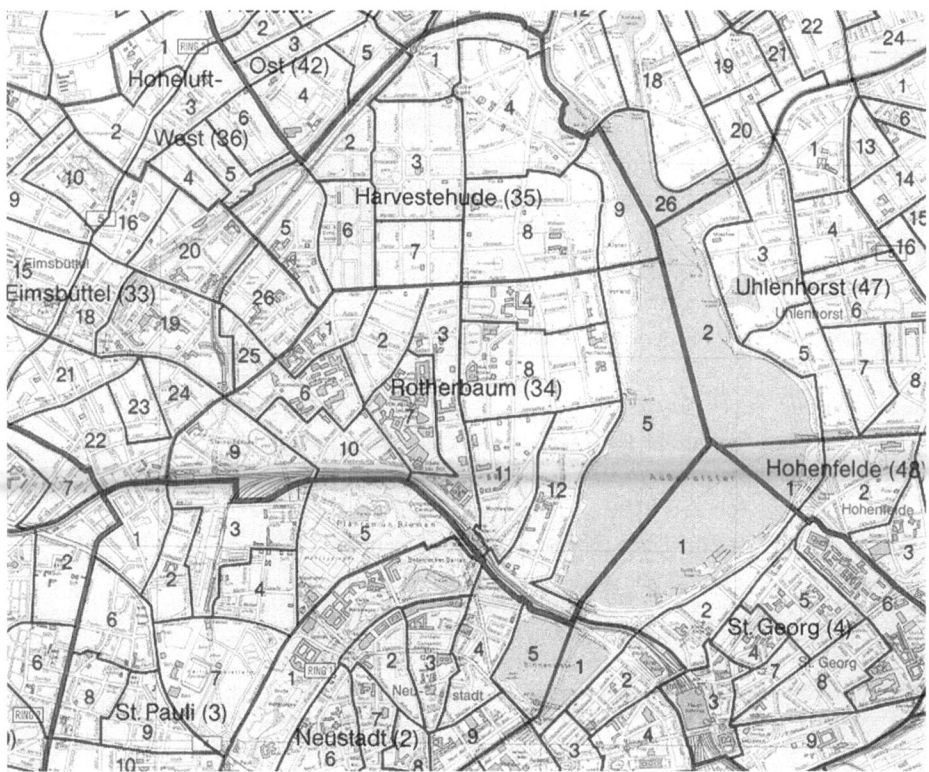

Fig. 5.2 The digitised road network

factual data can be systematically used for this purpose. These data include city district data, entries in telephone directories and so on.

Following a proven convention, for the ADM design from 2003 onwards 210 such sample points are selected for 1000 interviews to be carried out in the western part of Germany and 48 in the eastern part (since the territorial reform in 1996). These are designed to cover an average of 600 to 700 households.

In order to be able to conduct interviews in the sample points, a so-called network is required, i.e. a certain number of interviewers distributed over these areas.

Following our model developed above, here we have a design with stratification. The stratification criteria used in the ADM design are the federal states, the administrative districts and the counties in combination with a municipality typification.

"The cells in the sampling system are formed from the counties/county-free cities and the municipality size classes. The municipality size classes are used here in an expanded form in a division of 10 … The municipalities that do not belong to an urban region are assigned within a stratum (counties) according to their actual (political) size." (ADM 1999, p. 86 f.).

Second Step: The Selection of the Target Household
The aim of the second selection step is to determine the target households in the sample points in which a survey is to take place later. For this purpose, a random walk (also referred to as random route) of the sample point is carried out starting from a specific starting address. For this random walk, the survey institutes each create their own walk-through regulations, which the interviewers have to follow. The following list shows an example.

Address Listing Procedure
Instruction for the Collection of Addresses (Private Households)[4]

1. On the separately enclosed sheet "SAMPLE-POINT-Demarcation" we have described to you the constituency (Sample Point) in which the addresses are to be collected. Only the streets listed on this sheet belong to the Sample Point. **Only addresses located within the Point may be sampled**.
2. There are two different sample point demarcation:

A.	*Larger communities with listed street names*
	In larger municipalities, streets are indicated (with or without house number). From the sample point delimitation, one street has been ticked. This is your start street, i.e. this is the street where you begin the listing. The starting point is always the lowest house number specified for an identifier, such as G, U or D. This means that for identifier G you can only list households from even house numbers, for identifier U you can only list households from odd house numbers and for identifier D you can list households from all house numbers in the street. If there is an entry from 0 (1) to 9999 after the house numbers, all households are to be included in the inspection here. If you are unable to list the required number of households in the starting address, take the next street in the sample point boundary, and so on. When you reach the bottom, at the last address listed, you can start at the top, if your starting street was not the first street, until you reach your starting address again
B.	*Smaller communities without street names*
	In smaller municipalities (sometimes) no streets are specified. This is also not necessary for delimitation, since the entire municipality (or self-contained districts) represent a sample point.
	In this case, look for an address whose first letter matches the first letter of your surname and start the address survey there at the lowest house number. If there is no such street, look for a street with the next letter in the alphabet until you have found your starting street. Here you start at the lowest house number, follow the street side and collect addresses in ascending house number sequence. If this side of the street is not sufficient to collect the desired number of addresses, list the addresses of households on the opposite side of the street in descending house number order, then turn left into the next street and proceed in the same way as in the first street

---

[4] Source: (ADM 1999, p. 80).

3. For A. and B. applies:
- Only addresses **within the sample point boundary** may be collected.
- They list **all the** addresses (door signs) of private households.
- **No** institutional households, households in dormitories, in retirement homes, student housing, apartments in summer homes.
- **No** offices, practices or other commercially used space.
- If you do not find a name tag, please determine the name of the private household by asking.
4. In the case of multi-family houses, always start in the basement or on the ground floor, then go to the next higher floor and so on up to the top floor. If several private households live on one floor, first record the household on the left as seen from the entrance, then the one next to it on the right, and so on.
5. Fill in the address list carefully and legibly, using block letters.

In the Remarks column, please enter anything that may make it easier for one of your interviewer colleagues to find the addresses you have collected when interviewing at this Sample Point later. For example, side entrance on the left, back house, side house, access through courtyard driveway on the right, etc.

If you have any questions in connection with the address listing, or if difficulties arise on site, please call us immediately.

We thank you for your cooperation.

The interviewer, starting from the starting address given to him, walks a certain route according to the established rule and enters the selected households in an address listing sheet. Now the procedure can be continued according to three variants:

The *first* option is to send the address listing sheet to the survey institute, where a random selection of the households to be interviewed is then made (again) (cf. Häder and Häder 1997, p. 43 ff.). In the second and third options, the interviews are conducted immediately by the interviewers. In the *second variant,* the interviewer is allowed to continue the random walk until he or she has succeeded in conducting the number of interviews specified to him or her. In the *third variant, the interviewers are* given a maximum number of households to be contacted. Only with these households may the interviewer then attempt to obtain an interview. In the event that he does not reach anyone in the households or if he is confronted with a refusal, the number of interviews to be carried out by him is reduced.

It is very complicated for the survey institute to control the exact compliance with the – sometimes quite complicated – walk-in rule. Back houses, non-existent house numbers, roundabouts, reaching the city limits, private apartments occupied by caretakers in such institutions as hospitals or school complexes, communal housing, plots of land where dogs are warned against and a number of similar problems pose a certain challenge to the diligence and perseverance of the interviewers. In a seminar exercise at the TU Dresden, five students failed to list the same households starting from the same starting address in Dresden city centre and following the same walk-through procedure. This is all the more remarkable because the students in this exercise were able to complete the random walk relatively stress-free because they were only expected to list addresses and not to conduct their own interviews.

A control of compliance with the rules for the random walk would have to be carried out with the help of a costly follow-up visit by another interviewer. It would also be conceivable in the meantime to use an app to determine the geo-coordinates at which the relevant households are located.

Third Step: The Selection of the Target Person
Within the target households identified in the second step, the next step is to find the actual target person. Two different strategies are generally used in survey practice for this purpose. First, a systematic list is made of all household members who belong to the basic population and are therefore in principle eligible for a survey. This has to be done in such a way that the persons are entered in a list according to their age. Now (variant one) those persons can be asked to participate in the survey who had their birthday last. Another possibility (variant two) is to print random numbers on the address listing sheet and then determine the target person to be interviewed in the household.

On the one hand, this selection step is an easily understandable necessity, since it is not primarily those persons who are currently to be found in the home or who are particularly interested in participating in an interview who are to be interviewed. On the other hand, it poses a certain challenge for the interviewer to reject a person who may be willing to be interviewed and instead either contact the household again (in case the actual target person is not present) or to forego an interview altogether (in case the target person refuses to be interviewed).

The latter variant for selecting the target person is also called the Kish Table or Kish Grid (cf. Kish 1965) or the Swedish Key, after its developer. The interviewer's scope for possible deviations from the rule appears to be more limited than with the birthday method. For this reason, this procedure is more often preferred for face-to-face interviews. The following list presents the corresponding procedure.

While – at least theoretically – all elements of the population had an equal chance of becoming part of the sample in the first two selection steps, this was no longer the case in the last selection step. Here, members of smaller households have a greater chance than those from larger households. To compensate for this, weighting procedures must be used in the evaluation. These are dealt with elsewhere (cf. Sect. 5.8).

Finally, attention should be drawn to the limitations and some sources of error that can occur or are present in the application of the ADM design:

Above all, if a net number to be interviewed is specified to the interviewers (variant two), there is a risk that easily accessible persons are overrepresented in the sample and that the quality of the sample is thereby reduced. As a countermeasure, the number of contact attempts to be made by interviewers could be set much higher. This error can also be countered to a certain extent by the other two variants presented. The higher costs incurred in this way should be noted.

The selection of the person to be interviewed in the household according to the ADM design (compare ADM 1999, p. 84).

Selection of the target person (= person to be interviewed)

* Find out how many people of *survey age 14 and over* live in the household.
* Record the ages of these individuals in the boxes under the heading "Individuals of Interview Age," separately by:

Male = M, box no. 1–4

Females = W, box no. 5–8

In each case, start with the oldest person (box no. 1 = oldest male, box no. 5 = oldest female), then the second oldest, and so on to the youngest person of each sex.

(If there are ever more than 4 males or more than 4 females in a household, start with the second oldest person of that gender).

* Then go through the series of random numbers from left to right …:
* The person whose box number appears first in the row of random numbers is selected (to be interviewed).
* Curl that random number and finally Sue enter the target's first name on the line provided.

Example

There are 4 people of interview age living in the household:

The 49-year-old father, the 46-year-old mother, and a 19-year-old son and a 16-year-old son.

People of survey age: from 14 years random numbers

Personen im Befragungsalter: ab 14 Jahren				Zufallszahlen
M 1	2	3	4	
49	19	**16**		8 6 ③ 2 5 1 4 7
W 5	6	7	8	
46				
Vorname der Zielperson: **Björn**				

Vorname der Zielperson: Björn

The 16-year-old son is selected for the interview because his box number appears first in the series of random numbers.

In addition to compliance with the rules for the random walk, which is difficult to check, the correct implementation of the Swedish key or the birthday method also represents a source of error that is difficult or impossible to check. Thus, one or the other interviewer may succumb to the temptation and rather interview a person in the household who is willing to be interviewed than give up without having achieved anything. Under certain circumstances, the sheets shown for listing addresses or for determining the target person in the household do not pose an insurmountable hurdle for the interviewers.

Another problem is to systematically record the reasons for failure that occur in a study. From a methodological point of view, it is necessary to distinguish between sample-neutral and systematic nonresponse. When keeping statistics on nonresponse, the interviewer also has a relatively large amount of leeway, which can only be controlled in a complicated way. It is relatively difficult to check afterwards whether a target person did not participate in the survey because he or she does not speak sufficient German or whether there is another reason, such as an explicit refusal.

The ADM design used until 2003 was developed to cover the (German) population eligible to vote. This design was not originally intended for empirical surveys targeting other subjects. However, problems could result, for example, from the fact that – in the case that the aim was to survey not only Germans but the entire resident population – foreigners usually live together in relatively concentrated regions. In the case of a random route, it was then possible that such clumps were over-represented in the sample. Similarly, if a survey was to have a population consisting of people aged 16 and over, this was likely to be the case. These difficulties have, however, been overcome with the revised ADM design used since 2003.

In conclusion, the ADM design presents problems mainly in the control of field-work. It is common to check the interviewers, most of whom work on a fee basis, by means of telephone calls and postcards. The participants in a survey are asked whether an interviewer was in their (target) household at all, how the interviewer determined the person to be interviewed, how long the interview lasted, which topics were addressed, and the like (for information on checking the quality of the interviewer's work, see Chap. 10).

5.2.2   Register Sampling: The Sample of Population Registers

In view of the problems described above (cf. Sect. 5.2.1) in connection with the ADM design, the use of a register sample is currently regarded in Germany as the best possible methodic procedure for drawing a random sample for a population survey (cf. for example Koch 1997, p. 99 ff.; Albers 1997, p. 117 ff.; ADM 1999, p. 119 ff.). In Germany, the files kept by the residents' registration offices at the municipal level can be used as registers. However, these are only available for scientific research or research that is in the public interest. As an example of the use of register samples, the well-documented ALLBUS studies of 1994, 1996 and from 2000 onwards should be mentioned. The design of these studies was two-stage:

First Stage
In the first step, a selection of municipalities must be made. Similar to the ADM design, the selection is stratified by federal states, administrative districts and counties. As a rule, the municipalities are identical to the sample points. However, very large cities, such as Berlin, may well be included in the selection with several sample points. In the ALLBUS

studies, 111 sample points located in 103 municipalities were selected for the western part of Germany and 51 sample points located in 45 municipalities were selected for the eastern part of the country.

Second Stage
Now the actual target persons can be selected from the population registers. In the ALLBUS studies, 48 addresses are now randomly drawn for each sample point in a first step. These were increased by 17 more in the second stage of the field period. The main consideration here is that neutral dropouts may occur during the actual survey.

The population registers have some additional information about the inhabitants, which varies from federal state to federal state. This information can be made available for scientific studies. These are mainly the age, the gender as well as the nationality of the persons living in the respective municipalities. This information will still prove to be important for the control of the interviewers.

The use of register sampling is very time-consuming and expensive. Above all, cooperation with each individual registration office must be organised. In this respect, very different experiences are available in practice. It has also been shown that the files kept in the registration offices are not always up-to-date.

A positive aspect of this design is that the scope for the interviewers on site is greatly reduced. Any listing of addresses by the interviewers is omitted and thus all uncertainties associated with this step. It is also no longer necessary to select the target person to be interviewed in the target household. Instead, the interviewers are given the name and address of the person to be interviewed.

A control of the interviewer activity is additionally made possible by the fact that the age information, which was determined in the interviews, can be compared with that which is registered with the registration offices. In addition, the keeping of a non-response documentation is considerably facilitated. The information from the civil register is also available for this purpose. This means that the fieldwork can be described much more transparently than with the ADM design (see Table 5.5).

5.2.3 Telephone Sampling

In view of the growing importance of telephone surveys (cf. Sect. 6.1.3), it was also necessary to develop a methodologically sophisticated sampling strategy for telephone surveys. It was obvious to look for a register containing all elements of the population, i.e. all telephone connections. However, the telephone directory and also the commercially available electronic directories (CD-ROM) in Germany are not directories that can be used for social science surveys. As in the USA, for example, telephone directories in the Federal Republic of Germany no longer contain even an almost complete list of all subscribers since the abolition of compulsory registration in December 1991 (cf. Heckel 2002, p. 13). At present, the number of unlisted lines continues to rise and is estimated – with strong

Table 5.5 Result of the fieldwork in the ALLBUS studies 1996 and 2010 (register samples in each case)

	1996		2010	
	West	East	West	East
Original sample (gross)	4939	2246	5772	2652
Neutral failures, including: Wrong address, target deceased, moved and the like	509	188	539	200
Net sample (=100%)	4430	2058	6311	2852
No one found in the household	3.1%	2.2%	5.4%	4.3%
Target not found	1.6%	1.2%	1.5%	1.3%
Target person unavailable	3.2%	3.2%	2.2%	2.7%
Target not ready due to time constraints	3.1%	2.4%	–	–
Target generally unwilling to	32.5%	35.9%	52.5%	55.5%
Target person does not speak German	1.5%	0.6%	2.0%	0.4%
Interview not conducted correctly	0.8%	0.2%	1.0%	1.9%
Total failures	45.8%	45.8%	65.1%	66.6%
Evaluable interviews	54.2%	54.2%	34.9%	33.4%

regional variations – at up to 40%. Against this background, the random digit dialing (RDD) procedure has been developed in the USA. In this process, the telephone numbers of the households to be contacted are randomly generated as sequences of numbers. In view of the uniform structure of telephone numbers in the USA, this procedure is quite practicable. However, due to the different structure of telephone numbers within Germany, an adaptation of this procedure is not possible. If it were practiced, one would have to reckon with a large number of unsuccessful attempts, since a large number of the number sequences generated in this way do not have a connection. In Germany, the randomised last digit (RLD) method is still used relatively frequently. In this procedure, telephone numbers are taken from a directory and then randomly changed, for example by replacing the last digit with a digit determined by a random procedure. Then, in the second stage, the person to be interviewed in the target household is usually determined using the birthday method (cf. Sect. 5.2.1).

However, the best practice in telephone sampling is currently a different design (compare Gabler and Häder 1997, 2002; Heckel 2002). Here, all registered telephone numbers are downloaded from an electronic directory (CD-ROM). Afterwards, all connections that obviously do not belong to the population are deleted. In the case of general population surveys, this includes above all the business connections. Now the remaining numbers per local network area (of which there are 5200 in Germany) are arranged according to size. If you then form blocks of 100, for example from 0 to 99, from 100 to 199 and so on, you can place the existing telephone numbers in these blocks. For example, the number 06321 33703 is located in the block 33700–33799 of the local network area 06321. The basic assumption is that the unlisted telephone numbers are also located within the blocks occupied by at least one registered number. Therefore, all blocks in which no telephone number is listed in the directory are eliminated. In all remaining blocks, in which at least one telephone number is listed, all 100 conceivable digit sequences are generated, thus for example in the local network area 06321 the sequences 33700, 33701, 33702, 33703, ..., 33799. The set of all these digit sequences in all local network areas then represents the selection frame, from which the samples for telephone surveys can be drawn. This universe contains registered telephone numbers (for example 33703), unregistered telephone numbers and digit sequences that are not currently switched.

In the following selection of the target person in the household, the birthday method has so far mostly been used again. This means that the person who had the next (or alternatively the last) birthday in the household and who also belongs to the population to be surveyed is determined. This is then the target person to be interviewed on the telephone. It is also possible to use the Kish Grid to determine the target person.

Randomised birthday selection (RGA) represents a further development here (Schlinzig and Schneiderat 2009, p. 92 ff.). The persons contacted are asked which person has the last or next birthday before or after a randomised date of birth that was randomly assigned to the telephone number. The key advantage is that it takes into account the uneven distribution of births across the year and the resulting different inclusion probabilities of the elements of the population at the household level. This is particularly important for studies that survey in a particular cycle over a repetitive period.

At first glance, the elaborate procedure suggests that the persons contacted already refuse to participate in the precarious introductory phase of the interview. However, the opposite is not infrequently true. Using this selection procedure, the willingness to participate can even be slightly increased. It can be assumed that this comparatively complex identification of the target person has a cognitive stimulating rather than a deterrent effect on the contact person and can thus be effective as a kind of immaterial incentive insofar as the interviewee's confidence in their own competence to answer further questions is strengthened (cf. Meier et al. 2005).

The positive experience of a question already answered at the beginning of the interview also increases the commitment of the respondents and thus promotes the answering of further questions in the sense of consistent behaviour patterns.

It is also helpful if, in addition to the telephone numbers, an estimate is made of the characteristics of the region in which the respective connection is located. This makes it possible to stratify the sample. The basis for this is provided by the "Verflechtungs- bzw. Agglomerationstypen" (interdependence or agglomeration types) drawn up by BIK on behalf of the working group (v. d. Heyde 2002, p. 33; Häder and Glemser 2004, p. 157 ff.).

The described procedure offers several advantages:

- Compared to a random generation of the entire telephone number according to the RDD procedure, the number of generated number sequences with no connection is reduced. Since even after a call it is not always possible to determine unambiguously whether a dialled number sequence also conceals a connection, several contact attempts – sometimes up to twelve (compare v.d. Heyde 2002, p. 38) – are therefore made. An excessively large number of such failed attempts would considerably increase the effort involved in a survey and thus increase the costs. At the same time, however, it is in principle possible to contact unlisted telephone numbers in this way. Since incoming but unanswered calls are often registered at the called number and could be perceived as a nuisance, the number of contact attempts should not be increased further.

- Another general advantage of telephone surveys is the lower territorial clumping of the sample. Thus, a compilation and selection of arbitrary telephone numbers scattered all over the country can be realized without any problems. In contrast, in the case of a face-to-face interview according to the ADM design or also in the case of a sample from the residents' registration office, it must be ensured that the interviewers are on site in the selected region or can go personally to the corresponding addresses.
- The selection of digit sequences on the basis of an electronic directory such as a CD-ROM allows – similar to the face-to-face design of the ADM – the addition of regional background variables. For example, telephone directories exist that contain information on purchasing power or the number of unemployed within a certain area code.

However, some other problems remain:

- Thus, a strategy has to be found to deal with dialled fax and business lines, with answering machines and with multiple lines. Households with multiple connections are therefore a statistical problem, as they have a higher probability of inclusion than households with only one connection. A countermeasure to this problem would be to use a question to determine the number of lines in the household and to introduce a design weight (cf. Sect. 5.8).
- Gatekeepers pose a further problem. These are upstream electronic filters that ensure that only callers previously selected by the subscriber are put through. A solution to this problem has yet to be found.
- The selection of persons in the household must take place in a critical phase of the interview due to the high risk of dropout. While an interviewer who asks for an interview at the door of the home has various strategies at his disposal to convince the target person of the necessity of an interview, these possibilities are considerably reduced on the telephone. Tests have shown (compare Friedrichs 2000, p. 171 ff.) that letters of announcement can have a positive effect on the willingness to participate. However, sending such letters presupposes that the full addresses of the target persons are known to the organiser before the survey begins. However, this is not the case with such an approach, so that this technique cannot be used and the households usually have to be contacted "cold".
- The structure of the number blocks in which assigned numbers are located changes considerably. In the meantime, experiments have shown that there are also assigned connections in blocks in which no number is listed.

At present, around 12% of households no longer have a landline, but can only be reached via a mobile phone. This meant that a way had to be found to reach this group of people in telephone surveys. The solution is a dual-frame approach. In this approach, the surveys are conducted in parallel via the fixed network as well as via mobile phones. The synthesis of both sub-samples is then carried out using a weighting procedure (Häder et al. 2009, p. 21 ff.).

The selection frame here consists of the quantity of mobile numbers assigned by the Federal Network Agency (currently around 279 million numbers). However, the number of telephone directory entries for mobile numbers is only 2.3 million:

First, the actually valid number range is determined for each area code. It has proven useful to look at blocks of 10,000 numbers, which can be used to compile information. Firstly, these are the entries on the telephone book CD. Secondly, Internet searches are also successful. In addition, thirdly, one can also use special websites that give users clues about hidden costs. Fourthly, there are also blocks that are reserved for technical services of the providers or mailboxes and are therefore also ruled out for a survey. From all sources of information, a list sheet is generated for each provider at the 10,000 block level or, if the information is available on a finer scale, also at the 1000 block level. This list sheet provides information on which number blocks within an area code are actively used and which are still inactive. These lists form the basis for generating the telephone numbers (Häder et al. 2009, p. 43).

For more information on telephone sampling, see Häder and Sand (2019).

5.3   Sampling for Intercultural Studies

In addition to mathematical-statistical requirements, certain national circumstances shape the framework for the selection practice operated in the respective country. These include, for example, the existence of a central civil register or the possibilities anchored in national jurisdiction to draw samples from these registers for social science surveys. These boundary conditions result, for example, in the concrete procedures for face-to-face population surveys in the Federal Republic: the ADM design and register sampling. The situation becomes more complicated when it comes to developing sample designs in different countries that allow comparisons of the various estimators obtained from the data between countries.

In addition to the International Social Survey Programme,[5] the PISA studies[6] and the European Social Survey (ESS)[7] have recently received considerable attention.

The ESS will be used as an example to show how samples can be drawn for intercultural studies. In the various countries involved, there are very different technical-organisational

---

[5] For the ISSP, compare: http://www.gesis.org/issp/ last accessed 08/02/2019, the ESS is discussed in more detail in Sect. 6.1.4.

[6] For the selection design of the PISA study, see Sect. 5.5.

[7] For the ESS, see: http://www.europeansocialsurvey.org/ last accessed on Feb. 11, 2019, and Sect. 7.5.

and legal requirements for sampling. This in turn results in very differentiated practical experiences with sampling schemes at the national level.

A clear definition of the population is a prerequisite for developing a sample design. The ESS defines this as all persons with a minimum age of 15 years living in private households in the participating countries. An upper age limit is not specified. Furthermore, the selection is to be independent of nationality, citizenship and language.

The procedure followed by the ESS is further specified by the following points:

- Random selection should be used in all countries, i.e. it must be possible to determine the selection probabilities of the individual persons. Only this approach provides the possibility of making design-based estimates for the population on the basis of the data obtained.
- The second objective is to cover the entire population of a country by means of the selection made. This means that any minorities living in the countries must be taken into account by applying appropriate sampling strategies.
- Thirdly, a high response rate should be aimed for. The target is 70%.
- Fourthly, systematic failures that occur during fieldwork may not be replaced by other persons who are willing to take part in the survey. This rules out, for example, the use of the ADM design described above (see variant two in Sect. 5.2.2).
- Fifth, the design effects due to different selection probabilities and sample clumping should be kept to a minimum. Especially in countries with a relatively large territory, interviewers might have to travel quite long distances to reach all target persons if they were selected at random without restrictions. This would involve considerable cost. For this practical reason, spatial lumping is used in the selection of target subjects. However, these have the (negative) effect that the spatial proximity of the target persons generally makes the sample more homogeneous. This homogeneity can be measured by the intra-class correlation coefficient, which in turn enters into the calculation of the design effect (compare Gabler and Häder 2000). Finally, this leads to an increase in the confidence interval of an estimator (cf. Sect. 5.1) compared to that of a simple random selection, i.e. the estimate is less precise.

This can only be compensated by choosing a higher sample size in countries with more complex sample designs.

With his considerations, Kish provided the starting point for solving the problem of different sample designs in comparative surveys: "Sample designs may be chosen flexibly and there is no need for similarity of sample designs. Flexibility of choice is particularly advisable for multinational comparisons, because the sampling resources differ greatly between countries. All this flexibility assumes probability selection methods: known probabilities of selection for all population elements" (Kish 1994, p. 173).

Consequently, the selection strategies in each country are to be handled flexibly – however, under the premise that the best available design for random sampling is used in each case. This means that it would be wrong to try to apply (as far as possible) an identical sampling procedure in each country. Since the preconditions for sampling differ greatly from country to country, one would otherwise have to use the least suitable procedure – i.e. the lowest common denominator.

The required procedure is fulfilled by finding and using the most feasible selection strategy in each country. As in the realization of any other random selection, the question of an available inventory of all elements of the population is at the beginning and in the center. The ESS looks for a selection frame (a list or directory of all elements of the population) in each country and asks the question about the quality (completeness, timeliness, accessibility for survey research and so on) of this frame. Furthermore, the experience available nationally in sampling must be taken into account. Last but not least, the costs involved also play an important role.

It has been found that aggregated data on the respective demographic composition of the population exist in all countries participating in the ESS and can be used for the selection of the target population. It was also found that the frequency of updating such statistics ranges from a few months to a few years.

To illustrate, the situations found during the sampling will be described.[8] Thus there were:

- Countries with reliable population lists that are also useful for social research, such as Denmark, where there is 99.9% population coverage.
- Countries with reliable lists of households that can also be used for social research. An example of this is Luxembourg.
- Countries with valid address lists that can also be used for social research, such as the Netherlands.
- Finally, countries where the selection must be made without such lists because they do not exist or are not accessible. These include, for example, Portugal and Greece.

Sampling is particularly difficult in countries where no lists are available. Here, area samples have to be used, in which the municipalities are selected first and the households in a second step. The selection of households is particularly critical here. The following two methods have proven to be feasible:

First variant: First, a list of all addresses in a certain area is drawn up. Then the target households are drawn from this list. Here, a relatively strong clumping must be accepted. This procedure is practiced in Greece, for example.

In the second variant, random route elements are used for the selection. The problem here is the exact adherence to the prescribed random route by the interviewers (see also Sect. 5.2.2). This concerns both compliance with the specifications for the actual random

---

[8] Compare for example: Sampling for the European Social Survey Round VI: Principles and Requirements. Guide: available at: http://www.europeansocialsurvey.org/docs/round6/methods/ESS6_sampling_guidelines.pdf, last accessed 08 February 2019.

walk and the control of the interviewers during the survey itself. This is the case, for example, with the design chosen in Austria.

There are also issues to consider for countries with a valid selection framework. In Italy and Ireland, electoral registers can be used as a basis for selection. However, these only contain persons aged 18 and over. Thus, such electoral registers can only serve as an address frame for the addresses of the households and cannot be used for the determination of the individual target persons.

The procedure described helps to implement the best sampling strategy from a methodological point of view in each country. By calculating the respective national design effects, it is also possible to select the individual sample sizes in such a way that the data collected in the countries are of comparable quality.

5.4   Sampling for Access Panels and Internet Surveys

5.4.1   Selections for Access Panels

The term access panel is used to describe – in some cases very extensive – lists of persons willing to be interviewed. In principle, access panels (cf. Sect. 6.1.3) can be set up as a pool for participants in surveys by telephone or via the Internet, for surveys conducted by post or for surveys in the Mixed mode.

In the meantime, extensive methodological research has been conducted internationally on access panels (for an overview, see Bartsch 2012, p. 28 ff.). The background for such projects is, firstly, the dramatically lower response rates in general population surveys and the naturally increased uncertainties about the quality of the data obtained in this way, as well as the significantly increased costs of such studies. Secondly, social researchers have a need for panel data of the highest possible quality, since this allows them to work on causal hypotheses and to focus on the general population of the Federal Republic of Germany. Thirdly, both the possibility of rapid data transmission offered by the Internet and the high degree of accessibility of the target persons promise interesting possibilities for the collection of data.

Access panels have a number of other advantages. Particularly in the case of very extensive panels, it is possible to include in a survey people who are particularly mobile and who are relatively difficult to reach via other means. Furthermore, they offer market research in particular "pinpoint access" (Lamminger and Frank 2002, p. 95 compare also Hoppe 2000; Heckel 2003, p. 87) to specific target persons. If very specific segments of the population are required for a market analysis (people who wear glasses, buyers of baby nappies, wet shavers and comparable types of person), extensive and time-consuming screenings would otherwise have to be organised in order to identify them. Finally, access panels – once recruited – represent a very cost-effective survey mode.

Access panels are considered problematic because the participants have to be motivated to take part in the surveys with the help of incentives. Furthermore, non-coverage is expected, especially in the case of online access panels. This means that the part of the population that does not use the internet is excluded from such surveys. Panel effects due to repeated surveys must also be expected.

There are two main approaches to recruiting participants to such panels:

First, panel members can be assembled through self-recruitment. In this process, target persons are recruited for participation, for example, with the help of references on Internet pages. This strategy is not uncontroversial for various reasons and is rejected, for example, by the ADM (cf. Wiegand 2003, p. 68 and the ADM guidelines for online surveys).[9] For example, this approach raises the possibility that individuals may gain access to incentives primarily through multiple participation. Such people, also referred to as bargain hunters, professional respondents, or heavy users, compromise the quality of the data. Another problem results directly from the self-selection of the participants (compare Scheffler 2003, p. 37). Thus – as required for random selection – no information can be provided on the inclusion probabilities for the elements of the population.

Second, in surveys based on random sampling, screenings can be used to recruit people willing to be interviewed. Here, it does not matter whether these interviews are conducted face-to-face or by telephone. In such a screening, all participants are asked about their willingness to take part in a regular survey. However, the willingness to participate is problematic here. Success rates of only between three and a maximum of 25% are reported (Scheffler 2003, p. 37). In one case it is said to have been 14% (Heckel 2003, p. 91). However, once the willingness to participate has been declared by the target persons, response rates of around 70% are achieved with access panels.

In some cases, there is also talk of active and passive recruitment of participants. Which of the two approaches is the most appropriate depends on the purpose of the survey. If the aim is to test a particular theoretical model, as in the case of a psychological experiment to measure attention, or if a pretest is to be carried out, there is hardly anything to be said against using the less complex approach and employing a self-recruited panel. Access panels are also a suitable instrument for exploratory studies that are looking for initial clues to clarify a problem.

Special efforts in recruiting participants are being made by GESIS for a population-representative Omnibus Access Panel. Funded by the Federal Ministry of Education and Research, persons willing to be interviewed are first recruited on the basis of a register sample. To this end, face-to-face oral interviews will take place in the target households. Interested persons will then be given the option to be surveyed further either online or by mail.[10] Participation in this Omnibus Access Panel is made possible for social scientists free of charge or at low cost after an application procedure.

---

[9] Compare: https://www.adm-ev.de/wp-content/uploads/2018/07/RL-Online-Befragungen.pdf, last accessed February 08, 2019.

[10] Compare: http://www.gesis.org/unser-angebot/daten-erheben/gesis-panel/sample-and-recruit-ment/ last accessed on February 08, 2019.

5.4.2   Sampling for Intra- and Internet Surveys

Internet use already increased rapidly between 1997 (4.1 million users) and 2002 (28.3 million users) (compare Bandilla 2003, p. 76). In 2012, 79% and in 2017 over 90% of all households were equipped with an internet connection.[11] Thus, it is obvious to also ask the question about sampling strategies that are suitable to implement surveys that are mediated via this medium. By now, almost all large companies and many medium and small companies are likely to have intranets as well, so the problem of sampling for in-house surveys is also topical. While general population surveys via the Internet are not currently feasible, it is certainly possible to survey special populations, such as students, in this way. An example of this will be shown later.

The procedure for selecting participants for surveys on the Internet also depends above all on the intended purpose of the study. As with the recruitment of access panels (cf. Sect. 5.4.1), a distinction must again be made here between model tests and explorative studies on the one hand and parameter estimates, in which statements are to be made about the expression of a characteristic in the population, on the other.

Intra- and internet surveys for selected subgroups of the population are quite realistic. In principle, the first question to be asked is whether current and complete lists exist in which the members of the respective population are listed. It is then relatively easy to randomly draw survey participants from these lists. At the moment it is quite possible to conduct employee surveys via the intranet. Borg (2003, p. 158 ff.) presents the example of an employee survey conducted worldwide at SAP via the intranet.

However, studies with students are also conceivable. Here we are dealing with a population that is currently already likely to have a full supply of access to the Internet. The 2004 university ranking by the news magazine Der Spiegel,[12] the Internet provider AOL and the management consultancy McKinsey & Company can be cited as an example. This study is – according to the almost euphoric comment of a member of the scientific advisory board that accompanied the survey – "a real breakthrough in the field of online surveys." At the same time, another member of the scientific advisory board describes the response rate achieved as "simply sensational".

More than 80,000 students participated in the aforementioned study. The aim was to determine a subject-related ranking of universities, which included 15 subjects frequently chosen at universities. Via the internet portal, technical, methodological and content-related filters were already used to prevent incorrect and multiple answers and to achieve a high level of data quality. A broad-based, cross-media advertising campaign (for example, posters in the departments and print advertisements) was necessary to prepare for the

---

[11] Compare: https://www.destatis.de/DE/Publikationen/Datenreport/Downloads/Datenreport2018 Kap6.pdf?__blob=publicationFile, last accessed February 08, 2019.

[12] Compare: http://www.spiegel.de/unispiegel/studium/studentenbefragung-des-spiegel-die-me-thode-a-329.082.html last accessed on February 08, 2019.

survey; it ensured good acceptance and a corresponding response rate. This campaign made an average of four contacts with each student. Finally, the survey resulted in 50,000 evaluable questionnaires.

5.5   Lump Sampling: The Example of the PISA Studies

Cluster selection (the term cluster sample is also used synonymously, cf. Bortz and Döring 2002, p. 438) is a special case of multi-stage random selection. The procedure of cluster selection can be demonstrated relatively easily by means of a student survey. It may be of interest, for example, to what extent there are differences in the performance of students of a certain grade level in different federal states. In order to optimize the survey effort and to guarantee a comparable survey situation, the surveys could be conducted in the respective school classes. Assuming that a corresponding list of all schools is available, a selection of schools (also referred to as clusters) in the respective federal states would have to be made in a first step – at random – on the basis of this list. In a second stage, the school classes in which the pupils are to be interviewed are again randomly selected. Finally, at the third stage, all the elements of the selected clump – in this case all the pupils in the classes concerned – are examined. It is thus sufficient if a list of all schools is available; a list of all pupils (clusters) is not necessary.

Ultimately, however, the F2F design of the ADM (cf. Sect. 5.2) is also a variant of a lumped sample. Here, the areas selected at the first stage form the clumps. However, in the ADM design not all elements of the selected lump, i.e. all persons belonging to the population who live in the selected areas, are surveyed.

The sampling design used in the PISA study (cf. Baumert et al. 2001a, b, p. 34 ff.) will serve as an example to explain the procedure. The aim of the PISA study was to survey basic competencies among students in different countries. Specifically, reading literacy, mathematical literacy and basic scientific literacy were assessed.

To this end, the PISA study examined "students who were between the ages of 15 years/3 months and 16 years/2 months at the beginning of the testing period, regardless of grade attended or type of educational institution" (Baumert et al. 2001a, b, p. 34). This defines the population. The population did not include individuals who, for mental, emotional, or physical reasons, were unable to complete the tests independently, or foreigners who did not have a sufficient command of the test language and had been taught German for less than one year at the time of the test. The PISA target population defined in this first step comprised a total of 924,549 students in Germany.

It is interesting to note that in some countries, for example in Brazil and Mexico, a considerable part of the target population could not be covered by the chosen design because they had already left the schools again. In these cases, the chosen sampling design did not fully cover the population. As these are likely to be lower performing students, the estimators from the sample will be biased accordingly.

In Germany, 220 schools, in sampling terminology these are the clumps, were drawn at the first stage. The drawing probability was proportional to the size of the clumps. This followed the international guidelines of the organizers. The second stage made a random selection within schools (clumps). While all students in vocational schools and 35 students in integrated comprehensive schools were included in the sample, the remaining schools had 28 students each. The realized sample of the PISA study finally had a size of n = 5073 persons in Germany.

Unlike unrestricted random sampling, which uses a lottery drum approach, lumped sampling has a wider confidence interval. This means that an estimator (for example, mean or proportion) obtained using unrestricted random sampling is more precise than an estimator obtained using lumped sampling when both samples have the same size. In other words, a larger sample size is required to produce an estimate that is as precise from a lumped sample as from an unrestricted random sample (see also Sect. 5.4).

This becomes clear if one imagines that, firstly, the pupils in a class are taught by the same teacher, secondly, that their parental homes may be in a similar settlement area, thirdly, that the conditions in the social environment, such as the unemployment rate, purchasing power, political climate, are also similar. Error reduction (compare Bortz and Döring 2002, p. 440 f.) in lumped sampling becomes possible if:

• a larger number of clumps (in this case, school classes) is surveyed,
• smaller lumps are selected,
• these lumps are as heterogeneous as possible and
• the elements are as homogeneous as possible between the lumps.

The phenomena that can lead to a bias in the sample can again be shown in concrete terms if we return to the PISA study. The results show that schools in Germany do indeed differ greatly in terms of performance, or rather that one can already predict the expected performance of students relatively well on the basis of the type of school. Contrary to the above-mentioned demand for heterogeneous clusters, we are therefore dealing with very homogeneous clusters in the PISA study in Germany.

To determine the degree of homogeneity of a lump, the intraclass correlation coefficient is calculated. It is p = 0.60 in the PISA study for the reading performance of 15-year-olds. "In countries where the school system is not structured, the value is significantly lower. The difference to Sweden, where an intraclass correlation of p = 0.10 can be found based on the PISA study, is striking to begin with" (Sibberns and Baumert 2001, p. 511 f.). This means that in Germany a student's reading performance can be explained to a large extent by the school he or she attended.

Stratified lumped sampling can be used to counteract such drawbacks of lumped sampling. Thus, no unrestricted random selection was made when drawing the schools in the first stage. Rather, it has already been taken into account that

- there are different types of schools (Gymnasium, Integrated Comprehensive School, Secondary school, Main school, vocational schools and special schools),
- different numbers of persons from the population live in the individual federal states, and that
- private and public schools are available.

This structure is now exactly reflected in the sample drawn due to the stratification carried out during the draw.

As indicated, sampling errors are generally larger in lumped samples than in unrestricted random samples. This effect is countered by the stratification carried out.

"With the help of the design effect, the so-called effective sample size can be determined, which denotes the sample size that leads to equally precise estimates with a simple random sample. […] Thus, in the case of reading literacy, a random sample of approximately n = 2225 15-year-olds would yield similarly precise estimates" (Sibberns and Baumert 2001, p. 516). As a reminder, the actual sample size in the PISA survey was 5073.

5.6 Quota Selection

Above, the example of a coin toss was used to show the considerations on which random selections are based (cf. Sect. 5.1). It was assumed that the coin is always tossed in such a way that the result obtained is random. Possible systematic manipulations, false registrations and the like were excluded. This was the prerequisite for being able to make predictions with the help of probability theory. Only in random experiments can the characteristics of a sample be inferred from the population.

In the practice of survey research, random selection thus plays an important role (compare especially Sects. 5.2, 5.3, 5.4 and 5.5). In addition to these procedures, however, non-random procedures have also become established. These are, for example, quota selections, also known as quota design. First, the procedure for these selections will be described. Then the prerequisites of quota selections will be discussed and finally the debate in the social sciences about their advantages and disadvantages will be mentioned.

Quota selections are a certain type of deliberate and at the same time arbitrary selection. The interviewers are given certain characteristics (quotas) according to which they have to deliberately and specifically search for their target persons. The quotas are usually combined characteristics with such criteria as age, gender and level of education. The persons to be sought out for an interview are described on the basis of the above criteria and then specifically selected by the interviewers. If the interviewers adhere strictly to these specifications, the result is a sample whose structure corresponds exactly to the quota specifications. If these quota specifications in turn correspond to the structure of the population, this also applies to the sample. Figure 5.3 shows an interviewer instruction such as that used by the Institut für Demoskopie in Allensbach (cf. Schneller 1997, p. 16). It combines the criteria of municipality size, age, gender, occupation and household size.

In order to be able to make selections according to the quota procedure, a number of prerequisites must be fulfilled. Firstly, knowledge about the structure of the population must be available for the creation of the quota specifications. It is not sufficient to know only the distribution of a single characteristic, but – as shown in Fig. 5.3 – combined specifications from different criteria are required. Such information is usually provided by

Please return quota statement with completed interviews!

Name of interviewer:	Survey
Residence:	
A total of 10 interviews with people aged 14 and over	Questionnaire no.
In residence/in:	

Community size

Municipalities under 2000 inhabitants#	1	2	3	4	5	6	7	8	9	10	11
2000 to under 5000 inhabitants#	1	2	3	4	5	6	7	8	9	10	11
5000 to under 20,000 inhabitants#	1	2	3	4	5	6	7	8	9	10	11
20,000 to under 100,000 inhabitants#	1	2	3	4	5	6	7	8	9	10	●
100,000 to under 500,000 inhabitants#	1	2	3	4	5	6	7	8	9	10	11
500,000 and more inhabitants#	1	2	3	4	5	6	7	8	9	10	11

Age

		Male								Female				
14 – 17 Years	1	2	3	4	5	6	7	1	2	3	4	5	6	7
18 – 24 Years	1	2	3	4	5	6	7	1	2	3	4	5	6	7
25 – 29 Years	1	2	3	●	5	6	7	1	2	●	4	5	6	7
30 – 39 Years	1	2	●	4	5	6	7	1	2	●	4	5	6	7
40 – 49 Years	1	2	3	4	5	6	7	1	●	3	4	5	6	7
50 – 59 Years	1	2	3	4	5	6	7	1	2	3	4	5	6	7
60 – 69 Years	1	2	3	4	5	6	7	1	2	3	4	5	6	7
70 Years and older	1	2	3	4	5	6	7	1	2	3	4	5	6	7

Professionals

Farmers and family workers in agriculture and forestry (including horticulture and animal husbandry)	1	2	3	4	5	6	7	1	2	3	4	5	6	7
Manual workers (including agricultural workers, skilled workers, non-self-employed craftsmen and trainees)	1	2	3	4	5	6	7	1	2	3	4	5	6	7
Salaried employees (including trainees)	1	2	3	4	●	6	7	1	2	●	4	5	6	7
Civil servants (including soldiers)	1	2	3	4	5	6	7	1	2	●	4	5	6	7
Self-employed and assisting family members as well as liberal professions	1	2	3	4	5	6	7	1	2	3	4	5	6	7

(Non-working persons (also unemployed, for pensioners previous occupational status, for unemployed last occupational status, for housewives, househusbands, pupils, students, etc.). Occupational status of the breadwinner/main earner)

Farmers	1	●	3	4	5	6	7	1	2	3	4	5	6	7
Workers	1	●	3	4	5	6	7	1	●	3	4	5	6	7
Employees	1	2	3	4	5	6	7	1	●	3	4	5	6	7
Officials	1	2	3	4	5	6	7	1	2	3	4	5	6	7
Self-employed and Liberal professions	1	2	3	4	5	6	7	1	2	3	4	5	6	7

Conduct EXACTLY ONE of the 10 interviews in a household with 5 or more people and EXACTLY ONE in a household with 1 person

Note: Valid are the numbers before the stamp. If, for example, the number 3 were stamped in the line "Workers (professional) female", two female workers would have to be interviewed in any case. For the rest, please cross off the applicable statistic after each interview so that you can immediately overlook how many interviews still need to be conducted in the category in question

#The size of the entire municipality is decisive, not the size of districts or incorporated suburbs

Fig. 5.3 Example of an interviewer instruction for the identification of target persons according to the quota method

official statistics. Which criteria of a population are used to create quota specifications for a survey is the responsibility of the institute conducting the survey.

This also says that quota selections can come about in very different ways. So far, no generally applicable standard has been developed according to which quota specifications can be determined. The consequence is that each survey institute works with its own quotas and the label "quota selection" thus stands for very different strategies.

A second prerequisite results from the fact that the quotas given to the interviewer must be as easy as possible for him to recognise or query. It would be unrealistic, for example, to use political attitudes for a quota.

Quota selections are not considered particularly suitable when surveys are organised with only one topic. "In monothematic studies, there is a tendency for interviewers to preferentially select people who have an above-average interest in the research topic and are particularly able to provide information. A bias towards the 'expert' would be pre-programmed" (Schneller 1997, p. 8).

However, quota selections have a significant advantage: they are less costly than random selections and are therefore also offered by the survey institutes at a significantly lower cost. According to data, these cost advantages amount to 25 to 100%. This also includes the fact that with this method – for face-to-face interviews – relatively short field times can be realised. A time saving of 40% is given as a value here (compare Althoff 1997, p. 25; Koolwijek 1974, p. 84). This is contrasted by a number of criticisms which have led to quota sampling still being regarded as controversial.

The main objection is the fact that in quota selections the selection probability cannot be specified for the individual elements of the population. This has the consequence that mathematical-statistical estimation procedures, with the help of which one can infer parameters of the population from the structure of the sample, cannot be applied. Calculations of confidence intervals are not possible.

Furthermore, it can be assumed that the control of interviewers is particularly difficult in the case of quota selections. Deliberate falsification is easier than, for example, with register samples. It is conceivable, for example, that an interviewer addresses persons known to him and instructs them in such a way that they provide the necessary information, for example on age and qualifications, in the event of queries from the survey institute.

Another problem that is discussed in connection with quota selections is that the interviewers – without necessarily violating the rules given to them – look for easily accessible people in their circle of friends and acquaintances who are particularly willing to be interviewed and then ask them for an interview. This would mean that people who are particularly willing to communicate, sociable and involved in networks would be over-represented in the sample.

Finally, the quotas given to the interviewers must have a minimum degree of relevance to the problem under investigation. It is obvious that it makes little sense to compile a sample according to the criteria shown above if, for example, the development of foot sizes is to be determined in a study commissioned by the shoe industry.

In the context of the discussion about the efficiency of quota selection, its protagonists point out that such samples lead to valid results. The aforementioned study on the American presidential elections of 1936, in which quota sampling enabled the Gallup Institute to predict the correct outcome of the election (see also Sect. 5.1), has become particularly well-known. Today it is still possible to make particularly precise election forecasts on the basis of quota selections (cf. Schneller 1997, p. 17).

It should be borne in mind here that there is an immense reservoir of social science experience, especially in the prediction of electoral decisions. This enables the respective institutes to create particularly sophisticated quota specifications. Thus, a reference to a successful election forecast is not yet too strong an argument for the fact that it is also possible to implement this technique successfully for less well researched facts.

Furthermore, no information can be provided on the response rate for quota selections. After all, it is not known how many people the interviewers approached in order to carry out the number of interviews they were given.

If one chooses the quota specifications according to which the interviewers have to search for the target person so concretely that only a relatively small scope for decision-making remains for the interviewers, one approaches the procedure with a random selection. After all, in this way the interviewers are forced to go from door to door looking for suitable persons.

In view of response rates that are currently between 30 and 40% for the ALLBUS (cf. Sect. 5.8), a further argument for the cost-effective quota design seems obvious. After all, only those persons who are in principle willing to participate in surveys based on random selection do so. Ultimately, one could assume that the same elements are examined in a quota sample as in a random sample anyway.

Various designs of social science studies do not require random sampling anyway and can certainly make use of quota samples. This applies, for example, to pretests (see Chap. 8) and to experimental studies (see Sect. 7.1). Here the aim is to make model-based estimates, these "rely on the assumption that the experimental effects are uniform across all units of the relevant population." (Zins 2018) In contrast, as shown above, random samples do not require any further preconditions.

5.7   Sampling for Special Populations and for Qualitative Studies

In the social sciences, statements are sometimes to be made about very special populations. These can be drug users, single fathers in a certain age group, gamblers at vending machines or experts in a rare field. For finding participants in such surveys with numerically small and/or hard-to-reach groups of people, the *snowball method* has proven to be effective. This belongs to the group of ascending (compare van Meter 1990) selection methods and presupposes that the persons concerned are connected via networks or at least know each other.

The snowball approach is that one participant (an element of the population) completes the survey documents and also distributes further questionnaires to his or her circle of acquaintances, asking them to participate. It would also be possible for the person in question to simply arrange participation on behalf of the social researcher in question. In this case, the recruited target person would contact the survey institute himself.

This design is applicable when, firstly, there is no selection frame available, i.e. there is no list in which all elements of the population are listed. Secondly, the snowball procedure will have to be applied when the target persons are those who cannot be determined even by screening. In a screening, a general population survey – or better still, several different ones – asks about the specific characteristics of the target person that are being sought. For example, it might be possible to identify single fathers in a particular age group. If these characteristics are available, a corresponding survey can be conducted.

If, however, the characteristics sought are very rare in the general population or if a corresponding query poses major difficulties for survey research ("Are you possibly addicted to gambling?"), one is more likely to resort to selection by snowballing. Cost arguments may also speak in favour of such a decision.

The question of the generalizability of the results of a snowball survey to a population can in principle be answered negatively. Nevertheless, the snowball method can provide initial access to the study population of interest. The attempt to obtain estimates of population parameters with the help of sophisticated statistical procedures (cf. Gabler 1992; Kish 1988) will only be referred to here. It is assumed that the generalizability of the results depends firstly on the structure of the network and secondly on the selection of the initial sample through which the target persons are recruited.

Gabler (1992) and Biernacki and Waldorf (1981) respectively mention various problems of the snowball method. For example, people who are relatively isolated in their networks are likely to be underrepresented in the samples. Particularly challenging and consequential in the snowball procedure is finding an initial sample with which to begin the survey. It is also necessary to check the eligibility or suitability of the newly reported target persons and thus to identify persons who have already been interviewed or, if applicable, fee hunters.

An important prerequisite for the success of the selection is still the creation of trust among the persons involved. There is a particular need here to leave a trustworthy impression during the interview. Finally, the participants in the interview make themselves helpers of the investigator in forwarding the request.

Another possibility for interviewing special populations was already referred to in the presentation of selection procedures for access panels. Here, too, appropriate samples can be compiled with pinpoint accuracy, provided that a sufficiently large pool of people is available.

Qualitative studies also require principles for determining the target subjects (see Sect. 3.3).[13] Various approaches have become established:

Theoretical Sampling

Grounded theory (see, for example, Glaser and Strauss 1967; Strauss and Corbin 1996) – an approach guiding qualitative research – envisages the targeted selection of cases to be studied in such a way that they systematically expand knowledge about the object of study. It is an iterative procedure in which one or only a few units of inquiry are initially analyzed. On the basis of the results thus obtained, a search is then made for further cases that might be suitable for confirming the previous findings or for modifying them. This strategy is referred to as theoretical sampling.

Selection of Typical Cases

The selection of typical cases is a form of deliberate selection. A specific attempt is made to recruit particularly characteristic cases. Firstly, this requires knowledge about the nature of the population or about the problem to be addressed. Only if it is clear what a typical case is, can it be selected. Secondly, it must be assumed that certain cases can only be described as typical in very specific respects. This procedure is thus associated with numerous problems.

Selection According to the Concentration Principle

Here, those – weighty – cases are selected which, due to certain characteristics they possess, determine the distribution in the population. This selection also presupposes knowledge about the population. For example, the aim might be to investigate the propensity to violence among young people. If one knows that this propensity only applies to a very specific (small) part of the population and one is further informed that these are persons of a certain age, with a certain origin and of a certain gender, then an appropriate selection would be conceivable. Similar to the selection of typical cases, however, this approach is only suitable for dealing with very specific problems.

River Sampling

This strategy involves searching the Internet for people who are willing to participate in a survey study. For this purpose, corresponding search advertisements are placed on several thousand pages. It may be advantageous if these pages have quite different contents, which in turn are visited by different interested parties. As a result, the addresses of numerous users are available, ideally from quite different social milieus. This procedure is similar to that of a fisherman who casts his nets and then looks at his catch. This is also where the name for this procedure comes from. Samples can then be drawn from the pool obtained in this way. This is a form of self-recruitment of participants for surveys, which is of course not random.

---

[13] For an overview, compare For example, Akremi (2019b, p. 312 ff.).

5.8 The Non-Response Problem and the Possibilities of Weightings

5.8.1 Nonresponse

Nonresponse is firstly the failure of an element of the sample, for example a person to be interviewed. The term unit nonresponse is also used for such a failure. Secondly, nonresponse also refers to the absence of individual items of information about a selected unit. The term item nonresponse is commonly used for this situation. This form of nonresponse occurs relatively frequently, for example, in the context of a survey on income levels.

The situation in social research is different from that in a quality control department, for example, which also uses random sampling to determine the quality of a mass-produced article. A certain procedure is used to randomly select the products to be checked from those produced during a day's production. Now it should not be a problem for the company to subject exactly these products to a quality inspection. The situation is different in empirical social research. Here, the persons to be interviewed may have been determined on the basis of a theoretically well-founded selection procedure, but this does not guarantee that these persons will ultimately participate in the survey. Numerous reasons can lead to the fact that the elements of a sample to be investigated evade the investigation in the end. Furthermore, it cannot be guaranteed that all the desired information can be obtained from each element.

Table 5.6[14] shows a statistic that shows the result with which the originally almost 30,000 generated telephone numbers were processed. This is the study (cf. Sect. 3.3) on public attitudes towards self-defence, which was carried out by means of a telephone survey. The overall response rate in this study was 54%.

From the compilation in Table 5.6, various aspects of the fieldwork become clear. First, a distinction is made between neutral and systematic failures. In principle, neutral failures do not affect the quality of a sample – only the confidence interval in the parameter estimation is negatively affected by the smaller sample size. As can be seen, a neutral failure is, for example, when the household contacted by telephone does not belong to the population. In contrast, systematic failures are characterised by the fact that they can affect the quality of the sample. In particular, households that were not reached or persons who could not be contacted as well as refusals by the contact person or the target person are to be mentioned here. The distinction between target person and contact person is as follows: The person who answers the phone first is called the contact person. The contact person gives the interviewer a piece of information that the interviewer needs to identify the person to be interviewed. The target person is then randomly identified among the members of the household in question.

---

[14] There is no binding system for the compilation of failures in the context of field work. Here, each institute follows its own rules (compare Schnell 1997). The allocation to neutral or systematic failures can also be discussed further.

Table 5.6 Field report on the Dresden self-defence study 1999, collected and compiled by USUMA GmbH, Berlin

| | Outages | | | |
| | Quality neutral | | Systematic | |
	n	Percent	n	Percent
Gross/net sample	29,875	100	6376	100
Wrong phone number: No connection	14,509	48.6		
Wrong phone number: Computer, fax	1066	3.6		
Wrong phone number: Not identifiable	234	0.8		
Occupied	126	0.4	3	0.1
Not reached: Dial tone	2876	9.6	71	1.1
Not reached: Answering machine	956	3.2	24	0.4
Contact person re-call possible	198	0.7	5	0.1
Contact person indefinite appointment	161	0.5	4	0.1
Target person specific appointment	23	0.1	1	0.0
Target person recall possible	31	0.1	1	0.0
Target person not present during field time	908	3.0		
Connection is not part of the sample	1687	5.7		
Contact person denied			500	7.8
Target denied			382	6.0
Absolute refusal			1	0.0
Comprehension problems	724	2.4		
Termination without appointment			21	0.3
Termination with appointment			0	0.0
Interview not evaluable/quota fulfilled			1900	29.8
Quality neutral failures/interviews	23,499	78.7	3463	54.3
Interviews in the target group			3463	54.3

A distinction of three types of non-response has become established among survey researchers (compare Groves et al. 2004, p. 170 as well as Engel and Schmidt 2019, p. 385 ff. and Fig. 5.4). This differentiation refers to the form of nonresponse to an interview: first, people who cannot be reached; second, people who refuse to participate in the study; and third, people who are unable to participate in the survey due to various dispositions, for example due to illness.

Nonresponse would not be a problem for survey research if one could assume that refusals or failures would occur by chance. In this case, there would be no difference between the cooperative population on the one hand and the non-respondents, the refusers and the persons not suitable for a survey on the other. Admittedly, such an assumption seems untenable, not least because of various experiences of survey researchers. For example, it has been found that certain social groups are systematically underrepresented in the samples of social researchers. These include people who have no interest in survey research or are unable to see the point of it. This applies, for example, to people with less education. But fear of victimisation can also lead to a refusal to participate. This motive plays a particularly important role among older citizens. The group of inaccessible people,

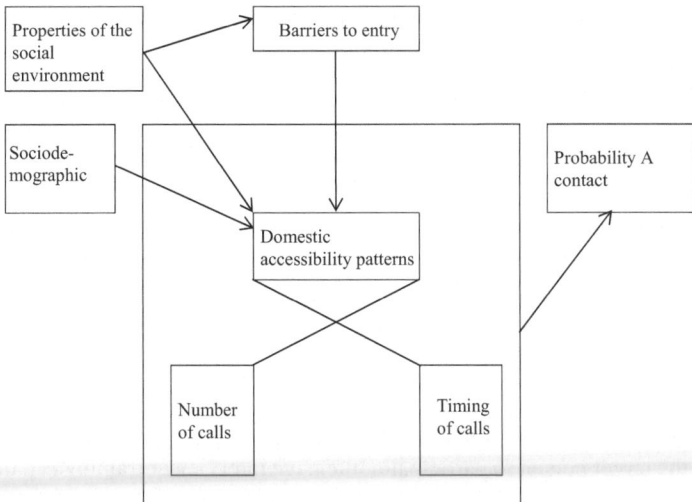

Fig. 5.4 Model explaining the inaccessibility of a household in the sample. (Adapted from Groves et al. 2004, p. 170)

on the other hand, includes people who are particularly committed to their profession. The participation situation therefore resembles a blind flight in some ways, in which the crew of the plane is largely unaware of their own position.

A whole series of authors, among them Schnell (1997); Dillman (2000); Stoop (2005); De Leeuw (1999); Groves and Couper (1998); Kalton (1983); Esser (1986b) and Koch and Porst (1998), have further studied the nonresponse problem and published their experiences on it. As a result, theoretical approaches are available that aim to explain the willingness or refusal to participate in surveys. These approaches are discussed in more detail below (see Sect. 6.1.2). At this point, reference should be made to strategies that have been developed to curb unit non-response.

Social research has various strategies to curb the rate of non-participation in surveys. However, the goal of achieving 100% participation is unrealistic for general population surveys. Sometimes 70% are aimed for (compare ESS), but about 30 to 40% are realized (as in recent ALLBUS studies). Reference should be made to the following strategies:

- Giving incentives to participants can increase the response rate. These can be financial incentives as well as such things as raffle tickets, vouchers, calendars, product samples and the like.
- The survey should be organised by institutions that have a high public profile and a particularly positive image. This usually applies, for example, to opinion research institutes and universities.
- The subject matter of a survey also plays an important role in a positive decision to participate. Particularly in the case of telephone surveys, where the decision on

participation or cancellation is made in the first few minutes or seconds (see also Sect. 6.1.3), questions stimulating the willingness to participate should be asked at the beginning.

- As far as the design of the survey allows, announcements should be made before the actual survey. In the case of face-to-face interviews, this can be done by post or by telephone.

- Social research has different strategies for conducting surveys. These include postal and telephone surveys, face-to-face oral interviews in the target person's home and surveys conducted online via the internet. Different people will find each of these forms differently sympathetic. So the idea is to let the target person choose in which form they would like to participate in the survey. These are then the mixed-mode designs.

- In face-to-face interviews, the interviewer can exert considerable influence on the motivation to participate. Even interviewers in telephone surveys still have some leeway to motivate the target person to participate. Intensive interviewer training can thus contribute to reducing non-response.

- In order to reduce the proportion of those – mostly mobile – persons who are not encountered by the interviewers, increasing the number of contact attempts as well as varying the contact times can bring success.

- In postal surveys (see also Sect. 6.1.3), follow-up or reminder campaigns are used to reduce the number of questionnaires still outstanding.

- The current situation in survey research is in a dilemma – this can be described by two phenomena: Firstly, as already described, failures in random sampling are increasing dramatically and at the same time a cost explosion is taking place here, so such studies are becoming increasingly expensive. Secondly, online access panels, some of which are recruited by means of river sampling, represent a clearly cost-effective variant for survey research. They sometimes compete with considerable sample sizes in order to substantiate their claim to representativeness. This gives rise to a certain need for research. This concerns the question of whether a combination of both approaches could not pave the way out of the dilemma. Of course, this requires the development of a theoretical foundation for such a design.

- Individual missing data can be supplemented with the help of imputation procedures. A rough distinction is made between *singular* and *multiple* imputation. In singular imputation, each missing value is replaced by a specific estimated value. In the simplest case, the mean value of the available observations is used. For the replacement of missing values, however, procedures can also be used which make use of the knowledge of correlations between different variables and the missing value. If, for example, a person's income level is missing but knowledge is available about the years of education he or she has completed, about his or her age and about weekly working hours, this knowledge can be used to estimate the missing value. For this purpose, hot-deck and cold-deck techniques as well as and regression methods were used. In multiple imputation, multiple values are estimated for each item for which nonresponse has occurred.

For more information on the imputation techniques, see Beretta and Santaniello (2016) and Lüdtke et al. (2007).

- If, despite appropriate efforts, the response rate remains unsatisfactory and a systematic bias is to be expected due to the non-response, then attempts are sometimes made to improve the quality of the survey by weighting the data sets. This instrument is discussed below.

5.8.2   The Weighting of Samples

As illustrated, nonresponse in social research is a more or less annoying fact that can reduce the quality of the sample, but cannot be completely avoided. One strategy to compensate for the effect of nonresponse is weighting.

In program packages for statistical data analysis, as used by social scientists, weighting procedures are relatively easy to implement (for SPSS compare, for example, Wittenberg and Cramer 2000, p. 95). These are intended to adjust the distributions of certain characteristics in the samples to the known distributions of characteristics in the population, such as age and gender. Here we will clarify what is meant by weightings. Afterwards, different strategies for weighting samples will be discussed by example.

Suppose a social scientist is interested in what the income is in a certain population. To determine this, a random sample is drawn from this population and the target persons are asked about their income. After an initial analysis, however, one now finds that – for whatever reason – the proportion of women in the realised sample is greater than that in the population. One reason for this may be that, for example, more women than men were willing to be interviewed. (The recognition that certain distributions – in this case gender – are different in the sample than in the population presupposes, of course, that the latter are known to the social scientist). Further, it should be assumed that there is a theory, or at least an educated guess, about the relationship between the parameter of interest in the population (mean income) and the structural characteristic of the sample (gender). Thus, there is probably a correlation between income on the one hand and gender proportion on the other, such that women presumably earn less than men. It can now be concluded that the income level estimated on the basis of the sample is lower than the actual level in the population – after all, there is a lack of higher-earning men. Equipped with this knowledge, it would now be possible to set up a model for the estimation, i.c. to weight the sample in order to eliminate this bias.

For such a simple form of weighting, the distribution determined in the sample (the gender proportion) is compared with that from the population (target-actual weighting, compare Rösch 1994 and Kiesl 2019, p. 405 ff.). In this way, a quotient is determined that is used for weighting. This weighting factor would be exactly 1 if the ratio between the proportion of men in the sample and the proportion of men in the population is identical. It will be greater than 1 if there are too few items with a particular expression in the sample. It is usual to normalize the weights to the number of cases. In our case, there will be values greater than 1 for men. Correspondingly, the value for women will be less than 1 because their proportion in the sample is greater than in the population. This value would then be the weighting factor. Now, in the data analysis, the income of the men and that of the women is multiplied by the respective weighting factor and thus – if the model represents reality well – the mean income is estimated better than without this correction.

There are numerous assumptions made in this chain of thought. First, it is assumed that the distribution in the population (in this case, the sex structure) is known to the researcher.[15] Further, it is assumed that there is a known relationship between gender and income. This is plausible, but it does not necessarily have to be so. If these assumptions were false, weighting could not improve the estimator.

Various weighting procedures are now distinguished in survey research. Reference should be made to the following two:

Design Weights
In the design of the ADM for face-to-face interviews (cf. Sect. 5.2.2), a random determination of the target person within the previously selected target household takes place in the last selection step. This results in differences in the selection probabilities. For example, a person living in a single-person household has a different selection chance at this selection stage than a person living in a larger household. One must therefore expect that persons living in smaller households will be more strongly represented in the sample than their share in the population. To compensate for these different selection probabilities, an appropriate weighting factor can be calculated using the inverse of household size.

Another example is provided by the 2001 Dresden self-defence survey, the purpose of which has already been described in more detail (cf. Sect. 3.3). One question concerned the individual experiences of the persons surveyed with self-defence. Among other things, they were asked to report whether they themselves had ever been in a situation where they had had to use force to defend themselves against an attacker.

A disproportionate sample design was used for this study. This was designed in such a way that more people were interviewed in eastern Germany than would correspond to the proportion of the population in the Federal Republic. The East German population (and consequently also the West German population) were therefore not included in accordance with their proportion of the population in Germany as a whole, but a disproportionate consideration of a certain group of people was made. Such a technique is also referred to as oversampling. This oversampling is justified by the fact that it is interesting from a social science perspective to make comparisons between West and East Germany. In order to be able to do this meaningfully for smaller subpopulations, a certain, relatively large number of East Germans and West Germans should be represented in the sample. If, however, the aim is to estimate a parameter of the totality of Germans, it is of course necessary

---

[15] In the practice of survey research, the results of the microcensus are mostly used here. For this survey, the target persons are obliged to provide information, so that the non-response problem only occurs to a limited extent.

to correct this oversampling again with the aid of an appropriate weighting factor, otherwise the East Germans would distort the result due to their overrepresentation The following Tables 5.7[16] and 5.8 show the corresponding figures.

According to Gabler (1994, p. 78), the weighting factors can be calculated as follows: For the West German respondents:

$$\text{Weighting factor}\left(\text{west}\right) = n / n_w \cdot N_w / N$$
$$= 3463 / 2450 \cdot 332,023 / 408,658$$
$$= 1.148$$

For the East German respondents:

$$\text{Weighting factor}\left(\text{east}\right) = n / n_o \cdot N_o / N$$
$$= 3463 / 1013 \cdot 76,635 / 408,658$$
$$= 0.641$$

To the data from each respondent, this weighting factor was added as an additional variable in the data set.

Table 5.7 The East-West weighting of the Dresden Self-Defence Study 1999 using the Microcensus 1997

	Microcensus 1997			Dresden self-defence study 1999		
	West	East	Total	West	East	Total
	(NW)	(NO)	(N)	(nW)	(nO)	(n)
Persons in private households, 18 years and older	332,023	76,635	408,658	2450	1013	3463

Table 5.8 Answers to the question "Have you ever been in a situation in which you defended yourself against an attacker?" (Dresden Self-Defence Study 1999)

		Unweighted		Weighted	
		Frequency	Percent	Frequency	Percent
Valid	Yes	940	27	930	27
	No	2520	73	2530	73
	Total	3460	100	3460	100
Missing		3		3	
Total		3463		3463	

---

[16] Microcensus data are from Koch et al. (2001, p. 52).

It can now be demonstrated (compare Table 5.8) whether it makes a difference if the data from the 1999 Dresden self-defence study are weighted. The question already addressed above is used for this purpose.

As can be seen, the use of the weighting variable hardly changes the marginal distributions of the variable under consideration. On the one hand, this is remarkable, since the unweighted sample shows a clear shift in the proportions of respondents in East and West Germany compared to the actual proportion. On the other hand, this shows that it is relatively meaningless in answering the question considered here whether the target person comes from East or West Germany. Nevertheless, design weights should in principle be used, as they are theoretically more correct estimates. A prerequisite for this, however, is that the selection probability for each element of the sample is known and documented.

Redressment

While design weighting neutralizes biases that have arisen due to the selected sample design – for example, in the third selection stage of the ADM design or in the case of disproportionate sampling approaches – redressment is intended to eliminate sample biases that have arisen, for example, due to refusals and dropouts. The procedure is otherwise basically the same as for design weighting. Cases that are too rarely present in the sample are weighted up and cases that are too frequently present are weighted down. In order to be able to use a redressment, various prerequisites must be observed:

1. The distribution present in the population must be known for all relevant variables. Assuming that such characteristics are education and gender, one would need to know for each level of education and for each gender how strongly it is represented in the population.
2. It must be sufficiently known that there is a relationship between the dependent variable on the one hand and the variables used for weighting on the other. It must therefore be possible to say which of the "relevant variables" mentioned here in 1. are at all. If, for example, the dependent variable were the voting decision and one were to weight the sample on the basis of blood groups, the success of this action would be at least questionable. In practice, information from the microcensus in particular (cf. Sect. 6.1.4) is used as the basis for calculating weighting factors. Interested parties can also find the information required for weighting in the joint publication of the ADM Arbeitskreis Deutscher Markt- und Sozialforschungsinstitute e. V., the Arbeitsgemeinschaft Sozialwissenschaftlicher Institute e. V. (ASI) and the Statistisches Institut für Sozialforschung e. V. (ASI). (ASI) and the German Federal Statistical Office. Here one can compare the distribution realized in a sample with that determined by the microcensus on the

basis of selected indicators. In some cases, the survey institutes supply their clients with the corresponding weighting factors for the redressment.[17]

3. In order to be able to weight, a realized sample must not be too heavily distorted. As explained, weighting increases the importance of certain cases, namely those that are too rarely represented in the sample. Now it is easy to see that there are limits here. For example, only a relatively small number of cases found in a sample may lead to the construction of numerous others.

Because of these preconditions, redressment weightings are not uncontroversial (compare Gabler et al. 1994 for a more comprehensive account). As mentioned above, such a weighting procedure can be helpful in principle, but this is not necessarily so. Weighting is particularly helpful when it is known that the parameters sought are related to the variables corrected by the weighting. This is conceivable if one thinks of election forecasts, for example. Here, a whole series of studies is available, which can then be evaluated on the actual election results. The sometimes precise predictions of election results also speak for the success of this approach. The situation is different, however, if such knowledge is not or not sufficiently available. In such cases, the use of the weighting procedure is not advisable.

Weightings based on the so-called recall question represent a special form. The recall question is directed at the last voting decision of the target person. In the ALLBUS 2002 the following question was asked:[18] (It is also interesting to note which logical controls or filters were built into the computer-assisted survey in order to guarantee a high quality of the data. See also Sect. 6.1.3).

Die letzte Bundestagswahl war am 27. September 1998. Waren Sie bei dieser Wahl wahlberechtigt?

1: Ja weiter Frage #S75

2: Nein weiter Frage #ID_K

9: Keine Angabe weiter Frage #S75

BEDINGUNG: #S74 Code 02 und Geburtsdatum ZP liegt vor September 1980 und #13 Code 01.

Sie haben angegeben, dass Sie bei der letzten Bundestagswahl am 27. September 1998 nicht wahlberechtigt waren. Ist diese Angabe korrekt?

1: Nein, war wahlberechtigt, Angabe korrigieren weiter Frage #S74

2: Ja, Angabe korrekt, war nicht wahlberechtigt

Haben Sie gewählt?

1: Ja weiter Frage #S76

---

[17] Compare: https://www.destatis.de/DE/Publikationen/StatistikWissenschaft/Band17_Demographische Standards1030817109.004.pdf?__blob=publicationFile last accessed on Feb 11, 2019.

[18] Compare: http://www.gesis.org/fileadmin/upload/dienstleistung/daten/umfragedaten/allbus/Fragebogen/ALLBUS_2002.pdf last accessed on Feb 11, 2019.

2: Nein weiter Frage #ID_K

9: Keine Angabe weiter Frage #ID_K

Welche Partei haben Sie mit Ihrer Zweitstimme gewählt?

INT: Bei Rückfragen: Zweitstimme ist die Parteienstimme!

01: CDU bzw. CSU

02: SPD

03: FDP

04: Bündnis 90/ Die Grünen

05: Die Republikaner

06: PDS

07: Andere Partei

08: keine Zweitstimme abgegeben

97: Angabe verweigert

98: Weiß nicht mehr

One can compare the result of this recall question with the actual result achieved in the last elections. This quotient is then used to weight the data set. Such a procedure is likely to be useful if the target persons are also asked the "Sunday question" about a possible upcoming election decision. The results of this Sunday question are then corrected with the help of the weighting factor, i.e. changed on the basis of the results of the recall question. The idea here is that, for example, an underestimation of the previous election decision for a particular party also meant an underestimation for the prospective election decision. If such an assumption is true, the result would then be improved. Weighting according to the recall question thus assumes a certain empirical correlation and only works if this is sufficiently validated.

Propensity Weighting

This approach attempts to estimate the probability of failure of a person to be interviewed and then use this estimation result – inversely – for weighting. Logistic regression models are used for this purpose. These must allow a relatively reliable estimation of the probability of failure or the accessibility of certain persons. This procedure has proven to be particularly useful for online surveys. It is known that participation in surveys via the Internet is strongly determined by self-selection. Accordingly, it would be problematic to transfer the results to the general population. With the knowledge of how likely participation in such surveys is among certain groups, an appropriate weighting can be applied. However, this strategy has also been used to compensate for the undercoverage of unlisted households in telephone surveys (cf. Cobben et al. 2012, p. 173 ff.).

In such propensity weightings, a different, randomly controlled survey is used to explain this propensity to participate in the respective survey technique. Of course, this type of weighting also assumes that the variables relevant to the problem in question are known and can be used for weighting.

In summary, it can be stated that weightings can in principle play a positive role in survey practice. They *can* significantly improve the quality of a sample. However, this only applies under the numerous conditions presented here. Finally, it seems important to point out that evaluations that are carried out on the basis of a weighted data set should always document this accordingly. This is the only way to ensure that the calculations can be traced and that the evaluation is scientifically legitimate. Gabler (2004); Sand (2018) and Sand and Gabler (2019) are recommended as further reading.

Survey Methods

6

Abstract

This chapter deals with the three methods used in empirical social research for data collection. In addition to the survey, these are observation and content analysis. In addition to the theoretical foundations of these methods and the diversity of their variants, the presentation also endeavours to show their application by way of example.

6.1 Surveys

6.1.1 Classification Options

Social science interviews are a survey method based on systematically controlled communication between people. The face-to-face oral interview, in which an interviewer sits across from another person and asks him or her questions, was long considered the standard instrument in empirical social research, its royal road, as René König wrote: "If it [...] is subject to methodological control, the interview in its various forms [...] will always remain the royal road of practical social research" (König 1957, p. 27).

Due to this outstanding position, a very differentiated methodological knowledge about surveys can be referred to in the meantime. This has been systematically gained in parallel with the frequent use of this survey method.

We also encounter surveys in everyday life and here they are quite generally certain social situations that are structured by communication. The salesman asks his customers about their wishes, parents ask their children about their experiences at school and so on. In everyday life, too, certain rules apply to such communications. Imagine, for example, travellers in a train compartment. In case they get into a conversation with each other, there are standards for such a communication like: One lets the other finish, one is polite and honest oneself, one assumes that the other communication partner is also telling the truth, one reports to each other, one reacts spontaneously to the other and laughs when joking, one tries to avoid redundancy in communication, one does not address topics that are socially taboo, and so on. The prerequisite for a conversation or a survey to take place is the willingness of the respective partners to cooperate. There is also a cultural context that defines various norms for such an interaction.

Surveys in the social sciences have some special characteristics. They are:

- according to plan, as they pursue a scientific goal,
- one-sided or asymmetrical, since the interview is ultimately guided only by the interviewer,
- artificial, as they do not occur in a natural way (such as in a train compartment),
- take place among strangers and
- they're without consequences .

Communication in social science surveys can take place in very different ways, be conveyed via different channels and serve differentiated purposes. The term social science surveys thus stands for very different approaches. A classification of the types of surveys should give an impression of the diverse possibilities that exist in the design of this method.

According to the ADM, more than 90% of the projects handled by these institutes use some form of survey to collect data. Table 6.1[1] shows which survey forms are used. The surveys conducted by ADM are both social science surveys and market-based studies. Pseudo-scientific experiments such as surveys via TED or campaigns staged by magazines are not counted here.

The table shows a clear dynamic that has taken place in recent years. According to this, the proportion of face-to-face interviews has fallen by almost two thirds in the last 25 years, while the proportion of interviews conducted by telephone has risen slightly. In the meantime, online surveys and surveys conducted using a mobile app have also been added. This compilation does not take into account the numerous surveys organised by non-commercial institutions such as universities.

Table 6.1 Quantitative interviews of ADM member institutions by survey type, figures in percent

	1990	1995	2000	2005	2010	2015	2017
Personal interviews	65	60	30	24	21	24	27
Telephone interviews	22	30	45	45	35	29	33
Postal interviews	13	10	23	9	6	5	8
Online interviews			2	22	38	38	34
Mobile app						1	1

[1] Compare ADM's annual reports at https://www.adm-ev.de/jahresberichte/ last accessed on 15 February 2019. It is important to note that the composition of ADM institutes has changed over this period. For example, distortions may have arisen as a result of institutes becoming new members of ADM, each specialising in particular survey methods.

For the presentation of the different strategies, the first step is to distinguish between surveys according to the way in which the communication between the respondent and the interviewer is mediated. This can be done in person-verbally, in writing, by telephone-verbally or using modern communication media such as the Internet via WebTV or via apps and mobile phones.

Oral Interzzasviews

In oral interviews, an interviewer usually goes to the home[2] of the person to be interviewed and conducts the interview[3] in person or face-to-face. Social contact is established between the two people during the interview, which is very important for the success of the interview. The assumption is that in this way the target person can be particularly strongly motivated to give the interviewer valid and reliable information. Commercial institutes offer their interview networks for conducting such surveys.

Oral interviews are almost universally applicable as far as the content of communication is concerned, i.e. almost all topics relevant to social science can be communicated in such an interview.

Three problems in particular must be taken into account with this approach: Firstly, the sometimes only low willingness to participate of the persons asked for an interview, secondly, the relatively high costs of such an approach and finally, thirdly, an influence of the interviewer on the result of the survey must be taken into account. Oral interviews can in turn be designed in various ways:

Oral Interviews with Paper and Pencil (PAPI)

Interviews can be conducted by an interviewer equipped with paper and pencil (hence the terms paper-and-pencil and PAPI interview). In addition to the actual paper questionnaire, template lists, card games on which the questions are printed, envelopes with questions for self-completion, samples of goods and so on can also be used. The interviewer notes down the answers in the respective questionnaire and forwards it to the survey institute.

Oral Interviews with the Help of a Computer or a Laptop (CAPI) or a Personal Digital Assistant (PDA)

In the technique of computer-assisted face-to-face interviewing, the computer replaces the paper questionnaire. The questionnaire is programmed and stored on the interviewers' laptops. These read the questions from the screen to the persons to be interviewed. Also,

---

[2] A special form of in-person oral interviews that do not take place in the target's home are those conducted at the exits of polling places. These are called exit polls. They are considered a relatively reliable method of estimating election results. In market research, interviews are also often conducted in shops or supermarkets (point-of-sale surveys). Such surveys can also be conducted at exhibitions and trade fairs.

[3] The term interview is used synonymously with oral questioning.

the answers are immediately entered into the computer and stored. There is no need to dispense with the use of template sheets, card games, samples of goods and so on.

Another variant is the self-completion of the questionnaire by the target person on the computer. Here, the respective interview partner then enters his answer into the computer himself.

The main advantages of this approach – which is initially more expensive due to the costs of equipping all interviewers with a laptop – are firstly the possibility of using particularly complicated questionnaires. The entire filtering is carried out automatically by the programmed questionnaire, so that only those questions are displayed to the interviewer which actually apply to the respective target person. Secondly, there is a permanent control of the data input. Certain input errors, for example values outside the valid range, the so-called wild codes, can be avoided. At the same time, the influence of the interviewer on the result of the survey is somewhat reduced. Thirdly, certain controls for the proper work of the interviewers can also be carried out with the help of this technology, for example by registering the required completion time or determining the location where the interview takes place by means of the Global Positioning System (GPS).

Oral Interviews with Hard, Soft and Neutral Strategy
Interviewers have various strategies at their disposal for dealing with the persons to be interviewed. A somewhat simplified distinction is made between hard, soft and neutral interviews. In soft interviews, the interviewer always agrees with the target person and signals his or her understanding. This can be appropriate in order to reduce inhibitions in the person being interviewed. In the hard interviews, on the other hand, the interviewer acts in an authoritarian manner. For example, he points out contradictions in the answers, expresses his distrust and so on. A well-known example of a hard interview is the Kinsey Report on sexual behaviour (compare Kinsey et al. 1948). Mostly, however, the neutral strategy is used for interviews. Here, the interviewer's task is mainly to register the interviewee's behaviour and not to comment on the answers on his own initiative. He adheres to the norms of conversation and reacts accordingly, for example, when he is told something entertaining by the interviewee.

Written Surveys
In the case of written surveys, the questionnaire is usually passed on by post to the person to be surveyed. However, written surveys also include questionnaires completed in the form of examinations, for example. The latter procedure is often used more widely in the context of creating an opinion of students about lectures.

Since written surveys encounter certain problems of oral surveys, some social scientists temporarily expect a comeback of this method. Above all, written surveys are characterized by the elimination of interviewer influence, by a lower overall effort compared to face-to-face interviews, and by a higher degree of anonymity. However, some fundamental difficulties with written surveys must also be noted. These concern firstly the question of who ultimately fills in a questionnaire sent by post. Secondly, the expected level of participation is almost unpredictable, but generally low (cf. Sect. 6.1.3). Thirdly, it is not uncommon for problems to arise in creating the address file required for mailing.

Telephone Surveys
In telephone surveys, communication between the target person and the interviewer is merely acoustically mediated. As shown, this type of survey was numerically dominant in quantitative empirical social research in Germany until a few years ago. Telephone interviews are currently the second most frequently used access to the target persons (compare Table 6.1).

The prerequisite for the use of this design is that almost all households are equipped with a telephone connection, which is now the case in Germany. With the help of the dual-frame approach, it is now also possible to contact mobile phone connections as well as landlines (cf. Häder et al. 2012). The use of this computer-assisted survey technique makes it possible to shorten the field time, tighten controls on the proper work of the interviewers and use sophisticated sampling strategies. Problems arise primarily due to the significantly shorter duration of a telephone interview (cf. Sect. 6.1.3).

Other Forms of Survey Mediation
Above all, the development of modern communication media has opened up new possibilities for the design of surveys, specifically for access to the target persons. Surveys have become established that are sent to the persons to be surveyed via the Internet (www) or (in the meantime already less frequently) via e-mail. The sending of a diskette programmed with a questionnaire (disk-by-mail) or the use of fax machines for surveys have also been tried out. New possibilities are also offered by WebTV as well as programs (apps) which have been installed on the mobile devices of the target persons (compare Sect. 6.1.3). Web surveys (see Weiß et al. 2019, p. 801 ff.) should also be mentioned in this context.

Surveys as Individual and Group Interviews, Group Discussions
Surveys can be addressed to individuals or to groups. The first form is the most common. As already described, communication takes place between an interviewer and a target person. An example of a group survey has already been given in the form of surveys directed at students, for example, in order to ascertain their impressions of courses. Here, an exam situation is used to obtain numerous individual judgments in a relatively research-economical way. For this purpose, a questionnaire with identical content is filled out by several persons at the same time and usually also under supervision.

Group discussions offer another strategy. Here, aspects such as the norms and values of a real cooperating social group or the dynamics within such a social group are addressed. As a rule, such a discussion takes place face-to-face. However, this is not exclusively the case. In Delphi surveys, for example, questionnaires are sent to a group of experts by post or via the Internet (cf. Sect. 7.3). The survey can therefore also be conducted in writing or on a PC. Finally, bulletin boards are a form of group survey that is conducted via the Internet (cf. Sect. 6.1.3).

Surveys with Varying Degrees of Structure
In order to differentiate between surveys according to their degree of structure, a continuum is to be assumed. At the poles, on the one hand, there is the fully standardised survey and, on the other, the completely open survey. In a fully standardised interview, not only is the entire procedure of the interviewer pre-structured, but the target person is also given all the answer options available to him or her. In a completely open-ended interview, the interviewer would simply present an initial stimulus and allow the target person maximum latitude in their response. In practice, however, these extreme forms rarely occur. Mostly mixed forms are used. In the meantime, it has become common practice to distinguish between three forms: firstly, less structured surveys, secondly, partially structured surveys and thirdly, standardised surveys.

Less structured interviews can be done face-to-face. Without using a pre-structured questionnaire, the interviewer asks his questions. The respondent's answers then determine the rest of the interview. In written ones, only an essay topic would be given and the interviewee would be completely free to decide how to respond. Concrete forms are the focused interview, the in-depth interview and the narrative interview.

In semi-structured interviews, the interviewee is asked pre-formulated questions in a guide. Finally, standardised interviews are characterised by the use of a questionnaire which strongly pre-structures and guides the entire interview situation. The interviewer, for whose behaviour certain rules are also fixed in the questionnaire, transmits the verbatim pre-formulated questions and thus collects comparable data from numerous target persons. Whereby a complete elimination of the interviewer's influence is not possible even with standardized surveys.

In an overview (compare Table 6.2),[4] the three main survey techniques will be compared on the basis of various criteria.

Table 6.2 Comparison of the three procedures from a technical-methodological point of view, modified representation according to Porst (2000, p. 17)

	By phone	In person-verbal	Written
Costs	Medium	High to very high	Low to medium
Sample size	Large	Medium	Very large
Sequence control	Very high (CATI)	Medium	None
Data accuracy, robustness against errors	High	Medium	Medium to small
Interviewer influence	Medium	High	None
Anonymity	Medium to high	Medium to low	High
Sample	Partially limited	Without restriction	Limited, address pool necessary
Questionnaire complexity	Low	High	Low
Exploitation	Difficult to determine	Medium	Medium to low
Length of the interview	Medium to long	Long	Short to medium

---

[4] The abbreviation CATI stands for Computer Assisted Telephone Interviewing. These are interviews conducted over the telephone, where the communication is controlled with the help of a computer.

6.1.2   Theories of Interviewing

For a long time, the elaboration of questionnaire questions and thus also the design of the entire questionnaire was regarded as an art (cf. Payne 1951). This was an expression of the fact that the formulation and composition of questionnaire questions is not easy to learn or to convey, but is strongly intuitive. This already indicates problems with regard to systematizability, formalizability and with regard to the justification of concrete rules for the development of questionnaires. Moreover, the latter are apparently not easy to put into words. At best, recommendations at the level of rules of thumb can be given. Theories play a subordinate role in this context. For some time now, however, mainly as a result of a fruitful collaboration between social science questionnaire designers and cognitive psychologists, the art of questionnaire design has been measured in more detail in numerous empirical experiments. Especially with the help of splits, in which different variants of a question are systematically tested, the answering process becomes (more) comprehensible. When designing a questionnaire, it is thus possible to draw on corresponding theoretical approaches. They make it possible to explain the occurrence of errors and thus to make the optimal design of a questionnaire intersubjectively comprehensible. The art of designing a questionnaire for a social science survey becomes – step by step – a structurable, comprehensible and thus learnable activity.

Among the theories that focus on behavior in the survey environment, it is important to distinguish between two issues. Firstly, theories attempt to explain the participation or non-participation of people in a survey. Secondly, these theories are devoted to explaining the way in which responses to individual questions are made. In a first step, theoretical approaches are presented that explain the willingness to participate in social science surveys. In the following, attention is then given to theories that explain errors that occur when answering individual questions.

Participation in Social Science Surveys as a Rational Choice

Esser (1986b) explains participation in surveys with a rational decision model. According to this model, respondents choose from the alternative courses of action participation or non-participation the one from which they expect a higher benefit (cf. Table 6.3).

To this end, they calculate the sum of the individually weighted incentives for participation and reduce it by the costs of participation, which in turn consist of the transaction and opportunity costs (cf. Esser 1986b, p. 41). For many target persons, this calculation yields only very small differences between the benefit and cost factors. They are thus often undecided when it comes to a participation decision. A decision is therefore ultimately often made, this approach concludes, on the basis of peripheral or random factors.

Consistent refusers can be people who cannot see any benefit from surveys. This group of people would then be systematically left out of surveys. If one follows this thesis, it cannot be ruled out that special efforts to increase the response rate can only motivate the undecided persons, but fail to reach convinced non-participants and thus ultimately

Table 6.3   Should I be interviewed?

Possible costs	Possible benefit
Time expenditure	Hope for a pleasant conversation
Invasion of privacy fear	Sense of accomplishment
Xenophobia	Desire for variety
Fear of misuse of information	Opportunity for positive self-expression
Fear of embarrassment and ignorance, avoidance of an examination situation	Fees such as lottery tickets, telephone cards, vouchers, samples, money and others
Fear of crime	Providing a basis for decisions
Lost due to non-participation	…
Benefit	
…	

Table 6.4   Agreement with various views on surveys, choice of "strongly agree" and "tend to agree" as percentages; the nonresponse rate was 35

Polls …		
–	Are useful and important for politics and society	86
–	Are generally important and useful	89
–	Are serious and are carried out responsibly	74
–	Guarantee anonymity and privacy	77
–	Are fun and bring variety	70

generate artifacts. Schnell (cf. 1997, p. 159 ff.) also formulates a theory of participation in surveys based on rational decisions.

A study on the opinions about surveys in the Mannheim area has shown (compare Table 6.4) that social science surveys are not viewed negatively by the participants in this survey.

Participation as a Heuristic Decision

Groves, Cialdini and Couper (cf. 1992, p. 480) also proceed from the basic assumption that the general population is generally not particularly interested in survey participation. Because of this indifferent attitude, potential targets invest little time and little cognitive effort in the corresponding decision-making process. Instead, they use heuristics to choose between alternative courses of action in a way that is efficient – convenient – for them. Heuristics are pragmatic decision rules that (nevertheless) lead to a sensible result in the majority of cases with little cognitive effort.

Groves et al. (1992, p. 489) caveat that their basic assumption cannot be sustained for all types of surveys. For example, in an intra-organizational survey or even in a graduate survey, individuals may be interested in the results and instrumentalize the data collection for opportune goal attainment. In this case, processes other than those under consideration play a role in deciding on participation behaviour (cf. also Bosnjak 2002, p. 60).

Participation as a Result of Social Exchange

Social exchange theory assumes that individuals base their actions on the benefits or rewards they receive from others as a result of their actions (cf. Gouldner 1960). This theory is thus also based on a decision-theoretical model in which the two components of costs and rewards are contrasted. However, there is an additional factor in the decision to participate in a survey: trust. This includes the probability that the promised rewards (A particularly interesting study awaits you here.) will actually outweigh the costs (compare Dillman 2000, p. 14).

Social exchange differs from a purely rational, economic decision-making model. While an economically oriented decision-maker chooses the alternative action that maximizes his individual utility, individuals in social exchange try to balance the individual utility.

Thoma and Zimmermann (cf. 1996, p. 141 ff.) were able to demonstrate this in an experiment on the reciprocity norm. They asked graduates of a vocational academy to participate in a postal survey. To this end, they first sent out a fairly comprehensive survey instrument and observed the response. Following the total design method, this was then followed up by sending out various reminders and follow-up actions. For the second follow-up action, Dillman proposes to enclose a questionnaire with the letter again. In this way, some of the people contacted received the original questionnaire again, while others were given an instrument that had been shortened by about 60%. This was clearly marked by means of printing ("short version").

"Surprisingly, the majority of BA students did not resort to the shortened version; instead, almost 60% opted for the original version of the questionnaire and returned this questionnaire completed. The BA students thus reacted much more positively than expected to the researcher's concession to only ask for a questionnaire that was significantly shortened in scope." (Thoma and Zimmermann 1996, p. 152).

Apparently, the people in the experimental group (for experimental designs, see Sect. 7.1) saw the completion of the questionnaire as an opportunity to repay the attention they had previously received by shortening the questionnaire and to rebalance the resulting imbalance of benefits – in this case to their disadvantage.

The theory of social exchange can also be used in survey research to explain the empirical findings on the effect of incentives as a participation-increasing measure (compare Goyder 1987, p. 163 ff.). For example, Diekmann and Jann (2001, p. 25) found that only promised rewards have a negative effect on the willingness to participate. Incentives given in advance with the questionnaire, on the other hand, lead to an improvement in willingness to participate (compare also Church 1993). The positive advance performance by the researcher creates a reciprocal obligation that induces the target persons to increasingly comply with the researcher's request to participate in the survey (compare also Bosnjak 2001, p. 90). If one were to assume that people decide purely rationally, different findings would have been expected here.

Social exchange theory is a universal theory for explaining human behavior, and survey researchers use it in special cases to study participation decisions in research contacts.

Action-Theoretical Approaches

In psychological attitude research, multivariate action theory models have become established with the Theory of Reasoned Action (ToRA) and the Theory of Planned Behaviour (ToPB). Across several levels, Ajzen and Fishbein (1980) model human behavior as a causal consequence of intentions, which in turn can be explained by behavioral attitudes and subjective norms. Figure 6.1 illustrates the postulated relationships with which the ToPB explains intentions to act.

The ToPB has been formulated in general terms and empirically tested several times in different areas (cf. Ajzen 1991, p. 187). It has thereby proven its explanatory power. With regard to survey research, Bosnjak (2002, p. 69 f.) cites two studies by Lockhart (1986) and Hox, de Leeuw and Vorst (1996), respectively, which also successfully predict willingness to participate in surveys.

In summary, it can be stated that a whole series of facets are important for participation in a survey. Groves et al. (1992, p. 477), like other authors (for example Porst and v. Briel 1995, p. 10), mention the following:

- Cultural-social factors, such as general attitudes towards surveys and the degree of over-surveying (this refers to the excessive frequency with which people are asked to participate in a survey),
- the characteristics of the instrument,
- the individual characteristics of the target person,
- the characteristics of the interviewer and
- the characteristics of the interaction between the interviewer and the interviewee.

In the following, theories will now be discussed that structure and explain the process of answering questions within a survey. First, the classical assumption according to Holm (1974) is presented.

The Classical Theory of the Interview

This approach assumes that four different influences play a role in answering a question. According to this approach, the answer in a survey is composed of:

- a true value that represents the actual target dimension of interest, for example, the income level of a target person,
- a random error, for example, the interviewer may have misheard the answer to the income question or made a mistake when taking notes,
- a systematic error, which can result from the fact that the target person forgets to state their holiday and Christmas bonuses when asked about their income. It is also conceivable that the target person fears that their details could be passed on to the tax office. This is the effect of an external dimension,
- an influence that arises from the effect of social desirability. The dress and/or appearance of the interviewer may cause the target person to round up or down their actual income.

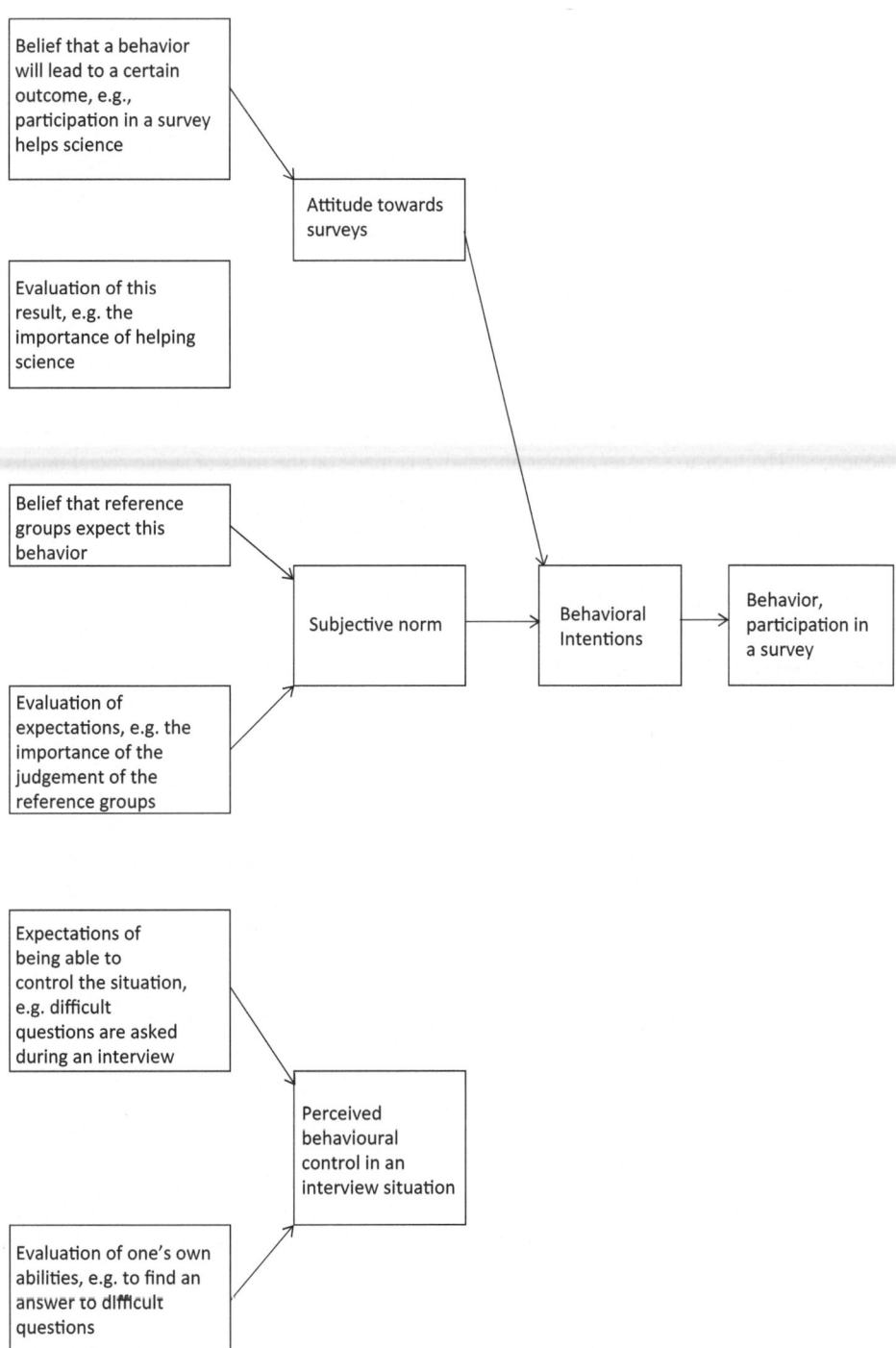

Fig. 6.1 Elements of the Theory of Planned Behaviour (ToPB) according to Ajzen

Any deviation from the true value is interpreted in the classical theory of the interview as an error that can be attributed to various causes, depending on the question and the research design. In numerous studies, survey researchers demonstrate a variety of influencing factors, such as social desirability effects (Silver et al. 1986; Aquilino 1993; Kane and Macaulay 1993), scale effects (Stadtler 1983; Schwarz et al. 1985; Bishop 1987; Böcker 1988), context effects (Cantril 1944, cited in Strack 1994; Schuman and Presser 1981, p. 47, and the review by Schuman 1992), and sequence effects (Payne 1949; Dillman et al. 1995).

Response Selection as Utility Maximization
Esser interprets (compare 1986a, p. 325) each answer choice as a decision situation, which he models analogously to his considerations on the participation decision in a survey (see above). To this end, he describes respondent behavior as a special case of a general theory of situational action whose core argument is a utility trade-off between positive and negative expected consequences. The decision for an answer alternative is made on the basis of stable social norms and personal preferences on the one hand, and situational objectives (these include, for example, the interviewer's dialect, clothing, badges worn, as well as the interviewer's apparent age) and risk perceptions on the other. From this perspective, the true value, namely an "attitude deeply rooted in the personal identity of the interviewee" (Esser 1986a, p. 321), is only a possible point of orientation in a person's reaction to a question stimulus. The question whether there is a true value at all, respectively whether interviewees can lie, must thus be answered with "no". "One can summarize that biases of responses according to social desirability can be explained as a combined result of motives, needs, and evaluations on the one hand [...] and certain expectations about the connection of an answer with certain consequences." (Esser 1986a, p. 6) A "true" attitude results from the comparison of different, possible responses.

Esser assumes a multi-stage decision-making process of the target person in an interview. The approach will be demonstrated in a schematic diagram. First, different goals of a person and different variants of action, which this person has, are distinguished:

$U_1, U_2, \ldots U_n$:	Be different goals of a person to be interviewed, for example, the pursuit of social recognition and the desire to help social science. They are each considered differently important by the respective person
$A_1, A_2, \ldots A_m$:	Be the alternative actions that exist in a situation, for example, to choose different answer options given in a questionnaire
$p_{11} \ldots p_{mn}$:	Represent expectations that a particular action, such as choosing a particular answer (A_m), will lead to a particular goal

The following assumptions are now made:

1. For each alternative action Aj (for each given answer option), the respondent weights the associated probability of achieving a certain goal. He determines the product of Uj and pij.
2. For the choice of the response variant, he finally forms the sum of the products $p_{ij} U_j$. This is also his subjective utility expectation. He chooses the action alternative with the greatest subjective utility expectation (SEU).

This is to be shown with a fictitious example. In a survey, the statement "I like going to Christmas parties" is to be evaluated by the target persons. For this purpose, they are given three variants:

A_1	= Agreement with this statement,
A_2	= indifference to the statement,
A_3	= rejection of this statement

Furthermore, it can be assumed that the target persons pursue different goals with their actions. For the sake of simplicity, it should be assumed that only the following goals are relevant for them (in reality, there will be many more goals):

U_1	= Confession of one's own Christian conviction,
U_2	= gaining social recognition towards the interviewer

Now the subjective probabilities (p_{ij}) are calculated. For this purpose, the target persons ask themselves the question with which probability, given the choice of each of the three answer variants (A_m), the two goals (U n) can be realized (for example, purely fictitiously):

A_1 and U_1	Positive relationship to Christmas celebrations and confession of a Christian conviction	0.80
A_1 and U_2	Positive relationship with Christmas celebrations and gaining social recognition	0.00
A_2 and U_1	Indifferent relationship to Christmas celebrations and confession of a Christian belief	0.40
A_2 and U_2	Indifferent relationship to Christmas celebrations and gaining social recognition	0.20
A_3 and U_1	Negative relationship to Christmas celebrations and confession of a Christian belief	0.00
A_3 and U_2	Negative relationship to Christmas celebrations and gaining social recognition	0.40

To be able to do this, the target must make a typing of the manifest characteristics of the interviewer. For example, she will estimate the clothes, the probable age and so on.

Now, the next step is to consider that the goals are pursued by the interviewee with a different intensity. It is assumed that demonstrating a Christian conviction is more important than gaining social approval from the interviewer. Specifically, be:

U_1	Importance of professing a Christian conviction	1
U_2	Importance of gaining social recognition from the interviewer	5

Now the subjective utility expectation of the action alternatives can be determined. For this purpose, the target person will proceed as shown in Table 6.5:

Finally, in the fictitious example described, answer option A_1 is chosen because it promises the highest subjective benefit (here 8).

Table 6.5 Procedure for determining subjective utility expectations, continuation of the fictitious example

	p_{i1}	p_{i2}	$p_{i1} U_1$	$p_{i2} U_2$	$p_{ij} + U_j$	=	SEU_i
A_1	0.80	0.00	(0.80) 10	(0.00) 5	8 + 0	=	8
A_2	0.40	0.20	(0.40) 10	(0.20) 5	4 + 1	=	5
A_3	0.00	0.40	(0.00) 10	(0.40) 5	0 + 2	=	2

According to Esser (1986a, p. 13)

Critics consider the proposed model to be insufficiently plausible, since respondents would have to complete an extensive utility calculation – presumably subconsciously and in only a few seconds – for each answer decision. This processing would involve a high cognitive effort and contradicts findings from psychology. According to these findings, people use simplified decision rules such as procedural schemata (Hastie 1981, p. 41) and heuristics precisely in decision situations that have no consequences for them, such as questionnaire processing (compare, for example, Bless and Schwarz 1999, p. 423).

The Frame Selection Model

The Frame-Selction Model is also a general sociological explanatory model for action decisions. It is based in particular on the work of Esser, Ajzen and Fishbein as well as Fazio. It was eventually developed further by Kroneberg (2005, 2011). His basic ideas are the following: In a decision situation – such as during an interview – the respective person first makes an interpretation of the situation. Thus, a first reduction of complexity takes place. This interpretation of the situation can be either automatic, spontaneous, or reflexive, calculating. In the first case we speak of the as- and in the second of the rc-mode. This interpretation then provides the basis for the subsequent action selection. Depending on whether the as- or the rc-mode is chosen, the decision is made for a certain action program or for a certain script. In the case of a decision for the rc-mode, there is conscious thinking about the action to be chosen, similar to what has already been described above in the SEU model. If, on the other hand, the as-mode is chosen, the decision to act is then based more on routines and emotions. In addition to other applications, this model has already been used to explain respondent behaviour with regard to response bias due to social desirability (Stocké 2004).

In his study and building on the frame selection model, Bär pursues the question "whether different situation definitions can be found as a *frame* for the opinions ultimately formulated in the interview situation. Based on the distinction between the two selection modes of *automatic-spontaneous* and *reflexive-calculative* imprinting, this raises the question of which influencing variables are at work in the corresponding mode and how they relate to each other" (2013, p. 8). For this purpose, Bär looks at the formation of opinion on security and defence policy as determined in a survey.

If this opinion-forming process takes place on the basis of a value-related frame, this leads via the as-mode of decision-making to a corresponding value-conforming judgment. The prerequisite for this is that the target person has sufficiently internalised a suitable value conviction and that this conviction does not have any gaps for the topic to be evaluated. If these preconditions do not apply, the interviewee begins a search for information in order to evaluate the topic and thus arrives at a decision in a reflexive-calculative mode.

Empirically, it has been shown that "a strong internalization of value beliefs such as discipline or sense of community that correspond to elements of military culture leads to a more positive opinion towards the armed forces, while values that run counter to this, such as personal autonomy, lead to a more negative opinion" (Bär 2013, p. 99).

The Cognitive Psychological Model

Survey researchers, together with cognitive psychologists, established a phase model with the help of which the answering process of a questionnaire question is also to be explained theoretically. In contrast to the behaviorist tradition in survey research, cognitive models focus on the psychological processes that take place between the presentation of the question stimulus and the delivery of a response (compare Sirken and Schechter 1999, p. 2). These processes are understood in the cognitive psychological paradigm as information processing processes. Information retrieval from memory plays a central role in these considerations (cf. Strack 1994, p. 44). Figure 6.2 illustrates the four characteristic sub-processes when answering questions within a survey study.[5]

First of all, the respondents have to understand the questions presented to them and to acquire the meaning of the question stimulus (comprehension). Then the information relevant to the answer must be retrieved from memory. This forms the basis for forming a judgement, the result of which is formulated in a final step and communicated on the basis of any answer variants that may be available (response).

Such a sequence makes it possible to draw on numerous findings from cognitive psychology. The alignment of various pretest strategies (cf. Chap. 8) is also based on such a cognitive psychological model. In the following, the facets of the model are illustrated in more detail.

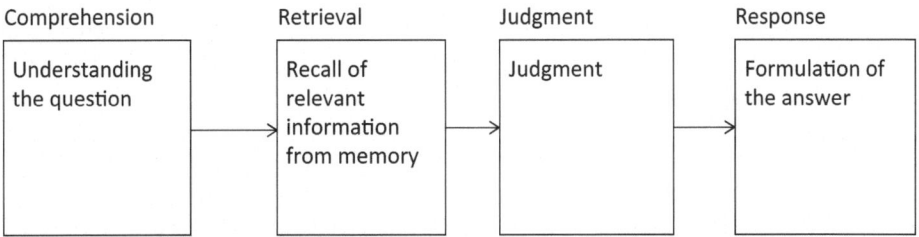

Fig. 6.2 Cognitive psychological processes when answering questionnaire questions

---

[5] In their account of this model, various authors (compare Strack 1994, p. 53; Sudman et al. 1996, p. 56, and Tourangeau et al. 2000, p. 8) agree in content, but sometimes use different labels. A comparative presentation of six other cognitive psychological process models can be found in Jobe and Herrmann (1996).

Question Comprehension

Understanding the question is an essential prerequisite for a meaningful answer. Survey researchers therefore traditionally urge the unambiguous and comprehensible formulation of questionnaire questions (compare Payne 1949; Noelle-Neumann and Petersen 1998, p. 101), although the realisation of both maxims sometimes contradicts each other: the formulation of unambiguous questions may require extensive and more complicated explanations, which in turn then leads to less comprehensible formulations (Fowler 2001, p. 56).

The respondents draw on all verbal and non-verbal stimuli available to them for the interpretation of the question content (compare Strack 1994, p. 70 ff.). In addition to the answer specifications (compare Dillman 2000, p. 33) and the scale labels (compare Schwarz et al. 1998), the preceding questions can also be a source of such interpretive aids. There is talk of context effects. Context effects always occur when the respective question is interpreted by additional information from the environment of this question (compare Strack 1994, p. 57; Sudman et al. 1996, p. 59).

A study by the Allensbach Institute for Public Opinion Research can be used to demonstrate how the response models presented to the test subjects affect their response behavior. The authors first asked the question: How successful have you been in your life so far? Please say it according to this ladder here! They used two different versions (see Fig. 2.1) and obtained the widely differing results already reported in Sect. 2.1.

Both versions lead to different findings. The reason lies in the way the scales are labeled, which the subjects used to interpret the question. The label 0 is obviously associated with a certain content, such as unsuccessful, and the negative scale labels could be understood as failures in life. In the corresponding alternative answer list, these cues to question comprehension were missing and accordingly resulted in different marginal distributions.

Information Procurement (Retrieval)

In order to retrieve information from memory that is relevant for answering a question, respondents form a retrieval strategy. This strategy causes further information to be remembered at the same time, namely that which is linked to the information sought. Tourangeau et al. (2000, p. 9 ff.) distinguish retrieval processes for questions about facts such as times, durations, frequencies, as well as for those questions that are directed at the opinions of the target persons.

The critical factor for the accuracy of the information is the availability of memory content. Such content then determines the extent of the information base available for assessment. The information acquisition process is also susceptible to context effects: Social psychologists speak of priming (Higgins et al. 1977) when certain information or frames of reference are activated in preceding questions and are thus available particularly quickly during a later retrieval. If this causes other information that is actually objectively just as relevant to be neglected and the answer is biased in favor of the additionally activated information, this context produces assimilation effects (Schwarz and Strack 1991,

p. 38). However, according to an inclusion-exclusion model of Sudman et al. (1996, p. 109), under certain conditions, opposite reactions may occur: If priming is very obvious, the available information is deliberately excluded from evaluation. The response is then biased in the other direction by contrast effects (compare Strack 1992, p. 34; Sudman et al. 1996, p. 105).

Schwarz and Bless (1992) present an illustrative example of the occurrence of such a context effect. In an experiment they asked subjects a question about their attitude towards the CDU: "All in all, what do you think of the CDU in general? Please answer using this scale!" (This example has already been seen in Sect. 2.1. in Tables 2.1 and 2.2).

Furthermore, they varied the design of the questionnaire and used three versions: In version A, before the question about the attitude towards the CDU, the question was asked: "Do you happen to know which office Richard von Weizsäcker holds that places him outside the party?" In version B, a preliminary question was omitted. Finally, in version C, subjects were first asked, "Do you happen to know which party Richard von Weizsäcker has belonged to for more than 20 years?" The mean values obtained in these three versions are shown in Table 2.1.

As can be seen there, the three versions produce very different findings. Version B is a result that could not be shaped by context effects. In version A, the CDU scores significantly worse in the ratings. Here, the preliminary question caused the particularly positive image emanating from Richard von Weizsäcker to be subtracted from the assessment of the CDU. In contrast, in variant C the CDU was rated particularly positively. In this case, the preliminary question resulted in the positive image of Richard von Weizsäcker being reinforced and now playing a special role in the assessment of the CDU.

Sudman et al. (1996, p. 73) differentiate between two groups of determinants of information availability. First, they count the described context effects among the temporary determinants; they arise from the sphere of influence of the questionnaire. Secondly, permanent (or chronic) determinants can also be identified. These vary with the characteristics of the respondent, for example with their experience or expertise in a subject area.

Formation of Judgement
Based on the information retrieved, the respondents formulate an initially provisional answer in the phase of forming a judgement. Depending on the individual power of memory, the information sought for this purpose is available to the respondents to a greater or lesser extent and, if necessary, is supplemented by interpolation processes. These processes differ according to the type of information and the question.

The empirically documented instability of answers to opinion questions (Strack 1994, p. 16, also Tourangeau et al. 2000, p. 13) can be explained with a cognitive belief sampling model. According to this model, attitudes and opinions are by no means prepared, for example as "true values", for retrieval by the respondent, but are articulated or formed by him only during reflection on the question (compare Strack and Martin 1988, p. 125 and also Schwarz and Strack 1991, p. 41). Judgment is formed on the basis of a large number of beliefs stored in memory, from which a selection was already made during the information acquisition phase.

An example will show how unstable such judgements can be. In the context of preliminary investigations for a study on television consumption, the probing technique was used (cf. Chap. 8). This involved asking the test subjects specific follow-up questions in order to shed more light on the various aspects of answering a question. Two examples from this study are shown.

First Example

Interviewer:	How long have you watched television in the past seven days?
Target:	Oh, dear, maybe six hours
Interviewer:	What makes you think that now?
Target:	Guessed, estimated – An hour every day. More every day. Every day two hours. Is 14 h. yes 14 h

Second Example

Interviewer:	How long have you watched television in the past seven days?
Target:	Ten hours
Interviewer:	And how did you come up with ten hours?
Target:	Well, the program …
Interviewer:	How do you get the ten hours?
Target:	Yes, so I have about two hours a day – No, it must be 14 hours. Two hours, maybe a little more. Every day I watch about two and a half hours in total

The two examples show how – merely provoked by a short question from the interviewer – the test subjects retrieve information again and form a new judgment based on it. In both examples, this judgment differs significantly from the answer given in the first attempt.

Response

According to the cognitive-psychological model, the response process ends with the submission of an answer. In mapping, respondents project their individual judgments onto the given answer options and check them in a second step (so-called editing) according to consistency and politeness criteria (compare Tourangeau et al. 2000, p. 13). But also effects of social desirability have to be taken into account when giving the answer.

Schwarz and Strack (1991) and Schwarz et al. (1998) consider mapping to be the cause of scale effects: Thus, on an eleven-point frequency rating scale with the verbalized end labels "rarely" and "often," individuals choose significantly higher response levels when these were numbered from zero to ten, rather than from one to eleven. Apparently, here too, absolute zero leads respondents to interpret the end of the scale as "never" (Schwarz et al. 1998). Similar findings are also cited by Dillman (2000, p. 33) and Tourangeau et al. (2000, p. 247).

Further data show that respondents also interpret the graphical arrangement and the size of the answer fields in a written questionnaire as indications of the response distribution expected by the researcher and adapt their response behaviour accordingly (Schwarz et al. 1998, p. 180).

A similar example was presented above (see Sect. 2.1). The duration of television consumption was surveyed. Different scales were also used. As shown in Table 2.3, a significantly higher duration of television viewing is indicated if the centre of the scale is shifted upwards, i.e. in the direction of a longer duration. Here, the scale is obviously interpreted in the sense that it is assumed that the average length of time spent watching television lies (approximately) in the middle of the answers given. This then explains the findings obtained.

Theories of Parallel Processing

Given the complexity of the response process, it seems unrealistic that all respondents actually complete every step of answering every question of a questionnaire precisely as assumed in the model just shown. Empirical evidence, such as the tendency to agree or the clusters of non-substantive responses or the choice of neutral response alternatives, suggests certain shortcuts taken by respondents along the path of the cognitive-psychological response process. Tourangeau et al. (2000, p. 254) note: "There's no reason why respondents should work hard to answer the diffcult questions posed in many surveys. The evidence indicates that many respondents may choose to take it easy instead." The authors conclude that in practice individual components of the cognitive-psychological response model are omitted or simultaneously overlaid by other processes (2000, p. 15).

Other authors transfer the model conception of parallel processing pathways frequently used in social psychology (compare Petty and Cacioppo 1986; Chaiken and Yaacov 1999) to the response process. Specifically, we now discuss Strack and Martin's (1988) considerations for answering opinion questions and Krosnick and Alwin's (1987) satisficing-optimizing concept.

The Concept According to Strack and Martin

Hastie and Park (1986) investigated the representation of opinion judgments and value attitudes in human memory and postulate a model according to which persons form and retain prefabricated opinion judgments on only a few factual issues. Strack and Martin (1988, p. 133) conclude analogously for the answering of opinion questions that the classical notion of true values, which are stored in mental drawers of the target persons as prefabricated judgments waiting to be activated in the context of an interview, is only *one* path of a complex model. In many cases, respondents form their opinion judgments only in response to a question they have just been asked. This is shown in Fig. 6.3.

The shortness of the response time – a typical opinion question is answered in less than five seconds (compare Tourangeau et al. 2000, p. 184) – is hardly sufficient for a comprehensive consideration of complicated issues and thus favours the use of heuristic strategies

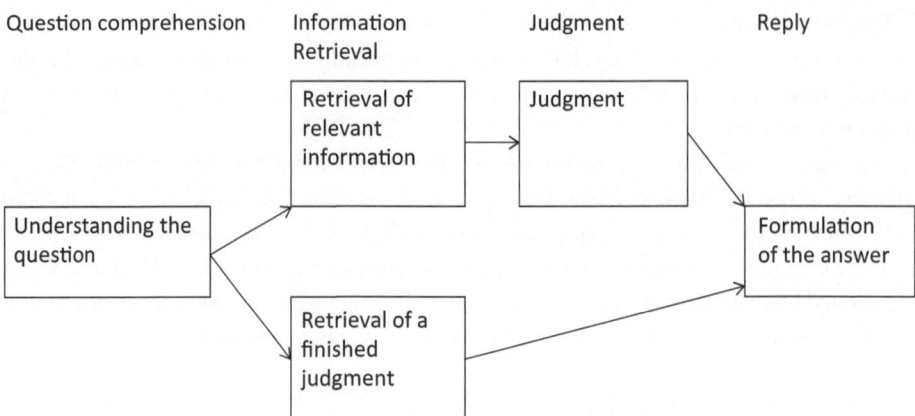

Fig. 6.3 Distinction between retrieval and computation of opinion judgments. (Source: Strack and Martin 1988, p. 125, for an illustration compare Böhme 2003)

to form judgements. Stocké (2002, p. 48) works out empirically that people who have preconceived opinions on a controversial issue report these more quickly in telephone interviews and generate more stable responses than people with less established opinions.

The Concept According to Krosnick and Alwin
The model of parallel processing paths developed by Krosnick and Alwin (1987) can be applied beyond the special case of opinion questions to all phases of the response process. According to the model, less motivated respondents perform the retrieval and processing processes only superficially, or even skip them completely. Such persons then select from the answers given to them an apparently sufficient one, for example a neutral middle category, or choose the white-not option (compare Krosnick 1991, p. 213).

This type of behavior, called satisficer, is contrasted with the optimizer, who completes all subprocesses with exemplary precision and a high degree of cognitive effort (compare Fig. 6.4).

The probability that respondents use the satisficing strategy increases with the difficulty of the task, with low motivation, and with low respondent ability (compare Krosnick 1991, p. 225 and Narayan and Krosnick 1996, p. 72). Optimizers, on the other hand, tend to be people who have a personal interest in the survey in question, expect favorable consequences from their participation, or in whom the personality trait Need for Cognition is strongly pronounced (compare Krosnick 1999, p. 549).

6.1.3 Forms of Questioning and Their Specific Features

After the different possibilities of surveys have been presented, the most important forms will now be examined in more detail. In addition to standardised face-to-face, postal and telephone surveys, we will also look at qualitative approaches to surveys. Finally, the

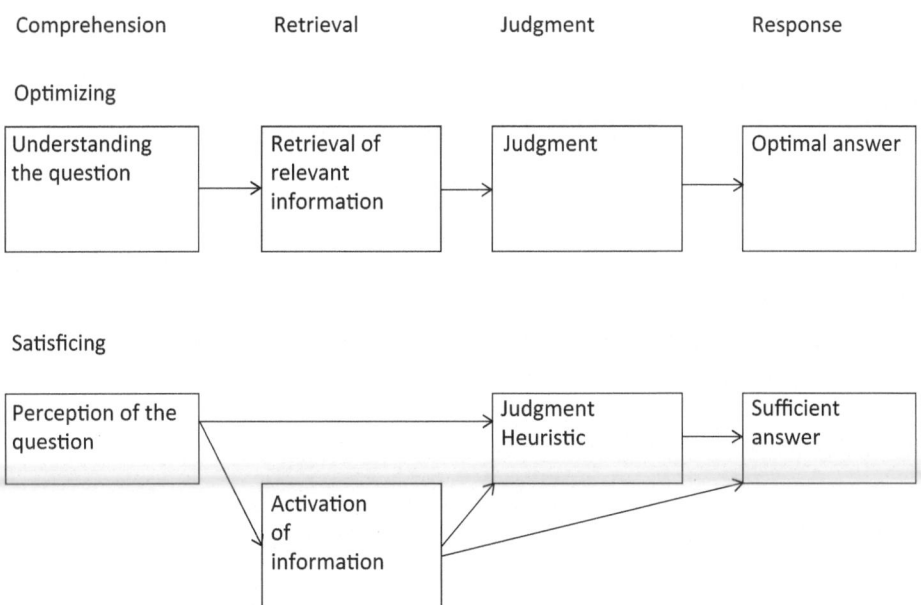

Fig. 6.4 Ideal-typical response strategies with high, respectively low, cognitive input, for illustration compare Böhme (2003, p. 205)

Internet and electronic media open up further possibilities for accessing the target persons. Attention is also given to these variants. Finally, group surveys and the survey of networks are discussed.

The Personal Oral Interview

The face-to-face interview is a communication between an interviewer and a target person designed according to special rules, whereby this form is of particular interest in social science. An essential element of this communication is the questionnaire used for it. The problem arises as to which sources of error occur in face-to-face interviews and what means can be used to counteract them. By way of introduction, the errors to be taken into account in face-to-face interviews will be discussed, after which the design of a questionnaire will be of interest.

Sources of Error in Face-to-Face Interviews

Errors occur when incorrect decisions are made[6] in the preparation of an interview design (compare Groves et al. 2004, p. 41). The sources of error that occur in face-to-face oral interviewing are distinguished according to the originators. These are firstly the respondents, secondly the questionnaire as well as thirdly the interviewers and finally fourthly

---

[6] At this point, selection errors that may occur in the sampling process are ignored, but compare Chap. 5

the situation or the conditions under which the interview takes place. These sources of error will now be discussed. At the same time, an attempt will be made to draw attention to strategies for avoiding the errors. The total survey error model also offers a comprehensive approach to explaining data quality in surveys (compare Faulbaum 2019 and von Hermanni 2019).

Source of Error 1: Respondent Characteristics

The target persons of a survey represent different views, come from different age groups, belong to different occupational groups, have differentiated views with regard to survey research, are each equipped with certain cognitive and communicative abilities – and so on. But all of them are expected to provide information about the same facts. It is therefore obvious that, in addition to the things asked, the personality traits of a respondent also have an influence on their answers.

These influences are manifold. A distinction is made between: First, influences that come about due to the tendency in the respondent to orient his answers to social desirability. Secondly, various so-called response sets when answering the questions. We speak of response sets when the target persons, for certain reasons, give their answers independently of the questioned facts, for example, they agree in principle with the statements presented to them. Thirdly, the existence of a supposed or actual lack of opinion.

Socially Desirable Response

The phenomenon of socially desirable responses was dealt with by Edwards (1957) in the context of personality research. The problem is that the greater the deviation of a questioned opinion from social desirability, the more unpleasant such a question is perceived by the target. Thus, there is an increased cost for them in answering and, consequently, a lower quality of the given answer can be expected. In the case of responses, it is necessary to take into account the tendency for the target person to present himself or herself better, reporting desirable characteristics more frequently than undesirable ones. It also depends on what is considered socially desirable in society according to the individual view of the target person.

If the question is about the duration of television viewing, it is necessary to assess whether the target person believes that there are social norms concerning the duration of television viewing. If a person holds this view, it is necessary to determine what he or she considers to be normal or what duration the target person believes others are likely to consider normal. If it is assumed that the interviewee in his own opinion consumes far too much television, he is in a conflict with the interviewer to report this to him and at the same time to maintain his own reputation. The result can be a reduction of the reported television time and thus a distorted answer.

Another example is provided by the question about voting intention. In the event that the target person prefers an extreme party positioned on the fringe of society, he or she also faces the problem of giving a socially desirable answer instead. Such a phenomenon has been empirically observed by American sociologists in the run-up to an election in

Nicaragua. They conducted the following experiment. A sample was randomly divided into three parts. The elements of all three subsamples were to be asked in a standardized way about their voting intention. For this purpose, the interviewers handed the target subjects a ballot paper on which the parties standing for election were noted. Furthermore, they handed the target persons a ballpoint pen with the request to mark their preferred party. In the first subsample, the ballpoint pen bore no inscription; in the second subsample, it bore the logo of the ruling party; and in the third subsample, the ballpoint pen inscription consisted of the abbreviation of a politically extreme party. The results were surprising. The subsample that was handed pens that had the abbreviation of the politically extreme party on them came closest to the later, actual results of the election. Their results were also superior to the design that used neutral pens.

This apparent paradox can be explained with the help of the bias of the answers in terms of social desirability. Apparently, there was a consensus in Nicaraguan society not to vote for extreme parties. Nevertheless, that extreme party had gained a certain number of supporters. The latter only gave a correct answer in election interviews in those cases in which the interviewer signaled that he also sympathized with the extreme party by handing over an appropriately labeled ballpoint pen. In the other cases, there was a distortion of the election prediction due to the phenomenon of social desirability.

To counteract such biases, social research has several strategies at its disposal:

- When formulating questions, neutral terms should be used that are associated with social norms as little as possible. For example, it is not recommended to use terms such as theft or tax evasion. Instead, it is advisable to ask the target person whether, for example, they have ever taken something from a shop without paying for it or whether someone has not told the truth in their tax return. Unfortunately, it is not always possible to tell whether a term has such a negative connotation (possibly only among certain social groups). Nor does the use of neutral terms completely prevent socially desirable response behaviour.

- It may be appropriate to use suggestively formulated questions in order to de-dramatize deviations from a social norm and to trivialize them in this way. Such formulations could be: "How old were you when you took hashish for the first time?" Or, "Everyone has ridden public transportation without paying. How many times have you done this?" Or, "In every partnership, there are arguments. How often do you argue?" Or, "Many people take things in stores without paying for them. How often have you done this?" The targets person are caught off guard in this way. However, such a strategy is also not without its problems. Under certain circumstances, it can greatly annoy the target person and thus cause him or her to abandon the interview.

- It is possible to increase the anonymity of the survey situation by suitable means. If this is successful, it can be expected that the answers given will be less biased towards social desirability. Two examples from the ALLBUS will be used to demonstrate this strategy. *First, the* use of a mix of letters makes it possible to promote the willingness to answer the income question. The ALLBUS first asks directly about the amount of income: "What is the total net monthly income of your household? Here, I mean the sum that remains after deducting taxes and social security contributions."

Table 6.6 Template sheet with income groups from ALLBUS

B	At	200 €
T	200 € until under	300 €
P	300 € until under	400 €
F	400 € to under	500 €
E	500 € until under	650 €
H	650 € to under	750 €
L	750 € until under	875 €
N	875 € until under	1000 €
R	1000 € until under	1125 €
M	1125 € until under	1250 €
S	1250 € up to under	1375 €
K	1375 € up to under	1500 €
Z	1500 € until under	1750 €
C	1750 € up to under	2000 €
G	2000 € until under	2250 €
Y	2250 € until under	2500 €
J	2500 € to under	2750 €
V	2750 € until under	3000 €
Q	3000 € to under	4000 €
A	4000 € to under	5000 €
D	5000 € to under	7500 €
W	7500 € and more	

In the event that a target person is not prepared to state his or her income in response to this question, the interviewer once again refers to anonymity and hands over a list on which income groups are recorded. What is unique about this list is that the income groups are labeled with letters and that the order of the letters appears to be random. The list with the income groups has the following appearance, shown in Table 6.6.

This means that a target person is no longer faced with the hurdle of having to state a specific sum; all that is now required is the mention of a seemingly innocuous letter code.

Table 6.7 shows that this strategy was used successfully with a whole series of target persons. After all, 15% of the respondents were persuaded in this way to give an answer to the question on income after all, even though they had initially refused to do so.

A *second* possibility to signal a higher degree of anonymity to the target persons is the use of envelopes within a face-to-face interview. The target persons are handed a separate sheet by the interviewers on which the sensitive question is printed. The instructions then stipulate that the interviewer should influence the target person as little as possible when filling it in, for example by leaning back. The target person then puts the completed sheet into an envelope and returns it sealed to the interviewer. There is also an example for the

Table 6.7 Response behaviour to the income question from the ALLBUS 2000, n = 3234, own calculations

	n	Percent
Open net income query answered	1746	54
No income	287	9
Open net income question denied	1195	37
No answer (the corresponding place in the questionnaire is empty)	46	0.2
Total		100
List query answered	497	15
List query denied	698	22
Total		37

use of this strategy from the ALLBUS. The corresponding text is shown in Fig. 6.5.[7] Here, the strategy described above was also used and the most neutral terminology possible was resorted to.

- The randomized response technique (RRT) can be used to minimize socially desirable response behavior. The RRT procedure involves several steps. For example, when asked a question about heroin use, the first step is to ask the target to draw a ball face-down. For example, in an urn there are two white balls and eight black balls. It is important that the interviewer does not see the color of the drawn ball. Then the subject is presented with two questions. One question is the sensitive heroin question (For example: Have you ever used heroin? Yes or No?) and the other is a completely harmless and innocuous question. It might read: Is your birthday in May? Yes or No? If the target person has drawn a white ball, he or she is asked to answer one question, and accordingly the other if a black ball has been drawn before. With the help of corresponding procedures (compare Warner 1965; Diekmann 2001, p. 488 ff.; Groves et al. 2004, p. 225; Schneider 1995) it is possible to calculate how many persons now have experience with heroin.

  Reference should be made to the relatively elaborate experimental design that will be required in the implementation of RRT, which is likely to ensure that its use is not all that widespread. It is also clear that in this way it will be possible to estimate a certain proportion in the sample. However, it is not feasible to attribute the responses to specific individuals.[8]

- It is possible to use factor analyses or comparable evaluation techniques (cf. Backhaus et al. 2000, p. 252 ff.) to determine the extent to which socially desirable answers were

---

[7] Source: https://www.gesis.org/allbus/inhalte-suche/studienprofile-1980-bis-2016/2000-papi/ last accessed on Feb 18, 2019.

[8] Recent findings from methodological research on dealing with sensitive questions can be found in a thematic issue of mda, methods, data, analyses Journal for Quantitative Methods and survey Methodology, Volume 13, 2019 | 1, edited by Ben Jann, Ivar Krumpal & Felix Wolter Social Desirability Bias in Surveys – Collecting and Analyzing Sensitive Data.

As you know, many citizens commit a minor violation of the law every now and then. Below are four such minor violations of the law. For each of these four behaviors, please mark with a cross how many times you have done this in your life

1. Uses public transport without having a valid ticket

Never .... ☐

1 Time .... ☐

2 to 5 times .... ☐

6 to 10 times .... ☐

11 to 20 times .... ☐

More than 20 times .... ☐

2. Driving a motor vehicle with more than 0.8 per mille of alcohol in the blood

Never .... ☐

1 time .... ☐

2 to 5 times .... ☐

6 to 10 times .... ☐

11 to 20 times .... ☐

More than 20 times .... ☐

3. Taken goods in a department store or shop without paying

Never .... ☐

1 time .... ☐

2 to 5 times .... ☐

6 to 10 times .... ☐

11 to 20 times .... ☐

More than 20 times .... ☐

4. Made false statements in the income tax return or in the annual wage tax adjustment in order to have to pay less tax

Never .... ☐

1 times....................................... ☐

2 to 5 times .... ☐

6 to 10 times .... ☐

11 to 20 times .... ☐

More than 20 times .... ☐

Fig. 6.5 Example of the use of the sealed letter technique in ALLBUS 2000

given for the individual questions. Afterwards, those indicators can be eliminated for which an excessively high proportion of socially desirable responses has been identified.

- Finally, the degree of social desirability of response behavior can be measured using specific scales (for example, Marlowe and Crowne 1960; for an overview, see Krebs 1991). "By affirming the positive and negating the negative traits, each respondent can be assigned a value for the need for social approval or the need to present oneself in a positive light. The summed value of these scales is used to statistically control for the extent of social desirability bias in response to attitudinal questions" (Krebs 1991, p. 3). It is possible to locate target subjects with a particularly strong tendency to give socially desirable responses. For example, if one uses regression models to identify the determinants of a particular response, the degree of socially desirable response behavior can be included in the model of explanatory variables and estimated.

Response Sets

In addition to the tendency towards socially desirable answers, certain personality traits of the respondent cause answers to be given more or less independently of the actual intended content of the question and thus systematically distorted. In this case we speak of response sets. Two tendencies in particular are relevant: *firstly,* the tendency to prefer the middle category or an extreme answer category in standardised questions and, *secondly*, the tendency to always answer questions in the affirmative or with "yes", regardless of the content. Such a tendency is also called akquiescence. In this way, the target persons make the answering process easier for themselves.

Response sets occur primarily in respondents with low ego strength. They also represent a personality trait in rather underprivileged persons that has been learned in everyday life and serves here as an assertion strategy. The following example (compare Table 6.8) illustrates this tendency to answer questions in the affirmative. In an experiment, an (apparently) identical question text was varied in such a way that agreement in the first variant corresponds to disagreement in the second variant. Thus, it would have been expected that both values would be very similar. However, the results show that different findings were obtained.

Antidotes include: The individual items of a question battery should be formulated in such a way that they express the target dimension both positively and negatively. This is referred to as opposite polarity. As an example of such an approach, a scale from the Dresden self-defence survey will be cited. A series of statements (compare the following list) were to be assessed by the target persons. The aim was to determine attitudes to self-defence. The questionnaire contained eleven (A to K) items for this purpose. Some

Table 6.8 Methodical experiment to prove a yes-say tendency

Variant	Result	n
A these days you really don't know who you can count on anymore	61% agreement	46
B nowadays you know who you can count on	10% rejection	48

Source: Carr (1971), quoted from Holm (1974)

expressed a positive polarity or a positive attitude towards self-defence (A, C, D, F, I and J), others a negative or a negative attitude (B, E, G, H, E and K). Now it can be assumed that the approval tendency can be neutralized in this way.

Question Formulations from the Dresden Self-Defence Study to Determine Attitudes Towards Self-Defence

A.	If someone attacks without reason, he deserves no protection in principle. It is his own fault if he comes to harm in the process
B.	Unexpected situations, such as a robbery, completely overwhelm you
C.	There are always ways to fight back, even if you are physically attacked
D.	You should carry irritant gas with you so that you are prepared for an attack and can defend yourself
E.	What happens to you in the event of an attack, you can not influence yourself
F.	In the event of an attack, you should defend yourself so as not to lose reputation with your friends and relatives
G.	For the sake of your family, if you're attacked, you should get to safety ...
H.	Even if you are capable of helping yourself in the event of a robbery, it is better to call the police so as not to put yourself in further danger
I.	When you catch a burglar in the act, it's easier to defend your property yourself than to rely on the police and the insurance company
J.	If everyone capable of doing so would defend themselves in the event of a robbery, there would be less crime
K.	If you fight back during a robbery, you may cause too much damage to your attacker

In standardised surveys, the target persons are given the options available to them for their answer. For example, how satisfied are you with the subject matter mentioned: firstly very satisfied, secondly satisfied, thirdly partly satisfied, fourthly dissatisfied or fifthly very dissatisfied. Another tendency in answering such questions is to prefer the middle category, in the case "partially satisfied". This assumes that the number of response specifications is odd, otherwise there is no middle category. An extensive and as yet undecided discussion has been devoted to the decision as to whether a scale without a middle category should be presented to the target subjects (as a counterstrategy) (compare Porst 2000, p. 53 ff.; Faulbaum et al. 2009).

The reason for not using such a middle category is the pressure on the target persons to make a decision or the more unambiguous answer that can then be expected. At the same time, however, the cognitive effort required by the respondents to answer the question is increased, or the costs for the target person increase. As a result, item non-response increases. Respondents who would have chosen a middle category in the other version now refuse to answer.

Opinionlessness

The actual or supposed lack of opinion on the part of the target person is another aspect that influences the quality of a survey. Either a "don't know" answer is given, although the target person has relevant knowledge for answering the questions, or an answer is given, although there is *no* opinion on the questioned issue (non-attitude). Both types of behaviour lead to a reduction in the quality of the data obtained during the survey.

In order to deal with this phenomenon, it is necessary to be clear about what facts the question is intended to clarify. The situation is relatively simple if the question is a knowledge question. This was the case, for example, in the Dresden self-defence survey with the following question (compare Fig. 6.6).

In the case of a knowledge question, it is absolutely necessary to provide an alternative answer (here: "don't know"). In the above-mentioned survey, this was chosen by 74 people (n = 3463).[9]

The situation is similar if, for example, an attitude question is to be asked in relation to a relatively unusual circumstance. In the case of such a question, a corresponding alternative category should also be provided. Here is an example from the ALLBUS 2000 survey (cf. Fig. 6.7).

The questionnaire developer must be aware of various risks if he proceeds as shown in Fig. 6.7. Firstly, one must reckon with people who think that they are always competent

---

A man gets into an argument with another man in a pub. After the other man has left the pub, the man takes the precaution of waiting an hour before leaving the pub as well. When he then steps outside the door, his opponent lunges at him with the words "I'll kill you". What do you think is permissible under the law?

Is it permissible under the law to fight back 1
or
Is it at best permissible under the law to avoid such an altercation by retreating to the pub,

for example? 2

Don't know 9

Fig. 6.6 Example of a knowledge question from the Dresden self-defence survey

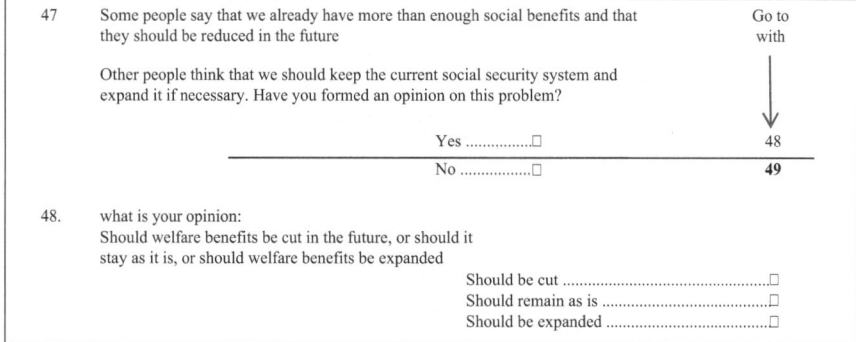

Fig. 6.7 Example of an opinion question with upstream filter question from the ALLBUS

---

[9] After all, 60% of respondents chose the wrong default, stating that they were required by law to avoid such a dispute here.

enough and can provide information on all questions in principle. Secondly, they are confronted with target persons who guess the function of the filter question (47) and deny knowledge (although it is available) in order to shorten the questioning somewhat in this way. Thus, a certain risk cannot be avoided when using such an approach (compare Fig. 6.7).

However, if the questions concern attitudes to general issues, the specification of a special category can be dispensed with. This applies, for example, to the question about one's own life satisfaction. Here it can be expected that there is sufficient understanding of this issue in the general population and that a target person is in a position to form an appropriate judgement. This means that it is possible to dispense with both the preliminary inclusion of a filter question and the explicit presentation of a white-non-answer option. It would even be conceivable to dispense with the specification of a middle category for such a question.

Source of Error 2: The Questionnaire

The entire section is devoted to sources of error that can occur in the context of a face-to-face oral interview. On the respondent's side, besides the tendency to socially desirable answers, response sets and the problem of lack of opinion have been addressed here. Another source of error is hidden behind the questions that – anchored in a questionnaire – are presented to the target persons. The aim must be to ensure that, as far as possible, all questions are understood in the same way by all respondents and answered on the same basis. Here, too, it is obvious to assume that certain question formulations, however, result in a different understanding of the question by the different target persons and thus in insufficiently comparable answers. The order in which the questions are presented to the target persons and the answer categories provided for answering the questions have an influence on the response behaviour.

Question Wording

The wording of a question, the position of the question in the questionnaire, as well as the categories provided for the answer have an influence on the response behaviour. Experiments have shown that even small changes in the formulation of questions lead to significantly different answers. An example from as early as 1940, which should not be missing from any methods textbook, is the following (compare Table 6.9):

Reuband (2003, p. 87 ff.) refers to a whole series of replications of this experiment (compare Schuman and Presser 1981; Hippler and Schwarz 1986; Waterplas et al. 1988;

Table 6.9 Results of an experiment on the consequences of a variation in question wording (compare Rugg 1941)

Variant A	Do you think the US should allow public speech against democracy?
Result	21% are in favour of "allow", 61% are against and the rest are undecided
Variant B	Do you think the US should ban public speech against democracy?
Result	39% are now in favour of not prohibiting (allowing) and 46% are in favour of prohibiting (not allowing)

Glendall and Hoek 1990; Loosveldt 1997; Holleman 2000; Häder 2009), with similar results in each case.

Against the background of these findings, it should be noted how sensitively the response behaviour reacts to apparently only slightly changed formulations in the question. This is particularly important in replication studies. Here, the complete equivalence of the instrument is a prerequisite for correctly interpreting the content of differences in the response distributions that occur between two studies.

In this context, Reuband draws attention to a second practical aspect. This concerns the answer alternatives that are not explicitly formulated in the question texts used. In principle, it is better to ask whether something should be "forbidden or not forbidden" or "permitted or not permitted". He succeeds in showing (2003; p. 93 ff.) that the effects that could be demonstrated in the above-mentioned experiment disappear as soon as the target subjects are presented with both answer variants in the question text.

In summary, the main purpose of this experiment was to show the sensitivity of the responses received in relation to the particular wording of the question.

Answer Specifications

In addition to the wording of a question, the guidelines provided to the target persons for their answers also influence the answering of the question. The answer specifications provide the target persons with a reference system for their answers, i.e. they interpret the answer specifications and draw further conclusions from them for their answer behaviour. An example is the question about the average frequency of television viewing in the last week, for which answer scales of different widths were used (compare Table 2.3 in Sect. 2.1).

This assumes that the target person does not know how long he or she spent in front of the television set during a certain period of time. In this case, answering the corresponding question involves a not inconsiderable cognitive effort on their part. The respondent has to reconstruct the television viewing time from memory. Thus, in order to find the corresponding value, he would have to remember all the days of the last week and process the individual hours into a sum. This would then have to be divided by seven. Further, he would have to consider whether the last few days were typical of his regular viewing behavior or whether there were any special features that would make it unwarranted to extrapolate a corresponding value. A simpler alternative is offered here. In this case, the middle answer specification is of particular importance. Thus, the target person could orientate himself on the presented answer categories. For the respondent, the conclusion is obvious that in the middle of the scale, the questionnaire developer has indicated the duration of television consumption that also applies to the average population. All that remains is for the respondent to make a judgement as to whether he or she (presumably) watches as much, more or less television than the average.

As a conclusion, it can be stated that an open question formulation should rather be used for such queries. This eliminates the influence of the answer specifications on the response behaviour. At the same time, however, the cognitive effort to be made by the target person increases. The question also arises as to alternative methods of determining television consumption. Electronic devices that are installed directly on the television set and lead to better results than a survey are an option here.

Question Order Effect

The answer to a question should, the developer of the question will hope, be oriented towards the subject of the question – i.e. the facts being asked. Within a survey, however, the answer can also be influenced by the fact that previous questions have an effect on the following ones. This so-called halo effect is the radiation of one question to the next(s). The information initially remembered by the target person is also used in the answers to the following question(s). The answer thus results from the position of the questions in the questionnaire. As an example, we can again refer to the question about the attitude towards the CDU. Depending on the previous question, very different results were obtained (compare Tables 2.1 and 2.2 in Sect. 2.1).

As a means of counteracting the halo effect, the question sequences can be varied within the survey. For this purpose, card games can be used in the face-to-face interview. The facts to be assessed are printed on cards. The interviewer shuffles these and presents them to the target persons in a different order. In the case of a telephone interview or a computer-assisted face-to-face interview, a random number generator can be used for this purpose. In this way, it is hoped that different sequence effects will occur in each case, which, however, neutralise each other in total.

Another means is the use of interstitial texts. With appropriate explanations, the respondents are specifically attuned to the questions that follow and it is assumed that this can reduce the halo effect. The insertion of questions on another subject can also be helpful in reducing the halo effect.

It is important to localise such effects with the help of pretests (cf. Chap. 8) and by using splits. Only then can appropriate measures be taken to counteract the radiation of questions.

Source of Error 3: The Interviewer

A face-to-face interview is a social situation involving the target person and an interviewer. Respondents react in different ways to the questions asked, to the interview situation and to the person of the interviewer. The response behaviour, as shown, is not only based on the deeply anchored ideas, knowledge and evaluations of the interviewee regarding the questioned facts, but is also partly a reaction to the instrument of questioning, to the respective situation and last but not least to the interviewer, which is only evoked in the course of the interview (compare Jacob and Eirmbter 2000, p. 53). With regard to the influence of the interviewer, three aspects should be distinguished: *First,* the influence on the interviewee due to manifest characteristics of the interviewer, *second,* the influence on the interviewee due to latent characteristics of the interviewer, and *third,* influences generated due to misconduct of the interviewer, such as falsification.

Manifest Characteristics of the Interviewer

The manifest behaviour of the interviewer, or behaviour that is visible to the target person, influences the course of the interview. The interview style, the extent to which the interviewer follows the instructions given to him, for example, reads out the question texts verbatim, adheres to the given order of the questions, comments verbally on the answers of the target person and/or accompanies them with facial expressions. The visible characteristics of the interviewer also include his or her skin colour, gender, probable age, possible physical disability, clothing, hairstyle, voice and much more.

The influence of the visible characteristics of the interviewer has been of particular interest to social science methodological research, although contradictory research results have been obtained here. Within the studies that have been able to confirm an interviewer effect (compare Biemer and Lyberg 2003, p. 171), however, there is a broad consensus that the external characteristics are particularly effective when they are related to the topic of the questionnaire; thus, the gender of the interviewer is likely to exert an influence on the respondents' answering behaviour above all when questions on gender roles are to be answered in the interview.

Furthermore, it can be assumed (compare Scheuch 1973, p. 99) that the interviewer is only insufficiently able to conceal his own views on the facts inquired about. The interviewee also already registers very subtle signs that signal the interviewer's attitude towards the facts inquired about. The consequence is that the interviewee reacts in the sense of socially desired answers and agrees with the supposed views of the interviewer. In psychological experimental research, these influences have become known as experimenter effects (compare Bortz and Döring 2002, p. 86). It is assumed here that the leader of an investigation also unconsciously communicates his expectations regarding the outcome of an experiment and that the subjects react to this in a hypothesis-confirming manner.

The effectiveness of manifest personal characteristics of the interviewer on the willingness to participate in telephone surveys was investigated by Hüfken and Schäfer (2003, p. 328). Their analysis shows that open, enthusiastic and friendly interviewers who spread a cheerful mood achieve significantly higher participation rates. Even in telephone interviews, such interviewer effects can be effective. Oksenberg et al. (1988, p. 106 f.) proved that a strong variation in voice pitch, loud and fast speaking, and accent-free American English contributed to a lower refusal rate.

A number of other analyses also confirm the importance of demographic interview characteristics for response behavior. For example, Groves and Fultz (1985, p. 47 f.) show that in telephone interviews male and female respondents expressed greater optimism about economic indicators to male interviewers than to female interviewers.

Latent Characteristics of the Interviewer

Non-visible or latent characteristics of the interviewer, which can also influence the course of the interview, are the interviewer experience, the social status of the interviewer, his education and his cognitive abilities and expectations, and finally also the non-visible attitudes towards the questions he asks.

Stember and Hyman (1949, p. 493) were able to show that interviewer attitudes operate in different ways. Thus, those respondents whose interviewer represented the majority opinion with respect to particular party preferences tended to respond in that direction. On the other hand, if the interviewer was a representative of a minority view, the respondents' answers tended towards the know-not category.

Further analyses (compare Reinecke 1991, p. 132) brought out that the opinions of the interviewer become effective in particular when the priority of the question topic is high for the interviewer and low for the respondent, however, and the interviewer has a consistent attitude. Interestingly, Finkel et al. (1991, p. 326) even demonstrated the influence of dark-skinned interviewers on white-skinned respondents in a telephone survey.

Here, too, survey research has techniques at its disposal to counteract the influence (conscious and partly unconscious) of the interviewer on the results of a survey. Thus, it is advisable to commission an experienced, heterogeneously composed interviewer staff with the survey.

If only one interviewer were used for a survey of 1000 people, a particularly strong influence of the interviewer would be expected. This is probably the worst solution. If 1000 interviewers with very different personality traits were assigned to the survey (each conducting only one interview), the influence of the different personality traits of the interviewers on the respondents would probably be only slight. Each interviewer leaves a different impression on the respondents and these numerous different influences eventually neutralize each other. For practical reasons (see also Sect. 5.2.2.), however, even such a model is not feasible. Therefore, a compromise must always be made when composing the interviewer staff.

It will never be possible to completely neutralize the influences emanating from the person of the interviewer. Therefore, the most important characteristics of the interviewer should be added to the respective data sets. Such characteristics are mainly the age, the gender and the educational level. But also the period of time during which an interviewer has already been working for the respective survey institute (as an expression of his or her experience) can provide helpful information in the data analysis. It should be considered whether the interviewer himself first answers the questionnaire that he is to collect later. Then the interviewer's attitudes could also be added to the data set and thus be checked during data analysis. Corresponding attempts have already been made (compare Krausch 2005). This information can then be used to identify the influence of these variables on the dependent variable within the framework of regression models.

Misconduct of the Interviewer

The possibilities of the interviewer to influence the data to be collected are manifold. In addition to the complete or partial falsification of interviews, an influence on the general willingness of the interviewee to participate, but also on the response behaviour of the target persons to individual questions cannot be ruled out.

Three types of forgery are distinguished among the forgeries made by the interviewers. Firstly, partial falsifications are likely to be the most frequent; secondly, total falsifications cannot be ruled out. Thirdly, a violation of the selection rules in the random walk or in the determination of the target person in the household (cf. Sect. 5.2) must also be feared.

Partial falsification is when the interviewer has visited the target household and the target person, but has not conducted the complete interview with them. Thus, it can happen that the interviewers mainly obtain information on the demographic characteristics of the target person and then end the interview relatively quickly. In this way, the interviewer avoids awkward questions and also gains time. The missing information is then subsequently entered by him into the questionnaire, i.e. falsified. Experiments have shown (compare Reuband 1990; Schnell 1991) that experienced interviewers are able to come relatively close to the expected marginal distributions in this way. It is relatively difficult to detect partial falsifications in face-to-face interviews. As will be shown in more detail, computer-assisted interviews (CAPI) can provide a way out of this problem.

In the case of total falsification, the complete interview is filled in by the interviewer without contact with the target person himself. The news magazine DER SPIEGEL (1994, issue 26, Ohrfeige an der Haustür, pp. 41–46.), for example, publicly presented such a practice on the basis of a single example and expressed the fear that sufficient validity of survey results is therefore not given in principle.

Finally, a third variant of falsification consists in not following the given rules correctly when selecting the target household or when identifying the target person in the target household. For example, interviewers on their random walk may not stop at properties that clearly warn of the dog or/and are too remote and located outside the city, although they should have been contacted. It would also be a violation of the rules, for example, if the interviewers do not inquire sufficiently whether, for example, there is not also a private caretaker's apartment in a hospital complex.

Survey research uses various means to prevent falsification. However, these work with varying degrees of effectiveness.

The postcard method of interviewer control is more or less routinely used by survey institutions. This involves contacting[10] a certain proportion of the respondents by means of a reply postcard. This is done by using the addresses from the contact sheets prepared by the interviewers. The questions to the target respondents contained on the postcards may vary. The obvious thing to ask is to provide information about whether an interviewer conducted an interview with the target person, approximately how long it lasted, what topics were addressed during the interview, how the interviewer identified the person to be interviewed in your household, and so on.

Assuming that the persons contacted are willing to answer such a postcard, it is in principle possible to detect total falsifications in this way. It is also possible to identify interviews that have been significantly shortened in time and possibly also to locate partial falsifications. Selection errors made in the determination of the target budget, on the other hand, can hardly be detected in this way.

---

[10] A certain quota can be agreed between the client and the survey institute. This is usually between ten and 20%.

A particular problem with this postcard method is the incomplete return. A non-response to the postcard can therefore in no way be equated with a certain total or partial forgery.

Another way of preventing falsification is better pay and more intensive training of interviewers. The interviewers usually work on a fee basis and are paid by the survey institutes if the interview is successful. As a rule, a particularly long random walk to find a remote target household is not remunerated. The interviewer is also not always paid for a repeated visit to a household that was unsuccessfully contacted in the first attempt – admittedly without knowing beforehand whether the repeat attempt will lead to the target. Extensive personal training of the interviewers, which informs them, among other things, about the importance of adhering to random selection, can also be helpful and motivate the interviewers to work carefully.

If necessary, it is in the interest of the quality of a survey if longer field times are planned for a survey. This gives the interviewers more time to contact the households, even repeatedly, and to arrange longer-term appointments with the target persons. In this way, falsifications made due to time pressure can be prevented.

For some time now, laptops have often been used in surveys (CAPI technique). On these devices the questionnaire is stored and the interviewers can enter the answers of the respondents directly into the PC. In such surveys it becomes possible to carry out a special control of the completion behaviour. Since the interviewer has to read the questions from the screen of his laptop, the clocks built into the computers can be used to identify extreme short interviews, i.e. partial falsifications.

In the context of the use of telephone interviews (CATI interviews), there are also special possibilities to counter all types of falsification. For example, a supervisor can monitor the extent to which the interviewer recruits the target persons in accordance with the rules. For this purpose, the supervisor intervenes in the interview during the contact initiation phase. Furthermore, due to technical controls, total falsifications are almost impossible. Attentive supervisors should not miss partial falsifications.

One strategy for detecting forgeries is to carry out checks on the basis of samples taken from population registration offices. The ALLBUS (cf. Sect. 5.2.1) has provided some interesting experience in this area. Most of the ALLBUS surveys conducted from 1994 onwards were based on samples taken from residents' registration offices, which allow the interviewers to be given the first name and surname as well as the address of the person to be interviewed. This eliminates the random walk and thus all possible errors in this context.

In addition, the gender and age of the target person can also be determined from the civil registers. This offers the opportunity to control the work of the interviewer in a new way. While the name of the target person, which is given to the interviewer, should actually indicate the gender of the target person, the age of the target person is unknown to the interviewer. In case of a possible falsification, such a falsification can be identified by comparing the data from the interview with the data from the population register. Koch (1995) reports on the results of the ALLBUS checks. The results will be briefly reported, not least to prove that a general mistrust in the findings of survey research is not justified.

Table 6.10 Results of the interviewer controls in the ALLBUS 1994, after Koch (1995)

Action	Cases
Total number of cases to be resolved	196
Of these were:	
Not ascertainable	50
Number of cases remaining	146
Of these were:	
Total counterfeits	45
Wrong target person interviewed in the household	51
Interviewer error during data recording	31
Census error	19

Initially, 15 out of about 3500 interviews were already identified as conspicuous due to the postcard method. In the subsequent more intensive control, 12% deviations in the age information and even 0.5% deviations in the gender information were determined on the basis of the information from the registers. These were 196 problem cases in absolute terms, which now had to be investigated with the help of further checks. Clarification of the discrepancies was sought by means of telephone and home visits. Table 6.10 contains the experiences made in this process.

It should be noted that all 15 total falsifications that had already been uncovered using the postcard method could also be detected using this strategy. Of the target persons interviewed incorrectly in the household, there were also cases in which mother and daughter or father and son had the same first name, so that here we cannot speak of interviewer falsification, but only of confusion. The overall result of this study is that 6% of the interviews can be classified as suspicious and only 3% as false. Of course, it must be taken into account that not all falsifications could be determined beyond doubt in the way described.

Source of Error 4: The Situation and Environment in which the Interview Takes Place

Now problems are addressed that can arise due to the situation in which a survey takes place or in the environment of the communication. *Firstly,* the sponsor of a survey plays a role and *secondly, it must be taken into* account that third parties may be present during the survey. In this case, it must be asked what effect this has on the result of the survey. The place where the survey is conducted is also important. It is assumed that this is usually the home of the target person.

Sponsorship

It is to be expected that the result of a survey will determine the facts of interest, for example the voting intention of a certain group of people. In this context, it should not matter on behalf of which party such a survey is organised. However, sponsorship has the effect that the results of a survey not only reflect the questioned facts, but are modified by the client of the survey.

When preparing for the interview, the target persons are also informed about the client of the study. This can be either a government agency, a media institution, a scientific institute or a private company. In the event that the initiator of the survey has a polarising public image, it is to be expected that the target persons will modify their answers. A loyal supporter of a trade union could be influenced in his answers by the fact that he is being interviewed on behalf of the business association. With the help of empirical studies (compare Stocké and Becker 2004), such an effect could be demonstrated.

A study conducted by the Institute of Sociology at the Technical University of Dresden aimed to determine the appearance of market economy surveys and scientific surveys. The research question was whether differences are seen in the general population between the two types of surveys and, if so, what they are. To this end, a randomly selected sample of n = 703 eligible voters and German speakers was surveyed.[11] One half of the study subjects were presented with questions related to *market research surveys*. The other half of the sample was asked (only) about their views on *scientific surveys*. An advantage of this approach is that the subjects did not have to explicitly comment on differences between scientific surveys and market research surveys, but were only asked to evaluate one form at a time. The distinction between the two survey forms is only made during the evaluation. Following the technique of semantic differentials, pairs of properties were presented to each of the target subjects. The task was for them to say which property tended to apply to the respective survey form. The result is shown in Table 6.11.

With the exception of the property pair unpopular versus popular – both market studies and scientific surveys are considered to be equally unpopular – the two survey types differ significantly from each other in all other property pairs. Scientific surveys are attributed above all with a higher degree of seriousness and greater thoroughness.

This small study (n = 860) shows that different views of the surveys of different clients exist in the population. It thus suggests that such differentiation also leads to differences in response behaviour.

Effective countermeasures that neutralise the influence of the client of a survey on respondent behaviour are difficult to find. Research ethics as well as data protection regulations require that the target persons are sufficiently and correctly informed about the handling of the answers they give, so that there is no alternative for an honest statement (compare Sect. 4.9).

Pretests (cf. Chap. 8) should not and must not be dispensed with in order to reveal at least the extent of the sponsorship on the response behaviour of the respondents. Replications of studies offer the possibility of determining the strength of the effect. Finally, when interpreting data from a survey, reference should also be made to the client of the respective study.

---

[11] The survey took place in 2004/05.

Table 6.11 Market research surveys (bottom) and scientific surveys (top), attributed characteristics

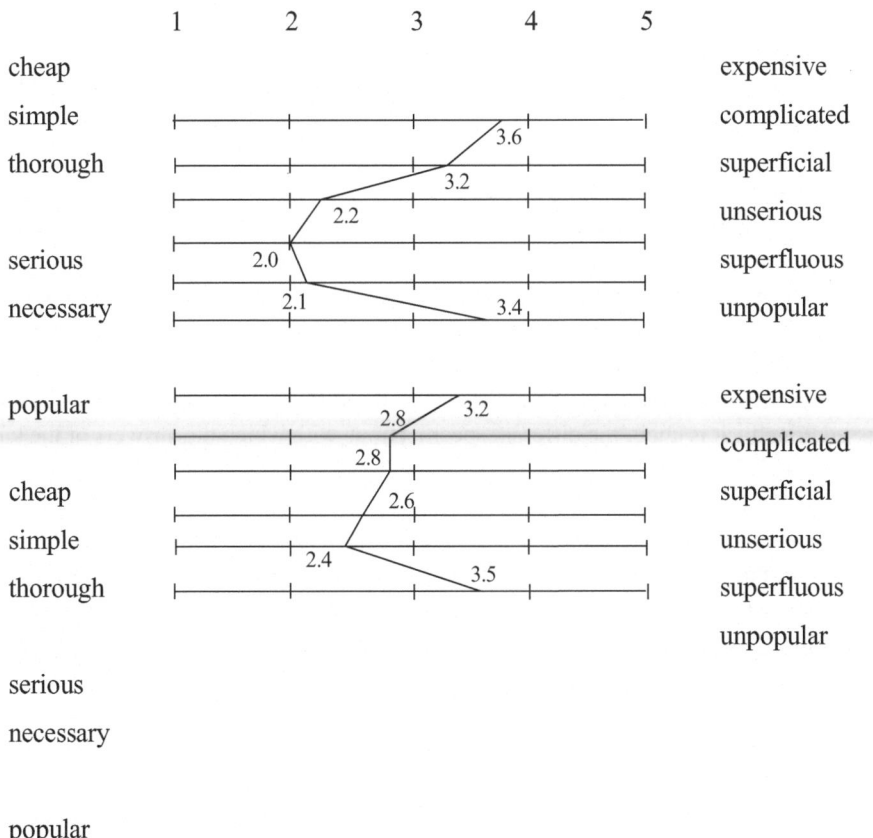

Third-Party Presence

If possible, face-to-face interviews should not be conducted in the presence of third parties. Even if persons present do not directly intervene in the course of the interview, they do bring about a special form of social control of the interviewee's response behaviour (compare Hartmann 1995; Reuband 1992). Although interviewers can be made aware of this fact in training courses and asked not to conduct interviews in the presence of third parties if possible, in practice it will only be possible to a limited extent to avoid family members living in the same household being present during a personal oral interview.

The following applies: If such a situation cannot be avoided, it should at least be documented. Here, too, reference should be made to the ALLBUS. Its data records contain information on whether a third person was present at the interview, who it was and also whether this person intervened in the course of the interview. In the 1998 survey, as many as 28% of interviews had been conducted in the presence of a third party. A very similar figure (26%) is reported by Mohr (compare 1986, p. 67) from the Welfare Survey conducted in 1984. However, these figures make it possible to control for the influence of third parties in the evaluation. At the same time, both figures illustrate the considerable extent of this problem.

A telephone survey can be regarded as a relatively effective countermeasure. Although the presence of third parties cannot be ruled out here either, their influence on response behaviour is likely to be less pronounced (cf. also Häder and Kühne 2009).

The Questionnaire Design for an Interview
A well-designed questionnaire makes the interviewer's job easier, leaves the respondent with a positive impression of the survey research and provides the social scientist with robust information. Questionnaire design will now be discussed in more detail. Among other things, the order of the questions has to be determined and the layout has to be designed when constructing the questionnaire. For these steps there are some hints and rules, which will be presented first. After that, the different question types will be discussed. Finally, it is about the different specifications with which the answers of the target persons can be registered.

All beginnings are difficult, as the saying goes. This may also be true for the developer of a questionnaire who is still sitting in front of a blank sheet of paper. However, for the target person to whom the questionnaire is submitted and who is now supposed to answer the questions, the proverb should not prove true, if possible.

Questions about Warming Up
At the beginning of a questionnaire, a good interview atmosphere should be created with warming-up questions. Suitable warm-up questions are those that are particularly interesting for the target person and at the same time easy to answer. This means that questions about age (too boring) or net household income (too difficult) are out of the question. Also, the questions asked first should already belong to the topic complex at which the survey as a whole is aimed. In this way, the interest and curiosity of the target persons are aroused or satisfied. This means that demographic questions are not suitable as an introduction to a survey. The respondent knows his marital status, but wants to know something about the information expected from him in the interview.

From a psychological point of view, questions for which the answer could be "no" should also be avoided when starting the survey. Such no-answers cause, as already shown, higher cognitive efforts for certain respondents and can thus have a demotivating effect for the entire survey.

Finally, only questions that are relevant to all interviewees should be asked at the beginning of an interview. For example, it does not make sense to start a general population survey with questions about partnership, even if this type of question could be interpreted as being of particular interest to the interviewee. For people living alone, one would provoke does not apply (TNZ) responses. These could also have a demotivating effect at the beginning of the interview and, for example, awaken in the target person the idea that the survey is aimed at an exclusive group of people to which they themselves (unfortunately?) do not belong.

The Dresden self-defence project was presented above (cf. Sect. 4.2). In the context of a telephone survey, the first question here was:

> At the beginning you will be presented with some situations in which people defend themselves with their own means. I would like to ask you to tell me in each case how you evaluate the behavior of the respective person.
>
> A woman is standing in the last free parking space to reserve it for her husband's car. A driver asks her to make room for him, otherwise he will drive towards her. The woman stops because she believes she has a right to the parking space. The driver then pushes her out of the parking space with his car. The woman suffers abrasions as a result. How do you assess the behaviour of the car driver. Do you think his behaviour was justified or not?

The developers of the questionnaire hoped that this introductory question would be easy for the target person to answer, and that it would already clearly lead to the further issues to be answered.

Challenging and Difficult to Answer Questions

Another maxim when compiling the questions of a questionnaire is to place the most important questions in the front part of the questionnaire, i.e. after a certain warm-up phase has been completed. This is based on the idea that a relationship of trust between the interviewer and the target person is built up with the first questions and at the same time the interview is not yet perceived as strenuous by the target person. In this way, one can prevent the interview from already being burdened by fatigue.

The curiosity, interest and attention with which a target person follows the interview are likely to be greatest at the beginning of the interview. A certain tension as to what further course the interview will take can be assumed. At the same time, the degree of fatigue and frustration at this point – it is to be hoped – is not yet particularly strong. The routine and also the fatigue then increase in the course of the interview – not least depending on the topic of the interview and on the constitution of the target person. After a certain time, curiosity is likely to be satisfied and interest in the further course of the interview also decreases.

Further Recommendations

- Sensitive questions that might provoke the target person to break off the interview should be asked at the end of the interview. For example, income questions are usually placed shortly before or directly at the end of an interview.
- The part of the questionnaire that identifies demographic information about the target person should be placed at the end of the interview. Demographic questions are not interesting for the target person. At the same time, answering them should not require cognitive processes that are too sustained, so that they can still be easily answered by the target person even if they are somewhat tired.
- Filter questions should be used in the questionnaire where necessary. These have the function of guiding the target persons over questions that do not apply to them. This

prevents the respondents from being asked superfluous questions and thus losing motivation to participate further too quickly.

- The questionnaire should be designed in such a way that possible question order effects (see above) are avoided. If there is a suspicion that there could be such effects in a questionnaire, split variants can be provided. In this case, the population of respondents is randomly divided. Both halves are then confronted with questionnaires whose succession differs. In this way, data analysis can determine whether such sequence effects have occurred. Another variant is the use of card games. The interviewer shuffles these before each interview. In this way, the target persons are presented with the questions in a different order for evaluation. In this way, systematic radiation is avoided.
- The questionnaire is to be grouped into content-related topic complexes. Similar to everyday communication, in which different topics are discussed successively and then changed, a questionnaire is also to be constructed. It would be wrong to arrange the questions randomly. Transitions to the next set of questions can be designed with transition sentences. Such transition texts are also suitable to reduce the radiating of questions (see above).
- The layout of the questionnaire should be interviewer-friendly. The questions should be arranged in an easily visible way, clear instructions should be given to the interviewer. A multicoloured print may increase the clarity of the questionnaire.
- The questionnaires should not be too long. Even if the target persons are prepared to complete a very long questionnaire, their attention wanes in the course of the questionnaire, depending on the subject and their condition. An empirically based guideline time cannot be given. Experience shows that a personal oral interview should not last much longer than one hour. Excessively long studies unnecessarily strain the patience of respondents and damage the public reputation of survey research.
- The questionnaire is to be subjected to a suitability test with the aid of pretests (cf. Chap. 8).
- Contact logs are to be kept by the interviewers. For control purposes, these mainly contain information on the target household and the target person. Also a characteristic of the respective interviewer, for example his gender, educational level and age should be added to the data set on the answers of the target person for the reasons already mentioned above.

Question Types and Answer Specifications

Some rules should also be presented for the formulation of questions. The English term wording is often used to name this work step.

- Short, comprehensible and sufficiently precise formulations should be used. This can be a contradiction, as comprehensible formulations are usually more detailed and therefore longer. It is therefore necessary to optimise between both requirements. Foreign words should be avoided in population surveys. Certain terms, such as net household income, should be explained to the target persons.

- Ingratiation with the target person as well as the use of dialect should be avoided in order to emphasize the scientific character of the interview.
- Double negatives should also be avoided wherever possible, both in the question text itself and between the question text and the answer specifications. They cause unnecessarily high cognitive costs for the respondent in understanding the question.
- The response categories should be disjunctive, as well as exhaustive and precise. Expressions such as "frequently", "sometimes", "somewhat" and "rarely" are fuzzy terms and thus not very suitable for surveys.
- Strongly value-laden terms should be avoided. These include terms such as freedom, justice, crime, communist, boss and bureaucrat.
- Multidimensionality should also be avoided in the development of answer specifications (see also Sect. 4.3). Questions containing lists of several items to be assessed simultaneously (How satisfied or dissatisfied are you with A, B and C?) are multidimensional and therefore unsuitable.
- Indirect questions should only be used in an interview under certain conditions. Indirect questions are those that ask about the actual issue of interest through the medium of another question. In some cases, however, it may be appropriate to ask indirect questions, for example to avoid socially desirable response behaviour. For example, the question about the number of birthday presents given per year provides an indication of the social isolation or integration of a person. In the event that the correlation between the number of birthday presents and social isolation is theoretically certain, however, a corresponding question can be asked.
- Suggestive questions are not suitable for surveys. Questions can already achieve a suggestive effect if only one answer option is named in the question text. This is shown by the following example contained in Table 6.12.

In a two-page article, the weekly newspaper DIE ZEIT reported on empirical findings on the national state of mind of the Germans (compare Bittner 2002, p. 10 f.). The article is positively distinguished by the fact that one is informed not only about the marginal distributions determined in the survey, but also about the exact question text. The following list shows the corresponding indicators in the foreigners complex:

Question Texts of a Survey (Bittner 2002)
- There are too many foreigners living in Germany.
- The many foreign children in school prevent the German children from receiving a good education.

Table 6.12 Two variants of a question

Variant		Consent
A	Do you think that in a workplace all workers should be in the union?	36%
B	Do you think that in a workplace all workers should be in the union, or should it be up to each individual to decide if they want to be in the union?	14%

Noelle-Neumann (1996)

- Foreigners living in Germany should choose their spouses from among their own countrymen.
- When labour becomes scarce, foreigners living in Germany should be sent back to their home countries.

From the point of view of the methodologist, a high level of agreement must be expected for all questions – due to the yes/no tendency to be expected regardless of the content of the question. This approval tendency is likely to be reinforced by the fact that the texts are not balanced, i.e. they contain both possible answer poles (compare the example shown in Table 6.14) and thus have a clearly suggestive effect. The results of the survey then turned out as expected. They did, however, come as a surprise to those with less methodological experience. The journalists felt compelled to ask the President of the Bundestag for a statement on the empirical findings. (compare: The elbow becomes too important, ZEIT conversation with the Bundestag president about the results of the study in the same 2002 issue).

- It is recommended that the individual items be polarised differently, especially in the case of questions about people's attitudes. The questions should be balanced in order to prevent a tendency towards agreement. As an example, a battery of items on the role of women from the ALLBUS 2000 is shown (compare the next list). Certain items (B, C and D) are aimed at a more traditional image of the role of women, others (A and E) at a more modern one.
- The target persons must not be overwhelmed in the answer search, otherwise they tend to use heuristics and just give a somehow plausible answer. The following could stand as an example of an inappropriate question: How wide and how long is your living room and how many square meters is this room?
- Questions about motives and facts that occurred a long time ago – recall questions – should also not be included in the question programme if possible due to the high cognitive demands on the respondents. As a rule, the use of such indicators only yields opinions or assumptions about one's own motivational situation or about the fact in question.

Statements of Question 4 from the ALLBUS 2000

A	A working mother can find as warm and trusting a relationship with her children as a mother who does not work
B	For a woman it is more important to help her husband with his career than to make a career for herself
C	A toddler will certainly suffer if his mother is employed
D	It is much better for all concerned if the man is fully employed and the woman stays at home and takes care of the household and the children
E	It is even good for a child if his mother is working and does not only focus on the household
F	A married woman should refrain from working if there are only a limited number of jobs available and if her husband is able to support the family

- If necessary, white-not categories and other special answers such as does not apply should be inserted (see above). After all, such answer options motivate respondents to then choose them.
- The problem of whether respondents should be presented with a scale with a midpoint, i.e. with an odd number of response options, or a scale without a midpoint, i.e. with an even number of response options, is difficult to answer and – as indicated – is the subject of controversial debate. It is known, for example, that some target persons interpret the middle response level as neutral or as a white-not category. If this specification is missing, an answer to the corresponding question is then more frequently refused (compare Porst 2000).

The ALLBUS 1998 included the following questions on personal attitudes, shown in Table 6.13.

Various scales were used for the answer. It can now be shown whether the claimed tendency can be proven. Table 6.14 summarises the number of questions per scale type.

Table 6.13 Questions of the ALLBUS 1998, which were provided with answer specifications with different numbers of scale points

Subject	Submitted list, cards	Variables	Scale type
Personal economic situation today	1	1	5
General economic situation in the future	1	1	5
Personal economic situation in the future	2	1	5
General economic situation future	2	1	5
Importance of different areas of life	Card game	8	7
Musical taste	5	5	5
Importance newspaper information	8	1	7
Importance TV for information	8	1	7
Importance of personal conversations to obtain information	8	1	7
Policy influence	Card game	11	7
Political disenchantment	9	6	4
Politically active	10	7	4
Satisfaction with democracy	11	1	6
Satisfaction with the Federal Government	11	1	6
Social differences	12	3	4
Situation old new federal states	15	9	4

Table 6.14 Scales used in the ALLBUS 1998 according to the number of given answers

Response levels	Number of questions
Five	9
Seven	22
Number of questions with odd number of answer levels	31
Four	25
Six	2
Number of questions with even number of answer levels	27

Table 6.15 Mean values of the indices formed

	N	Mean value	Standard deviation	Minimum	Maximum
Straight scales	3234	0.95	0.11	0.11	1
Odd scales	3234	0.99	0.05	0.48	1

Furthermore, all variables were recoded. If the target person answered the respective question, i.e. if he or she was classified in the scale, the value one was assigned. If the question was not answered, the value zero was assigned. Zero was also assigned if White Not or No answer or Refused was given. The following index formation involved the addition of the recoded variables divided by the respective number of questions. The index – arithmetic mean – was formed for all variables with even or odd scale points (compare Table 6.15).

The Wilcoxon test (cf. Brosius 2002, p. 827 f.) now yielded a statistically significant result, according to which the question is not answered more frequently when an even number of categories is specified than in comparison with questions in which an odd number of answer options is provided. This confirms the assumption made above.

- Many of the rules mentioned apply only under certain conditions. Ultimately, the pre-test must decide whether the questionnaire works.

The GESIS Leibniz Institute for the Social Sciences has compiled a compilation of social science items and scales which contains more than 250 instruments in 2019 and is available online.[12] These as well as the book by Faulbaum, Prüfer and Rexroth (2009) can be a helpful support for the construction of own questions.

The Written Survey

In this section, an online search is used as an introduction to show the many possible applications of written surveys. For this purpose, the results of a search in the database SOWIPORT[13] are presented. This database is a cooperative offering of the Bertelsmann Foundation, the German Center for Gerontology, the German Central Institute for Social Issues, the Friedrich Ebert Foundation, GESIS, the Cologne University and City Library, and the Social Science Research Center Berlin gGmbH. Here one can obtain information on current research contexts in which, for example, written surveys are used. Subsequently, the annual reports of the ADM also provide information on the use of written surveys. Afterwards, the different variants of written surveys are presented, the possibilities and limitations of this approach are discussed, and rules and recommendations for the use of this instrument are pointed out.

---

[12] Compare https://zis.gesis.org// last accessed on Feb 25, 2019.

[13] Compare http://www.leibniz-transfer.de/einrichtungen/wirtschafts-und-sozialwissenschaften-raum-wissenschaften/gesis-ao-leibniz-institut-fuer-sozialwissenschaften/sowiport/ last accessed on Feb 25, 2019.

SOWIPORT Research

With the help of an online SOWIPORT search, projects can be identified that have been reported and for which the keyword "postal survey" is included in the project description. Such a search yielded 148 hits in spring 2009, with 20 studies dating from the period since 2006. This shows that postal surveys are currently used to investigate very different topics. This is the result of an overview of those social science research contexts in which postal surveys are used:

1. Postal surveys can be addressed to *private individuals as well* as to *institutions*. An example of a survey of individuals is the survey of 14 to 17 year old young people living in Filderstadt on their leisure time behaviour, what offers they use or miss, what problems young people have and how they assess the work of the Youth Community Council (Freiburg Institute for Applied Social Science). The survey of 581 senior housing facilities with a focus on assisted living concerned the problems of very old people at risk of developing dementia (Gerontological Research Group at the Research Centre for Educational Psychology and Developmental Psychology) It represents an application of postal surveys of institutions.
2. Postal surveys appear to be well suited for *graduate studies*. Examples include a survey of public health graduates on their career prospects and professional situation (TU Berlin) and a survey of graduates on the factors influencing academic success, job search and income of social scientists (University of Erlangen-Nuremberg).
3. In a number of projects, the postal survey also addresses other *populations,* some of which are very *special*. Examples include: a survey of former DFG scholarship holders on science and career, living situation and supervisee (University of Kassel, Scientific Centre for Career and University Research) and a survey of nursing service employees on changes in outpatient care (Institute for Health Analyses and Social Concepts Berlin).
4. The postal survey method is used for surveys of the *general population in spatially delimited areas*. The survey of inhabitants of various cities on municipal crime prevention (University of Heidelberg) as well as a partner survey on the social situation of mixed-denominational married couples in the area of Lower Saxony with a focus on Weser-Ems (University of Oldenburg) are examples of this.

ADM Annual Report

According to the annual reports of the ADM institutes, the share of this form of administration fluctuates considerably. A precise overview was provided by Table 6.1 in Sect. 6.1.1. According to this, at the beginning of the century about 20% of quantitative interviews were conducted by post by the institutes in Germany; thereafter the trend is clearly declining. This is attributed to the substitution of this form of survey by online interviews.[14]

---

[14] Compare https://www.adm-ev.de/wp-content/uploads/2018/08/ADM_Jahresbericht_2017_Web-6. pdf; p. 15, last accessed February 24, 2019.

Written surveys are said by some authors to be making a comeback in the arsenal of methods used in empirical social research (Diekmann 2004, p. 439). Thus, it can at least be stated that postal surveys at least enjoy a constant attention. The reasons for this situation lie in some special possibilities that postal surveys offer. For example, due to the increased fear of crime – especially in large cities – there has been an avoidance of contact with unknown persons (as interviewers are) and thus a decline in the response rate. This applies both to face-to-face interviews and to telephone interviews. The overall increase in costs for such surveys can also be blamed for this trend.

The ADM Annual Report for 2001 draws attention to another aspect: "With a share of 28%, the importance of written interviews continued to increase in 2001. This is probably primarily due to the numerous customer satisfaction surveys, which are often conducted as written interviews." (ADM 2002, p. 10 f.).

According to this source, postal surveys also play a conspicuously important role in the social science projects funded by the German Research Foundation (DFG) (57%), and individual interviews are used in 53% of the research projects (quoted from Klein and Porst 2000, p. 4).

Variations of Written Surveys
The procedure for written surveys is manifold. One should always speak of a written survey when the questionnaire is filled out (in writing) by the respective target person himself. This can be done with a certain variety. Thus, the following procedures are possible:

- The questionnaire is handed over to the target person by an interviewer and either collected again by him or sent by post to the organiser of the survey. The great advantage of such a procedure is the possibility to check the questionnaire for completeness when collecting it, the disadvantage results from the immense effort involved in this strategy. Examples of this approach, which is not used too often, were the censuses of the Federal Statistical Office, the last one in 1987.
- The questionnaire is sent by post to the target persons or target households, who then complete it themselves and return it to the survey institute by post. This procedure is also called mail survey or postal survey. Such a strategy is sometimes wrongly used as a synonym for written surveys in general.
- The questionnaire is completed in a situation similar to an examination, i.e. by a group under supervision. Here the advantage lies mainly in the control of the survey situation. The evaluation of teaching within the framework of a student survey, which takes place in a lecture hall, is a relatively frequently practised example of such an approach. The PISA study was also surveyed in this way with students in classes.
- A combination of different forms of administration is sometimes used in the written survey. The target person can first be visited and questioned by an interviewer. Following the face-to-face interview, the interviewer hands over the written questionnaire and tries to motivate the target person to complete it as well. Such an approach is also referred to as drop-off. As an example of this combination, the ALLBUS and the

surveys within the framework of the International Social Survey Programme (ISSP) can be cited as a self-completer.

- Another type of combination of written and oral questioning provides for individual sections within a PAPI or CAPI study to be completed in writing by the interviewee himself. The hoped-for advantage of this strategy lies in the higher degree of anonymity of such an approach. This type of combination is mainly used for sensitive questions, whereby the part of the survey completed in writing by the target persons is then usually handed over to the interviewer in a sealed envelope.

An essential prerequisite for all the above-mentioned procedures is a high level of motivation on the part of the target persons to participate in the survey. This applies above all to those variants in which such an attitude cannot be generated by an interviewer in a personal conversation, i.e. above all in the case of mail surveys.

Advantages and Disadvantages of Written Surveys
Due to the wealth of variants of written surveys, their advantages and disadvantages cannot be presented in general, but must be considered in a differentiated manner. It is attempted to name the main possibilities and limitations of written surveys. The advantages lie above all in the following points:

- Written surveys are more cost-effective than face-to-face oral surveys (whether conducted with paper and pencil or with a laptop) or telephone surveys. Above all, the required personnel expenditure is lower. Hippler gives a guideline, according to which written surveys incur only a quarter of the costs that become necessary in a PAPI study (1988, p. 244). However, this still does not make postal surveys a cheap affair.
- If one excludes telephone studies and studies conducted with the help of online access panels (cf. also Sect. 5.4), written surveys are a survey variant that can be implemented in a relatively short period of time.
- In the event that the addresses of the target persons are available in the written survey, this procedure is characterised by a sample that, in contrast to the ADM design, for example, is not territorially clumped. It is well known that questionnaires can be easily transported to all parts of the country by post.
- Especially in mail surveys, the lack of interviewer influence can be seen as an advantage. Numerous studies prove that the person of the interviewer has an influence on the answering of the questions. However, the absence of an interviewer can also become a disadvantage of this approach.
- Various authors expect more honest answers from the target persons in written surveys (compare Schnell et al. 2005, p. 359). Since the interview situation is not controlled by an interviewer, the tendency of the target persons to give socially desirable answers is eliminated. However, other authors express doubts as to whether this is actually the case (cf. Schnell 2001) and call for (further) empirical evidence of this tendency.
- Mail surveys in particular offer the target respondents the opportunity to complete the questionnaire at the time that is most convenient for them. At the same time, respondents have, in effect, any amount of time to think through their answers. This advantage

is especially true compared to surveys in which an interviewer personally conveys the questions on site.

- Questionnaires completed independently by the target person are likely to have been answered with a higher degree of anonymity than questionnaires for face-to-face interviews, where this is done by an external interviewer. This also applies to telephone interviews.

The various possibilities of written surveys are confronted with various limitations and problems that put the charm of this approach into perspective. These include:

- One of the main objections to the use, especially of mail surveys, is the hardly predictable level of the non-response rate. The fears tend to be that a too high non-response rate prohibits a generalisation of the findings of written surveys. Diekmann (2001, p. 441) considers 20% but sometimes only 5% response rates typical for mail surveys. Hippler (1988, p. 244) relies on Richter (1970) and places the response rate between 10% and 90%. For the level of item non-response, however, Hippler cites much lower values. According to this, they lie between 1 and 8%. Following other authors, the response rates are typically between 30 and 50% (compare Hopkins and Stanley 1981, cited in Klein and Porst 2000) but can be increased to 60 to 75%. Dillman (1978), on the other hand, claims that, if his design proposal is implemented accurately, at least 60% response can always be expected (compare also Berger 2006, p. 82; Dias de Rada 2005).
- Another major objection to mail surveys is the possible distortion of the sample due to a strong self-recruitment of respondents. At the same time, it remains largely uncontrollable who fills in the questionnaire, under what conditions and at what location. There is no effective control of the data collection situation. It is not known how seriously the survey is taken by the target persons. Especially in comparison to telephone surveys where a relatively high degree of control is possible, the deficits of mail surveys become clear at this point.
- The advantage of written surveys of obtaining more considered as well as less socially desirable answers from the target persons becomes a disadvantage if the organisers of the survey are concerned with obtaining spontaneous answers to the questions posed. Thus, written surveys are largely unsuitable for eliciting knowledge questions. Respondents will usually skim through the questionnaire first. Under certain circumstances, they will then also endeavour to give consistent, i.e. non-contradictory, answers to the various questions in the questionnaire.
- Postal surveys do not offer the target persons the opportunity for (direct) queries about the study or about individual questions from the survey standard. However, this can be countered by providing telephone numbers or e-mail addresses for queries in the cover letter.
- The lack of an interviewer results in another specific feature of written surveys. The target persons cannot be presented with questionnaires that are too complicated, for example with numerous filter guides or with ranking questions, and also with open questions only to a limited extent. There is also the demand that the questionnaire

should be particularly well thought out in this form of administration (compare Scheuch 1967a, p. 167 f.). Whereby also a traditional PAPI questionnaire must be sufficiently comprehensible and must not be made accessible to the respondents only with the help of the explanations of an interviewer. The interviewer as a "traditional crutch of badly constructed questionnaires" (Dillman 1978, p. 119) is finally missing in postal surveys.

If one assumes that the questionnaires should also be self-explanatory in the case of a face-to-face or telephone survey, i.e. that they do not first require interpretation by the interviewer, then this reservation about written surveys becomes relative. What remains, however, is the omission of the interviewer as a motivator for participation in the survey.

- A prerequisite for the organisation of mail surveys is the existence of an address pool. This can be customer files, addresses from local registration office registers or member lists. These can serve as a basis for sampling. For general population surveys a problem arises, since no central civil registers are available for social research in the Federal Republic of Germany. A way out can be offered by listing addresses with the help of a random walk. In this case, however, a further problem arises: the determination of the target person to be interviewed in the household by means of a random procedure (compare Sects. 5.2.1 and 5.2.2).
- Certain segments of the population, such as homeless people, illiterate people and people with severe visual impairments, cannot be reached through written surveys.
- The need for research in order to be able to use mail surveys in nationwide studies is currently still immense. Schnell (2001), in particular, mentions some concrete further questions here that still require clarification.

The Procedure for Postal Surveys: The Total-Design-Method and the Tailored-Design-Method

In the meantime, an immense amount of research has been done on the design of mail surveys and especially on optimizing the response rate to such surveys. Dillman with his Total Design Method (TDM) is particularly worthy of mention. The underlying idea is that all aspects of the survey should be optimally designed, since a higher response rate can be expected primarily due to the interaction of all sub-aspects. In addition, a relationship of trust must be created between the target persons and the organisers of the survey.

With Dillman's work, a certain methodological paradigm shift takes place. Whereas initially it was mainly the incapable and/or unwilling target persons who were held responsible for the problems with the response rate, it is now hoped that a better design by the survey organiser will provide a remedy. Another basic idea lies in the application of the cost-benefit concept to postal surveys. Here, Dillman (1983) assumes that people are most likely to answer a questionnaire if (first) the perceived costs of participation are minimized for them and (second) the rewards are maximized, and if at the same time (third) the respondent can trust that he will actually receive the promised rewards or the expected benefits.

The costs incurred by the target persons include above all the time required as well as suspected difficulties in answering the questions. On the benefits side, for example, the entertainment value of a survey, the belief that they are doing something positive for science, identification with the topic of the survey and/or with the institution conducting it.

In order to build up a relationship of trust between the target person and the researcher, it is a good idea to stick stamps on the return envelope instead of labelling it "Recipient paid". As the survey organiser, you are signalling that you trust the target person and assume that they will use the stamp for this very purpose. Incentives should also be enclosed directly with the initial mailing and not just offered as a reward for participation (cf. Berger 2006).

Dillman (1983, p. 360) claims that if the total design method is applied consistently, response rates can be achieved that are never below 60%. Accordingly, in an evaluation of a total of 48 written surveys, the response rate was as high as 74% on average.

In his further development of the total design approach, Dillman (2000) proposes to apply the tailored design method. According to this, surveys should be conducted in mixed-mode design, following the spirit of the twenty-first century. This approach for surveys is based on the same theoretical model as the Total-Design-Method. The intention is now additionally to compensate for the respective weaknesses of one method by using different methods. This technique, known as mixed-mode, provides for each target person to be interviewed according to his or her individual preferred survey mode. One problem that might arise in this process is mode effects. These are influences that come from the different survey techniques on the results of the survey. Thus, not only is the answer to the questions based on the facts asked, but it is also influenced by the survey technique used, as has been pointed out several times. In other words, an identical question may generate different answers when asked in a postal interview or when asked in a face-to-face interview.

The following aspects should be considered when designing a written survey:

The Cover Letter
It has proven useful to use a cover letter for postal surveys. Aspects that should be included in such a text concern the benefits of the study, the necessity of the target person's participation in this study, the introduction of the institution organising the study, a reference to confidentiality in dealing with the results of the survey and an explanation of the way in which the target person got into the study group.

In many cases, an identification number (ID) is used to check the response or to specifically request the outstanding questionnaires. This ID is usually printed on the questionnaires as a stamping number. As a result, the anonymity of the survey may appear to be reduced from the point of view of the target person. It is therefore necessary that the objectives pursued with the printed numbers are explained in the cover letter.

There are different experiences with an alternative procedure. In this case, separate, non-anonymous reply postcards are sent out together with the questionnaire to identify the respondent. This is combined with the request to return the questionnaire and the reply postcard separately to the survey institute in order to ensure a response control. According to Linsky (1975, p. 85), such an approach serves its purpose. Schnell and colleagues et al.

(2005, p. 358 ff.) contradict the usefulness of this strategy, admittedly without naming their own studies.

Finally, when writing the cover letters, some other aspects should be emphasized. These include a personal (title, first name and surname) salutation of the target person, a handwritten signature of the organiser of the survey in blue ink, a request to make comments and remarks on the subject of the survey or the questionnaire, the use of an official letterhead, the indication of a deadline by which the questionnaire must be completed, the offer to be informed of the results of the survey. Finally, an acknowledgement to the target person for his or her participation in the survey should be considered.

The necessity of specifying the time fund required for answering the questions in the cover letter is discussed. However, this depends on a whole range of factors, such as the number of indicators filtered out and thus not to be answered by the respective respondent and the reading speed of the target person. If too much time is given, potential respondents might refrain from completing the survey. The deliberate indication of too short a period of time should be avoided for reasons of research ethics. Experience has shown that respondents tend to underestimate the time needed to complete an interesting questionnaire anyway (compare Brückner 1985, p. 69; Kühne and Häder 2009, p. 185 ff.). For these reasons, specifying the time required to complete the questionnaire in the cover letter seems to be rather dispensable.

Questionnaire Including Cover Sheet
Some hints can also be given for the design of the questionnaire and the cover sheet as well as the last page of the questionnaire.

The questionnaire should be bound or stapled, i.e. presented to the target persons in the form of a brochure. White paper – not grey environmental protection paper – should be used for this purpose. The format of the booklet should be A_5. The last page should be explicitly reserved for open comments and messages from the target persons to the interviewing institution, i.e. it should be left blank.

The cover sheet of the questionnaire brochure should contain the title of the study, the organising institution including its address and an original logo. The aim is to arouse the interest of the target person.

Also on the second cover page no questions should be printed yet. It should be used to give instructions on how to fill in the questionnaire. An example should be used to explain how to tick the answers, where to put numbers, and so on.

Although less attention needs to be paid to the order of questions in written surveys than in face-to-face interviews – it can be assumed that the target persons will leaf through the questionnaire anyway before they start filling it out – the first question should also be chosen carefully here. It should be easy to answer, related to the topic of the overall survey, and applicable to all target respondents. The respondent's expectations of the entire study are largely based on this first question. The hint to put a throwaway question at the beginning if necessary is also given by some questionnaire developers.

The question section on demographics should be located at the end of the question-naire, as is usual in surveys. Such questions are not interesting for the target persons and they usually do not require any special concentration from them when answering.

It is recommended to use filters sparingly in written surveys. Where filters cannot be avoided, they should be designed in a visually unambiguous manner.

Some open-ended questions should also be used in written surveys. This is not so much recommended for reasons of content – after all, the evaluation of open questions requires immense effort on the part of the researcher and answering them also requires increased costs on the part of the respondent – but rather to allow the target persons to make remarks and comments that they would otherwise address to the interviewers.

The length of the questionnaire should be designed in such a way that it supports the respondents' confidence in the survey. According to various recommendations (compare Hippler 1985, 1988; Heberlein and Baumgartner 1978), twelve pages are probably opti-mal here. If the questionnaire is only one page long, this would make the survey seem too insignificant; if it is 30 pages long, participation in the survey becomes an imposition.

The design of the questionnaire should be as uniform as possible throughout the entire survey. It is recommended to divide the questionnaire from top to bottom and to refrain from jumping from a right to a left column on a page.

Figure 6.8 shows the design of a cover sheet for a written expert survey in an exam situ-ation. In this case, a separate cover letter could be dispensed with. The cover sheet con-tains all the essential information about the study.

Shipping and Follow-Up

A number of recommendations have been drawn up for sending out the questionnaires, compliance with which is important for the success of the survey. Above all, the following principles should be referred to:

In addition to the cover letter and the questionnaire, a stamped and addressed return envelope – preferably with a charity or special stamp – should be enclosed with the mail-ing. This constitutes the confidence-building measure required above. The envelopes should not be provided with address labels in order to distinguish them from advertis-ing mail.

The initial mailing of the instrument should be done in such a way that the question-naire reaches the target persons towards the end of the week. Since questionnaires are mostly filled out over the weekend, this prevents the survey from being forgotten too much. This does not apply to surveys of companies or institutions. Here, the survey docu-ments should rather be available to the target persons at the beginning of the week, so that the response can be planned for the following working days.

To increase the response rate, it is necessary to start various reminder or follow-up campaigns. The first reminder campaign should take place after one week. In the form of a postcard to all those involved, an expression of thanks is made for their cooperation and at the same time an appeal is made to all those still in default to now (finally) complete the questionnaire.

TU Dresden	Center for Surveys,	TU Dresden
Faculty of Law	Methods and Analyses (ZUMA)	Faculty of Philosophy
Chair of Criminal Law,	P.O. Box 12 21 55	Institute of Sociology
Criminal Procedure & Theory of Law		Chair of Methods of
01062 Dresden		Empirical Social Research
Prof. Dr. Knut Amelung	68072 Mannheim	01062 Dresden
Ines Kilian	Sabine Klein	Prof. Dr. Michael Häder

Written survey of lawyers on attitudes of the population to self-defence

As part of the third-party funded project "Attitudes about and potential behavioural intentions in self-defence in the general population of the FRG", funded by the VW Foundation, a telephone survey was carried out this year to ascertain the attitudes of the population towards self-defence. With the help of a random selection, around 3000 people aged 18 and over were interviewed. The survey described situations in which people defended themselves. The situations were mainly based on cases from case law. The respondents were asked to indicate whether they considered the behaviour of the respective person to be justified or not

The aim of **this** survey is now to determine to what extent lawyers are able to estimate the assessments of the population. Therefore, we would like you to give us an estimate of how many respondents chose "justified" and how many chose "not justified" in each of these cases. We realize that this can only be approximate. Please estimate in percentages, assuming for simplicity that there were no "I don't know" responses

At the end of the survey, you will of course have the opportunity to comment on the survey if you wish

The survey is of course anonymous. If you are interested in the results of the survey, please let us know on a separate sheet

Thank you very much for your participation!

Fig. 6.8 Example of a cover sheet for a written expert survey

The second reminder campaign was to be launched three weeks after the initial mailing. This is now only addressed to target persons who have not yet participated in the survey. In addition to a reminder, the mailing should also contain a renewed questionnaire. This serves to minimize the effort on the part of the respondents, since it cannot be ruled out that the questionnaire from the initial mailing has been misplaced in the meantime. A prerequisite for this strategy is that the return of the questionnaires can be checked with the help of the identification number.

The total design method provides for a third reminder seven weeks after the first mailing. Here, it is advisable to request all target persons who are still in default to participate in the survey with particular urgency. If this reminder is also accompanied by a new questionnaire, this also minimizes the costs for the respondent. In the USA the documents are sent by registered mail. In Germany, this procedure does not appear to be unproblematic, as registered mail is mainly used here for official reminders and sometimes has to be collected personally from the post office. This appears to be counterproductive, as it increases the costs of participation again for the target persons. As an alternative, telephone reminders can therefore be used more advantageously.

In his further development of the total design method, Dillman suggests a fifth contact in order to increase the response rate. The extent to which such a recommendation should be followed depends not only on the financial and time resources available, but also on the success of the previous reminder campaigns. There is a rule of thumb which states that each reminder campaign will have approximately the same success among the still out-standing participants as the initial mailing (compare Hippler 1988, p. 245). This would make it relatively easy to calculate the effort required for a fifth contact and the benefit to be expected from it (compare Klein and Porst 2000, p. 15 as well as, with a specific view of older persons, Kunz 2010, p. 127 ff.).

Use of Incentives
Another instrument for increasing the response rate are tangible and intangible incentives. These can be (small) monetary gifts, vouchers, lottery tickets and the like. When giving such incentives, it should be emphasized that it is a (small) recognition for participating in the survey. The impression should be avoided that the aim is to reward the target persons for the effort they put into completing the questionnaire. The enclosure of materials about the research project, for example in the form of newspaper cuttings, has also proved useful (for a detailed description of the literature, see Berger 2006).

It should be noted that the effect of incentives is good if the corresponding incentive is already enclosed with the survey documents when they are first sent out. It is less effective to offer these incentives only if the questionnaire is also completed by the target persons. On the one hand, such a strategy would be at the expense of the anonymity of the survey and, on the other hand, this procedure would contradict the maxim of approaching the target persons with a leap of faith. Incentives cannot replace reminders, for example.

The Door-in-the-Face Technique
The Door-In-The-Face technique is a variant of this approach. The basis for this is the norm of reciprocity, according to which positive behaviour on the part of others is in turn responded to positively (cf. Diekmann and Jann 2001). When implementing this technique in the context of postal surveys, the organiser first makes a maximum demand on the target person. This can be, for example, a (particularly) long questionnaire that is to be com-pleted. Those persons who refuse to fill in such a complicated questionnaire are then con-fronted with more moderate demands. This means that a much shorter and/or easier questionnaire will be sent to them. As a result of this concession on the part of the survey organiser – to be satisfied with a shortened questionnaire – the researcher now also hopes for a reward from the target person, i.e. participation in the survey.

An Experiment
A number of experiments were dedicated to increasing the response rate of postal surveys. One such experiment – representative of numerous others – will be presented.

The aim of this test was to determine the effects of various individual design aspects of a postal survey (compare Thoma and Zimmermann 1996). In terms of its content, the survey was designed to ask graduates of the Baden-Württemberg University of Cooperative Education to evaluate their studies there and to provide information about their own professional development.

Thus, the list of graduates of the University of Cooperative Education was available to the organisers for determining the sample. As has been shown, graduate surveys are a relatively frequent field of application for postal surveys. The easy access to the sample or to the addresses of the persons to be surveyed is probably one reason for this.

Finally, 2840 graduates were selected from this list. The design included different splits in order to assess different elements of the experimental design. First, different cohorts of graduates were asked to participate in the survey. The assumption here was that this would allow different levels of attention to the subject of the survey to be tested. If it is true that the more interesting a survey topic is for the target persons, the higher the participation rate in the survey, then graduates from earlier years, for whom their studies took place some time ago, should participate less in the survey.

Secondly, the accompanying letter varied. Some of the respondents were confronted with a letter that was particularly personal and signed by the head of the institution. The assumption here was that such a covering letter signalled a higher level of authority and contributed to a correspondingly more extensive response.

The third split version concerned the reminder letter. A subset of the respondents received the reminder letter in conjunction with a shorter questionnaire, while another subset was not sent a new questionnaire. It was expected that, due to the norm of reciprocity (see above) with the help of the shortened questionnaire, there would be a positive influence on the response rate of the survey. Some results of this test:

It turned out that the older the graduation years, the lower the response rate. The assumption that a decreasing attention value in relation to the concern of the survey leads to a lower response rate was confirmed. However, other influences, such as particularly extensive professional commitments among older graduates, cannot be ruled out as a cause for the lower response rate.

Contrary to expectations, the personal letter from the director of the institution did not have a positive effect on the response. A certain concern about data protection or the anonymous treatment of the responses may explain this.

Finally, sending out a shorter questionnaire in the second wave produced a particularly interesting result. The associated emphasis on the importance of the project now ensured a higher response to the original – i.e. the significantly longer – questionnaire.

Summary

In recent times, extensive methodological efforts have been made to further develop the design of postal surveys. Thus, a change in the evaluation of postal surveys in the context of methodological textbooks can be pointed out. While Atteslander and Friedrichs were still quite reserved about the possible use of postal surveys, a more differentiated view can be stated in the meantime, for example in Diekmann. The following quotations are intended to substantiate this assessment:

The written interview is suitable only for ascertaining simple facts that pose minimal psychological problems for the respondent. (Atteslander 1984, p. 115)

One will work with written surveys if no other method than the survey yields the necessary
information, but interviews are not possible for reasons of time and cost. [...] If it is [...]
about an exact hypothesis testing including the calculation of the security limits of the statement [...], a written survey is not advisable. (Friedrichs 1990, p. 237)

The "written, postal survey [is] an alternative to personal and telephone interviews that is
definitely worth considering." (Diekmann 2001, p. 520) If one follows the latest methodological findings, postal surveys are also suitable for representative, i.e. randomly recruited,
samples. The assessment of the performance of postal surveys is even more optimistic
among the outspoken advocates of this approach. According to Goyder (1985) it is even an
optimal method for the post-industrial society and if one follows de Leeuw (1992), this
survey technique is characterized by a particularly good data quality, especially for sensitive questions. Finally, various other authors suggest that sensitive questions in particular
would be answered more honestly (compare Krysan et al. 1994; de Leeuw 1992, p. 33;
Nederhof 1985, p. 296).

Important prerequisites for the decision to choose a postal survey as a survey mode are
above all the existence of an address pool, a (probably) good motivation of the target persons to participate in the survey and – related to this – a relatively homogeneous target
population. Accordingly, its use is reported relatively often, especially for graduate surveys, customer satisfaction studies, for teaching evaluations, for member surveys, for
regionally limited studies and for expert surveys. Experiences with the conception and
implementation of nationwide mail surveys, in which the general population represents
the target population, are accordingly not yet available.

Furthermore, a strict orientation towards the rules of the total design method presented
above should be followed. In addition, the response rate for interesting surveys is 12%
higher than for surveys on uninteresting topics (cf. Hippler 1985). Variations in the anonymity of the survey have an effect of plus or minus 15% (compare Wieseman 1972). The
use of environmental protection paper instead of white paper reduces the response rate by
four percentage points (compare Hippler and Seidel 1985).

Finally, it should be noted that research on improving the design of postal surveys is
probably not yet complete. Further efforts are required to remedy the still existing, sometimes "serious inadequacies" (Blasius and Reuband 1996, p. 35).

The Telephone Survey

Again, by way of introduction, the prevalence of telephone interviewing in the practice of
survey research should be considered as an indicator of the importance of this survey technique. Table 6.1 showed the corresponding figures. Thus the share of telephone interviews
conducted by the ADM institutes in all surveys in 2005 was 45%. The number of CATI
workstations at these institutes increased from 1529 in 1995 to over 4000 in 2001. The

survey institute FORSA[15] in alone had telephone studios with over 300 interviewer work-stations at which 1200 interviewers were employed in Frankfurt am Main, Dortmund and Berlin in 2019 and offers daily multi-topic surveys with sample sizes of n = 1000 persons. Other institutes offer comparable services. In the USA, 90% of all surveys are now said to be conducted as telephone interviews.

Similarly, methodological knowledge about telephone surveys has increased. Initially, there were reservations about this survey approach. It was assumed that this method would only be suitable for reference date surveys, that a restriction to surveys in which house-holds and not individuals represent the unit of analysis had to be accepted, and that only short and uncomplicated questionnaires, which at the same time do not place too high demands on the target persons, could be used (compare Scheuch 1967a, p. 172 f.). At pres-ent, these restrictions apply only to a very limited extent or can be regarded as having been largely resolved.

In older and also in less old methodological textbooks, telephone interviews are either not described at all (compare Atteslander 1984) or only as special forms or as a variant of oral interviews (compare Kromrey 1998; Friedrichs 1990). In more recent editions, how-ever, these points of view have been revised.

Information about the rapid development of the framework conditions relevant to the field of telephone surveys is also provided by the rapid moral attrition in which knowledge about this survey technique is no longer up to date. As an example, only the density of entries in telephone directories shall be mentioned here. As already shown (see Sect. 5.4), this information – which is not unimportant for sampling – was still described as almost complete in 2001 (Diekmann 2001, p. 433). Only a short time later it was established that – with a further upward trend – 30% of telephone owners in Frankfurt, 32% in Hamburg and 35% in Berlin are not (or no longer) registered in a telephone directory (cf. Deutschmann and Häder 2002, p. 76). It is now assumed that more than 40% of telephone subscribers are not registered.

Thus, first of all, an already high frequency of use and secondly, rapid changes in the preconditions for telephone surveys describe the current situation of this survey approach. However, there are also reservations in principle about telephone surveys. Noelle-Neumann and Petersen draw a rather cautious conclusion: "Compared to the telephone interview, the face-to-face interview is the more sensitive, 'more susceptible to interference' but also the considerably more differentiated and variable method. […] Survey research that relies predominantly on telephone interviews is regressing to its primitive initial forms" (Noelle-Neumann and Petersen 2000, p. 198). It cannot be discussed further here to what extent such an assessment is (actually still) true. In the context of the discussion about the pos-sibilities of telephone surveys, it does at least represent one facet.

For a detailed account of the current state of research on telephone surveys, compare Häder et al. (2019).

---

[15] Compare: https://www.forsa.de/methoden/ last accessed 02/25/2019.

The Procedure for Telephone Interviews

For telephone surveys, the following procedure is usually followed: The questionnaire is programmed so that the entire survey can be controlled by a special CATI system (for example CI3[16] or VOXCO),[17] that is, with the help of a computer. The interviewers are each gathered in front of screens in a laboratory and conduct the interview on the telephone from there. At present, such laboratories are also frequently operated by university institutions, for example, as part of their methods training. The survey system enables automatic sampling and provision of telephone numbers (not necessarily the dialling itself, but this is also possible), control of the interviewers by a supervisor, call administration, data recording, filtering in the questionnaire, encoding of open questions, permutation of questions, recording of audio samples and permanent production of interim results. The system can also be used to account for the work performed by each individual interviewer.

The ADM has drawn up a set of rules (last revised version: January 2008)[18] for the procedure of telephone surveys, which is mainly dedicated to data protection and quality assurance. These guidelines can be followed by other institutions. The most important rules concern:

- The differentiation of scientific studies from commercial advertising and sales promotion or the renunciation of advertising in the context of scientific surveys.
- Strictly maintaining the privacy of the target persons, for example, by observing the unannounced calling hours between 9:00 am and 9:00 pm.
- Compliance with general data protection rules similar to face-to-face interviews.
- The intermittent listening in of telephone calls by a supervisor for quality assurance purposes.
- Finally, the stipulation that tape recordings of telephone calls may only be made if all persons involved agree.

Advantages and Limitations

One background to the high frequency of use of telephone surveys is the growing problems with face-to-face surveys, here above all the decline in response rates. Surveys in recent years in particular show a dramatic decline in the response rate to ALLBUS surveys (cf. Table 6.16). In addition, costs have risen and the time required to carry out the surveys has increased (cf. Koch 2002). Telephone surveys initially promised to remedy this situation, but – as will be shown – they cannot solve the problem.

Against this background, the following advantages can be claimed for telephone interviews compared to face-to-face interviews:

---

[16] Compare: http://www.sawtoothsoftware.com/support/downloads/download-ci3 (last accessed Feb 25, 2019).

[17] Compare: https://www.voxco.com/survey-solutions/voxco-cati/ (last accessed Feb 25, 2019).

[18] Compare: https://www.adm-ev.de/wp-content/uploads/2018/07/RL-Telefonbefragung.pdf (last accessed Feb 25, 2019).

Table 6.16 Reported exhaustion and case prices of the ALLBUS surveys of 1996, 2000 and 2012.

Study	Reported exhaustion rate (%)	Case price, without value added tax in DM
ALLBUS 1996	49.1	Approx. 250 DM
ALLBUS 2000	54.2	Approx. 180 DM
ALLBUS 2012	37.6	k. A.

Sources: Koch 2002, p. 33 and http://www.gesis.org/allbus/studienprofile/2012/

- The fieldwork is computer-assisted and computer-controlled. This allows for better field control and better standardisation of interviewer behaviour and, as a consequence, more objective results. In the interests of survey objectivity, the scope for interviewers to determine the target persons in the case of population registration office samples and, above all, ADM samples (cf. Sect. 5.2) is considerably restricted.
- Due to the supervision, targeted (follow-up) training of all or individual interviewers can be scheduled. This is largely impossible with the decentrally conducted face-to-face interviews.
- The costs of telephone surveys are significantly lower than those of face-to-face interviews.
- The interview situation on the telephone is more impersonal than in a face-to-face oral study. Thus, it is to be expected that the tendency to socially desirable answers occurs less.
- The interviewees perceive a higher anonymity of the interview situation. On this basis, it can be assumed that their answers are less oriented towards the person of the interviewer. Thus – also due to the more impersonal interview situation – an overall tendency towards better answers is expected. As an example, a study will be cited that contained a question about one's own voting behaviour. This issue was better identified in a telephone survey than in a face-to-face study (compare Diekmann 2004; Rogers 1976). In some cases, however, such findings are vehemently contradicted (cf. Noelle-Neumann and Petersen 2000, p. 189).
- According to Frey et al. (1990, p. 198), today's society demands particularly fast data that is available "at the shortest possible notice". Survey projects can be completed extremely quickly with telephone surveys. FORSA, for example, provides its clients with the results of a survey of 1000 people within 24 h. The results are then made available to the client. For numerous studies in the social sciences, however, this supposed advantage may be irrelevant or even disadvantageous. This will be discussed later.
- In the event of queries, interviewers can contact the supervisor in a relatively straightforward manner.
- For target persons who want to be sure about the identity of the organiser of the interview, the possibility of a call-back can be offered as a confidence-building measure.
- Even complex indicators have proven to be collectable in telephone surveys – thanks to the computer-assisted presentation of the questionnaire on the interviewer's PC.
- The primacy effect, i.e. the tendency for respondents to choose one of the first categories in questions with several answer categories, can be reduced in telephone surveys. Especially when the target persons are presented with the answer specifications in writing, a tendency

can be observed according to which one of the first specifications is chosen, regardless of the content of the question. This is naturally not the case in telephone surveys.

- It becomes possible to play audio samples into the interview by means of special *.wav files (compare Häder and Klein 2002). This can serve both to further standardise interviewer behaviour and to loosen up the conversational situation on the telephone.

The disadvantages of telephone surveys over face-to-face oral studies are as follows:

- One consequence of the more impersonal interview situation is greater uncertainty for the interviewer when making contact. In order to motivate the target person to participate in the interview, the interviewer needs scripts. In the case of an interview conducted on the telephone, the interviewer only has the voice of the person he is talking to at his disposal (cf. also Meier et al. 2005, p. 37 ff.). The situation is different in a personal oral interview. Here, for example, the furnishings of the apartment and the impression left by the residential area, the clothing and the appearance of the target person can be used to create such a script.
- Another consequence of the impersonal interview situation on the telephone is more frequent abandonment and a higher item non-response. It is easier for the target person to break out of the role of the interviewee and end a telephone interview prematurely by hanging up. In contrast, a respondent in a face-to-face oral interview still has to ask the interviewer to leave early.
- It is not possible to use visual and other aids on the telephone, such as lists, pictures, samples of goods and the like.
- Telephone interviews should be designed for a shorter duration. A guideline value of 30 min and an upper limit of 45 min are given. However, there are also reports of interviews that are said to have lasted up to 1.5 h (cf. Diekmann 2001, p. 503).
- Older people who have completed their socialisation without a telephone are more likely to have problems with this type of questioning. For example, they are more likely – for whatever reason – not to answer individual questions.
- The effect of an ISDN display, which signals in advance to the called party who the caller is, on the target's willingness to participate is still largely unknown. It is conceivable that in the case of unknown callers, as with a survey institute, acceptance of the call is refused.
- The continuation of face-to-face interview series with the help of telephone interviews can also be problematic. Here again, the fashion effects already mentioned must be taken into account. A corresponding change requires more extensive tests (cf. Wüst 1998; Häder and Kühne 2009).
- Open-ended questions are answered more unimaginatively in telephone studies. Recall questions, in which events from the past are to be recalled, are also more difficult to use. In general, a lower quality of answer must be expected on the telephone to questions that require a higher degree of information gathering to answer. The causes here are the same as for answering open-ended questions. People who are interviewed on the telephone arrive at their answers more quickly and this speed is at the expense of the information activated in the memory for finding the answer.

- The recency effect occurs. If the target persons are read answer specifications, there is a tendency that, regardless of the content of the specifications, one of the latter variants is preferred.
- Additional information on the credibility of the respondent is missing, such as a reconciliation between the reported household income and the home furnishings.
- For respondents, telephone interviews are considered boring to a higher degree.
- Although initially higher response rates were reported for telephone surveys (up to 70%), in the meantime they have been brought into line with the response rates achieved in face-to-face surveys (cf. Diekmann 2001, p. 502; Hüfken 2000, p. 11), and in some cases they are even lower (cf. Schlinzig and Schneiderat 2009).

Methodology

The use of telephone interviews for surveys among the general population is linked to several preconditions. The first concerns above all the degree of coverage with telephone connections. This network density has reached a level in Germany that has allowed us to speak of full coverage since about 1997. This now also applies to eastern Germany. Full coverage means that at least 95% of households are equipped with telephone connections. The following Table 6.17[19] provides concrete information on this.

In telephone population surveys, it has so far been assumed that the telephone connections installed in the households include at least one fixed network connection. At present,

Table 6.17 Equipment of private households with information and communication technology in Germany, results of the current economic accounts (total households)

Phone type	1998	2003	2008
Equipment level in %[a]			
Phone	–	98.7	99.0
Stationary	96.8	94.5	89.7
Mobile	11.2	72.5	86.3
Equipment stock per 100 households[b]			
Phone	–	234.7	268.3
Stationary	109.5	120.5	114.5
Mobile	12.2	114.2	153.8

Status: 1 January of the respective year. Excluding households of self-employed persons and farmers and excluding households with a monthly net household income of €17,895/€18,000 (as of 2002) and above

[a]Number of households with corresponding durables in relation to extrapolated households in the respective column

[b]Number of consumer durables present in households, based on extrapolated households in the respective column

---

[19] Compare http://www.destatis.de/basis/d/evs/budtab2.php (last accessed 15/02/2005 and 29/06/06 respectively). The data were updated on 15.10.2004 and 02.03.2006 respectively. The data for 2001 are from the 2001–2003 economic accounts, the rest from the 2003–2005 economic accounts. The data for 2008 are from the source indicated in the following footnote.

it can be observed that households are increasingly foregoing the installation of fixed-network connections in favour of mobile telephones. This poses new challenges for survey research (compare Häder 2000; Häder and Glemser 2004; Häder and Häder 2009). In 2008, about 12%[20] in Germany only (still) had a mobile phone. "TNS Infratest determined in its Face-to-face-BUS 2005 on the basis of about 30,000 interviews still a share of 8.3% of households that can only be reached via mobile phone. Moreover, due to additional telephony providers (e.g. Voice over Ip), the problems of sampling are becoming increasingly complex" (Gabler and Häder 2006, p. 120). Thus, it was necessary to look for new ways. The solution is a mixed-mode design in which the telephone survey is conducted in parallel via the fixed network and mobile phone. This instrument is described in detail in Häder and Häder (2009) and Häder et al. (2012).

Surveys of specific populations who, for example, have landline telephone connections for professional reasons – such as doctors, lawyers and shopkeepers – can also be carried out when network density in the general population is or becomes lower.

Furthermore, it must be possible to draw random samples for such telephone surveys. Since no even approximately complete lists of all telephone connections are available, special methodological efforts must be made for this purpose. The telephone directory is no longer suitable as a sampling frame. At present, about 40% of telephone numbers in Germany are no longer listed, i.e. entered in the telephone directory, and this proportion is rising. In the USA, it has become common practice to generate telephone numbers randomly. However, this procedure – Random Digit Dialing (RDD) – requires a certain structure of the telephone number system and is therefore not applicable in this form in Germany.

The current practice of drawing telephone samples was discussed in detail in Chap. 5.

Some problems that may arise in the application of this technique should be briefly mentioned:

- Fax and business connections as well as answering machines and unassigned telephone numbers are also dialled with this procedure. However, this cannot be avoided. However, the use of automatic dialing machines – predictive dialers – minimizes the resulting effort.
- Multiple connections are overrepresented in this design, as they have a higher probability of inclusion. If a corresponding demand is made in the survey for the number of existing connections in the household, this problem can be solved more or less well by means of a design weighting. This assumes that the persons called can reliably provide information about these aspects.
- Announcement letters also have a positive effect on the response rate in telephone surveys (compare Friedrichs 2000, p. 171 ff.; Hüfken 2000, p. 11 ff.). However, this presupposes knowledge of the addresses and a telephone directory selection and cannot be realised with the RLD technique mentioned.

---

[20] Zuhause in Deutschland – Ausstattung und Wohnsituation privater Haushalte, http://www.destatis. de/basis/d/evs/budtab6.php last accessed 23/03/2009, now unfortunately no longer available.

From a sampling perspective, it is also important to point out several advantages of telephone surveys:

- Above all, the lower spatial clumping in comparison to ADM – or also residents' registration office samples is to be mentioned as an advantage.
- A CD-ROM-based RLD technique allows the addition of background variables. Thus, electronic telephone number directories are available which also contain information on the economic situation, the unemployment rate and other socially relevant aspects within the respective area code.
- Finally, as mentioned above, the supervisor ensures permanent control of the interviewers. While the selection of the target person in the target household is largely beyond the control of the survey institute in an ADM sample, this critical phase of the interview can be well controlled on the telephone. However, the problem of falsification and partial falsification of interviews can also be countered.
- Whereas in the case of face-to-face interviews, each additional contact attempt is associated with a high level of effort, a telephone survey can proceed more effectively. The survey system automates the control of call times and sets callbacks to different times of day or days of the week, for example. It also becomes possible to have the appointment management controlled by computer when a callback is requested by the target or contact person at a specific time. Finally, it is possible to continue an interview that was interrupted due to time constraints at the appropriate point at a later time.

Reference has already been made to some technical aspects of telephone surveys. Here, too, a whole series of advantages can be named compared to face-to-face interviews. For example, filtering can be considered to have been solved in telephone surveys. Consistency checks of the answers or of the data entered into the PC by the interviewers can be programmed and carried out during the interview. The separate data entry or transfer into a machine-readable data file is no longer necessary. Intermediate evaluations are thus easily possible. Furthermore, the survey program enables an automatic and random permutation of answer specifications or/and questions. Sequence effects can thus be neutralized more easily. Of course, this also applies to face-to-face interviews conducted with the help of a laptop (CAPI).

Finally, telephone surveys – as already mentioned – are considered to be a particularly fast way of obtaining information. However, easily accessible persons are likely to be overrepresented in speed samples surveyed in this way. This is all the more true the faster a survey is conducted. The design of a simple flash survey eliminates the need for callbacks on different days of the week and at different times of the day. It thus forgoes a key advantage that this design contains.

In order to further illustrate the methods of the telephone surveys, some aspects will be discussed:

It has been shown that telephone surveys have to contend with the same problems that occur in face-to-face surveys. These are, for example, the education bias; telephone surveys also have too low a participation rate among people with elementary, secondary or no school-leaving qualifications.

The design of questionnaires in telephone surveys follows the same basic rules that apply to other surveys. This concerns, for example, the order of questions, the need for clear wording of indicators and so on. Rating questions can be used in both modes without any problems. However, it is advisable to change overly differentiated answer scales in CATI surveys. As is well known, the use of template sheets, which visually present the answer scales to the respondents, must be avoided on the telephone. Possible scales here are those that are constructed in analogy to school grades and those that the target respondents can easily imagine. Thermometer scales ranging from zero to 100, which can be used to determine the degree of agreement with a certain statement, can also be used in telephone surveys. So that the cognitive abilities of the target persons are not overtaxed, no more than five verbalised answer categories should be read out to the target persons for selection.

In order to enable comparisons with scales that contain more extensive specifications or to ensure the continuation of survey series, the unfolding technique can be used. This technique is used, for example, for the left-right or the top-bottom question. Both indicators originally use far more than five specifications. The solution would be to first ask whether the target person would classify himself politically more to the left, more to the right or more in the middle. This is then followed by another question that asks how far to the right or how far to the left a respondent would classify themselves.

In contrast, ranking questions are more difficult to use in telephone surveys. Here, the visual support provided by lists to be presented or card games is particularly lacking. But even this questioning technique does not have to be completely dispensed with in telephone surveys. What is important is the scope of the facts to be put into a hierarchy. Pair comparisons, which are easier to master, are also suitable in order not to make the cognitive demands on the target persons too great. In pair comparisons, each fact to be evaluated is contrasted with every other fact. The use of this technique is limited by the number of comparisons required, which increases rapidly with the number of items to be assessed.

Open questions can and should also be used in telephone surveys in order to take away some of the artificial character of the communication. It is important to note that the answers to be expected are likely to be the result of only a small amount of thinking on the part of the respondent (compare Noelle-Neumann and Petersen 2000, p. 183 ff.). After all, the CATI technique offers better opportunities for field coding of open-ended questions. Whereas in a PAPI interview the answers given have to be noted down (verbatim) by the interviewer, the CATI interviewer can search for a classification of the answer on the computer.

Of particular importance in telephone interviews is the design of the introduction and the decision on the most appropriate initial questions. As recommendations can be mentioned:

- The first question should be a closed one to signal to the respondent that the interview will be easy.
- The second question could be asked openly. This gives the interviewee the opportunity to articulate himself relatively freely and to build up a relationship with the interviewer. It also gives the target person the opportunity to possibly express displeasure and thus reduce it.

As an example, the Politbarometer can be cited, which is organized by the Forschungsgruppe Wahlen in Mannheim (FGW). The procedure looks like this:

- The first question is about the federal state in which the target person is eligible to vote. Here the answer can be found easily, at the same time a reference to the content of the survey is established. The rule according to which the first question of a survey may not be answered in the negative is also taken into account.
- The second question is directed at the number of inhabitants of the place of residence in which the respondent lives. As a rule, this is not known exactly. This opens up the possibility for a conversation with the interviewer. The anonymity of the interview also underlines this question, as obviously the caller does not have this information. If necessary, it is a good idea to explain the random selection made in more detail to the target persons. Psychologically, such a question builds on a special identification with one's own place of residence.
- The third question in the Politbarometer is aimed at the most important problems in Germany at present. It is an open question. It is intended to give the respondent the feeling of being an equal partner to the interviewer. At the same time, the content profile of the entire survey is thus already well addressed.

Data protection in telephone surveys entails some special features. The question must be answered as to whether a person who is not listed in a telephone directory can be expected to be called unexpectedly and asked to participate in a survey. In the meantime, there is a judicial decision on this.[21] A balancing of interests was carried out. On the one hand, there is the freedom of research and the use of an attractive survey technique. On the other hand, there is the protection of the privacy of the target person who may feel harassed. The view of the court is that a person can be expected to at least accept such a call and, if necessary, signal a refusal to participate. It is important that no commercial purposes are connected with this questioning, for example advertisement for a certain product is operated. In case the target person or the target household refuses to cooperate, the resulting costs for them are relatively low: the telephone connection is only blocked for a short time. In contrast, not generating telephone numbers – this would be the consequence if it were forbidden to contact unlisted persons – would make the use of telephone surveys pointless. The costs on the part of survey research would be correspondingly immense.

ADM has issued a policy on telephone surveys. The main aspects of this guideline concern:

- The generation of random numbers is not prohibited.
- Hard deniers are no longer allowed to be called.
- In the case of repeat interviews, the target persons must be given appropriate information on data protection.
- The (currently still rare) use of voice computers for the interview is not permitted without further ado. The consent of the target person must also be obtained for this.

The following text opened the Dresden Self-Defence Study, a telephone survey:

---

[21] Compare: http://www.kanzlei-prof-schweizer.de/bibliothek/neu/index.html?datum=2005-11 and http://www.kanzlei-prof-schweizer.de/bibliothek/urteile/index.html?id=12951 each last accessed on February 25, 2019.

Hello, my name is … I am calling on behalf of USUMA GmbH, a market, opinion and social research institute, in Berlin. We are conducting a survey on behalf of the Technical University of Dresden and ZUMA in Mannheim on critical situations that one can get into every day. In order to make the survey representative, I would like to interview one person in your household who is at least 18 years old. If more than one person aged 18 or over lives in your household, I would like to speak to the person who had their birthday last. May I ask you a few questions about this topic?

INTERVIEWER: If the target person is not on the phone, please have him/her put on the phone or if the target person is not present, make an appointment!

ONLY IF NEEDED: Your phone number was selected through a scientific random process. The survey is voluntary, but it is very important that as many selected people as possible participate in order for the survey to provide a true result. The analysis is anonymous, i.e. not in connection with your name, address or telephone number.

Summary
The most significant problems with telephone interviews are: firstly, the non-binding interview situation, which entails the risk of the target person abandoning the interview, and secondly, the necessity of using a relatively simple questionnaire and having to do without a varied – also visually appealing – design of the questions. This is offset by various advantages. First of all, the significantly lower costs, secondly, the possibility of supervision, thirdly, the absence of spatial clustering of the sample, and fourthly, the meanwhile good methodological experience.

Telephone surveys have a firm place in the canon of methods of empirical social research. With regard to the criticism by Noelle-Neumann and Petersen (2000) cited above, the following should be noted: First, the limitations of telephone surveys mentioned above must be acknowledged by social researchers and taken into account when interpreting the findings. Second, methodological research can help to reduce remaining deficits. Possibly, technical developments in the field of telecommunications can open up new possibilities.

Qualitative Surveys
Surveys play an important role in both the quantitative and qualitative research paradigms. There has been a lively debate about the relationship between and the performance of quantitative and qualitative methods (compare, for example, Hopf 1993, p. 11 ff., and Schnell et al. 2005, as well as the descriptions in Sect. 3.8). The question of whether and to what extent qualitative methods are also suitable for testing theories or hypotheses is always disputed. In order to understand this "sometimes tense" (Bortz and Döring 2002, p. 295) debate, the analytical-nomological and interpretative approaches to social research will now be discussed. Afterwards, some qualitative techniques of interviewing and the corresponding strategies of data evaluation will be presented in more detail.

The understanding of analytical-nomological science is based primarily on the assumption that the ultimate goal, also of social scientific research, is the knowledge of laws.

Laws are characterized by the fact that they describe relationships that are realized independently of the acting individuals. In hypotheses, scientifically justified assumptions about such relationships are formulated in the interest of the knowledge of laws. In testing them, such techniques must now be employed in which individual influences are largely avoided. The ideal of an objective measurement, in which the measuring subject has no influence on the result of the measurement, is pursued, even if not completely achieved. The survey strategies presented so far have been designed accordingly. They are characterized above all by a high degree of standardization. This standardisation appears necessary in order to minimise the subjective influences which, for example, emanate from an interviewer during a survey. Such a paradigm is admittedly not without controversy.

The journalist and historian Joachim Fest takes a different view in an essay. He deals, for example (compare 2003, p. 38), with the role that particular personalities such as Adolf Hitler and Mikhail Gorbachev have played in history. In this context, Fest even encounters what he calls a basic fallacy of history: it is about the "utterly false understanding of the nature of man […]. All the causal bravura that is attributed to events is nothing but presumption. What is actually going on is the eternally confusing drama of ignorant, erring, proving themselves or failing. […]. But history is much less science than its academic spokesmen like to make us believe, because all science operates with regularities. But precisely these do not exist in historical processes, but only turbulence, whims, convulsive states, and the discharges that repeatedly throw all conditions into confusion."

If one is convinced that history is determined by the subjective arbitrariness of individuals – and not by objective laws or peer pressure – and if one works on corresponding questions (How could a profile-less average type like Adolf Hitler come to power? How could a single individual bring about the dissolution of a world empire that was oriented towards the consistent maintenance of power?) Such problems can only be dealt with by means of open, i.e. qualitative methods. Standardizations to increase objectivity are no longer helpful at this point.

The basic assumptions of interpretative social science, on which qualitative forms of questioning are based, are correspondingly different. Here, the premise is that there are no objectively existing structures in society. Rather, these are constantly being constructed anew by the acting individuals. This has far-reaching consequences for the methods to be used. Thus, a special degree of openness is required in the survey in order to be able to respond to the acting individuals and to determine their individual interpretations of reality. For qualitative social research it is thus important to take into account all possible boundary conditions of the survey, because these boundary conditions can also lead to certain actions. Thus, there is a special need for flexible methods here, because only they are able to do this. A standardisation of the methods in order to obtain the most objective results possible, on the other hand, is insignificant or even a hindrance.

The focused interview, the narrative interview and the problem-centered interview will be presented. Subsequently, some techniques for the evaluation of such interviews will be addressed. The basic discussion about qualitative and quantitative methods will be discussed again at the end of the section.

Characteristics of Qualitative Interviews

The survey techniques have already been divided into standardised, partially standardised and non-standardised approaches. This classification refers to the situation of the interview, to the questions used for it and also to the answer specifications presented to the target persons. The open, non-standardised concept of a qualitative survey is characterised by features that distinguish it from standardised and partially standardised approaches. The following are particularly striking:

In structured surveys, a standardized questionnaire is used which makes it possible to summarize the answers obtained from many people and then to generalize. Qualitative interviews have more the character of a casual conversation. This is characterized by a stronger influence of both the target person and the interviewer. Thus, the interviewer is expected to behave differently. Special training must teach the interviewer to adapt his question formulations to the respective situation in the interview. This counteracts artificial (standardised) interview situations. Pre-fixed questions are replaced by free, partly spontaneous, situation-dependent formulations, resulting in answers that cannot be structured beforehand. In contrast to quantitative surveys, it is not uncommon for the interviewers and the researchers interested in the content of the questions to be identical persons.

Qualitative interviews can therefore be characterised in particular by the emphasis on subject-relatedness. It is not the questionnaire that determines the course of the interview in advance, but successively the reactions of the interviewee. This can only be achieved by keeping the situation of the interview open, i.e. not determining it in advance. The aim is to make the investigation seem less artificial and to carry it out in a way that is close to everyday life.

Such surveys are also characterised by the fact that they are more in-depth, allow the target persons to have their say more, require more intensive evaluation, are often directed at subcultures, social fringe groups and similar populations, and are particularly suitable for pretesting.

In various qualitative approaches, theoretical sampling is used to determine the person to be interviewed. This involves the theory-based, i.e. conscious, selection of the persons to be studied or the cases to be analysed. In most cases, particularly typical cases or, on the contrary, conspicuous exceptions are used.

The Focused Interview

One form of qualitative interviewing is the focused interview, developed and illustrated by Merton and Kendall (1979, p. 171 ff.). The focused interview makes use of the following procedure:

- First, a certain stimulus is given to the interviewee. This can be a film, a commercial, a product sample or something similar. However, a situation experienced in everyday life, such as the Elbe flood, can also be made the starting point. This stimulus then provides the basis for the subsequent questioning. The focused interview has also proved particularly useful for interpreting unexpected events as well as exceptional cases.

- Now a guideline is used to analyse the effect of the stimulus. Content analyses, for example of the film shown or the press coverage of the Elbe flood, can be used to develop this guide. Assumptions about the effect of the stimulus on the target persons can also be included in the development of the guideline. On guidelines and expert interviews, see also Helfferich (2019, p. 669 ff.).
- This is followed by the elicitation of the target person's reactions using this guide. The aim is to determine the subjective experiences of the interviewee in relation to the stimulus in order to validate and further develop the assumptions made on this basis.
- Since the interviewer is familiar with the situation, he or she can ask the target person more or less prepared follow-up questions in addition to the guide. The aim of focused interviews can be both an initial hypothesis check and the discovery of new clues.

A good example of the useful application of focused interviews is an experiment on voter turnout (compare Merton and Kendall 1993, p. 171 ff.). The starting point was provided by the fact that the mailing of certain materials increases participation in elections. In order to investigate the questions of why this is so, what role the sender of such materials plays in this, what the concrete content of the materials does, and so on, two solutions are possible in principle: Either one resorts to a relatively complicated experimental design, varies the senders in a split, forms different control and comparison groups, repeats the survey (compare Sect. 7.1) or – much simpler – one uses focused interviews with the procedure described here.

Reference has already been made to the special function of the interviewer in qualitative interviews. In the case of focused interviews, the interviewer must observe the following principles in particular:

- The maxim of non-interference applies. The interviewer should limit himself to a minimum of instructions, he should above all appear non-directive. The target person should say things that are important to him or her and not primarily report on facts that the interviewer would like to hear. In principle, all information provided by the target person is of interest. Recommended formulations for an unstructured question are: "What impressed you most about this film?" "What struck you most about this film?" Examples of successful semi-structured questions could be: "What new things did you learn from this leaflet that you did not know before?" "How did you feel about the part where … is described?"
  The interviewer should ask the questions from the guide in a restrained manner so as not to interrupt the target person's flow of speech. The fact that he may be confronted with the problem of irrelevant digressions must be accepted. Transitions to other complexes should, if possible, be made by the interviewee – and not by the interviewer.
- Furthermore, the principle of specificity applies. This implies that the reactions of the target person are to be found out even on details. Suppose a person, after viewing a propaganda film, makes an utterance such as, "The sight of marching soldiers in this film creates fear in me." However, this statement is not sufficient to determine the cause of the fear. The question would then be, for example, "Were certain faces of people, any comments in the film, or something else responsible for the fear described?" The

subject needs skillful cues from the interviewer to provide the appropriate information. Such aids are *retrospective introspection* and *explicit reference to* the situation at hand. In *retrospective introspection, the* past situation is brought to mind. For example, photos from the time in question can be shown (again) or radio broadcasts can be played. A useful example of a question could be: "When you think back, what scared you about this film?"

In the method of *explicit reference to* a situation, the situation is brought to mind again and coupled with an unstructured question. In the case of the propaganda film that created fear, the question could be: "Was there anything that gave you this impression?" In this way, it can then be determined that it was primarily the goose-step parade and the Sieg Heil shouts that generated fear in the viewer. These aspects thus have a special symbolic value in relation to fear.

The definition of the situation should be "fully and specifically enough expressed" in the focused interview (Merton and Kendall 1993, p. 178).

- In the focused interview, the aim is to record the entire spectrum of the target person's reactions to the stimulus. For this purpose, not only the expected reactions but, if possible, all reactions should be registered. The triggering stimuli and the reactions produced are to be comprehensively determined. In the case of taciturn informants, the interviewer must intervene helpfully.

- Another principle concerns the depth of the interview. Through targeted inquiries, it is important to sound out what affective evaluations the interviewee is making. The aim must be to gather the maximum amount of "self-revealing commentary from the informant about how they [experienced] the stimulus material" (Merton and Kendall 1993, p. 197). Focusing on feelings, for example, can be done with questions such as, "What did you feel when ...". The repetition of implicit or expressed feelings by the interviewer has also proved useful. In this way, the target person is encouraged to express further feelings and at the same time the interviewer signals his understanding of the expressed emotions.

Indicators that may suggest such emotionally charged facts include "prolonged pauses, self-corrections, trembling in the voice, incomplete sentences, embarrassed silences, slurred pronunciation" (Merton and Kendall 1993, p. 193).

The Narrative Interview

The narrative interview (compare Schütze 1977, 1983 as well as Küsters 2019, p. 687 ff.) is a form of narration that can be used above all to obtain particularly experiential, subjective statements. Here, it is mostly about biographical processes that are described by the target person in the context of the interview. The general aim here is also to understand the views and actions of the actors.

In a narrative interview, the interviewer simply gives a topic and then encourages the target person to tell the story. After that, the important thing is to avoid interrupting as much as possible. The interviewer takes on the role of the listener.

For the narrative, the target person is explicitly left free to decide what and how to report. A certain momentum is assumed. In his narration, the interviewee unconsciously has to solve a number of tasks. Three are considered particularly central:

- Figurative exploration: The narration must be done in such a way that it is understandable for the listener. For example, the people involved must be introduced. Special highlights are to be worked out by the narrator.
- The compulsion to condense: Within the framework of the narrative, the interviewee must set focal points. It is to be expected, for example, that when describing a catastrophe experienced, certain details will be dealt with in particular detail. Other details, on the other hand, can be skipped or only briefly touched upon.
- The compulsion to detail: The narrator must shape the narrative in such a way that it also makes connections clear to the listener and the motives of the actors become comprehensible.

The narrative phase is followed by an inquiry phase. In this phase of the narrative interview, any points that remain open are clarified. Finally, in the concluding summing-up phase, the target person is given the opportunity to respond to follow-up questions. Narrative interviews can be up to three hours long. They are recorded and then transcribed for analysis, i.e. written down according to predefined rules (see Sect. 9.1).

The Problem-Centered Interview
Another form of qualitative questioning is the problem-centred interview (compare Witzel 1992). In this technique, the interviewer asks questions according to a partially structured guideline. In contrast to the narrative interview, the interviewer exercises a relatively active role here. This is particularly necessary if the target person lacks narrative competence. At the end of the interview, there is usually a separate request for demographic data or other information.

It is often suggested that problem-centred interviews be combined with other methods, such as case studies (see Sect. 7.2), content analyses (see Sect. 6.3) or group discussions.

The Evaluation of Qualitative Interviews
Qualitative surveys also have their own typical procedure for evaluating the information collected. Three basic approaches will be referred to; further principles will be reported on later (cf. Sect. 9.1).

The Grounded Theory
Among the qualitative approaches, grounded theory has come to play a special role in data evaluation (and at the same time in their collection). Grounded theory is considered the classic, theory-discovering method. It is an approach that develops a "theory grounded in the data" (compare Brüsemeister 2000 and Strübin 2019, p. 525 ff.).

Grounded theory is characterized by the use of various data sources. The information obtained in this way is then attempted to be generalized step by step.

As an example[22] a study on the Elbe flood will be used, which was collected at the Institute of Sociology of the Technical University of Dresden. In a first step, it is observed how the flood was experienced by the people. All possible contexts are of interest, such as occupation, family, damage suffered, help experienced and so on. Based on this, secondly, more detailed investigations can be carried out, for example, comparisons can be made of the way in which the activities of different aid organisations (the Federal Agency for Technical Relief, the Federal Armed Forces, the professional fire brigade and the voluntary fire brigade were deployed) were perceived, or what role social differences – then still – play in a disaster. In a third step, the data obtained is sorted into patterns (sub-processes). This also involves drawing on existing theories for explanations, for example of social structure (pool of ideas). This procedure is characterised above all by a particularly flexible use of different data.

The Ethnomethodological Conversation Analysis
Ethnomethodological conversation analysis was founded in the USA in the mid-1960s. Social groups (ethno-) are examined with this approach to see which methods are used by them "as a matter of course in the conduct of their everyday affairs" (Bergmann 1991). The actors are metaphorically regarded as blank sheets of paper which, together with others, must first establish social reality, i.e. fill the sheets.

Typical examples that have been studied using the approach of ethnomethodological conversation analysis include gossip, counselling sessions at the social welfare office, saying goodbye, the communication of an airplane crew, or emergency calls to the fire brigade by telephone (cf. Bergmann 1993).

The evaluation of such analyses is then data-driven rather than theory-driven. The focus is again on the respective situation. The effects of certain utterances on the progress of the interaction are considered. It is thus purely an evaluation procedure.

Returning to the example of the analysis of fire brigade emergency calls, the following phases could be distinguished: First, the greeting, here already a concise intonation makes it clear to the caller that "he is dealing with an institution which – even without a concrete alarm – is already in the highest state of readiness, is, as it were, sitting in the starting blocks and is only waiting for a suitable occasion to spring into action." (Bergmann 1991). Secondly, the description of the concern, thirdly, a clarification phase, fourthly, the assurance of the mission, and finally, fifthly, the end of the conversation.

---

[22] The example of the lady in red has also gained notoriety: In an expensive restaurant, someone in the kitchen observes a woman wearing a red dress. The observer's mind signals skepticism and curiosity to him. Thus begins further investigation and reflection. The lady in red is specifically observed, for example, while working in the kitchen, during various communications. The results of the inquiries are then classified, for which purpose theories about the service society can be used, for example (compare Strauss and Corbin 1996).

The Objective Hermeneutics

According to the literal meaning of hermeneutics, it is the: "the study of the evaluation of texts as well as of non-linguistic cultural expressions". Objective hermeneutics now seeks an objective, representational meaning of signs, devices, social forms and the like (compare Freyer 1928). This approach was further developed for the social sciences by Ulrich Oevermann in the 1970s and by Hans-Georg Soeffner (1989) in the 1980s. Here, too, it is purely an evaluation procedure. It serves the reconstruction of latent structures of meaning of everyday actions with the help of hermeneutic text interpretation. The means of objective hermeneutics are sequence analyses. For this purpose, (first) all possibilities of meaning latently contained in a text are determined, then (second) it is empirically verified which of the possibilities is chosen (compare also Kurt and Herbrik 2019, p. 553 ff.).

The following example is intended to demonstrate this analysis technique. Let the following text be given: "A woman, 39 years old, married to a business graduate ten years older, has two schoolchildren and lives in very good economic circumstances. She is taking up a course of study." Now the search for possible justifications for this behavior begins. The following interpretations are conceivable (compare Oevermann et al. 1984, p. 28 f.):

1. Studying was always planned, it was just postponed because of starting a family.
2. The woman realizes that she is no longer fulfilled in the family alone and tries to do something socially recognized.
3. The woman thinks she is not educated enough in relation to the man, she wants to have more say.
4. The wife suspects the marriage will end soon and therefore wants to stand on her own two feet.
5. The former profession no longer fills the woman, she wants to continue her education for a more qualified one.
6. Concepts of emancipation and the desire for self-realization cause women to begin their studies.
7. The study corresponds to the woman, her previous interests in other life.
8. The woman has too little to do.
9. Only now do the external possibilities exist (for example, through distance learning) to take up a course of study.

This strategy looks for contexts in which an observed action – taking up studies – makes sense, regardless of the possible intentions of the actor. Only then does the second phase occur, the questioning of the person in question.

Problems of Non-Standardized and Partially Standardized Surveys

The condensed treatment of qualitative survey methods (for a detailed description see Denzin and Lincoln 2000) also documents the principle justification of this approach in the arsenal of methods of empirical social research. Although there is still a discussion about qualitative and quantitative designs, there is hardly any claim to exclusivity for only one of the two approaches, nor is there any doubt about the justification of the other strategy.

Finally, some problems of qualitative methods should be pointed out:

- Sample selection: this is usually done deliberately in qualitative studies. Thus, certain "types" are sought and then studied. This raises the question of the generalisability of such findings.
- Reliability and validity are basic problems of quantitative surveys. Here, tests have shown how sensitively target persons react when even seemingly insignificant changes are made in the survey standard, for example in the wording of a question (cf. Sect. 2.1). The findings of qualitative studies are likely to be similarly sensitive to such influences.
- Other keywords are interviewer influence, selective perception and the self-fulfilling prophecy effect (see Sect. 4.6). Here, too, sources of systematic bias in the answers are to be suspected in qualitative surveys.
- A considerable amount of material is available for the data analysis of qualitative studies. 40 interviews of two to three hours each result in approximately 2000 pages of transcribed text.
- Finally, reference should be made to the high demands placed on the interviewer in qualitative studies, to the necessary linguistic competence of the target person, and finally to the not inconsiderable amount of time required to conduct qualitative interviews.

Group Discussions

Conjectures are made as to whether group discussions can be understood as a specific technique of interviewing and should therefore be placed alongside postal, telephone and personal oral interviewing techniques, or whether this procedure should be understood as a method in its own right. Here Lamnek (1998, p. 25, compare also p. 31 f.) is followed: "Ultimately, no decision needs to be made as to whether group discussion is an independent method of data collection, or whether it is rather considered a specific questioning technique that exchanges questions and answers orally in the group situation in the exchange of all participants. It is arguably both, and more likely to be one or the other depending on methodological orientation and research design."

The group discussion is a mostly non-standardised oral form of questioning groups. Lamnek defines the group discussion as "a conversation of several participants on a topic that the discussion leader names [...] and that serves to gather information (Lamneck 1998, p. 11)." The two supporting elements of this procedure are thus firstly the interaction in a group and secondly the subordination of the communication to a scientific interest, which is concerned by the researcher or discussion leader.

While survey research is often concerned with constructing a statistical average opinion from (many) individual answers collected in a standardized way by different interviewers, group discussions are intended to capture the individual opinions in their contextual and social conditionality. In contrast, survey research abstracts from the various social contexts in which the interviews take place and summarizes all answers to a question into one value at the end. The group discussion, on the other hand, is different: the consideration of the social situation in which information is given is an essential specificity of this approach. After all, a whole series of social behaviours take place in the context of a more or less public everyday life and thus under conditions that are quite comparable to those in a group discussion.

It is irrelevant whether group discussions take place in groups artificially created for this purpose, which is more often the case, or in real existing and cooperating groups (real groups).

Due to their specificity, group discussions are more related to qualitative than to quantitative research. It is claimed, especially by their protagonists, that the importance of group discussions has increased in recent years. The consensus is that this instrument still receives relatively little methodological attention overall.

A basic methodological idea of group discussions is that the participants decide to make more detailed statements in the group discussion situation than they would in a personal oral interview. Due to a specially created group situation, it is expected that here the participants will also report on deeper-lying contents of consciousness. In view of a corresponding situation, it occurs – in contrast to a personal-verbal interview – that the discussants reduce psychological controls and decide to make spontaneous and uncontrolled statements. "It is also often easier in group discussions than in individual interviews to discuss questions of a very private, intimate nature with the interviewees. This assumption, which at first seems paradoxical, is justified by the fact that under the impression of frank, open-hearted contributions to the discussion by some participants, even more inhibited subjects are encouraged to make 'open' statements on such questions" (Mangold 1973, p. 230).

Group discussions were first used in 1936 by Kurt Lewin and Ronald Lippitt (1938) and by Dorwin Cartwright and Alvin Zander (1953). The names of Robert F. Bales (1950), who used group discussions primarily for small group research, and Robert K. Merton are also closely associated with the method. The latter became famous for his work on focus groups (compare 1987). The two approaches are relatively closely related.

In Germany, the Frankfurt School began to use group discussions in the context of studies on political consciousness (compare Pollock 1955). In addition, Mangold (1960) and Nießen (1977) were responsible for the further development of the methodological foundations of group discussions. Each of the three authors has left his own mark on the further development of group discussions:

For *Pollock*, the objective of group discussions focuses on non-public opinion. In a discussion and criticism of individual interviews, he states that this non-public opinion is not equal to the sum of all individual opinions recorded anonymously in individual interviews. Rather, opinions should be adequately collected in the real social environment. Non-public opinions thus become ascertainable above all in group situations.

Mangold, on the other hand, is concerned with informal group opinions that are independent of the situation. These, too, can only be determined in a group situation. They come about as a "consensus that emerges on a particular topic through mutual influence of the individual participants and the group within the discussion group" (Lamnek 1995, p. 143). Thus, what in quantitative social research tends to be considered an annoying disturbing factor, the influencing of the interview by the presence of third persons, is deliberately wanted in group discussions and placed at the centre of the procedure.

The relationship of the individual to the group and the way in which a group influences individual action are therefore typical questions that can be addressed with the help of group discussions.

According to *Nießen,* group discussion consistently belongs to the qualitative methodological paradigm. He assumes two prerequisites: firstly, only existing real groups should be interviewed and secondly, the participants in a group discussion should also be personally affected by the topic of the discussion in some way. While Mangold is interested in situation-independent group opinions, Nießen is concerned with situation-dependent group opinions. If one assumes, as the interpretative paradigm does, that social reality is generated and defined by the acting actors in dependence on the situation, then group discussions lend themselves to adequately capture such group opinions.

Finally, Lamnek (1998, p. 57 f.) refers to the lively interest in group discussions in market research. He attributes this to the fact that this method can be used to determine opinions, attitudes and behaviour in a relatively simple and quick way. The group processes themselves, on the other hand, are of only marginal interest in market research, for example in relation to the opinion leadership of certain individuals in group situations.

Cognition Intent
Lamnek (cf. 1998, p. 59 ff.) presents a good summary of possible cognitive intentions that can be pursued with the instrument of group discussion. The eleven directions identified by him are to be reproduced in abbreviated form and with slight commentary. According to these, group discussions are suitable for:

1. The uncovering of processes and procedures internal to the group. In this way, a certain dynamic is triggered in the group during the survey, from the observation of which insights can then be drawn.
2. The identification of group opinions. Certain objects and contents relevant to the research are discussed in the group and a corresponding opinion is then developed collectively.
3. The provision of information in market and opinion research. Here, unlike under (1) and (2), it is about the individual opinions, attitudes, behaviours and so on of the participants in a group survey.
4. Exploration. At an early stage of a research project, in addition to studying relevant literature, group discussions can be used to generate hypotheses. Such discussions to find ideas can be conducted with experts but also with members of the population of interest.
5. Pretests. As will be shown in Chap. 8, draft questionnaires, for example, can be dealt with in group discussions. Thus, group discussions also play a helpful role in the development of instruments for social science surveys.

6. As a corrective and/or complement to a questionnaire survey. For example, following or in the evaluation of a standardised survey, group discussions are intended to question the findings. In view of the low objectivity and reliability, however, this is probably a rather marginal function of group discussions.

7. For the plausibility and illustration of empirical findings. Collections of quotations created in group discussions can be used to support quantitative (statistical) findings, to make them more vivid, to give them life.

8. In the context of assessment centers. "Assessment centres are nowadays standard methods in almost all corporate sectors in personnel selection and personnel development for senior or even management positions" (Lamnek 1998, p. 63). Here, group discussions are intended to identify the management skills of individuals and thus to select the appropriate applicants for management positions (compare Obermann 1992, p. 142, but also already Mangold 1973, p. 228).

9. For method triangulation. In order to explore the possibilities and limitations of different social science instruments in more detail and to make empirical findings less vulnerable to attack, different methods are used in parallel to explore the same issue. Here, group discussions also play an important role.

10. As a therapeutic tool in clinical psychology, group discussions have significance.

11. Evaluation research. This seeks to determine the extent to which interventions undertaken in practice actually achieve their intended purpose (see Sect. 7.4). However, here too the usefulness of group discussions is likely to be relatively limited, as it is difficult to make generalisations on the basis of group discussions.

Advantages and Limitations of Group Discussions

On the basis of these cognitive possibilities attributed to group discussions, the advantages and limitations of this approach can now be summarized well. The particular attractiveness of group discussions lies in their use as a qualitative method. Above all, they fulfil the methodological premise of the qualitative paradigm (cf. Lamnek 1998, p. 78): group discussions collect data with openness, flexibility and closeness to everyday life. In addition, group discussions, probably like no other method, allow the study of group dynamic processes. If certain social objects of investigation or social behaviours are of interest, it can be assumed that the findings of group discussions are more relevant to behaviour, i.e. closer to reality, as they are collected in a natural – social – environment.

Finally, group discussions are afflicted with all those limitations, especially in comparison to quantitative approaches, which have already been presented and which apply to all other qualitative approaches. This concerns above all the question of the generalizability of the findings. For example, it hardly seems realistic to recruit the discussion participants by means of random selection and to generalize the results for a defined population. Furthermore, a strong influence of the discussion leader on the results of the study is to be expected. A check of the reliability of a group discussion session, i.e. the repeatability of the results, also seems to be hardly or not at all given. Also, imponderables arise as a result of the uneven participation of the various discussion participants in the conversation. The necessary information about the silent in such discussion rounds is naturally lacking. In contrast, the influence that comes about as a result of recording the discussion with the appropriate equipment is considered to be relatively small.

Conducting Group Discussions

The literature reports on group discussions in which between three and 20 people were involved (cf. Mangold 1973, p. 229), with group sizes between six and ten people being considered optimal. The problem with groups that are too small is that the individual characteristics of the participants carry too much weight here, while the informal character of the conversation is disturbed in groups that are too large.

In order to reduce the phenomenon of silent discussion participants, the group should be put together in a targeted manner. In principle, two variants are possible for the composition of a discussion group: A rather homogeneous or – the opposite – a heterogeneous composition of the circle. Both variants have certain advantages. For example, socially homogeneous groups promote the joy of discussion among the individual participants. Experience has shown that when people are among themselves, certain inhibitions are reduced during the discussion. On the other hand, there are also experiences according to which an inhomogeneity of the participants is able to enliven the discussion. In this case, it is particularly interesting for the discussants to receive the views of members of other social classes.

A discussion leader opens and ultimately leads the discussion, giving the floor to individual speakers. In the course of the discussion, the facilitator should clarify whether individual views are shared by the whole group or whether there are subgroups that hold a different view. It is also possible for the facilitator to present opposing views to the participants and then ask for opinions. Finally, the facilitator can add further information to the discussion and monitor the reaction of the group.

The discussion leader should encourage individual, less inhibited speakers; in this way he loosens up the atmosphere and subsequently tries to get other people to participate actively in the discussion. It is particularly advantageous if it is possible to establish a group norm in this way, in which the expression of one's own opinion becomes an obligation.

According to Spöhring (1989, p. 223), certain phases can be distinguished in discussions in groups where the participants have been recruited specifically for this conversation. Spöhring has further developed Mangold's (1973, p. 240) three-phase model for this purpose. These phases are shown in Table 6.18.

Evaluation

The discussion should be recorded with the help of a recording device, whereby a video recording of all interactions of the participants is also favourable. In principle, the participants in the discussion should be guaranteed full anonymity. To this end, it has proven useful to assign aliases for the participants. Mangold suggests the following further evaluation steps for a group discussion, shown in Table 6.18.

Table 6.18 Tabular progression model of group discussions

Discussion phase	Appearance	Hypothesis about causes
I. Foreignness	Cautious turns of phrase; non-commitment; reinsurance	Uncertain, because foreign situation; negotiating the definition
II. Orientation	Sensing; stimulating and provocative expressions	Desire for certainty; search for common ground
III. Adaptation	Consideration of previous remarks; slander	Need for approval; pleasure in confirming one's own opinion dispositions; group as an 'objective authority'
IV. Familiarity	Statement to other group members; concurring statements; supplementary heckling; consensus	Familiarity with the attitudes of group members; comfort with the collective; fear of isolation
V. Conformity	Uniform group opinion; no deviation of individuals; monologues; falling back on certain topics; taking sides against outsiders; defending against attempts at leadership; covering up aberrations	Contagion; group suggestion; identification; concern for group cohesion
VI. Fading away of the discussion	Decrease in tension; decrease in intensity of discussion; inattention; repetition	Sufficiency to the manufactured conformity; fatigue

After Spöhring (1989)

- In the first, qualitative phase of the evaluation, the issues raised throughout the discussion are worked through and the group's opinions on them are presented.
- Secondly, it is then possible to compare the discussion processes that come from members of the same social groups. In doing so, similarities and differences are to be uncovered.
- Third, "a comparison is made between the 'typical' group opinions of discussion groups of different social structure studied in the context of a sociological question" (Mangold 1973, p. 252).

Lamnek (1998, p. 162 ff.) provides two criteria according to which the evaluation of group discussions can be differentiated. Firstly, it is a question of whether a thematic evaluation, a group-dynamic evaluation or a synthesis of both approaches should be undertaken. Secondly, the evaluation can be described according to whether it should be descriptive, reductive or whether it should be explicative/intensive.

In a *descriptive* analysis, the main content of the discussion is summarised, typical quotations are presented and the course of the discussion is described. A *reductive* analysis focuses on the information and knowledge gained. For this purpose, the transcript produced is reduced in scope. Finally, an explicitly qualitative, hermeneutic interpretation (compare Oevermann et al. 1979) can be carried out.

For a detailed description of analysis strategies for group discussions, see Lamnek (1998, p. 162 ff.). Here you will also find the following overview of basic analysis strategies (see list below).

Basic Analysis Strategies for Group Discussions in Overview According to Lamnek (1998, p. 167)

- *Group dynamic analyses* are less significant and less frequent than the *content evaluations.*
- Therefore, group discussions are usually evaluated by *content analysis.*
- Group discussions can be evaluated *descriptively, reductively* and/or *explicatively.*
- The most common form of analysis is likely to be *descriptive,* followed by *reductive.*
- *Reductive* content analysis attempts to reduce the abundance of data material in such a way that *information* is *gained.*
- *Statistical-quantitative* analyses can be used for this purpose, but this is not necessarily required.
- A *qualitative-typological* evaluation is also a reductive analysis.
- *Interpretative-explicative-extensive* analysis procedures only occur in the scientific community and are therefore rather rare in group discussions.
- *Reductive* content analyses – regardless of whether they are statistically or typologically oriented – require the *formation of categories.* The diverse, conceptually differently conceived units
  - Encoding Unit
  - Context Unit
  - Evaluation unit
  - physical unit
  - Reference unit
  - thematic unity and so on

indicate different associated analytical strategies and cognitive goals.

Sociometric Surveys

In surveys, as described in the previous sections, information is mostly obtained from (many) individual persons. This information usually concerns the attitudes and behaviour of precisely these target persons. However, social science interest can also be directed at facts that are more closely related to people's social relationships. For example, it could be hypothesized that the party preference of a particular person is related to the party preferences of those people with whom that person has closer social relationships. In order to test such a hypothesis, it is clearly not sufficient to simply ask the target person about his or her own party preference. It would also be important to find out with which people this person maintains social relationships and then which parties these people favour.

According to another hypothesis, the opinion leadership of a person in a group could be linked to certain characteristics possessed by the person who holds this opinion leadership. In order to work on such a hypothesis, it is also not sufficient to only collect information concerning an individual person. Rather, it would be necessary to determine who is the opinion leader in a group in the first place and, at the same time, which characteristics are attributed to this person by the other group members.

Empirical social research uses sociometric surveys and – as a more advanced form – network analyses to address such and similar questions. The term sociometry is used to describe analyses of interpersonal preferences among group members. According to Bjerstedt (1956), it is a "quantitative study of interpersonal relations from the standpoints of preference, indifference, and rejection in a choice situation." Sociometric surveys use standardized interviews in real cooperating small social groups, such as school classes, work groups, army units, and similar social entities. Thus, this approach is characterized by the fact that here not only (many) atomized persons are considered, but complete groups.

In principle, it would also be possible to use the method of observation (see Sect. 6.2) in order to identify certain behaviours in a group. For example, the seating arrangement in a school class could be observed by the teacher and evaluated according to various aspects.

Jakob L. Moreno (1934) is considered the founder of sociometry. His intentions went far beyond the sociological level of analysis under discussion here. Moreno's visions consisted, in addition to the targeted integration of members of society into groups, also in the therapy of people and ultimately aimed at a reorganization of society along the lines of scientific socialism. Admittedly, these far-reaching goals proved (so far) to be impracticable.

The basic idea of the sociometric approach is that in social systems certain social structures are also based on informal relationships. These are made the subject of the sociometric surveys. Starting from a continuum ranging from affection to dislike, each group member is asked to evaluate his or her relationships with the other group members. In this way, sociometric configurations can be made visible. These structures, formed on the basis of affection and dislike, can be contacts, friendships, enmities, advisors, confidants, helpers, isolates and so on. Always the individuals in the group form the origin and the goal of such contacts. Different relationships are distinguished: pairs, triads, symmetrical and asymmetrical bridge relationships and so on.

For the representation of such findings, a sociomatrix is drawn up. In this matrix, the cells contain the reported contacts. The marginal totals correspond to the sum of the elections received. Table 6.19 shows a (fictitious) example of a result of a sociometric survey with eight (A to H) persons. Person A, for example, voted for person C and was also voted for by C.

Table 6.19 Example (fictitious) of a sociomatrix, above the diagonals are the votes cast and below the diagonals are the votes received

	A	B	C	D	E	F	G	H	Σ
A	–	0	1	1	1	1	0	0	4
B	0	–	0	0	1	1	1	1	4
C	1	0	–	1	0	0	1	1	4
D	0	0	1	–	1	1	1	0	4
E	1	1	0	0	–	1	1	0	4
F	0	0	1	1	0	–	1	1	4
G	0	1	0	1	0	1	–	1	4
H	1	1	0	0	0	1	1	–	4
Σ	3	3	3	4	3	6	6	4	32

Another possibility of evaluation is the calculation of indices and coefficients. There is a wealth of variants available for this (for an overview, see Nehnevajsa 1973, p. 270 ff.; Ardelt 1989, p. 184 ff.):

* Sociometric status: This is the sum of the elections received.
* Social status: This is the number of relationships of a person (column plus row sum).
* Cohesion index: The proportion of positive mutual elections in all possible elections represents this index.
* Identify sociometrically isolated individuals: These group members receive neither elections nor rejections.
* Sociometric outsiders: These are group members who receive more rejections than the average of the group.

Friedrichs (1990, p. 261 ff.) presents a number of other indices that can be used to evaluate sociometric surveys. He names, for example: the probability of choice, the interest quotient, the attraction quotient, the integration into one's own or into the other subgroup, the group entropy and the expansion.

Another possibility for the evaluation and presentation of the results of sociometric surveys are the sociograms. These are the directional graphs of the identified interactions. The following are listed as examples in Fig. 6.9: the triad (A, B and C), the outsider (E), the star (D receives numerous elections), the three-party cliques with (I, J and K or K, L and M) and the bridge relationship (person K is integrated into two cliques) as well as the grey eminence (person F is only elected by two stars of a group).

Fig. 6.9 Different group configurations, each shown with the help of sociograms

Triad A, B, C

A   B

C

Outsider (E)   E

Star (D)   D

Cliques of three with bridge relationship (K)

Grey Eminence (F)

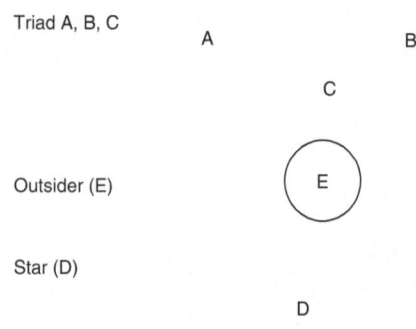

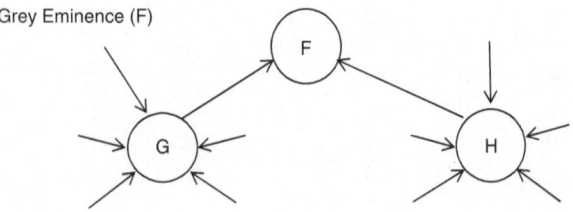

Northway's (1940) target charts provide another option. This is done by first determining the quartiles of the elections received. "Usually the individuals with highest quartiles (the, of course, estimated, 25% of the most frequently voted individuals) are placed in the inner ring of the diagram, etc., so that the outer ring includes the lowest quartile – the least voted individuals in relation to the voting pattern of the group" (Nehnevajsa 1973, p. 269).

The following methodological requirements should be met if one decides to conduct a sociometric survey:

- The group in question should not be too large. Each group member must know every other group member and be able to interact with them, at least in principle.
- It is necessary to derive criteria according to which the group structure is to be studied. Thus, activities that produce those sociometric configurations must be identified. Typical questions in a sociometric survey are: "Who would you most like to work with?" "Who would you most like to prepare the Christmas party with?" "Who would you most like to ask for advice and help on a personal matter?" "Who would you most like to invite to your home sometime?".
- Finally, a certain susceptibility of the findings to current group dynamic processes should be noted.

In addition, other aspects must be taken into account: For example, it has to be decided whether only positive or also negative choices (rejections) should be allowed in the survey. The latter provide a more complete picture of the group situation. However, it might be embarrassing for the group members to express such rejections to the investigator. What remains to be decided is how many choices per person are permissible. Here, too, in the interest of providing as complete a picture as possible of the group situation, it would be beneficial to obtain an assessment of all the members of the group. This would be possible, for example, with the help of a rating scale that asks the target persons the following task:

"How much do you like or dislike working with each member of your group? Please rate each group member using the following scale, ranging from very like to very dislike!" Other variations could include:

- How much would you like or dislike to do a seminar paper together with [names of all group members]?
- How much would you like or dislike doing something together with [names of all group members] in your free time?
- How much would you like or dislike asking [names of all group members] for help with a personal problem?
- Who would you most like to invite to your birthday party and who can't you imagine doing that with?
- To what extent do [names of all group members] possess the following qualities: ambition, confident manner, helpfulness, professional competence?
- Who in your group represented the ideal boss (the ideal colleague, the ideal employee) for you?
- Who do you think from your group will have indicated that they would like to work with you?

Sociometric surveys are usually organised as a written survey, which mostly takes place in the context of an examination, i.e. under supervision and guidance. Only rarely are socio-metric surveys conducted in person-orally, as an interview.

During the evaluation, the personality structures of the owners of certain relationships can now be studied. For example, what characteristics have people who form a clique together, people who are isolated, and so on.

Furthermore, it becomes possible to determine the social self-perception of individuals. For this purpose, the group is also asked about the assumed (positive and negative) choices. From this, the assumed and actual status of a person within a group can be determined. This in turn can be related to other characteristics of the respective personality.

Another way of examining group structures is to ask the members of the group who they would most like to see in the group as a leader, as a colleague or as an employee. If this questioning is combined with the assessment of various personality traits of the group members, such as competence, honesty, confident demeanour and so on, information is obtained about which traits are attached to the holders of the aforementioned functions.

Other facts, such as the preference for certain political views, can be predicted on the basis of sociometric structures (political preferences of the reference persons, their class affiliation and so on).

Sociometric surveys are used primarily in the educational sciences to examine the structures of school classes, for example.

In brief, we will present some of the substantive results that were obtained with the help of such sociometric surveys:

- Irregular distributions of sociometric choices are always to be expected. Thus, certain hierarchies exist in all social groups (compare Nehnevajsa 1973, p. 285).
- A high sociometric status correlates with insight into the relational relations of the respective group, a low status accordingly tends to express a false perception (compare Lemann and Solomon 1952; Gronlund 1955).
- There is a strong correlation between sociometric status and a person's level of intelli-gence. Achievement, mental health, socioeconomic status, and the degree of participa-tion in group life are interdependent (see Wardlow and Greene 1952; Grossman and Wrighter 1948; Thorpe 1955).
- Sociometric elections are usually directed upward in status hierarchies (compare Lundberg and Steele 1937; Proctor and Loomis 1951; Longmore 1948).
- Individuals with high social status are seen as holding the values of the group (compare Hallworth 1953; French and Mensh 1948; Powell 1951).
- Sociometric friends are mostly homogeneous in terms of educational level and in terms of general personality traits. They also live spatially closer together (compare Urdan and Lindzey 1954; Maisonneuve 1952).
- Groups assembled on the basis of their sociometric choices work together more effec-tively than others (compare v. Zelst 1952).
- The efficiency of group performance correlates with the sociometric cohesion of the group (compare Chesler et al. 1956; Fiedler 1954).

Nehnevajsa (1973, p. 285 ff.) presents further results.

Network Analyses

For some time now, the importance of network analyses has been growing strongly in the social sciences. This is a special type or a further development of sociometric methods. Network analyses go beyond the use of surveys (within groups). They also take into account that not only individuals, but also companies, organisations and even states can form networks.

While some authors regard sociometry as a "largely undemanding collection of simple survey techniques and index formations" (Schnell et al. 2005, p. 169 f.), this is by no means true of network analyses (compare Jütte 2002; Trappmann et al. 2005; Jansen 2003). First of all, it is important to distinguish between complete and ego-centered network analyses. *Complete network analyses include* direct information from all members of the network in the analysis. As the network is relatively extensive, this is a particularly time-consuming method of investigation. In many cases, therefore, *ego-centred network analyses are* produced as an alternative (cf. Wolf 2004, p. 244 ff.). These are based on a specific person to be examined. This person is asked about the persons to whom he or she relates. For this purpose, the method of ego-centered network analysis makes use of network or name generators. For example, the target persons were asked in the ALLBUS in 1980 and 1990:

"We now have some questions about the people you frequently hang out with in private: Please think for a moment about the three people with whom you are most often in private. These can be relatives as well as unrelated friends or acquaintances, just not people who live in the same household as you. For simplicity, let's call the three people "A," "B," and "C." So that they are not confused, please write down the first name or a special keyword for identification on this sheet. Always think of person A first in the next questions."

(Interviewer: hand over sheet and pen – first ask questions V755 to V761 for "A", then for "B", then for "C" and enter answers under the corresponding letter. If no persons at all are named as friends or acquaintances, then continue with next question).

Indicators that applied to the reference persons were, for example, "Is A male or female?" and "Can you tell me which party A voted for?"

With the help of such questions, problems can now be dealt with such as: Is it a heterogeneous or a homogeneous network, an open or an integrated network? What connections exist between the characteristics of the members of the network and those of the ego? Do characteristics of the network have an influence on the ego, for example on its political voting decisions?

However, there are also examples (compare also Baur 2019, p. 1281 ff.) of the use of complete network analyses. Three are presented below:

Schweizer and Schnegg (1998) analyse the social structure of a novel with the help of network analysis.[23] The basis for this is Ingo Schulze's book: "Simple Storys. A novel from the East German province" (1998). Here the changes of the people in the small town

---

[23] Text from 20.05.1989 from the Internet: http://www.uni-koeln.de/phil-fak/voelkerkunde/doc/simple.html, unfortunately however in the meantime no longer callable.

of Altenburg in East Thuringia are described. In the process, 38 people play an essential role. The narrative is laid out in such a way that in each case a particular person and his or her views are placed at the centre. In the individual sections the other persons appear again – related to each other. It is therefore a description of a complete social system consisting of a certain number of actors and relationships.

"An essential task of any ethnographer is to discern how persons operating in a social field are intertwined kinship-wise, economically, politically, and communicatively, and what subsets of persons (cliques, groups, positions) we can distinguish on the basis of close interactions and/or equal location in that system." (ibid.)

The analysis uses the following procedure. In a first step, all positive (e.g. marriage, kinship, friendship, love) and all negative (e.g. hatred, divorce, dismissal) relationships between these 38 people are registered (cf. Fig. 6.10). In a further step, a list is compiled that provides information about who is in which relationship with whom. Further analysis is carried out using computer programs. These make it possible to visualise the relationships that have been established. A graph illustrates all the relationships between these people. Further, only the positive relationships can be shown in a graph. Also, the existence of "*subgroups of* structurally similar actors […] who maintain similar relationships among themselves and with third parties can be determined." (p. 3).

As a result, network analysis proves to be a useful "research direction of ethnology and related social sciences (sociology, economics, psychology, communication studies)" to continue and further develop the old structuralism (Levi-Strauss 1967).

A second example of a complete network analysis is presented by Fürst et al. (1999). This is a study of the regional network of the city of Hannover. The network analysis is based on a regional science perspective.

The idea is that regions are considered particularly successful "where well interacting networks exist" (p. 1). In examining actor networks, the research group is interested in "whether such networks exist, who they include, and how they can influence topics of regional development, among other things, but they can also be less 'productive' for regional development if they exclude important actors, if they are very selective in their choice of topics, or if the actors themselves consider them to be relatively unimportant for their own actions." (p. 1).

The procedure was two-stage. In the first step (autumn 1997) all significant persons and organisations in the Hannover Region were recorded. This was followed by an initial reduction to 628 actors and 459 organisations, the criterion for the reduction being the influence on regional development which the actors are able to exert. In the second step (1998) this reduction was systematically continued, so that finally 179 persons remained in the pool.

The next step was to determine the dimensions to be studied. The decision was made on: (a) interorganizational relationships, (b) interpersonal relationships, (c) informal relationship options as well as on (d) group-structured relationships. A telephone survey

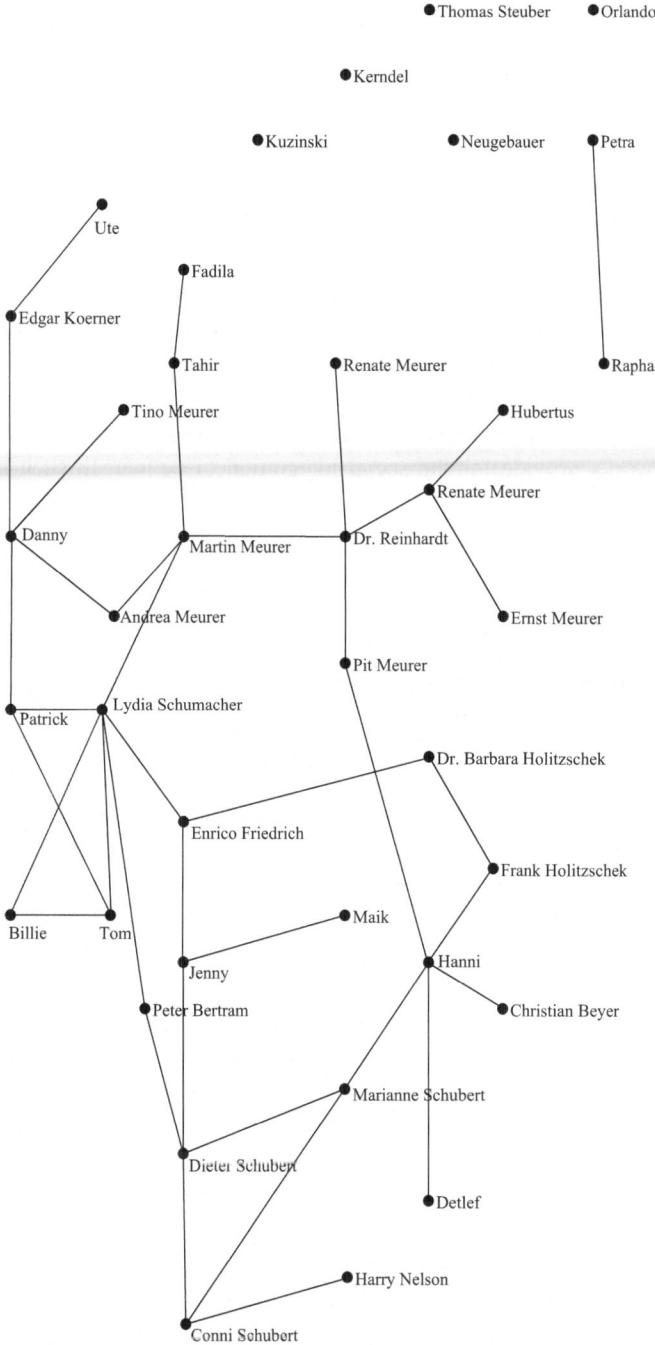

Fig. 6.10 Representation of positive relationships (marriage, kinship, love, friendship)

(following a postal announcement) was then used to collect the relevant empirical information. These now allow a description of the network based on its socio-demographic structure (p. 3), its age and occupational structure, according to sectors of activity, according to the strength of influence, according to the internal and external orientation of the actors, according to the field of engagement, as well as according to the nature of the ties between the actors.

The authors summarize the benefits of their study as follows: "The interaction of decision-makers in the region was not transparent until now. With the study of the regional actor network, the interplay has been mapped for the first time – like a photographic snapshot. The actors can recognize themselves from a distance as members of a network and evaluate the structure of the regional network itself." (p. 14).

Finally, a third example, which chose a somewhat different survey method, will be presented. It was about explaining fear of crime and finding a new job with the help of networks. Networks provide for the social integration of people. In such networks the exchange of information, goods, ideas and last but not least influences takes place. Granovetter (1973, 1995) first distinguishes between strong and weak social relationships in such networks. Thus, in close relationships – the strong ties – such as close friendships, more frequent interactions take place. Consequently, the members of such networks also know each other well. This is therefore a network of close social contacts. In weak ties, an exchange between the members takes place correspondingly less frequently; we are therefore dealing with loose relationships.

Empirically, it can now be demonstrated that it is primarily the weak relationships and not the close relationships that lead to success in the job search. This can be explained by the bridging function of weak relationships. Whereas in close relationships there is in any case a largely identical stock of information among the persons involved, it is above all the weak relationships that enable access to other circles and thus to other information.

However, the approach described has also acquired explanatory power for the occurrence of fear of crime. Firstly, close relationships can bring about a particularly high degree of social control in the neighbourhood and thus promote a high feeling of security. Secondly, however, the opposite is also conceivable, according to which such relationships lead to an isolation from other information, and so it happens that the foreign is perceived as hostile. In contrast, multiple weak relationships can contribute to a stronger exchange of information, to higher trust and thus to a weakening of fear of crime. On the basis of a secondary-analytical study with data from the 2009 SOEP, it has been possible to deal with this problem empirically. According to this study, "the fear-reducing effect of weak ties is preserved in the model of personal fear of crime ... The strong ties, on the other hand ... exhibit a fear-increasing effect" (cf. Gühne 2013, p. 93).

A question about the frequency of mutual visits in the neighbourhood served as the data basis here; it is considered an indicator of close social contacts. The SOEP also asked about the ratio of people in the neighbourhood; the answers here are considered an indicator of cohesion in the neighbourhood. Finally, the homogeneity of the neighborhood was to be surveyed via a question about foreigners living there.

Online-based social networks have now become enormously widespread. This offers special opportunities and challenges for social network research. Henry Polte (2017, p. 2 f.) warns, however: "The high availability of information about diverse interactions and relationships between individuals on the *world wide web,* however, also carries the risk that their analysis will deteriorate into a mere end in itself and lose the usefulness of sociological knowledge discovery."

Newer Forms of Questioning
In 1999, the Federal Statistical Office, together with the Arbeitskreis Deutscher Markt- und Sozialforschungsinstitute e. V. (ADM) and the Arbeitsgemeinschaft Sozialwissenschaftlicher Institute e. V. (ASI). (ASI) a conference devoted to survey instruments in social research at the beginning of the new century. At this conference, the then managing director of the ADM, Erich Wiegand, elaborated on five trends which, from the perspective of the time, describe the development and use of new survey instruments and influence the quality of research results (compare Wiegand 2000, p. 12 ff.). In doing so, he named:

- The pluralisation of interview forms. Whereas in the past there was a clear dominance of face-to-face interviews, according to the annual reports of the ADM institutes, various other forms of interview are now increasingly being used.
- Computerisation and mechanisation of data collection is taking place. Specifically, reference is made to the use of central CATI laboratories. These make it possible, for example, for speech to be recorded automatically, so that the interviewers no longer have to take notes of the answers, even in the case of open questions. Furthermore, the use of PCs or laptops to support interviews, CAPI interviews, is gaining in importance. In the business-to-business sector, the use of fax machines for sending questionnaires is increasing. There is also the sending of floppy disks on which the questionnaires to be processed are stored, the disk-by-mail method (DBM). Finally, various forms of online surveys are gaining importance. The course of group discussions is documented with the help of video recordings.
- For some time now, there has also been a noticeable tendency towards increased standardisation of the survey instruments. Particularly in the case of questions on the demographic characteristics of the target persons, corresponding standards, a so-called standard demography, have been developed. These allow, above all, a high degree of comparability of the results. However, the evaluation presentation in market research also uses such standards, for example, to show customer loyalty or the positioning of goods.
- Furthermore, a stabilization of the research process can be observed. Such continuous monitoring is now replacing ad hoc research. It enables market research to continuously observe product cycles, for example. Another important keyword at this point is the establishment of respondent pools, the access panels.
- A number of methodological problems arise from the increasing internationalisation of surveys. The problem of the comparability of the results is central to this.

In the following overview (compare Fig. 6.11), the variety of survey instruments now available to the social sciences is presented from a different angle. The three traditional

Paper   **Computer**

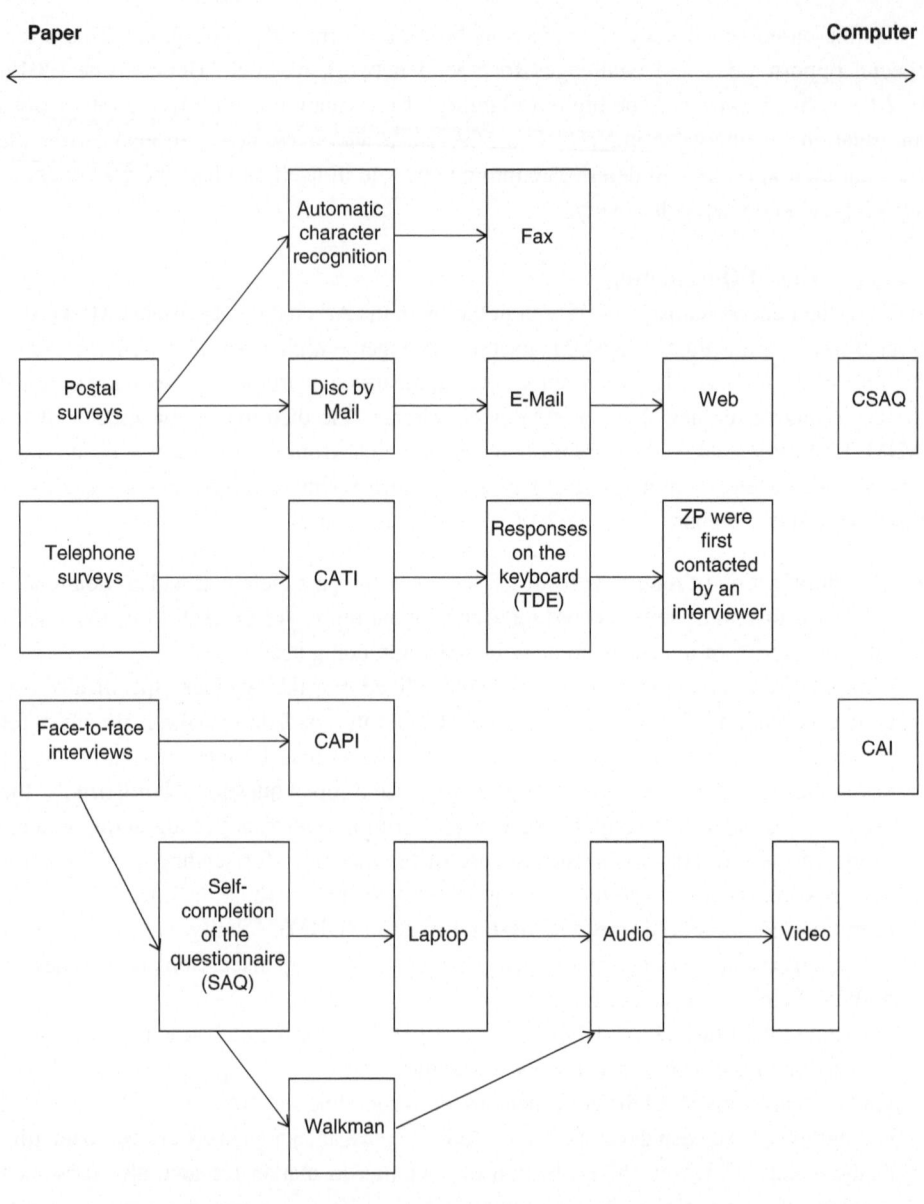

CSAQ: Computer Assisted Self-Assessment

CAI: computer-assisted interviewing

Fig. 6.11 The technological evolution of survey research. (Compare Groves et al. 2004, p. 140)

techniques (postal survey, telephone interview and face-to-face interview) are taken as a starting point and their computer support, which has meanwhile taken on many forms, is shown.

Caja Thimm and Patrick Nehls currently (2019, p. 974) identify four perspectives to describe digital methods. Thus, research can turn to (1) digital data that is process-produced, such as from Facebook, on Twitter, in blogs, or by means of online surveys. There are (2) methods of analysis available, for example in television research, to explore "digital phenomena". Such methods of analysis can (3) also be applied to digital objects such as networks. Finally, the authors cite (4) user studies as a separate focus of digital research.

In the following, only some of the tendencies mentioned will be dealt with in more detail. For example, the computer-assisted oral interview (CAPI), surveys conducted via the Internet or the Web (www), the use of fax machines or mail for the delivery of survey disks (disk-by-mail) and the recruitment of a fixed group of persons willing to take part in the survey (access panels) are considered.

CAPI: Computer Assisted Personal Interviewing
The special feature of face-to-face interviews is that the interviewer and the target person sit opposite each other and thus establish a social relationship. In conventional interviews (PAPI), the interviewer writes down the answers of the target person on a questionnaire made of paper. Now there are several advantages of replacing the interviewer's questionnaire with a personal computer. On the computer both the questionnaire and the answers of the interviewed person are stored. It is still possible for the interviewer to carry and use additional tools for the interview. These can be such things as template cards, product samples, magazine dummies or lists.

The ALLBUS has also been using the CAPI technique since the 2000 survey, although not without using a split to determine possible methodological effects in the transition from one survey mode to the other (cf. Koch et al. 2001).

Pfleiderer (2000, p. 57 ff.) discusses various motives that can cause such a switch to CAPI technology. He mentions four arguments: firstly, the costs incurred, secondly, the quality of the survey data, thirdly, the speed or the time required for the survey, and fourthly, the new possibilities offered by computer technology.

The cost discussion associated with CAPI technology goes in two directions: CAPI causes additional costs and at the same time saves other costs. Costs are incurred primarily due to the initial equipment of interviewers with PCs and the necessary software, due to the associated training of interviewers and due to the greater programming effort required. On the other hand, the elimination of the printing of questionnaires and the checking of data, the no longer necessary mailing costs (postage) and the overall time saved have a cost-saving effect.

The use of the CAPI technique can influence the quality of a survey in various ways. With regard to the questionnaire, positive influences can be expected due to the automation of filtering and the permutations of questions and/or answer specifications. In addition, there is the possibility of database-supported queries directly during the interview.

During pretests, the PC's internal clock can be used to obtain an accurate picture of the processing speed. In this way, for example, questions can be identified that require a particularly high time fund for their processing. A possibly negative effect could be the almost complete loss of flexibility with which the CAPI interviewer and thus also the target persons are confronted.

The realization of the sample can also be influenced both positively and negatively by CAPI. The stronger control of the interviewer makes at least partial falsifications more difficult, in which not all questions are asked properly by the interviewer. A corresponding control can be programmed and prevent the skipping of questions. In the future, it would also be conceivable for the interview to be Internet-based, mediated via a mobile phone. The questionnaire would then be retrieved from the network during the interview and the results would be stored immediately. This would make falsification even more difficult. The return control is also greatly simplified. With a daily delivery of the survey results to the institute via modem, it would always be possible to carry out appropriate checks. It is still largely unclear whether a lack of acceptance of PC technology is not to be found among certain persons, which could then generate systematic failures or response distortions. Reference should also be made to the problem of the power supply for the PCs and, finally, it cannot be completely ruled out that technical errors could cause a loss of data.

The quality of the data is in turn positively influenced by the use of computers, when apparently illogical answers are specifically questioned and wild codes (input errors) are avoided. The subsequent writing down of answers given in open questions, which is necessary in conventionally conducted interviews, is also omitted. The technical prerequisites exist for recording the answers given by the target persons during an interview to an open question via a microphone built into the laptop (compare Biemer et al. 2003). Such recordings can also be used to control the interviewers.

Another aspect to be discussed at this point concerns the time fund required for a survey. The opinion is expressed that the time fund required for CAPI surveys is only "minimally reduced at best" (Pfleiderer 2000, p. 67) compared to paper-and-pencil studies. The experience of Infratest Burke shows, however, that due to various other work steps, CAPI surveys save an average of five days compared to PAPI surveys (Pfleiderer 2000, p. 68).

A final line of argument concerns the many new possibilities that the use of multimedia PCs brings to CAPI. For example, commercials can be shown on the screen and audio samples can also be played.

Surveys Via the World Wide Web (www)

The expansion of the circle of people using the Internet is taking place at great speed (compare Groves et al. 2004, p. 151). According to one source, it is an increase of about 0.5% per month.[24] This raises the question of using the resulting new opportunities for social science surveys as well. "The advantages of such an endeavor are obvious, since

---

[24] Compare also: http://www.clickz.com/stats/sectors/geographics/article.php/3468391 last accessed on Feb 28, 2019.

surveys via the Internet, in comparison to postal surveys, require less financial effort and surveys with extremely high case numbers can be realized within short periods of time" (Bandilla and Bosnjak 2000, p. 166). In addition, there is the elimination of interviewer influence, problem-free data collection, the possibility of designing the questionnaire in a relatively sophisticated way, and the enormously increased costs of alternative survey techniques. However, reservations of principle should not be overlooked either. For example, it is also suspected that such surveys can hardly be used for serious scientific purposes in the medium term (Schnell 2012, p. 287 ff.). The assessment of Wagner-Schelewsky and Hering (2019, p. 797) is more positive.

In view of the great speed at which the Internet user base is expanding, it is extremely difficult to make a judgement on the possibilities that present themselves for the organisation of surveys. It currently appears that Internet-based surveys will be used in the near future primarily in the following areas (Bandilla and Bosnjak 2000, p. 116):

- In employee surveys, which can take place both locally and in internationally networked companies.
- For the survey of special populations (for example, experts and students), a large percentage of whom have access to the Internet. Delphi studies (cf. Kirsch 2000; Florian 2000) can also be conducted via the Internet.
- The survey of registered Internet users is characterized by the fact that the population is known and thus appropriate selection procedures can be developed.
- Internet-based studies have already proven their suitability for pretests in preparation for general population surveys (compare Bandilla and Bosnjak 2000, p. 166).

Furthermore, such instruments appear to be suitable when it comes to estimating models. This has been demonstrated, for example, using models to explain voting decisions and voter turnout (cf. Bieber and Bytzek 2012, p. 185 ff.; Zins 2018 and our comments in Sect. 5.6). However, such surveys are not a suitable instrument if parameters of the population, such as an election result, are to be predicted.

However, one should be warned against too much euphoria and too much authority, which apparently comes from large numbers of participants in surveys, such as this statement: "How is Germany? This is the question addressed by a large online survey [...] At the end of last year citizens were asked to give information about what they think, what they feel, what they demand. The response was overwhelming: 450,000 Germans filled out the questionnaire on the Internet. This makes 'Perspective on Germany' the most informative socio-political survey we have ever had" (Wintzenberg 2004, p. 54). It is certainly possible to achieve high numbers of participants quickly and cheaply in this way, but this alone is not a quality criterion for a survey (cf. Sect. 5.2).

Furthermore, it is problematic with this survey mode to motivate the potential target persons to participate. Experience has shown that this results in a high proportion of nonresponse. Furthermore, there are still strong social differences in access to the Internet. For example, older people, people in households with a lower income and people who are not in employment are much less likely to have access to the Internet. Undercoverage is the consequence here (see Schnell 2012, p. 288 f.).

The use of the Internet for surveys can be done in different ways. Currently, three main strategies are implemented: the survey is conducted by e-mail, the survey is conducted by means of WWW questionnaires or online panels are used for this purpose.

E-mail surveys are the simplest variant: the questionnaires are sent to the target persons in this way cheaply, quickly and asynchronously. The recipients process the questionnaire and send it back in the same way. This procedure has been used since the 1980s (compare Kiesler and Sproull 1986).

The Hypertext Markup Language (HTML) software is used for WWW surveys. This offers numerous good possibilities for the questionnaire design, for example a support of the survey by sounds and (moving) pictures. All persons who have access to the corresponding pages on the Internet can participate in such surveys. Such survey campaigns can be publicised via advertising banners. Thus, a controlled strategy cannot be assumed when assessing the sampling. Accordingly, the findings of such self-recruited studies should be interpreted with caution.

Online panels are a special form of access panels. Ideally, they can be recruited via random selection – for example, by telephone. Interested persons agree to participate (more or less) regularly in surveys. This group of people provides their demographic information. This enables the organiser of an online survey via the Internet to target a specific group of people to be surveyed.

Finally, reference should be made to the possibility of correcting the data obtained via the Internet by means of a propensity weighting. The possibilities and limitations of this tool have already been mentioned elsewhere (cf. Sect. 5.8.2).

Fax Surveys and Disk-by-Mail

In addition to the possibilities described above for communicating the questionnaire, there are two other possibilities: The questionnaire can be sent to the target person by fax, filled out by him/her and sent back to the survey institute by fax as well. Another variant is the disk-by-mail method. This consists of programming the questionnaire, putting it on a floppy disk or on a CD and then sending it to the target person. Both procedures are mainly used in the business-to-business area (compare Wiegand 2000).

The first use is said to have been in March 1986 in the USA (COMPAQ). The prerequisite is that all participants have access to a PC. In such cases, company surveys (such as here with hardware and software producers) are then possible, for example. It is important and must be assumed that there is a strong motivation to participate, for example due to an expected high own benefit from the results of the survey. The advantages are then obvious:

- There is no need for a separate transfer to data media and thus a time saving and error minimization can be assumed.
- Plausibility and consistency checks are possible during the survey.
- The filter guidance is automated.
- A random permutation of the questions becomes possible, thus sequence effects can be avoided.

- A relatively good response rate of around 50% was achieved for the relatively infrequent applications, which in turn indicates a relatively high level of acceptance of this technology.

However, this is confronted with the following limitations and problems:

- An extensive test phase is necessary, similar to CATI and CAPI surveys, in order to be able to use the survey program.
- Open-ended questions are of limited use, as there is a reluctance to answer them.
- A possibility to go back through the questionnaire and correct answers already given should be provided.
- The appropriate handling of categories such as "no answer" or "don't know" must be determined. If there is an explicit requirement, the choice of such categories is increased; if there is no requirement, the item non-response rate is increased. A possible solution is to set a time limit. If this time limit is exceeded, the respective question can automatically be declared as a special level (item nonresponse).
- Furthermore, various technical aspects need to be clarified. These include the operating system, drive types, taking advantage of color monitors, compatibility with other installed programs and the like.
- Queries should be made available to the target persons in as simple a manner as possible.
- The questionnaire must be easy to handle by the respondents. For this purpose, help windows, the F1 key, the ESC key and the like could be provided.
- Attention should be paid to virus protection. This can be done via a separate notice, the disk can be sealed in a foil and so on.
- Again, telephone follow-ups should be made.

Attempts to determine the performance of fax surveys are reported by Schneid (compare 1995, 1997); Bathelt and Griebel (2001) and Walker (1994). Here, the focus is again primarily on cost and benefit arguments. In the case of postal dispatch, postage and time are the main cost factors. Here, a fax survey proved to be faster and cheaper in each case. Thus one needed for a world-wide study only 40 seconds per side and/or 1.29 €. The example results from a survey of 291 market research institutes in 52 countries. The focus was on the use of computer-assisted survey programs. Problems were encountered, for example, in the surveys in Cuba, Lebanon and Nepal, where it was not possible to determine FAX numbers. 65 institutes could not be reached by fax even after three attempts. However, there is no comparative data to evaluate this finding. The very fast response (on average one to two days) as well as the favourable costs are also to be noted positively.

Another experiment concerned a fax survey in Europe (compare Schneid 1997). Here the costs for one page were only between 0.33 € and 0.52 €. In addition, the cheaper night dispatch via modem, i.e. after 6:00 pm, could be used.

Walker (1994) reports on a study in which participants were initially surveyed by post. The questionnaire was sent by fax to all those target persons who did not answer the questionnaire sent by post. This procedure gave the target persons the impression of a special urgency and led to a response rate of 83% (compare also Babbie 2002, p. 265 f.).

In summary, it can be assessed: Computer-assisted fax programs can also be used for sending questionnaires. Here, for example, the number of attempts (dialing) can be determined. However, there are numerous technical problems. For example, a relatively long transmission time as well as more complex programming work must sometimes be accepted.

For general population surveys, both approaches (fax surveys and disk-by-mail) are currently not suitable strategies. Reference should also be made to possible legal problems with (unsolicited) fax surveys.

In view of the ever-increasing use of the Internet and the possibilities it offers for electronic transmission, both fax surveys and the disk-by-mail technique are unlikely to become too important in the future.

Access Panels

Access panels do not necessarily have anything to do with the use of computers in surveys, but they are often associated with them. However, they should also be presented here as a strategy that has recently become increasingly important. Access panels are characterised in the literature as follows (compare Hoppe 2000, p. 147):

- It is a number of pre-recruited households or target persons.
- For a survey, a certain group of participants is selected from this pool.
- Such a selection can be made in a relatively targeted manner, since a great deal of information is already available about the individual participants.
- The survey is then conducted on an ad hoc basis.
- It has been established that this technique is particularly suitable for exclusive target group surveys.

The technique of access panels was developed against the background of and as a solution to a complicating situation in survey research. This can be described above all by an increasing reservation in the population towards interviewers, by a solidified scepticism with regard to the protection of personal data as well as by drastically increasing costs for the actual data collection. Furthermore, there is an increased fractionalization of target groups in the market economy, i.e. there is an interest in people who have very specific characteristics. These are correspondingly difficult to find in surveys. Last but not least, considerable time pressure is likely to have contributed to the particular attractiveness of access panels (cf. Hoppe 2000, p. 148).

The members of an access panel can be interviewed by telephone, by post or via the Internet. A certain pioneering role has been played here by the Netherlands with the telepanels (cf. Saris 1998) in which television sets play a decisive role. Here, a sample of households is regularly surveyed via videotext on screens. For this purpose, they are connected to the computer of the survey institute via modems or telephone lines. This equipment is made available to the households free of charge in order to prevent distortions. It is also likely that even older people are experienced in using the remote control.

Bandilla and Bosnjak (2000, p. 168) refer to a project attempt undertaken in the USA. Here, the mediation between the target person and the survey institute is carried out via the Internet. For this purpose, 30,000 households are provided with Internet computers free of charge. The aim is to obtain samples that reflect the general population.

A central problem in the use of access panels is the question of the extent to which the results obtained can be generalised. As a rule, it is not assumed that access panels are suitable for conducting surveys whose results can be transferred to the general population. Advocates of this procedure are of the opinion – probably rightly – that it is at least possible to obtain information about certain target groups, for example the consumers of a very specific product, in this way (cf. Hoppe 2000, p. 147).

Bandilla presents an experiment that addresses the question of the extent to which the results of a face-to-face oral general population survey and the results of a survey obtained via an Internet-based access panel can be compared. As a result, it turns out that this is not possible at the level of the general population. However, if we restrict ourselves to the subpopulation that was also represented in the access panel and compare these findings with the findings for the same subpopulation from the general population survey, we see that there are no longer any fundamental differences in the results here. This experiment supports the assumption that there are no differences in principle in the results simply because of the different survey mode.

Access panels sometimes contain a large number of potential respondents. The market research institute IPSOS, based in Germany, says that a panel size of 35,000 households and 85,000 people is appropriate for the research carried out at this institute (cf. Hoppe 2000, p. 162).

Schnell's fundamental criticism of access panels should not go unmentioned. His reservations concern above all the recruitment strategies for such panels (Schnell et al. 2005, p. 291 ff.). He expressed doubts about the usefulness of access panels for a long time to come, especially because of the lack of possibilities to draw random samples.

Also worthy of mention are efforts to install, at relatively great expense, a random sampling-based access panel for science. It is called the Open Omnibus Access Panel for Social Science.[25] It takes into account that certain individuals cannot be surveyed over the Internet. Therefore, a mixed-mode design is envisaged. Approximately 65% of the target respondents will each be interviewed via the Internet and the remaining 35% will be interviewed by mail. The panel is to comprise 4000 persons between the ages of 18 and 70. Interested scientists can apply to join the panel here.

6.1.4 Examples of Survey Studies

A whole range of data sets is available to empirical social research and an interested public. These are mainly processed by the GESIS Leibniz Institute for the Social Sciences, Department

---

[25] Compare   http://www.gesis.org/unser-angebot/daten-erheben/gesis-panel/   last   accessed 03/01/2019.

of Data Archive and Data Analysis, and can be used for a relatively small fee[26] or in some cases free of charge via the Internet, especially for scientific and educational purposes. The research deposited here was mostly publicly funded. Thus, a wide use of these resources represents an optimal use of spent funds. It is also important for replications and for reanalyses that these can be traced using the data set. The following surveys are described in more detail: the General Population Survey of the Social Sciences (ALLBUS), the European Social Survey (ESS), the Microcensus (MZ) and the Socio-Economic Panel (SOEP).

Numerous other datasets can be found in the dataset catalogue using online searches[27] and can be obtained from the GESIS Leibniz Institute for the Social Sciences.

The General Population Survey of the Social Sciences (ALLBUS)

Various sources of information on the ALLBUS are available to users. A number of these can also be accessed online. The following are particularly helpful:

- *Codebooks* documenting the questionnaires used in the ALLBUS are available as ASCII files, on DVD[28] or on paper.
- The *datasets of* the individual surveys as well as some cumulative datasets containing all indicators that have been replicated at least once.
- The *method reports,* which provide information on the origin of the questions used, for example, are available by post, online or also on a CD-ROM.
- A *bibliography* of all (known) papers that have used ALLBUS data and much more information on ALLBUS is also available.

The ALLBUS is a survey series on attitudes, behaviour and the social structure of the population in Germany (cf. Mayer and Schmidt 1984; Braun and Mohler 1991; Terwey 1998; Koch and Wasmer 2004). It has been conducted since 1980 with the help of face-to-face interviews at two-year intervals. In 1991, due to German unification, an additional survey was conducted in which persons living in East Germany were included for the first time. Since 2000, the CAPI technique has been used for the survey. The ALLBUS attaches particular importance to a methodologically sophisticated implementation as well as to a transparent presentation of the methodology used.

The ALLBUS questionnaire has the following structure: As a rule, one or two main topics (for example, deviant behavior, religious orientations, attitudes toward social inequality) are addressed in each survey. In addition, batteries of items and smaller individual indicators are used for various other areas. The main topics are replicated every ten years, the individual indicators and the item batteries every two or four years.

---

[26] The GESIS Leibniz Institute for the Social Sciences fee schedule for ordering datasets and codebooks is available online at: https://www.gesis.org/fileadmin/upload/dienstleistung/daten/umfragedaten/_bgordnung_bestellen/Gebuehrenordnung.pdf last accessed on 2019-03-01.

[27] Compare: https://dbk.gesis.org/dbksearch/?db=d last accessed 03/01/2019.

[28] Compare: https://www.gesis.org/allbus/download/ last accessed 03/01/2019.

Table 6.20 Overview of the main topics of ALLBUS

1980	Politics and authorities (attitudes towards different political aspects, party sympathy, assessment of social conflicts, contacts with authorities and attitudes towards authorities) Friendship relations (characteristics of friends)
1982	Religion and worldview
1984	Social inequality and the welfare state
1986	Education and cultural skills
1988	Attitudes towards the political system and political participation
1990	Sanction and deviant behavior, politics and authorities, friendship relations (partial replication of the 1980 focus)
1991	Family, career, inequality, politics, problems of German unification
1992	Religion and Worldview (partial replication of ALLBUS 1982) Problems of Unification
1994	Social Inequality and the Welfare State (Replication of the ALLBUS 1984)
1996	Attitudes towards ethnic groups in Germany
1998	Political participation and attitudes towards the political system
2000	No main topic, instead replications of questions from the entire previous ALLBUS programme
2002	Religion, worldview, and values (partial replication of the 1982 and 1992 religion emphasis)
2004	Leisure activities and media use, social inequality and welfare state, political attitudes, technological progress and computers, health
2006	Attitudes towards ethnic groups in Germany (replication ALLBUS 1996)
2008	Political participation and political culture (replication ALLBUS 1988 and 1998), social capital; as ISSP focus: leisure and sports, religion III (replication of 1991 and 1998)
2010	Replication from the entire question program, friendship relations as ISSP focus: Social inequality IV; environment III
2012	Religion and worldview (replication of 1982, 1992 and 2002) as ISSP focus: Health; family and changing gender roles IV

Table 6.20[29] contains an overview of the varying main topics of the individual ALLBUS surveys.

Each ALLBUS contains questions on the respondent's left-right classification, voting intention ("Sunday question"), materialistic or post-materialistic values (Inglehart items), political interest, subjective class classification, trade union and party membership, religious denomination and frequency of church attendance. The survey is supplemented by very detailed demographic information on the respondent.

Up to and including 1990, all persons eligible to vote in the (old) Federal Republic and in West Berlin living in private households constituted the basic population of the ALLBUS surveys. Since 1991, it has been the adult resident population (i.e. Germans and foreigners, provided they have sufficient command of the German language) in West and East Germany.

---

[29] Compare: https://www.gesis.org/allbus/inhalte-suche/frageprogramm/ last accessed 03/01/2019.

Until 1990, the sample size comprised about 3000 respondents; in 1991, about 1500 interviews each were conducted in West and East Germany; and since 1992, the net sample size has been about 2400 interviews in the Old and 1100 interviews in the New Länder.

The sampling strategy used from 1980 to 1992 and in 1998 was the ADM sampling design. In 1994, 1996 and since 2000, municipal samples with the drawing of personal addresses from the population registers were used. The ALLBUS is therefore not a panel study.

The ALLBUS is a GESIS project that is realized in cooperation with a scientific advisory board. The ALLBUS working group at the GESIS Department of Continuous Monitoring of Society is responsible for the conception, design, questionnaire and implementation (in cooperation with the private survey institute commissioned) of the surveys.

The European Social Survey (ESS)
The European Social Survey (ESS) is a series of studies launched in 2002 with the aim of identifying and explaining, in a methodologically particularly sophisticated way, the relationship between the changing European institutions on the one hand and the attitudes and behavioural patterns of national populations on the other. The second objective is methodological: to develop a sophisticated instrument that can be used in an intercultural context.

Similar to the ALLBUS, the materials and documents of the ESS – including the data file – can be obtained free of charge via the Internet.[30] The ESS also strives for a high degree of transparency and accordingly provides users with extensive materials.

The surveys will be repeated at two-year intervals. Participating countries in the first wave are: Austria, Belgium, the Czech Republic, Denmark, Finland, France, Germany, Greece, Hungary, Ireland, Israel, Italy, Luxembourg, the Netherlands, Norway, Poland, Portugal, Slovenia, Spain, Sweden, Switzerland, Turkey and the United Kingdom.

While the coordination of the study is funded by the European Science Foundation and supported by a separate group, the individual national studies are funded by national research funding agencies. At the national level, national coordinators are responsible for the work on the ESS.

Data collection uses face-to-face oral interviews lasting about one hour. These interviews are followed by a short supplementary questionnaire to be completed by the target persons themselves. The questionnaire consists of a core module. This is to be kept largely constant. Furthermore, there are two rotating modules. They are used at certain intervals. Each module is dedicated to a specific topic. The rotating module aims to take an in-depth look at a number of individual matters of academic or political interest. The core module aims instead to observe the change and continuity of a wide range of socio-economic, socio-political, socio-psychological and socio-demographic variables.

In order to be able to present a study that is as methodologically unassailable as possible and to ensure the principle of equality or equivalence, a number of conventions have been drawn up that have been declared binding for the individual national studies. Such

---

[30] Source: www.europeansocialsurvey.org, last accessed 01 March 2019.

conventions mainly concern the selection strategy, the translation of the questionnaires and all other materials as well as the fieldwork. This also ensures that all results are documented in a comparable manner.

Special attention is paid, for example, to achieving a high response rate, permanent reporting by the field institutes on the progress of the work already during the field phase, interviewer training on the basis of a uniform guideline, the use of standardised announcement letters and appropriate contact protocols.

The study carried out in this large number of countries has to face facts that are typical of international and intercultural studies:

- Ensuring comparability of national samples through consistent use of random sampling.
- Achieving a high participation rate among all nations.
- The compilation and translation of tested, functionally equivalent questionnaires.
- Ensuring consistent fieldwork and appropriate coding across all countries involved.
- Harmonizing measurement tools such as the left-right scale and the liberal-authoritarian continuum.
- Ensuring easy and quick access to documented data.

Specifications to be followed by all ESS participating countries concerned, in addition to the *selection strategy,* above all:

Fieldwork
All interviews were conducted in face-to-face oral mode by trained interviewers, who were required to make at least four attempts to contact the target subjects on different days and at different times. The number of interviews to be conducted by one interviewer was limited. In addition, a central quality control was carried out (see also Sect. 7.5).

Translation
The original ESS questionnaire was annotated for its translation. A separate translation of the questionnaire had to be made for each linguistic minority in a country that makes up more than 5% of the population. All translations have been assessed by third parties and each step taken in the translation has been documented for further work.

Observation of Events
All responses to a survey can be influenced by current events. The proximity of an election, economic or political unrest or even a natural disaster can change the response behaviour. For future evaluations of the ESS, such influences must be taken into account. Each participating country therefore documents important political, social and economic factors during the field period. Processed in a standard, these reports become components of each national data set.

Reliability and Validity
The ESS uses a variety of instruments to reduce and correct measurement errors. Thus, alternative approaches were tested in various experiments. Preliminary investigations took place in all countries.

Access

The fully documented data sets of the multinational rounds are available – since summer 2003. Scientists, journalists and politically interested people in Europe and beyond have access to these data. A download of all important documents from the internet is possible.

The Microcensus (MZ)

A great deal of information on the Microcensus (MZ) can also be found on the Internet.[31] The Microcensus is the official representative statistic on the population and the labour market of the Federal Republic of Germany, in which 1% of all households participate each year (compare Lüttinger 1999; Müller 1999, p. 7 ff.). A total of about 370,000 households with 820,000 persons participate in the survey, including about 160,000 persons in about 70,000 households in the new Länder and in East Berlin. The Federal Statistical Office is responsible for the organisational and technical preparation of the microcensus. The implementation of the survey and the processing are then the responsibility of the Land Statistical Offices.

All households have the same selection probability in the Microcensus (compare Hartmann and Schimpl-Neimanns 1992). A single-stage stratified area sample is conducted, i.e. areas are selected from the federal territory in which all households and persons are interviewed. One quarter of all households in the sample are replaced annually. Consequently, each household remains in the sample for four years.

The questionnaire programme of the microcensus consists of a fixed basic programme and a supplementary programme with annually recurring facts, most of which are subject to an obligation to provide information. This makes participation in the survey compulsory for the target persons. In this way it is possible to achieve a response rate close to 100%. At the same time, this obligation to provide information results in an obligation on the part of the organiser to monitor compliance with data protection regulations particularly carefully. In addition, there are supplementary programmes every four years which are partially exempt from the obligation to provide information. For reasons of data protection, the original data of the microcensus are anonymised in a special way (cf. Wirth 1992, p. 7 ff.) and can then be made available to scientists as scientific-use files for their own analyses. A simple download of the data sets from the Internet is therefore not possible.

The microcensus serves to provide statistical information on the economic and social situation of the population as well as on employment, the labour market and education. It updates the results of the population censuses. It also serves to evaluate other official statistics, such as the sample survey on income and consumption.

The annual basic programme of the microcensus includes, among other things, personal characteristics (such as age, sex, nationality), family and household context and, in addition, the characteristics of main and secondary residence, employment, job search, unemployment, inactivity, child of pre-school age, pupil, student, general and vocational training qualification, sources of subsistence as well as information on statutory pension insurance, nursing care insurance and the level of individual and household net income.

---

[31] Compare: https://ergebnisse.zensus2011.de/ last accessed 03/01/2019.

In the annual supplementary programme, among other things, additional questions are asked about employment and information is collected on previous employment as well as on vocational and general education and training. In the four-year supplementary programme, for example, information is collected on commuting for work and education, on the housing situation, on health insurance and on health and disability status.

At present, the 2006 microcensus – as well as the data sets of previous surveys – can be made available as scientific-use files. All files are distributed on a CD-ROM in ASCII format.

The provision of the de facto anonymised individual material is carried out by the Federal Statistical Office. The Federal Statistical Office, in cooperation with the GESIS Leibniz Institute for the Social Sciences, provides support and information for users. Detailed information documents on available variables and used classifications can be obtained from the Federal Statistical Office. In cooperation with GESIS, further information on the data, such as questionnaires, data set descriptions, setups for reading the microcensus files into SPSS, are also made available on the World Wide Web and can be accessed there.

The Socio-Economic Panel (SOEP)

The Socio-Economic Panel (SOEP) is a repeat survey of private households in Germany that claims to provide "representative findings". It has been conducted annually since 1984 among the same individuals and families in the Federal Republic. In June 1990, i.e. before monetary, economic and social union, the survey was extended to the territory of the former GDR. To adequately capture social change, an "immigrant sample" was introduced in 1994/95. Other additional samples were integrated into the current survey in later years.

With the help of the SOEP, political and social changes can be observed and analyzed. The data set provides information on objective living conditions as well as on the subjectively perceived quality of life, on changes in different areas of life and on the dependencies that exist between different areas of life and their changes.

The strengths of the SOEP consist above all in its special analysis possibilities through:

* the longitudinal design (panel character, cf. Sect. 4.6),
* the household context (interviewing all adult household members),
* the possibility for intra-German comparisons,
* the disproportionate sample of foreigners (The SOEP is currently the largest repeat survey of foreigners in the Federal Republic of Germany. The sample also includes households with a head of household of Turkish, Spanish, Italian, Greek or former Yugoslavian nationality. The questionnaire was translated for this purpose),
* the survey of immigration (the SOEP is currently the only sample of immigrants who came to West Germany between 1984 and 1995).

The SOEP is characterized by a high stability in participation. In 1984, 5921 households with 12,290 successfully interviewed persons participated in the SOEP-West survey. After 23 waves in 2006, there are still 3476 households with 6203 persons.[32]

---

[32] For SOEP comparisons: http://www.diw.de/deutsch/soep/uebersicht_ueber_das_soep/27180. html#79566 last accessed 03/01/2019.

The SOEP covers a wide range of topics. It continuously provides information on, among other things: household composition and housing situation, employment and family biographies, labor force participation and occupational mobility, income patterns, health, social participation and life satisfaction. The SOEP also focuses on different topics each year.

Since the survey year 2000, youth-specific biographical data have also been collected from 16- to 17-year-old household members. Since 2003, mothers of newborns have been answering questions about central indicators that have a high explanatory power for the developmental processes of children. And finally: Since 2005, the parents of two- and three-year-old children have also been surveyed in particular; since the 2003 birth cohort, the SOEP has thus also been a birth cohort study.

The anonymized SOEP micro-data set is made available to universities and other research institutions for research and teaching purposes for a small user fee. For reasons of data protection, the use of the data requires the conclusion of a data transfer agreement with the German Institute for Economic Research in Berlin (DIW).

The SOEP data are distributed as raw data and in various formats on CD-ROM. Training courses on the use of SOEP data are held annually. The SOEP-NEWSLETTER[33] regularly informs all data users about innovations. The SOEP is also conducted and developed in the form of a service facility for Leibniz Association research (WGL) based at DIW Berlin. The SOEP group disseminates the data to interested professionals and produces its own analyses. The fieldwork is carried out by Infratest Sozialforschung (Munich).

6.2   Social Science Observations

Scientific observation, along with interviewing and content analysis, represents one of the three basic data collection methods in the social sciences. In this section, some methodological assumptions associated with observation design are discussed. Then the various forms of scientific observation are presented and finally the application of the method is demonstrated by means of an example.

6.2.1   Basic Problems of Scientific Observations

There are numerous similar expressions for observing in colloquial language: for example, to look at, to stare at, to look at, to keep an eye on, to fix, to spy, to eyeball, to notice, and so on (see also Bortz and Döring 2002, p. 263). Observing, then, is a multifaceted everyday affair. Thus, one might assume that there is a certain kinship between scientific observation and everyday observations. Indeed, what both forms have in common is that they require certain decisions. For example, it must be decided what exactly is to be the focus

---

[33] Compare https://www.diw.de/de/diw_02.c.222316.de/soepnewsletter.html last accessed 03/01/2019.

of attention. Furthermore, in everyday life as well as in science, observing requires interpretations in order to evaluate what has been observed. By way of introduction, the term and the criteria for scientific observation will be clarified.

In the social sciences, a narrow and a broad meaning of the term observation is used. Following the *broad* version, observation is a method that social research uses in all forms of empirical inquiry. An interviewer asks a question and records – he observes – the answer of the target person. A content-analytical researcher observes – he evaluates – the entries in old diaries and interprets them. A market researcher discusses the advantages and disadvantages of a new product with a group and registers – he observes again – its marketing possibilities (compare Mees and Selg 1977; Cranach and Frenz 1969).

In a *narrower sense,* observation means the direct, immediate registration of facts relevant to a research context. The observed facts can be language, certain behaviours, also non-verbal behaviour (facial expressions, gestures), social characteristics (type of vehicle, clothing, badges and similar symbols) and coagulated behaviour (shop window displays, doorbell signs and similar things). "Compared with ordinary perception, observational behavior is more planned, more selective, determined by a search attitude, and directed from the outset toward the possibility of evaluating what is observed in terms of overarching intention" (Graumann 1966, p. 86).

As already indicated, observations and perceptions also take place in everyday life, so that it is appropriate to distinguish scientific observation from everyday observations more precisely. The following criteria must be fulfilled by an observation in order to earn the designation scientific (compare also Greve and Wentura 1997, p. 12 ff.):

- Hypotheses form the basis for scientific observation. They specify which relationships are to be clarified with the help of observation (cf. Feger 1983, p. 3). König (1973, p. 32) speaks of the principle of purposefulness of scientific observation. The observer makes a conscious distinction. Here, the everyday routine of observation on the one hand is contrasted with a systematic processing of certain questions with the inclusion of theoretical preliminary discussions in scientific observations on the other.
- Scientific observations must be subjected to some form of control. This can be done, for example, with the help of several observers whose results are compared with each other. The comprehensibility of the findings of a scientific observation can also be guaranteed with the help of a strict system. König speaks here of the control principle (König 1973, p. 31).
- Another criterion of scientific observations is their purposefulness. The selection of the units to be observed must be justified according to scientific criteria. Implicitly, this also determines the facts that are not to be observed.
- Finally, scientific observations are designed from the outset in such a way that they can be subjected to systematic, intersubjectively comprehensible evaluation and replication.

Some journalistic and ethnological attempts – for example Günter Wallraff's descriptions of his self-observations as a supposedly foreign worker in Germany – may be located here in the border area to scientific observations. However, they do not fulfil all of the criteria mentioned.

Also, there is no question that. s as a result of random observations quite interesting scientific findings can be drawn.[34] Merton originally coined the term "serendipity pattern" for this (1957, p. 104 ff.). This is "the chance observation of an unexpected, abnormal but strategically important datum which becomes the occasion for a new theory or the extension of an old theory […] it is the ability to preserve naive observation in a scientifically trained observer who is also able to recognize the implications of what chance has brought to him that is meant" (König 1973, p. 25).

Scientific observation has to deal purposefully and systematically with two problems. A first problem consists in selective perception. In principle, it is not possible to observe a social situation in its totality. For example, so many stimuli emanate from a group interaction that it would not be possible – even with modern technical aids – to register them all. Therefore, it is necessary to fix in advance those facts that are essential for the processing of the hypotheses and consequently to focus and structure the observation accordingly. For example, it may seem interesting to an observer under certain circumstances how often a speaker promises himself on his campaign speech or how often he touches his nose. However, when the focus of scholarly consideration is political strategies to combat unemployment, the observations mentioned here are most likely very beside the point. From another point of view, however, it may well be useful to also observe a speaker's slips of the tongue and nose-grabbing.

The second problem concerns the interpretation of observed behaviours, of symbols and of the coagulated behaviour, especially in foreign social milieus. Here, the difficulty lies in correctly identifying the meaning associated, for example, with a particular act of behaviour by an actor. While nodding one's head in Germany and in many other countries signals approval, in Bulgaria this behavioural act means the exact opposite. The same applies, for example, to the correct interpretation of symbols that, for example, young subpopulations use to signal to others that they belong to a certain culture.

König (1973, p. 7 ff.) refers to the "error of first sight". He describes the controversy between Redfield and Rojas (1934) and Lewis (1953) in this regard. Both conducted observations in a particular Mexican community. Redfield was impressed by the capacity for integration and also by the mutual accommodation of what should have been conflicting interests in this community. Lewis observed somewhat later in the same community. He reported, on the other hand, a community corroded by mutual distrust, in which the total underprivileging of certain classes had occurred. The reason for these different assessments lay in the "unteachability of the position once taken by the first observer, that is, a typical error of the first glance" (König 1973, p. 7). The first glance was therefore not followed by a second, and this was precisely the error here.

Bortz and Döring (2002) list five modeling rules to follow when making scientific observations:

1. Selection: From a universe, as already shown, a specific selection of certain stimuli must be made, which are then to be registered during observation.

---

[34] Such observations were made by Wilhelm Röntgen when he 'accidentally' discovered the rays later named after him.

2. Abstraction: The event to be observed is detached from its concrete environment and restricted to the significance that is essential in the research context. If the aim is to observe social status symbols in a residential area, for example, it is unlikely to be of interest how often interactions between residents take place in a given unit of time.
3. Classification: This is followed by an assignment of the observed event to certain event classes. To stay with the example of the observation of status symbols, it would have to be decided here whether a certain car brand indicates a low, medium or high social status.
4. The systematization: The observed individual events – it can be about doorbell signs, the condition of the front gardens, the number of shops with cheap offers or just about car brands – are compiled and condensed into an overall protocol.
5. The relativization: Finally, it has to be examined to what extent an integration of the observation material into a theoretical framework can take place or whether possibly the ascertained facts have the status of exceptions in some way.

In general – especially in contrast to surveys – the following *advantages of* social science observations can be identified:

- In observations, no verbal self-disclosure is requested, but what is to be observed is fixed unaltered by subjective reflections of an observing person. The influence of an interviewer on the subject of the investigation does not exist. This is particularly important in the case of facts that are socially strongly standardised (cf. Sect. 6.1.2). Instead of asking "As a pedestrian, do you ever cross at a red light?", it is better to observe how things are in real life.
- Observations are a good way of systematically investigating new phenomena. Even without too much differentiated prior knowledge, such facts can initially "only" be observed and explored in this way.
- Observations – especially in contrast to written protocols – are particularly suitable when it is a question of (further) interpreting an action also by observing facial expressions, gestures and the like.
- Observations offer good advantages for the study of unconscious behaviour and often represent the only possible approach (cf. Schaller 1980).
- Observation is more immediate, because it takes place at the same time as the event. This not only eliminates interviewer influences, but also errors that occur due to faulty memory performance.

Three facets, which in turn set certain *limits to* the use of observations, should also be pointed out:

- The first facet concerns the costs and time required. Here, surveys are likely to be significantly cheaper and less time-consuming when it comes to obtaining comparable amounts of data on facts.
- A second facet is the potential subject areas that could be investigated with the help of a survey. They are more diverse and universal than those that can be collected through observation. While communication can be established within the framework of a

survey about almost any issue, it is only possible to a very limited extent to conduct observations on issues relevant to social science. For example, the so-called Sunday question about voting intentions will still have to be asked in surveys.

- The methodological state of research, i.e. the knowledge about the functioning of surveys, represents a third facet. This is much broader and more differentiated than that on observations.

6.2.2 Forms of Observation

A scientific observation can be carried out in many different ways. Above all, the respective role that the observer takes on in the field can vary. The various forms of observation are presented in more detail below.

Participant and Non-Participant Observations

The criterion for this distinction is the role that the observer takes. The distinction already goes back to Lindeman (1924). If the observer becomes a component of the group to be observed, this is referred to as participant observation. If the observer does not participate in the life of the group, but limits himself to his role as an observer, we speak of non-participant observation. Participant observation enables the observer to gain more detailed insights into group life. However, it can become a problem that the observer, as a result of his activity in the group, influences the actual object to be observed. Thus, under certain circumstances, certain processes would have proceeded differently if the observer had not been active in the group. The discretion and research ethics expected of an observer are also sensitive aspects of participant observation. Finally, participant observations are particularly confronted with the phenomenon of "going native". This is a loss of distance of the observer from the objects of observation. Such a loss of distance can occur due to the social interaction between the observer and the observed. It leads to the fact that the observer identifies too strongly with the persons to be observed and as a result does not perceive certain phenomena (any more). Participant observation became particularly well known through a study conducted by Whyte (1951) on street corner gangs.

Field and Laboratory Observations

Field or naturalistic observation is when the observation takes place under natural conditions. Laboratory observations, on the other hand, use an artificially created environment. The advantage of laboratory observations is the possibility to control all boundary conditions or to prevent the influence of disturbance variables. The object to be observed is shielded from all unwanted influences in the laboratory. This makes it easier to determine the effect of the independent variable on the dependent variable. The price of this shielding, however, is an artificial situation that may be remote from practice. It leads to the question of the generalizability of the findings obtained. The situation is correspondingly different with field observations (compare Glück 1971).

Overt and Covert Observations

The differentiation of observations according to the criteria overt and covert results from the knowledge of the target persons to be the object of an observation. Covert observations – for example behind a one-way mirror – have a special charm due to their non-reactive character. It is well known that, simply because an investigation is taking place, the object being investigated changes, i.e. that the situation in which the data are collected affects behaviour (cf. Sect. 4.7). This disadvantage does not occur with covert observations. However, ethical problems of the research design must be taken into account in the case of covert observations. Particularly in the case of covert participant observation, a number of other problems are encountered. Apart from the change of the object to be observed due to the participation of the observer in the group events, the necessity to get a practical access to the object of investigation[35] should be pointed out above all. In addition, the covert observer faces the problem of also having to (covertly) record his observations. In most cases this will happen after the fact and is thus again prone to error.

Self- and External Observations

As a rule, specially trained observers are used in observations. However, there are also examples of self-observation – this form is also called introspection – in which the observer is not used and the person observes him/herself. Self-observation has proved useful, for example, in determining time use, i.e. in time budget studies (cf. Sect. 7.6). Empirical research on income and household consumption also relies to some extent on self-observation (see Zapf 1989 for another example). Of course, the criterion of verifiability and comprehensibility of the results demanded for scientific studies must be regarded as problematic in this approach. Thus, self-observations are only likely to be scientifically meaningful if they are conducted with the help of a highly standardized instrument. For a historical sketch of introspection, see Greve and Wentura (1997, p. 14 ff.).

Standardised, Partially Standardised and Non-Standardised Observations

The degree to which the procedure for an observation is fixed in advance determines whether it is a (fully) standardised, a partially standardised or a non-standardised or unstandardised observation. In the case of standardised observations, appropriate observation schemes are used. Comparable to the standardized questionnaire in a personal oral interview, all facts to be recorded are determined in advance. The selection of the categories to be determined is therefore already decided in advance of the observation. Similar to standardized interviews, estimation scales can also be used here. Thus it is possible to

---

[35] The covert participatory observation of a radical party by so-called V-men of the Office for the Protection of the Constitution raised the question in the Federal Republic to what extent such results could be used legally to bring about a ban of this party. In order to gain access to the field of investigation, the undercover agents had to actively participate in the party's activities. In the end, it turned out that such a ban could not be imposed, not least because of the influence the informers had exerted on the party.

equip different observers with the same observation scheme and in this way to determine the reliability of the observation. As already indicated, the standardization of the procedure can be regarded as a prerequisite for the verifiability and the control of the results of an observation and thus for the scientificity of the approach. This in turn gives rise to a certain need for legitimacy for non-standardised (scientific) observations.

Direct and Indirect Observations
In the case of direct observations, the facts to be investigated, for example an interaction in a group, are recorded during the execution. This can be done with the help of technical recording devices or by an observer. Indirect observations, on the other hand, focus on behavioural traces. These are the tangible results of behaviour, such as litter, wear and tear, or pizza consumption (see Sect. 4.7). The latter approach has already been described in more detail.

Technically Mediated and Technically Unmediated Observations
The criterion for the distinction here is whether a technical aid is interposed between the facts to be observed and the observer. This can be a video system, sound recording devices and the like. Above all, the repeatability of the observation as well as the possibility of using the slow-motion technique speak for the use of the corresponding technical devices (compare Manns et al. 1987, p. 25).

6.2.3 Observation Errors

Standardised observations must meet similar quality criteria as those already demanded for standardised surveys. These include above all reliability and validity.

Reliability refers to the reproducibility of an empirical finding. For example, a survey provides reliable results if these can be determined repeatedly after a certain time interval (test-retest design). The same applies to observations: "In general, reliability describes the reproducibility of observations under conditions that are theoretically equivalent for the occurrence of the observed" (Feger 1983, p. 24).

If different observers, looking at the same facts, arrive at the same results, then one can also speak of a reliable observation. If an observer is able to observe the phenomenon to be observed again, for example with the help of a video recording, it is in principle also possible that these results can be compared and thus the reliability of the observation can be determined on the basis of the observations of even only one observer.

The validity of a survey instrument refers to its suitability to determine exactly the facts that are to be measured by the instrument. The comparison with a target in archery may clarify the reliability and validity. If an archer always hits the same spot, his hits are reliable. However, this does not mean that this shooter always hits the bull's eye. This is what is meant by validity. The hits are valid if they are placed exactly where they are supposed to go.

There are several ways to determine the validity of an observation. One variant is to compare the result of an observation with a criterion that is determined independently of this observation. It is important here that this criterion actually contains the same facts that are to be determined with the help of the observation. Furthermore, it is possible to combine different observations into a so-called construct. Now a theory is needed that makes a statement about the relationship between the individual observed facts. Observations that lead to results that contradict this theory are therefore not valid.

The observer acts as an essential part of the measuring instrument. His tasks consist of selecting the necessary information during the event and recording it, assessing it, coding it if necessary and finally recording it. This range of tasks of the observer represents in principle a first source of error. The second complex concerns errors that can occur due to the chosen access to the object of investigation, i.e. here specifically the observation. Thirdly, it must be taken into account that observation errors can occur as a result of unfavourable external conditions.

Greve and Wentura have developed an approach for the systematization of concrete observation errors. This is shown in Table 6.21.

Error Incurred by the Observer
The Perception Error
This type of error occurs primarily when the object to be observed is not correctly recognized and interpreted by the observer. Some examples of this have already been mentioned above. These include further:

Table 6.21 Overview of the most important sources of error according to Greve and Wentura (1997, p. 60)

1	Error incurred by the observer
(1a)	Perception
	1. Consistency effects
	2. Influence of existing information
	3. Projection
	4. Expectation effects
	5. Emotional involvement
	6. Logical or theoretical errors
	7. Observer drift
(1b)	Interpretation
	1. Central tendency
	2. Personal tendencies or dispositions
(1c)	Reminder
	1. Capacity limits
	2. Memory distortions and selections
(1d)	Playback
2	Error in observation
	1. Reactivity and expectation effects
	2. Observation and study conditions
	3. Problems of the observation system
3	Error due to external factors

- The halo effect: Here, the observation is guided as a result of an overall impression that the observer has obtained. This then radiates to all subsequent observations. The observer strives to perceive what is being observed consistently, i.e. without contradiction (cf. Schaller 1980). The halo effect has also been observed in interviews. Here, the answer to a question radiates onto the answer to the subsequent questions.
- Errors that occur due to the influence of the temporal sequence: In this case, the events to be observed can be evaluated too strongly on the basis of the first impression. In the following, the events are then perceived selectively and quasi aligned with the first impression. But also information that comes from other observers and is reported in advance can influence the expectations of the investigator and other observers and thus cause errors.
- Observers can be prone to projections: In this case, more facts are observed and reported that the respective observers individually perceive in particular or do not want to perceive. This can be remedied by using different observers.
- Expectancy effects: This refers to the tendency of observers to make assessments that conform to hypotheses. According to the motto, he who seeks, finds, those things are perceived above all – partly unconsciously – that correspond to one's own expectations.

This includes the leniency effect. This is an overly generous assessment by the observer. The leniency error occurs, for example, in participant observation as a result of particularly close social contact between the observer and the observed or due to excessive emotional involvement on the part of the observer. The problem becomes particularly evident when "the observer shares the values of the observed, because he then lives in the illusion that he can gain a direct insight based on direct observation without mediating hypotheses" (König 1973, p. 12).

The use of an observation logic by the individual observers also plays a role in this error category. In the case that the observers do not evaluate the events strictly on the basis of the observation protocol as intended, but, deviating from this, according to their own logic of expectation, the observations of different observers would no longer be comparable.

- The emotional involvement of the observers or too strong identification with the object of observation can lead to errors. Particularly in the case of participatory observations that are conducted over a longer period of time, the observers may identify too strongly with the object of observation. The necessary distance is lost. However, the opposite effect is also conceivable: the observer distances himself too much from the object of observation.
- Logical and theoretical errors have to be taken into account: This is thinking of influences that come about due to implicit theories (compare Sect. 3.6) of the observers. Implicit theories are used by people in everyday life to deal with perceived facts. Prejudices are also among these implicit theories. In scientific observations, however, they are out of place.
- Observer drift refers to the "gradual change in an observer's standards" (Greve and Wentura 1997, p. 64). In the process, the observation rules initially learned can be forgotten again over time, fatigue can set in, the observer's motivation can wane, and so on.

The interpretation or interpretation errors. This type of error includes all those that influence perception.

- The errors of central tendency occur in particular during the delivery of assessments or evaluations of the objects to be observed. For this purpose, the observers are usually given scales on which they have to give their assessment. It would be conceivable, for example, to evaluate the degree of well-keptness of a front garden in a residential area. Tests have shown that in such situations observers tend to shy away from choosing extreme evaluation categories.
- Personal tendencies and dispositions of the observer can negatively influence the observation. For example, people may tend to make either particularly severe or particularly lenient judgments as a matter of principle. This would also include the tendency to perceive and register above all socially desirable facts.

The Memory Errors
In the case of participant observation, and especially in the case of covert participant observation, it is hardly possible to record the observed facts directly during the event. Thus, especially memory errors can occur if the observation protocol is created only after a certain time interval. Two types are to be distinguished:

- At some point, people reach the limits of their cognitive capacity in terms of their individual receptivity.
- Systematic biases and selections can occur when the observed facts are stored and have to harmonize with other, already stored, information.

The Reproduction Errors
In addition to deliberately falsified observations, there can also be distortions of the observations due to conformity pressure. Such errors could also be empirically proven with the help of corresponding experiments.

Error in Observation
In addition to the errors discussed here, which are at the expense of the observer, errors must also be taken into account, which are at the expense of the observation. Especially in the case of overt (not covert) observations, reactive measurements occur. These are measurements that are biased due to the fact that a measurement is taking place. An observed person knows that he or she is being observed and therefore behaves differently than in an everyday situation.

One antidote here is blind and double-blind trials. These are mainly used in pharmaceutical research to assess the efficacy of drugs. In this way, all target subjects are administered a preparation. However, only a portion of the drug given to the subjects is the drug being tested. The comparison group is given only a placebo. It is important that it is not known to the test subjects (blind test) or to the investigator (double-blind test) who is counted in the experimental group and who is counted in the comparison group.

There is an interaction between observer and observed. This can influence the result of the observation. This is especially – but not only – the case with participant observation. Here, the situation to be observed is distorted by the fact that another person, the observer, is also acting in the field.

Error at the Expense of External Conditions
Especially in the case of laboratory observations that take place in an artificial atmosphere, the external conditions can make the result of the observation erroneous (cf. Sect. 7.1). This is the case when the experimental conditions are so tightly controlled and thus take on an artificial character that generalizations (external validity) are no longer possible.

6.2.4   The Development of the Observation Design

In standardized observations, an observer classifies actions and other relevant facts according to a specific instruction. In the interest of objectifying these classifications as much as possible, an observation guide is developed. Such a guide can contain different criteria and instructions for the observation:

- Certain sign systems can be registered. These can be, for example, yawns, interjections, requests during a lecture and the like. It is possible to log both the frequency and the duration of such events.
- Classifications can be made. Observation tools are used to classify actions with the help of category systems. When observing group discussions, for example, it may be necessary to decide whether a contribution is constructive, neutral or destructive in terms of problem solving.
- Assessments of intensities are possible. In the context of standardized observations, the use of estimation scales offers the opportunity to record, for example, the intensity of agreement and/or disagreement with a discussion contribution on a multi-level scale.
- Sequences can be observed. The multi-moment method was developed in order to observe rapidly occurring processes, such as communications on a playground. This method provides for the relevant facts to be recorded again at certain time intervals.

Sample
In determining the sample to be observed, a decision must be made as to whether an event sample or a time sample should be used to select the units to be observed. The basis for this decision is provided by the manner in which the event to be observed occurs. In an event sample, a record is kept of whether a particular event occurs and how often it occurs. It would be conceivable to observe how often each participant speaks in a discussion and how long they speak. In a time sample, the period of observation is divided into time segments. When observing the discussion round, it would then be recorded – approximately every three minutes – which participant is currently making a contribution to the

discussion. Relatively evenly proceeding events can be observed expediently with such a time sample. If this is not the case, an event sample is recommended.

Time samples are "more suited to describing the whole event and event samples more suited to documenting specific behaviors" (Bortz and Döring 2002, p. 272).

6.2.5   The Social Prestige of a Residential Area: Example of a Standardised, Non-Participatory External Observation in the Field

Now the development and implementation of an observation design will be demonstrated by means of an example. To this end, a project will be presented that involved the observation of the social prestige of various residential areas in Dresden. This observation was standardized, non-participatory and designed and implemented as a field observation in 2004/2005. However, the main aim of this study was to familiarize students with the method of observation, to teach them the steps to be followed, and to make them aware of possible mistakes in the process.

In order to be able to carry out a standardised observation, an observation plan must be worked out on the basis of hypotheses (compare Bortz and Döring 2002, p. 269 ff.). In doing so, a number of decisions have to be made which – step by step – specify the layout of the survey. These include:

Theoretical Concept
The observation was intended to reveal a possible clumping of similar lifestyles in residential areas. Two hypotheses were created for this purpose:

1. In a residential area live mainly people with the same social status.
2. The higher or lower the social status of people, the more homogeneous the residential area in which they live.

These hypotheses should now be empirically processed with the help of an observation.

Operationalization
In order to be able to test hypotheses empirically, the concepts they contain must be operationalised. Obviously, such facts as social status or homogeneity of a residential area cannot be read directly on the basis of (only) one particular criterion. Instead, several observable facts must be named, which then allow conclusions to be drawn about these latent facts in the context of an observation. The following things have to be taken into account:

• Are the chosen indicators actually empirically observable? For example, the owner of a house (the resident himself, a cooperative or a non-profit association) may indicate the social status of the people living there, but it is likely to be difficult to determine this owner by means of observation. This makes such an indicator unsuitable.

- Do the indicators established point with sufficient certainty to the social facts sought?
  Here, one could observe the vehicles parked in the residential area and estimate their
  value. It must be assumed that the vehicles also belong to the residents and have not
  been parked here by strangers.

The decision was made to use the following indicators: The level of redevelopment of
the buildings, the landscaping, possible property security such as cameras and the like, dog
excrement on the footpaths, the type and condition of the letterboxes, the condition of the
rubbish bins and their cleanliness, a possible vacancy in the building – whereby presumably a low vacancy rate indicates more attractive apartments – vehicles parked in front of
the houses and their value, finally the type and condition of the doorbell signs, whereby
academic titles as well as references to certain professions should also be given attention here.

It was also decided that there should be a "house" observation sheet, which is used to
record all the flats in a house, and a "residential area" observation sheet, which is used to
record the area surrounding the house.

Coding Scheme

The observation plan contains instructions on how to register and record the facts to be
observed. For this purpose, the facts to be observed are further broken down into elements,
which then become the object of observation. Here it is helpful to provide the observers
with anchor examples as an aid. These give examples of which vehicle types are to be classified in the upper, middle and lower price categories.

The "House" observation sheet had the following appearance, shown in Fig. 6.12.

Determination of the Units to Be Observed

To address the two hypotheses, five districts were specifically selected, taking into account
the capacities available for observation in the context of a seminar, they form the population. After that, the observations were to take place randomly in these districts. The following two-stage procedure was followed.

First Step: Systematic Determination of the Residential Area

At the time of the survey, there are 99 city districts in Dresden. Five neighbourhoods with
as many different characteristics as possible were to be selected. The conscious choice was
made for a district that is mainly built up with villas (Gruna), a district inhabited by the
upper middle class (Laubegast), a district with a mixed population (Innere Neustadt); a
district with prefabricated slab buildings (Gorbitz-Süd) and a district with industrial settlement (Friedrichstadt). Each district has a number in the administration, namely Gruna: 57,
Laubegast: 62, Innere Neustadt: 13, Gorbitz-Süd: 95 and Friedrichstadt 05.

Second Step: Random Selection of Streets from the Street Directory

The 2003 street register, published by the Municipal Statistics Office of the City of
Dresden, is arranged alphabetically and contains, among other things, information on the

1.1 Which of the following applies to the house? (only one cross)
One-family house ( )
Two-family house ( )
Apartment house ( )

1.2 Style of the house (only one cross)
Detached house (4 sides open) ( )
Semi-detached house (3 sides open) ( )
Terraced house (2 sides open) ( )
Prefabricated building ( )
Rear building/side wing ( )

1.3 Number of floors (Please enter! If there is only one ground floor, then enter zero!)
_____

1.4 Number of households per house (Please fill in! Determine on the basis of the nameplates on the letterboxes! Vacant dwellings are counted)
_____

1.5 Degree of remediation (only one cross)
New building ....( )
Renovated (cleaned facade, new roof, new windows or in good condition, new doors)...................................( )
Partially refurbished (new roof or façade, no rising damp, one element missing from "refurbished")
....( )
Unrenovated (everything in old condition)..( )
Dilapidated/ruined ....( )
Unclassified ....( )

1.6 Securing of land (tick all that apply!)
Camera ( )
Alarm system ( )
Dog / Warning about the dog ( )
Motion detector (also connected to a light switch) ( )

2. Car parking spaces (motorcycle parking spaces etc. are not meant here!)
2.1 Are there parking spaces for cars directly in front of the house or the property?
yes ( )   no   ( ) continue with: 2.9

2.2 Type of pitches (tick all that apply!)
Garage on the property ( )
Separate parking facility/collective parking ( )
Underground car park ( )

2.3 Number of parking spaces better Number of parking spaces (please enter)
_____
not recognizable/not visible   ( ) continue with 2.9

2.4 Are the parking spaces reserved?
yes, all   ( )   yes, some   ( )   no, none   ( ) continue with: 2.6
                                        Not detectable   ( ) continue with 2.6

2.5 Type of reservation (tick all that apply)
Registration number of the cars ( )
Barrier ( )
Parking bollards ( )
Parking control device, e.g. parking meter ( )

2.6 Condition of garages (only one cross)
Good condition (no defects on facade or entrance gate)..( )
Medium condition (minor defects to facade or driveway gate)...( )
Poor condition (major defects on facade or driveway gate), plaster is missing or badly
        Damaged; damaged entrance gate, graffitied facade)...( )
No garages available ....( )

Fig. 6.12 "House" observation form for the standardised observation of social milieus in urban districts.

2.7 Paving of the access road to the parking facilities on the property (only one cross)
Unpaved (soil, gravel)..(   )
Paved (paved, asphalted)..(   )
No parks available on the property ....(   ) continue with 2.9

2.8 Condition of the parking facilities on the property (only one cross)
Good condition (no defects)   (   )
Medium condition (small defects, small cracks, small bumps)   (   )
Poor condition (major defects, large holes)   (   )

2.9 Is there a garden or property boundary (fences, etc.)?
yes   (   )   no   (   ) continue with 2.16

2.10 What is the limitation? Is it ...? (tick all that apply)
Fence/grid   (   )—continue with 2.11
Wall   (   )—continue with 2.12
Hedge   (   )—continue with 2.13

2.11 What is true about the fence, the grid? (Only one cross)
Particularly elaborate (i.e. artistic, extremely stable, expensive, very high)   (   )
By default   (   )
Kept emphatically simple   (   )

2.12 Is the wall: (Only a cross)
Unplastered   (   )
Plastered   (   )
Cancelled   (   )

2.13 Is the hedge well-maintained (   )   untended   (   ) or is there no hedge (   )?

2.14 Is there any damage to the fencing?
Undamaged: no visible damage..(   )
Slightly damaged: minor visible damage that does not affect the function or aesthetic appearance of the product.
             Only slightly restrict perception of the property boundary (e.g.: missing
             Garden door, chipped plaster, rusty chain link fence) .... (   )
Severely damaged: large visible damage, which severely impairs the function of the boundary
             Restrict. (e.g.: broken fence slats, toppled pillars) .... (   )

2.15 Are there any ornaments on the fencing which appear to increase its value?
                                                  Yes (   )   no   (   )

2.16 Is a waste storage area available?   yes   (   )   no   (   ) continue with 3.1

2.17 How many bins are visible? (Please enter the number)
_____

2.18 Is the rubbish storage area demarcated from the surrounding area or are the rubbish bins enclosed?
                          yes   (   )   No   (   ) continue with 2.24
2.19 What is the demarcation made of? (tick all that apply)
Wooden boundary   (   )
Metal demarcation   (   )
Vegetal delimitation   (   )
Wall   (   )

2.20 Is the waste storage area lockable? yes   (   ) no   (   )

2.21 How many dwellings is the bin site responsible for? (Please specify)
_____
Unrateable   (   )

Fig. 6.12 (continued)

3.1 Number of windows per house (please enter all, limit to the front of the houses, per entrance) or house)

_____

3.2 Size of the windows (Please enter in each case!) Attention: the sum must be exactly the same as specified under 3.1 (Number!)
_Number of small windows (e.g. toilet windows): _____
_Number of middle windows: (standard size): _____
_Number of large windows (e.g. picture windows): _____

3.3 How many windows have external window decoration (e.g. painting/ornaments)? (Please number enter each)
_____ _Windows are decorated_
_____ _Windows are not decorated_
_____ _Windows are not visible_

3.4 How many windows have window decorations (e.g. flowers) on the inside?
_____ _Windows are decorated_
_____ _Windows are not decorated_
_____ _Windows are not visible_

3.5 How many windows have been refurbished?
_____ _Windows are refurbished_
_____ _Windows are not refurbished_
_____ _Windows are not visible_

3.6 Are there any special window shapes, e.g. round windows?
                              yes   (   )   _no_   (   )

3.7 How many windows have curtains etc. on the inside?
_____ _Windows with curtains_
_____ _Windows without curtains_
_____ _Windows are not visible_

3.8 How many windows have blinds, shutters, etc. on the outside?
_____ _Windows with blinds_
_____ _Windows without blinds_
_____ _Windows are not visible_

3.9 How many windows are damaged?
_____ _Damaged windows_
_____ _Windows without damage_
_____ _Windows are not visible_
3.10 How many windows are boarded up (with cardboard, plywood, etc.)?
_____ _Windows are not boarded up_
_____ _Windows are boarded up_
_____ _Windows are not visible_

3.11 How many doorbell signs are there in the house?

_____

3.12 How many doorbell signs (apartments, not offices, etc.) are labeled?

_____

3.13 How many mailboxes (flats, no offices etc.) are there in the building?

_____

3.14 How many of the mailboxes are taped?

_____

3.15 Are there signs such as "For Rent" "For Sale" and the like?
                         yes ( ) no ( )
3.16 Is there any other damage (including graffiti) to mailboxes, lighting, etc.?
                         yes ( ) no ( )
4.1 How many cars are parked on the property area in total? (Please specify number)
_____ _Cars_
Not clearly identifiable ( )

Fig. 6.12 (continued)

4.2 Number of cars in the ... (Please enter in each case)
Upper price range (from 35,000 €): Mercedes E-, S-, M-, G- class, Mercedes SL, SLK, CL, CLK, AMG, BMW 5 series, 6 series, 7 series, Z3, Z4, Z8, Audi A6, A8, Audi TT, Opel Signum, Peugeot 607, VW Phaeton, Touareg, Porsche, Ferrari, Maserati, Lamborghini

———————————

Middle price range: (~ €16,000 to €34,000): Mercedes A-, C- class, BMW 1-series, 3-series, Audi A2, A3, A4, VW Golf, Passat, Peugeot 206, 307, 407, Opel Astra, Zafira, Vectra, Alfa Romeo, Ford Mondeo, Focus, Toyota Avensis, Corolla, Nissan Almera, Primera, Volvo; larger Citroen, Fiat, Honda.

———————————

Lower price range: (up to €15,000): Opel Corsa, Ford KA, Fiesta, VW Lupo, Polo, smaller Citroen, Fiat (Panda, Punto), Honda, Nissan (Micra), Toyota, Seat, Peugeot

———————————

4.3 Number of cars that ... (Please enter the number in each case)
Are clean, well-kept: Car looks like freshly cleaned, clean body, clean windows (tidy inside)

———————————

Well-kept, but slightly dirty (slightly untidy): overall. well-kept impression, only a few splashes of dirt, inside only few things lie around- max.5

———————————

Are dirty, unkempt: many dirt splashes (mud, bird droppings, tree sap), inside many things are lying all over the place

———————————

4.4 Number of cars with damage (please enter number in each case)
without external damage: no scratches, no dents

———————————

Minor damage: minor scratches, no dents

———————————

Medium damage: scratches and minor dents

———————————

Big damages: big scratches and big dents

———————————

4.5 Number of cars with ... (Please enter the number in each case)
Expensive equipment: leather seats, noble instruments, high-quality radio, possibly navigation system

———————————

Normal equipment: seats, fittings, etc.

———————————

4.6 Number of cars at the age of about: (Please enter the number and estimate the age)
Under 5 years

———————————

Older than 5 years, under 10 years

———————————

Older than 10 years

———————————

Enter the time of the observation (date, day of the week and time) here:

——————————————————

Enter the location of the observation here (street and house number):

———————————————

Fig. 6.12   (continued)

city area in which a particular street is located. In a first step, all streets were marked in the register for each city area (each district with a different colour). The following number of street names was now determined for the five selected city districts: Gruna 53, Laubegast 58, Innere Neustadt 54, Gorbitz-Süd 15, Friedrichstadt 37. Now four streets were to be drawn at random from each city area. When deciding on four streets per city district, there was the possibility to have each street recorded by several observers. With a higher number of streets this would have become problematic.

A special drawing procedure was used to draw the streets per district. This drawing resulted in the following: Gruna: 50, 04, 53, 06; Laubegast: 58, 27, 05, 51; Innere Neustadt:

35, 34, 01, 31; Gorbitz-Süd: 14, 12, 03, 02 and Friedrichstadt: 12, 07, 03, 11. Thus the following streets were determined for the observation:

- Gruna: Wiesenstraße, Am Grüngürtel, Zwinglistraße, An den Gärten,
- Laubegast: Zur Bleiche, Klagenfurter Straße, Azaleenweg, St. Pöltener Weg,
- Innere Neustadt: Melanchthonstraße, Löwenstraße, Albertstraße, Köpckestraße,
- Gorbitz-Süd: Wilsdruffer Ring, Tannenberger Weg, Espenstraße, Ebereschenstraße,
- Friedrichstadt: Friedrichstraße, Bremer Straße, Bauhofstraße, Flügelweg.

Now the starting point and the length of the observation could be determined. The observations were to take place on Saturdays or Sundays between 10:00 and 14:00.

The following were observed: firstly, the houses or the housing units and secondly, the residential area surrounding the housing units.

When *observing the houses* or the *housing units,* the smallest house number of the street formed the beginning. Observations were made in the direction of the larger house numbers. Thus, the observation started at the first housing unit or at the smallest house number, the observation was to be conducted for 50 housing units, after the 50th housing unit the observation was canceld. A residential unit was considered to be a household or a doorbell. It had to be taken into account whether the house to be observed was a single-family house (corresponding to only one housing unit) or a multi-family house. Vacancies were also counted in the housing units. However, ruins or completely vacant houses did not count. From the starting point, all housing units on the same side of the street were observed. If the street was not 50 housing units long, a right turn was made at the next opportunity and observation continued there. If this was not sufficient, it was now necessary to turn left. In the event that a turn in the respective prescribed direction was not possible, the turn was to be made in the opposite direction. In the case of dead ends, the other direction was to be taken and here the next opportunity to turn right or left was to be used. Pure commercial buildings, offices, surgeries, shops were not considered residential units. Rear houses, side wings and the like (also houses accessible only via courtyards) were also to be observed.

For the *observation of the residential area, it* was determined that the area around the residential units to be observed should be observed. Here, the area that was within five minutes walking distance was considered as residential area.

Pretest
The planned procedure and the observation protocol were tested during a preliminary investigation in another residential area. Some revisions were necessary, for example, various descriptions were additionally included in the observation protocol in order to ensure that all observers have the same understanding of the facts to be recorded. In particular, the observation of the vehicles should thus be somewhat more objectificd.

Main Survey
The main survey was conducted by about 30 observers at the beginning of 2005. The project was part of the methods training at the Institute of Sociology at the Technical University of Dresden.

Data Preparation and Evaluation

A total of 195 housing units were observed in this way in the residential areas mentioned. The individual observation protocols were transferred into an SPSS data file and checked for plausibility.[36] This was followed by the data analysis. At this point, only index formation will be discussed as an example. While the complex issue of social prestige had been broken down into individual dimensions during operationalization, the task now was to combine this breakdown into an index. This index then expressed the issue in question as a numerical value. Specifically, an additive sum index was to be formed[37] (compare Sect. 4.3.3). The index was formed in such a way that a higher value indicates a higher social prestige of the residential area. The following variables were included in this index:

Hau:	The size of the house, a 2 was coded for a single-family house and a 1 for a two-family house
Hhz:	Households per house, with a 1 coded for up to five households per house
San:	The degree of renovation of the house, new buildings and renovated houses were coded with 2 and partially renovated houses were coded with 1
Zau:	The condition of the fence, if no damage was visible a 2 was coded and if minor damage was present a 1 was coded
Mül:	The condition of the garbage area, for a clean garbage area a 2 was given and for an only slightly dirty garbage area a 1 was given
Fe1:	The window renovation, if all windows were renovated, a 1 was coded
Fe2:	Window damage, if there was no window damage, a 1 was coded

The index was checked by calculating Cronbach's alpha and then improved step by step. As a result, the following indicators were included in the index formation, which then had the reliability measures shown in Table 6.22.

It was now possible to determine this index for each of the residential areas studied and thus to test the hypotheses. The corresponding result is contained in Table 6.23.

In fact, the expected result could be more or less confirmed. Thus, the villa district and the district inhabited by the upper middle class show the highest values. The area characterised by prefabricated buildings performs even slightly worse than the industrial settlement.

The degree of homogeneity of the residential area can be assessed on the basis of the dispersion of the index. It was expected that the higher the social prestige or the lower the social status of a residential area, the more homogeneous it would be. In our small study, the areas rated best (Laubegast and Gruna) and worst (Gorbitz Süd) also had the lowest dispersion.

---

[36] The corresponding file can be ordered and downloaded from the author via the Internet: michael.haeder@tu-dresden.de

[37] Missing values have been replaced with mean values.

Table 6.22 Reliability determination for the social prestige index using SPSS reliability analysis (alpha), n = 195

Variable	Corrected item-scale correlation	Cronbach's alpha if item omitted
Hau	0.756	0.557
Hhz	0.735	0.564
San	0.401	0.679
Zau	0.209	0.714
Mül	0.334	0.697
Fe1	0.452	0.686
Fe2	0.239	0.712
Alpha = 0.71		

Table 6.23 Social prestige (index) of different residential areas in Dresden

Residential area	Index	Standard deviation
Gorbitz south	6.1	1.51
Friedrichstadt	6.3	2.22
Inner new town	6.7	1.84
Gruna	10.5	1.75
Laubegast	10.6	1.76
Total	9.5	2.49

6.3 Content Analyses

With the content analysis, following the survey and the observation, the third basic social science method for data collection is presented. In everyday life, the analysis of content is directed at a variety of containers. Conceivable are, for example, the contents of a barrel of wine, a newspaper, our treasury, a test tube, but also stomach contents can become the object of analysis. Obviously, the concept of content analysis needs to be defined more clearly here.

According to one definition, content analysis is understood as a research logic for the systematic collection and processing of communication content in texts, pictures, films, records and the like (compare Diekmann 2001, p. 576 ff.; Früh 2001; Schulz 2003; Rössler 2005; for other definitions, see Berelson 1952; Merten 1995). Thus, firstly, not only texts are the object of social science content analyses, but in principle all carriers of relevant information. Secondly, according to this provision, content analyses present themselves as open to the investigation of communicated manifest as well as hidden, latent facts. Thus, a newspaper article can be examined not only for what messages it contains about a fact, but also for what information it does not contain, but which would have been expected in the context addressed. If in a song "Telephone, gas, electricity – unpaid that also works" is sung, (human by Herbert Grönemeyer) then one can also assume such a latent fact here. A content analysis could also start at this point and look for the meaning of this sequence.

In order to give an impression of the diverse use of content analyses, some examples are presented. Afterwards, the essence of this method is dealt with in more detail and finally the procedure is described using a concrete example.

6.3.1   Examples of Content Analysis

A number of examples (compare Mohler et al. 1989 for even more illustrations of quantitative content analyses compare Mayring and Fenzl 2019 for examples of qualitative content analyses) from the social science environment will be used to show the wide range of applications that content analysis enables. Among others, the following have been examined so far:

- 5030 Exercises from 17 history, geography and social studies textbooks of different grades of the USSR. Due to the lack of empirical materials on the ideological situation in the Soviet Union, such an approach was appropriate. It was found, first, that history, rather than social studies, was the main source of the transmission of Marxist ideology and, second, that from grade four onward there was a marked increase in the transmission of ideology in Soviet schools (compare Cary 1976).
- Party platforms and election campaigns of the Republicans and the Democrats in the USA in the period from 1844 to 1964. The frequencies of the words used in a certain category were determined. The contents of the party programmes were then subject to two cycles: firstly, a 152-year period and secondly, a 48-year cycle. If at a certain point in time certain words are used particularly frequently, then half a cycle later there is a marked abandonment of these terms. Thus, in 1844, 1932, and 1980, terms describing the nation's prosperity dominated; in 1848, 1896, and 1966, political disputes took center stage; in 1812, 1860, 1908, and 1956, a cosmopolitan phase followed; and finally, in the 1872, 1920, and 1968 election campaigns, a conservative phase ensued (compare Namenwirth 1986).
- The subject of further analyses were the Speeches from Throne in Great Britain, which were delivered at the opening of the sessions of Parliament in a period from 1689 to 1972. These analyses confirmed the possibility of transferability of Namenwirth's American model of cultural change to Great Britain. However, two cycles of different length emerged: one lasting 72 years and one lasting 148 years. Here, in addition to an integrative and an instrumental phase, an expressive and an adaptive phase have also been identified (compare Weber 1986).
- To explore gender roles in the context of small group research, transcripts of various group sessions were analyzed. It was found that men in a male-only group express their feelings more in the form of stories, emphasize relationships with each other through laughter, and that more aggression topics are discussed than in a gender-heterogeneous composition of the group. Women in female-only groups discuss themselves and their families more. In mixed-gender groups, women talk less about home and family and

are more likely to represent traditional roles. Men take on the leadership role here, tell fewer stories, and reflect feelings more (compare Aries 1973, 1977).

- 816 marriage advertisements published in the weekly newspapers DIE ZEIT and "Heim und Welt" in June and July 1973. The main difference was the external image of the partner sought conveyed in the advertisements. While the former newspaper is read by the upper middle class, the readers of the latter newspaper are members of the lower middle class. For the analysis, among other things, the Type-Tock-Ration was determined. This is a measure of the richness of the vocabulary used in the texts. This is derived from the ratio of the number of different words to the total number of words used. The attributes advertisers in "Home and World" wanted in their partner were primarily young, vital, healthy and intelligent. The attributes reiselustig and musisch, on the other hand, were desired in partners preferred by ZEIT readers (compare Kops 1984). An analysis of personal ads of homosexual men was presented by Marscher (2004).

- 22 letters from suicides, 18 letters from simulated suicides, and the letters of eight doomed persons (compare Henken 1976). The simulated suicides were test subjects who wrote such a letter from the perspective of a suicide. The aim was to find out typical personality characteristics of the different groups. It was found that there is only a stereotypical knowledge of suicidal motives in the population. The differences in writing styles between the suicidal and the deathly ill were smaller than between the population and the suicidal.

- As part of the study "The Unemployed of Marienthal", the team of researchers led by Paul Lazarsfeld firstly examined school essays to see how expensive the Christmas presents they wanted were. Then these values were compared with those from a place where unemployment did not play such a formative role as in Marienthal. Secondly, we counted how often the grammatical form of the subjunctive – as an indicator of resignation and hopelessness – was used in the essays (compare Jahoda et al. 1960).

- Viktor Klemperer (1968) analysed the propaganda language of the Third Reich (LTI), which was accessible to him primarily in daily newspapers, and drew conclusions from it about the political and military situation of the country. He also assigned an indicator function to certain terms in the propaganda language: "In December 1941, Paul K. came home from work beaming. He had read the army report on the way. 'They are miserable in Africa,' he said. 'Do they really admit that,' I asked – they usually only report victories. 'They write: Our troops fighting heroically. Heroic sounds like obituary, count on it.' Since then, heroic has sounded like an obituary many, many times in the bulletins and has never been deceptive" (Klemperer 1968, p. 15).

- Supplier information from car manufacturers on car purchasing was analyzed to determine how purchasing risks are dealt with. For this purpose, 82 such risks were identified by experts, for example a too small boot, high susceptibility to rust damage and high fuel consumption. It was found that in the brochures and advertisements, verifiable information was more frequently given in the form of small print that was difficult to read. Information on fuel consumption is found most frequently, while rust problems are dealt with least frequently (compare Grunert 1981).

6.3.2   Specifics of Social Science Content Analyses

Similar to observation and interviewing, content analysis is a technique related to certain activities from everyday life. Social science content analysis is a procedure that follows certain rules. While the everyday reading of the content of a daily newspaper can be done with the arbitrariness left to the respective reader, for example from back to front, the social science content analysis is structured by a set of rules. Such a rule determines the selection of texts to be analyzed, it specifies the units to be coded, and so on. Certain methodological quality criteria, such as that of reliability and validity, must also be guaranteed in content analysis. For this purpose, the results of a content analysis can be checked, for example, by using several coders. At the same time, this set of rules makes it possible to trace the procedure chosen for an analysis and thus to check which conclusions are drawn on the basis of which instruments. Furthermore, social science content analysis is distinguished from everyday content analysis by its goal-orientedness. As a rule, it is hypothesis-driven. Scientific objectives are the triggers for the application of this technique.

6.3.3   Classification Possibilities of Content Analyses

Social science content analyses, similar to the other survey methods, exhibit a considerable methodological wealth of variants. They can – as already demonstrated – use a wide variety of sources and can also be distinguished in a correspondingly wide variety of ways. Some examples of sources that can be tapped using content analysis have already been presented above (compare Merten 1995, p. 16). Three approaches to structuring the procedure of content analysis should be referred to. One approach goes back to Früh, another one to Berelson and finally another one can be found in Diekmann.

A general model of content analysis illustrating its possible aims has been presented by Berelson. It is shown in Fig. 6.13.

The model contains the producer (Kr) of a text, he encodes his request in it in order to finally send it. Furthermore, the model contains the addressee (Rt), who receives this request or this text, decodes it and must understand it. The text itself is also part of the model. Finally, the social environment in which the text was created is integrated into the model. Based on this, the possible goals of the content analysis are now:

- *First,* conclusions are drawn about the producer of the text.
- *Secondly,* the addressee, the recipient of a text, can be the focus of interest.
- *Thirdly,* it is conceivable that a purely formal description of the text in question could also be given.
- *Fourthly, it is* possible to draw conclusions about the social context or the social situation of the communication. This takes into account the effect of the environment, which has an influence on the processes mentioned.
- Finally, it is also conceivable that the results of qualitative interviews are examined with the help of content analysis (see below).

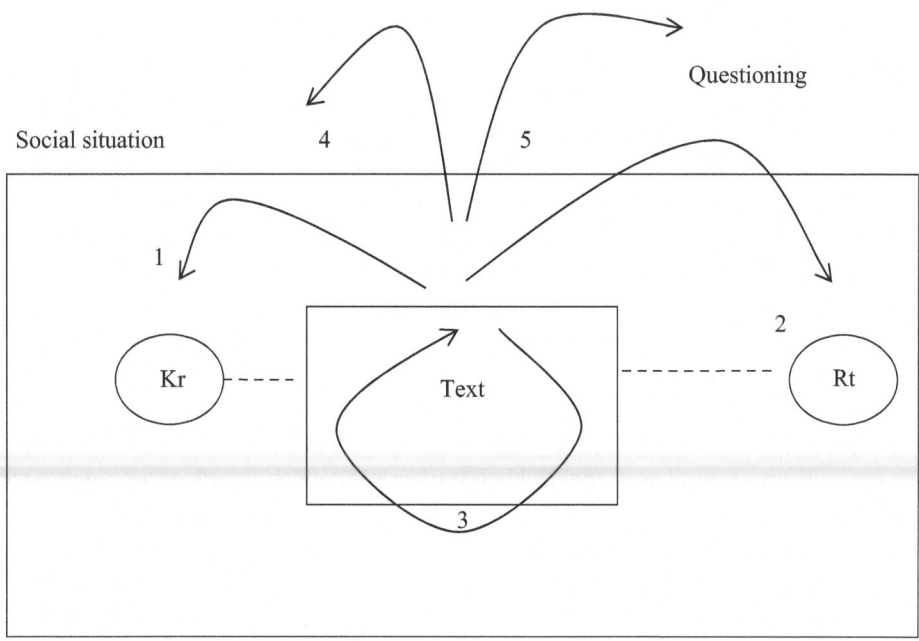

Kr = communicator, Rt = recipient

Fig. 6.13 Description and inference as goals of Berelson's content analysis. (According to Merten 1995, p. 56)

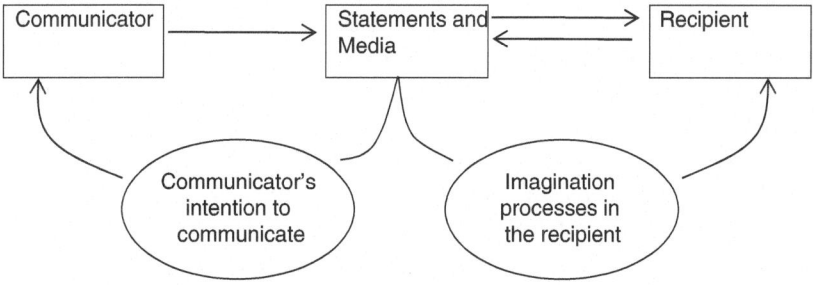

Fig. 6.14 Communication model. (According to Früh 2001, p. 41)

According to Lasswell (1952; quoted from Silbermann 1974), the various goals of content analysis can be summarized in a nutshell: "Who says what through which channel to whom and with what effect?"

Früh (2001, p. 41 ff.) also distinguishes five directions of content analysis in his own approach. All of them are based on a similar basic model as the one presented earlier. However, here it consists of the sender, the text and the receiver. This approach is shown in Fig. 6.14.

Früh also has a communicator at the beginning of the communication model. The communicator wants to get in touch with the recipient and send him a message. However, as already stated, the content of consciousness cannot be transferred directly from person to person, so that the communicator must encode his request. To do this, he uses certain sign systems (for example, language) and a medium (for example, a newspaper or television). These sign systems must be reconstructed by the recipient – he must read this newspaper or watch the television programme and also understand it. With this, the different tasks of a content analysis can now be described – further on in the following:

Content analysis can be used to study various text features. For example, it is possible to draw conclusions from the numerical ratio of verbs to adjectives, from the frequency of the use of the subjunctive and from the richness of the vocabulary used. This is the *formal-descriptive approach* (Früh 2001, p. 41 f.). Here, only purely external and not, for example, content-related features of a text are examined. The type-token ratio already presented above would also be classified here.

Furthermore, on the basis of content analysis, one can infer the author of the respective messages. Each source has a certain history of origin, which the *diagnostic approach* addresses. Here, conclusions or inferences are drawn from the messages to the author of these messages. Such decoding of the content is considered to be the main problem of content analysis. The diagnostic analyses aim at the relation producer and message. Here questions are dealt with such as: What does the author want to say with his text? Out of which situation did the text arise? What effect does the author want to achieve with his text? For example, marriage advertisements can be examined to see which self-image is conveyed to the reader by the advertisers.

Finally, messages are always addressed to a specific addressee. The recipient of a message reacts to certain characteristics of this message. It is to these effects that the *prognostic approach of* content analysis turns. Prognostic analyses say something about the effect of a text on the recipient, for example, about what future behaviour of the recipient can be expected. This is a central problem of content analysis, especially in the context of research on propaganda.

In addition, there is another direction of content analysis: the *communication-theoretical analysis*. The interrelationship between the sender and the receiver is the subject here. How does television consumption affect children, for example? Especially with such questions, an extension of the design by other methods of data collection makes sense.

Finally, the characteristics made by Diekmann (2001, p. 597 ff.) will be presented. He distinguishes between *frequency analyses*, *contingency analyses* and *valuation analyses*.

Frequency analyses represent the simplest form. Frequency analyses determine the frequency of occurrence of certain units in the text. This can be individual words, certain expressions, topics and so on. Frequency analyses are suitable for the study of television films, photographs in magazines and the like. It is advisable to carry them out with the help of computers. It can become a problem to recognize the correct semantic meaning of certain terms, remembering again the term "tiger", which – as seen – can experience very different interpretations. A hypothesis which claims that the meaning of a certain fact can be observed on the basis of the frequency with which certain terms are mentioned in the media could be investigated with a frequency analysis.

A second form is the *contingency analysis*. In addition to frequencies, this also takes association structures into account. Association structures are used to uncover the links between two terms such as "environment" and "dictatorship" or "tax reform" and "ruin" or "freedom" and "equality". Contingency analyses are interested in whether such terms occur particularly frequently together in a text.

Third, *evaluation analyses* or *intensity analyses are* used. They are used to measure the intensity of evaluations made by the author in the text. Certain attitude objects evaluated in a text are examined to determine the intensity of the evaluation.

6.3.4 Advantages and Limitations

Content analysis offers several advantages over other social science research approaches:

- The possibility of investigating changes in social values over the long term. While it is relatively difficult to address issues that lie further back in the past in surveys, historical texts can also be examined with the help of content analysis. As shown, the Speeches from the Throne from different centuries, for example, lend themselves to identifying possible cycles of social change.
- The results of content analyses are not burdened by an interviewer influence, they are thus not reactive. Of course, this only applies if the production of the texts was not influenced by the researcher.
- A falsification due to possible memory errors can be excluded in content analyses.
- Content analyses – again in contrast to surveys – can in principle be repeated as often as desired without changing the object of the analysis. This makes it possible to check the reliability of the data collection form (coding scheme) or to modify it if necessary and then use it again on the same text. The time factor thus plays a subordinate role.
- Statements can be made about people (recipients and communicators) who cannot be reached in any other way. Also, again in contrast to surveys, it is not necessary to rely on the willingness of other people to cooperate.
- Content analyses are considered less expensive than other methods.

This is countered by some problems that can affect the application of content analysis:

- Quantifying content analyses work with a coding or classification scheme. The reliability of these classifications must be ensured. For example, errors can occur in the analysis due to the ambiguity of terms and due to inaccuracies in the scheme.
- The encoders have an important function in content analysis. Their task is to assign certain text features certain characteristics in the coding scheme. With coders, both learning processes during coding and fatigue must be taken into account. Both can negatively affect the quality of the results.
- Content analyses should face the demand for stability (one encoder encodes the same text again in the same way), for repeatability (several encoders process the same text)

and for accuracy (for training purposes, comparisons are to be made with a standard encoding). This increases the necessary research effort for content analyses.

- In order to be able to carry out content analyses, the existence of and access to relevant sources must be given.
- In order to interpret the texts, a large number of assumptions, the auxiliary hypotheses, must be made. For example, if one wants to draw conclusions about the crime rate on the basis of the number of police reports, it is necessary to take into account that certain cases are only reported to the police if there is insurance cover. This may be particularly the case with bicycle thefts. While it may be possible to ask follow-up questions to the target persons in interviews, this possibility does not exist in content analyses.

6.3.5 Survey Research in the Mirror of the Press, an Example of Content Analysis

Using the example of an investigation into the image of survey research in the media on the basis of newspaper articles, the various phases of a content analysis will now be presented.

In the *first phase, the aim was* to determine the objective pursued by the content analysis as concretely as possible. For example, the declining response rate represents a particular challenge for survey research (cf. Sect. 5.8.1 and Stoop 2005). One of the determinants of willingness to participate is the reputation that survey research has in society. This image is likely to come about in a number of ways. It cannot be ruled out that one's own experience – as a target person – with surveys determines the reputation attributed to this method. But surveys are also judged and evaluated in the media. This gives rise to various questions that should be addressed by content analysis:

1. How do the media report the results of surveys? Which aspects of a survey are primarily communicated to readers and which are not?
2. How is survey research reported? For example, is the reliability of the findings questioned, is the anonymity of the respondents doubted, is the willingness to participate addressed as a problem, and so on.

Both questions ultimately formed the focus of the content analysis.

In the *second phase, the* relevant texts had to be identified. In other words, the population from which the texts to be analysed were to be selected had to be determined. This was to include all articles published in the following newspapers in 2004: DIE ZEIT, Der SPIEGEL, Frankfurter Allgemeine Zeitung, Süddeutsche Zeitung and Berliner Zeitung Online. After that, a sampling strategy could be determined. It had to be decided whether a complete analysis of every umpteenth issue of a newspaper should be carried out or whether a total survey of all newspapers was preferable. In the present case, issues were randomly selected from each publication series and these were then fully analysed as a cluster. In the case of the online newspaper, on the other hand, certain days of the week were randomly drawn and then considered in their entirety. When determining the population and deciding on a selection strategy, the available resources played a role above all.

Table 6.24 Articles included in the content analysis per medium

Medium	Absolute frequency	Percent
The MIRROR	65	20.8
THE TIME	77	24.7
Berliner Zeitung online	80	25.6
Frankfurter Allgemeine Zeitung	5	1.6
South German newspaper	84	26.9
(information missing)	(1)	(0.3)
Total	311	99.7

Table 6.24 shows how many relevant articles were finally found in the individual media using this strategy.

The *third step* served to determine the units of analysis or counting. It is conceivable to provide for newspaper areas, broadcast minutes, words, headlines, verbs, foreign words, word combinations and so on as counting units. The decision was made to analyze complete articles in each case. These are texts which have now been divided into the following categories: (1) letters to the editor, (2) short notes, (3) articles and (4) longer reports (with a length of more than one page).

In the *fourth phase,* the category system was developed. It contains the characteristics to be coded and forms the core of the content analysis. The category system is comparable to an interviewer's questionnaire, except that it is not addressed to a target person but to a text. In the category system, the expressions of the characteristics of interest are determined within the framework of operationalization. This is based on a number of theoretical assumptions.

The category system can be designed in different ways. Three types can be distinguished:

- The one-dimensional system, in which a distinction is made only between the occurrence or mention versus the non-occurrence/non-mention of certain terms or facts.
- The synthetic category system that captures multiple dimensions simultaneously.
- Finally, the relational category system, which goes beyond manifest content and encodes relations between features.

The subsequent *fifth phase of* a content analysis involves the establishment of coding rules. Similar to the instructions given to the interviewer in a standardized survey, instructions are compiled for the coders to follow when processing the text in question. In the planned study, ratings that were conveyed in the text via surveys were to be coded. To do this, the following options were available to the coders:

Positive	The tendency in the article is to present surveys in a clearly positive light
Negative	The trend is clearly negative
Both	There are both positive and negative evaluative reports about surveys in the article
Neutral	Reporting is done without an evaluation of the survey research
Not applicable	There are no statements in the article that would indicate an evaluation

If the article contained latent evaluations of surveys, it had to be decided how these turned out. In the second part of the coding scheme, for example, it had to be decided whether certain aspects of knowledge (information and facts about surveys) were included in the text.

In this context, it was necessary to further operationalise the knowledge aspects. One indicator was information about the way in which questions were formulated in surveys. This was understood to include comments such as *"From a methodological point of view, the questions were asked poorly"* or *"Care was taken in the formulation of the questions"*.

A synthetic category system was developed for the analysis of the image of surveys. It is shown in Fig. 6.15.

In the *sixth phase,* the pretest took place. In the course of this pretest, the coding instructions were checked for clarity, the selection of materials to be examined was questioned, the required number of coders was determined and, finally, the time required for coding could be estimated. For the pretest, it was expedient to use editions of the aforementioned media that did not belong to the population, for example, because they came from a different year of publication. Afterwards, the survey documents and the selection concept could be revised on the basis of the pretest results and the strategy of the actual survey could be reconsidered.

The *seventh phase of* the content analysis is the field phase. It provides for the coding of the material by various coders using the coding sheet. This forms the basis for the data matrix to be created subsequently. Similar to a survey, the data must be prepared and transferred into a data file. Statistical analyses, such as frequency counts and comparisons, then form the main components of the following *eighth phase,* the evaluation.

6.3.6   Special Forms of Content Analysis

Computer-Aided Content Analyses

Some special features have computer-assisted content analysis. This form can be done at very different technical levels. To facilitate the interpretation of texts, it is already possible to work with conventional word processing programs. With their help, certain terms can be counted. They can be found and easily marked with a text program. In addition, there is also the possibility of semi-automatic and automatic content analysis. The TEXTPACK program is capable of listing identical units (words) after the texts have been entered, identifying ambiguous terms in the process and processing them manually, as well as performing statistical data analyses.[38]

An example of the automatic processing of texts is available in the context of occupational coding with the use of the ISCO classification (International Standard Code of Occupation) of the International Labour Organization (ILO). In surveys, it makes it

---

[38] Other programs – for qualitative content analysis – are for example AQUAD 7 (compare http://www.aquad.de/ last accessed on 03/05/2019) or ATLAS (compare Muhr 1997).

Source: Issue: Page: ID:

Categorization:
Letter to the editor
Short note (less than 1 page)
Articles (1 page)
Comprehensive report (more than 1 page)

Graphics/tables   available yes/no
Study mentioned in the   title yes/no

A)   Evaluation of surveys (subjective feelings)
Positive: clearly positive presentation
Negative: Clearly negative representation
Both: Both positive and negative evaluation
Neutral: Naming without rating
Does not apply: no statement included

1. Duration of the interview (*positive: refreshingly short; interview was just the right length; negative: lengthy, tiring long*)
positive   negative   both   neutral does not apply

2. Clarity of the question wording from the respondent's point of view (*clarity = unambiguous, comprehensible, precise, without Technical terms*)
positive   negative   both   neutral does not apply

3. Entertainment value of the survey for the respondent (*positive: fun, entertaining, varied; negative: boring, monotonous*)
positive   negative   both   neutral does not apply

4a) Assessment of reliability of survey results (*positive: unambiguous results; feeling that results are true/valid; negative: feeling that results are untrue/ambiguous/outdated/obsolete*)
positive   negative   both   neutral does not apply

4b) Factual presentation of survey results (*survey results are presented like facts*)
positive   negative   both   neutral does not apply

5. Feeling informed about surveys (*positive: people feel well informed, think they know how surveys are done; negative: people have no idea about surveys; the most important things are kept quiet*)
positive   negative   both   neutral does not apply

6. How sustainable is the approach to people during interviews? (*positive: unobtrusive, pleasant; negative: intrusive, annoying*)
positive   negative   both   neutral does not apply

7. Belief in anonymous handling of answers (*positive: people believe in anonymous handling of their answers; trust in data protection in surveys; negative: uncertainty about who receives the answers; fear that answers are passed on to third parties without permission, surveys are only used to sound out consumers for advertising purposes*)
positive   negative   both   neutral does not apply

B) Knowledge about surveys
8. Information about the way questions were formulated in surveys. (*"The questions were asked in a methodologically poor way" or "Careful work was done on the question wording"*)
Briefly mentioned described in detail none mentioned   not assessable

9. Information on the benefits of surveys for the population (*statements on the entertaining nature/information gain from survey results*)
Briefly mentioned described in detail none mentioned   not assessable

10. Information on the fact that surveys must be conducted according to scientific methods (*statements on sample size and sampling, training of interviewers, naming of scientific institutes*)
Briefly mentioned described in detail none mentioned   not assessable

11. Information on the way the interviewer contacted the respondent (*interviewer repeatedly tried to reach the target person*)
Briefly mentioned described in detail none mentioned   not assessable

12. Information on misuse of data (*information obtained through surveys was passed on to third parties for advertising purposes against the will of the respondent*)
Briefly mentioned described in detail none mentioned   not assessable

13. Information on the prevention of data misuse (*data is treated confidentially, data is not passed on to third parties without authorisation*)
Briefly mentioned described in detail none mentioned   not assessable

14. Information on the handling, retention and use of the survey results (*the results of the survey were used to improve product X*)
Briefly mentioned described in detail none mentioned   not assessable

Fig. 6.15 Category system for coding survey coverage in the media

15. Information about the reasons why media report on surveys.
(*Increase circulation of the medium, inform the population about trends*)
Briefly mentioned described in detail none mentioned   not assessable

C) Survey behaviour (*intentions to act, participation behaviour: why to participate/not to participate*)
16. It is natural to participate in a survey
briefly mentioned described in detail not mentioned   not assessable

16a. Other reasons to take part in a survey (*entertaining surveys, making a personal contribution, financial incentives, doing one's civic duty*)
briefly mentioned described in detail not mentioned not assessable

17. It is a waste of time to participate in a survey.
briefly mentioned described in detail not mentioned   not assessable

17a. Other reasons for not participating in an interview (*fear of crime, privacy not being respected, fear of not knowing enough, invasion of privacy, time cost*)
Briefly mentioned described in detail not mentioned not assessable

Fig. 6.15 (continued)

possible to automatically assign a corresponding code to 30,000 occupations and to divide them into 390 categories for occupational activities (cf. Hoffmeyer-Zlotnik et al. 2006, p. 101 ff.). It is also possible to identify synonymous terms such as butcher, slaughterer and butcher by computer. In addition, it is possible to expand the system with each application and improve it in this way.

Table 6.25 shows the ISCO code of some of the persons interviewed in the Dresden self-defence study.

Above all, the possibilities of access to electronically stored texts, such as newspapers or chat rooms, given by the Internet and with the help of CD-ROMs, grant new possibilities to computer-assisted content analysis. Finally, the use of scanners makes it possible to process conventional texts electronically and to analyse them with computer support.

Qualitative Approaches to Content Analysis
Finally, the more qualitatively oriented, non-standardized or partially standardized content analysis will be discussed. For the purpose of hypothesis-generating research, it can be useful to examine a document relatively openly at first, without a previously determined system of categories. Firstly, the method according to Mayring (1993) should be mentioned (for further information, see Kuckartz 2005 and Mayring and Fenzl 2019, p. 637 ff.):

1. Summary. In this process, the essential statements of a text are elaborated (summarised) and reduced to individual categories. For example, the results of an open interview can be reduced to a few aspects.
2. Explication. This step involves a meaning analysis of text passages that appear problematic. It is possible and recommended to consult further material on the subject of the study.
3. Structuring. This involves filtering out structural features of a text using a category system.

The comparison with the observation of a conspicuous boulder found in a landscape illustrates the procedure according to Mayring: First, the part itself is observed from a distance (first step: summary) and its conspicuous features are highlighted. Then, it is a matter of analysing particularly interesting details (second step: explication) and activating other

Table 6.25 Occupations (ISCO 88) of persons interviewed in the Dresden self-defence study (extract)

Soldat	7	Uni.- HS-lehrer	5	Datenverarbeitungsfachkräfte	19
Leitende Verwaltungsbedienst.	5	Wiss. Lehrer Primarber.	5	Datenverarbeitungsassistenten	2
Leitende Bedienst. Interessenorg.	1	Wiss. Lehrer Vorschule	25	Bediener Datenverarbeitungsanl.	8
Leitende Bedienst. pol. Parteien	2	Wiss. Sonderschule	4	Photographen	7
Leiter Arb.geber-Arb.nehm.verb.	2	Schulräte	2	Fernsehanlagenbed.	1
Geschäftsleiter	2	Sonst. Wiss. Lehrkräfte	1	Bediener med. Geräte	1
Direktoren Hauptgeschäftsfüh.	4	Sonst. Wissensch.	1	Deckoffiziere	1
Produktionsleiter Verarb. Gewer.	2	Unternehmensberatungsfachkr.	1	Flugverkehrslotsen	1
Prod.leiter Groß- Einzelhandel	1	Wirtschaftsrechnungssachverst.	10	Baukontrolleure	1
Übrige Prod.- Operationsleiter	3	Unternehmensberatungsfachkr.	12	Sicherheitskontrolleure	7
Sonstige Bereichsleiter	2	Juristen	4	Biotechniker	8
Finanzdirektoren Verwaltungsl.	2	Anwälte	7	Hygienetechn.	1
Personalleiter Sozialdirektoren	4	Richter	2	Augenoptiker	4
Verkaufs- Absatzleiter	8	Sonstige Juristen	1	Physiotherapeuten	8
Werbeleiter	1	Bibliothekare	1	Pharma. Ass.	4
Versorgungs- Vertriebsleiter	4	Wirtwissenschaftler	20	Mod. Med. Fachberufe	3
Leiter EDV-Abteilung	4	Soziologen	1	Nicht-wiss. Krankenschw.	95
Sonstige Bereichsleiter	10	Philologen	4	Nicht-wiss. Hebammen	2
Betriebsleiter	25	Psychologen	4	Heilpraktiker	3
Betriebsleiter Landwirt.	1	Sozialarbeiter	31	Nicht-wiss. Lehrkr.	22
Betriebsleiter verarb. Gewerbe	5	Autoren	13	Sonstige nicht-wiss. Lehrkr.	11
Betriebsleiter Baugewerbe	1	Bildhauer	1	Effektenhändler	5
Betriebsleiter Groß- Einzelhand.	26	Komponisten	12	Versicherungsvertreter	16
Betriebsleiter Restaurant Hotel	10	Choreographen	1	Immobilienmakler	7
Betriebsleiter Transportwesen	2	Geistliche	10	Tech. Kaufm. Handelsvertr	15
Betriebsleiter gewerbl. Dienstl.	7	Techniker u.ä.	21	Einkäufer	1
Betriebsleiter Körperpflege	2	Physikal. ing.wissen. Techniker	15	Finanzfachkr.	2
Sonstige Betriebsleiter	3	Chemo- und Physikotechniker	21	Handelsmakler	1
Wissenschaftler	9	Bautechniker	5	Verwaltungsfachkräfte	56
Physiker, Chemiker	1	Elektrotechniker	3	Verwaltungssekretäre	27
Astronomen	1	Elektronik- Fernmeldetechniker	10	Fachkräfte Rechtsangel.	22
Chemiker	3	Fleischer	6	Buchhalter	32
Mathematiker	1	Bäcker	12	Verwaltungsfachkräfte	8
Informatiker	5	Möbeltischler	27	Zoll u.ä.	9
Systemplaner ,-analytiker	8	Korbflechter	1	Steuerbedienstete	3
Programmierer	9	Weber	2	Sozialverwaltungsbedienst.	9
Informatiker	1	Schneider Hutmacher	19	Polizeikommissare	5
Architekten, Ingenieure	47	Näher	1	Sozialpfleg.	50
Stadtplaner	8	Polsterer	3	Dekorateure	7
Bauingenieure	10	Schuhmacher	1	Fernsehsprecher	1
Elektroing.	2	Schlosser	26	Berufssportler	1
Elektronik- Fernmeldeing.	7	Elektroniker	3	Bürokräfte	310
Maschinenbauing.	14	Aufsichtskräfte Produktion	16	Stenotypisten	5
Chemieingenieure	3	Anlagen- Maschinenbediener	34	Datenerfasser	1
Bergbauingenieure	3	Metallschmelzer	1	Sekretärinnen	58
Kartographen	5	Glasschmelz- kühlofenbediener	1	Buchhaltungsang.	21
Architekten	14	Bediener Papierherstellungsanl.	1	Finanzangestellte	31
Biowiss. Mediziner	1	Bediener chem. Verfahrensanl.	7	Lagerverwalter	2
Biologen, Botaniker	2	Bediener Dampfmaschinen	3	Fertigungsplaner	2
Agrar- u.ä. Wissenschaftler	1	Bedien. automat. Montagebänder	1	Speditionsang.	11
Ärzte	23	Maschinenbediener	4	Bibliotheksange	1
Zahnärzte	2	Maschinenbautechniker	9	Postverteiler	7
Tierärzte	3	Chemiebetriebstechniker	2	Kodierer	1
Apotheker	2	Bergbautechniker	1	Sonst. Büroang.	12
Wiss. Krankenpfl.- Geburtshilfe	1	Technische Zeichner	13	Büroang. Kundenkontakt	6
Wissenschaftliche Lehrkräfte	91	Physikalische ing.wiss. Techn.	6	Kassierer	8

sources of information for this purpose. Finally, the inner structure of the boulder is explored, it is symbolically broken open in order to gain deeper insights (third step: structuring).

Another qualitatively oriented technique of content analysis is *hermeneutics* (compare Oevermann et al. 1979 and Kurt and Herbrik 2019). This involves a qualitative text interpretation in which understanding the meaning of a text is the goal. Whereas in the standardized approach a system of categories with the indicators worked out in advance is the prerequisite, in hermeneutics the text is first to be interpreted, according to open rules. The document is to be opened up in its entirety. This does not mean counting the frequency of individual words, as in frequency analysis. For hermeneutic analysis, the researcher should put himself in the place of the author of the text and dissect his text. While in standardised content analyses it is expected that the results are independent of the respective coder, in qualitative content analyses strongly person-dependent results are to be assumed (on qualitative techniques compare Sect. 3.8).

As an example of content analysis using hermeneutic methods, the evaluation of a narrative interview on the topic of "Traumatized life courses in Israel" will be presented. The starting point is the idea that decisions in the life course have an objective meaning, independent of the original intentions (compare Oevermann et al. 1979) of the person acting. In a first step, the chronological order of the life data, as obtained from a narrative interview, takes place. This is followed by a sequential examination of the individual stations of life in terms of their latent meaning. This sequential analysis of biographical data interprets the hard biographical data independently of the further biographical course. This is then followed by information about the life path actually taken. Afterwards this step is repeated again (compare Fischer-Rosenthal 1996). Here is a concrete example:

a.	Martin Janker (pseudonym), born in Saarbrücken in 1911 to Jewish parents,
b.	Studied medicine in Heidelberg for four semesters,
c.	In 1932/33 he met his future wife in a Jewish organization,
d.	1934 three months work in a car repair shop

In a first sequence, it is now asked what step (d.), the three-month work in a car repair shop can mean. The following readings are possible for this. The work in the garage is taken up:

(1)	Because Jews were forbidden to study medicine in Nazi Germany,
(2)	Because his future wife was expecting a child and he wanted to support the family,
(3)	Because he was working a student summer job,
(4)	Because he was planning to emigrate to Palestine, where car mechanics were needed …

Now we have to ask how the following sequence (e.) would look like in the course of life, with the individual interpretations. In (3) this would be the continuation of studies, in (2) the birth of a child, and so on. Now the actually reported station in life is considered:

e.	1934: Janker emigrated to Palestine; here he worked as an unskilled construction worker

For this sequence, it is again asked what its meaning can be. What readings for (e.): would be possible? It would be conceivable that the emigration and the activity in construction are motivated by:

1. For every immigrant in this time a pluck and a commitment to Palestine were indispensable. The behavior was borne of patriotism.
2. Janker was arrested in the illegal immigration case and sentenced to hard manual labor by the Mandate.
3. The person concerned only carried out the work temporarily and then went back to study.
4. He didn't work for a Jewish state, but for Arabs in an Arab settlement/city there because he switched to a different faith.

Now it can be fathomed what the reported sequence (f.) would have to look like depending on (1), (2), (3) and so on in each case. Thus, one could assume that if (1) were true, the person would resume his studies. However, the next life station actually reported then is:

f.	Nerve complaints and medical treatment ensued. Since then and up to the present there is both chronic diarrhea and regular use of tranquilizers or mood elevators

This means first of all that reading (1) cannot apply, likewise (3) should not be possible, while (2) cannot be excluded. The other sequences would be treated accordingly.

The next step in the evaluation concerns the assessment of the alternative actions chosen (or not chosen) by Janker. In this way, even the life not lived can be reconstructed with the help of corresponding structural hypotheses.

Complex Designs

7

Abstract

In the previous section, the basic social science methods for the collection of empirical information were presented in the form of interviews, observation and content analysis. Repeated reference has been made to the variety of techniques that can be observed in the application of the methods mentioned. The following section deals with complex designs. They are characterized by the fact that they make use of the basic methods and arrange them in a particular way in each case. For example, social experiments use observations and interviews, Delphi studies use postal surveys, and time budget studies use self-observations. Due to the special arrangement of the basic methods, each results in its own possibilities of knowledge. These will now be presented and discussed.

7.1 Social Experiments

7.1.1 Nature and History of Social Experiments

Before considering the history of the experimental method, which can be traced back relatively far, a variety in the use of the term experiment must be noted. If one follows Schulz (1970, p. 22), at least the following variants can be identified:

- Experiment means a procedure that is used experimentally and that is characterized by groping around and trying things out. Its main purpose is to gain experience.
- Experiment is a synonym for a specific empirical scientific approach.
- Experiment refers to a type of evidence that is primarily capable of uncovering causal relationships.

- The term experiment is used to designate a particular experimental arrangement. This is characterized above all by more or less strong artificial elements.
- Finally, particularly daring undertakings (such as the founding of the GDR or even the whole of socialism) are also referred to as experiments.

Petersen (2002, p. 11 ff.) describes the history of experiments in more detail. He points out that the original content of the term (coined by the Franciscan monk Roger Bacon, who lived from about 1210 to 1294) can rather be paraphrased with the maxim to make one's own observations. This is based on the view that for proper human knowledge of the world, mere reflection is not sufficient, as this can easily lead astray. But also to experiment without thinking, that is, to observe *exclusively,* Bacon considered useless. The content of such a concept of experiment thus corresponds roughly to today's understanding of empiricism. According to this, theory and empiricism form a unity: theory has a strong reference to empiricism and thus empiricism is also inconceivable without a reference to theory.

The demand for the introduction of experiments into science goes back to Francis Bacon (he lived from 1561 to 1626). With this demand, Bacon appealed to researchers to work empirically themselves and to draw scientific conclusions on the basis of their own observations. In doing so, he broke with a tradition of thought that primarily or exclusively relied on the authority of classical values and regarded them as accepted dogmas in science.

Finally, the name of John Stuart Mill (1806–1873) is associated with the development of the concept of experiment. To him is owed a change from the passive to the active concept of experiment. This expresses the fact that in experiments one no longer merely observes from the outside, but actively intervenes in reality. In this way, it finally becomes possible to uncover causal relationships.

Social experiments are considered the crown of knowledge. They are a particularly rigorous form of hypothesis testing that can help uncover cause-and-effect relationships. The prominent position of social experiments results from various reasons, which will be presented in more concrete terms below.

In the following, the essence of experiments is described and then the different forms of experiments are presented. Experiments can be characterized by the following properties:

- They are characterised by the parallel use of a comparison and an experimental group, whereby it must be ensured that both groups have an equivalent structure.
- In experiments, a change is made in the experimental group and the subsequent observation of the effect thus produced.
- Finally, reference should be made to the use of multiple measurements. Thus, surveys are conducted at different times, at least once before and at least once after the introduction of the experimental stimulus.

While a certain change is made in the experimental group, conditions in the comparison group are to be kept as constant as possible. It also stands to reason that only if the composition of both groups is comparable can the differences between the experimental and comparison groups determined at the second time point of the investigation be attributed to the effect of the experimental change. For this purpose, the procedure of randomization is used. Randomization is a form of control of the study situation. With the help of a controlled random, the subjects are assigned to one of the two groups. This can be done in a simple way according to the principle of the lottery drum. Furthermore, it is important to control the result of this random assignment. For example, the result of such simple random selection could be that there are unintended differences between the two groups. As shown above (in Sect. 5.1), the outcome of a random experiment cannot be predicted with certainty. Such an outcome can be countered by stratifying the allocation procedure. For this purpose, it is specified, for example, that the age and gender proportions should be identical in both groups. This proportion is then controlled and thus guaranteed when the test subjects are randomly assigned to one of the two groups.

Thus, identical experimental conditions in both groups and a comparable composition of the groups are crucial for the success of a social experiment.

In principle, social experiments offer the opportunity to conduct causal analyses. While a relationship between two characteristics can be determined in simple cross-sectional surveys, the direction of this relationship can also be identified in experiments. Since an effect (E) always occurs with a time lag after the cause (C), such cause-effect relationships (C → E) can be uncovered. Experiments in which a measurement is made before and after the introduction of an experimental stimulus can attribute changes determined in the process to the introduced stimulus. This is not true for cross-sectional studies.

It must also be pointed out that strict ethical standards must be applied to experiments in the social sciences. If, for example, one were to proceed methodologically consistently in an experiment testing the effect of a new unemployment promotion programme (= experimental stimulus), all eligible test subjects would have to be randomly assigned (= randomisation) to the experimental or the comparison group. In practice, however, socio-political or ethical criteria are more likely to play a role when it comes to implementing such a support programme. It is conceivable, for example, that support programmes could be used for people who have a special need for support. However, this would not make it possible to determine the possible effect of this support programme on all unemployed people in an experiment.

7.1.2   Types of Experiments and Their Sources of Error

Different types of experiments can be distinguished (compare Zimmermann 1972). The most common are laboratory experiments, field experiments and quasi-experiments.

- Laboratory experiments take place under controlled conditions, i.e. under artificial conditions (i.e. in the laboratory). For example, the test subjects may be invited to a test studio in which constant conditions prevail, i.e. no mobile phones ringing, for example. Here they are introduced to different versions of a new product and are asked to evaluate it. It is conceivable that all reactions of the test person can be precisely observed (by means of one-way mirrors, cameras and so on). Laboratory experiments also have the advantage that it is possible to monitor almost all conditions (the so-called boundary conditions) or to keep them constant. Thus, for example, disturbances caused by interventions of third parties can be avoided. The disadvantage to be accepted for this is the remoteness from reality of such a procedure.
- Field experiments, on the other hand, take place under natural circumstances. Here, firstly, the control of all conditions is more difficult. Secondly, this also applies to the recording of all reactions of the test subjects. However, there is a high degree of realism and correspondingly better conditions for generalising the findings.
- Quasi-experiments are those in which the above characteristics of social experiments are not completely fulfilled. Quasi-experiments modify the described (complete) experimental design. Thus, randomization may be absent and/or only the effects of an influence quantity are determined retrospectively – omitting the measurement at the initial time t_0. Quasi-experiments partly make use of quite normal processes in reality. These mostly do not take place on the basis of a scientific question, but are caused in a different way. If one were to understand the GDR (and its collapse) as an experiment of immense sociological interest, it would be obvious that neither the necessary before-and-after measurements nor a random division of the test subjects into the two groups (East and West Germany) could take place here. However, quasi-experiments are not scientifically worthless.

Two other types of experiments should be mentioned:

- *Experiments with an ex post facto design*: This refers to experiments in which the assignment to the two groups has not already been made before the experimental change. This is often the case with quasi-experiments. As a result of what is usually a relatively rare event, the social changes encountered in the setting are then studied in social science terms. Measurement before the experimental change is dispensed with.
- *Experiments with an ex ante design are* characterised accordingly by the fact that the formation of the two groups has already taken place before the experimental change. This is assumed in the classical design. Ex ante experiments thus have a significantly higher informative value.

In order to clarify the different types of experiments, a schematic representation can be used. The groups involved, the measurements made and the experimental change are illustrated in it. The first representation (see Table 7.1) is an incomplete design or a quasi-design.

Table 7.1 The procedure for incomplete experimental designs I

	Time		
	t_0	t_1	t_2
Experimental group	–	Experimental change	Measurement
–	–	–	–

Table 7.2 The procedure for incomplete experimental designs II

	Time		
	t_0	t_1	t_2
Experimental group	Measurement	Experimental change	Measurement
–	–	–	–

In Table 7.1 it can be seen that there is neither a comparison group nor that a measurement took place before the experimental change. Diekmann (2001, p. 332) describes an example to demonstrate the inadequacy of such an approach. Then one could look at which speed (= experimental change) more traffic accidents occur at, 50 km/h or 200 km/h. Since the frequency of accidents is likely to be greater at 50 km/h than at 200 km/h, a corresponding fallacy could be suggested. This would be that a higher speed leads to fewer accidents. However, this design lacks a comparison group that is characterised by the fact that it is *not* affected by an accident. One would then also have to measure the speeds driven here and could thus avoid the corresponding erroneous conclusions.

The following design (compare Table 7.2) is also an incomplete experimental design.

This design also lacks a comparison group. Such an experimental design would be present if, for example, persons with a cold, who are known not to have always suffered from this illness (= measurement at time t_0), were given a lolly (= experimental change at t_1) and the effect of this intervention was then measured. To attribute an improvement in health status determined in this way to the experimental stimulus would also be a fallacy in some circumstances. Here, too, the effort of a comparison group (persons who also have a cold, but who are not exposed to the lozenges) would lead to more differentiated findings. Under certain circumstances, both groups would be healthy at the same time and an effect of the lozenges would thus have to be negated. Such a design would also be an oversimplification of social experiments, which would then not lead to valid conclusions.

The situation is somewhat different with the survey design (cf. Table 7.3). Although two groups exist here, an experimental change has only taken place in one of them. (At this point, no further consideration should be given to the extent to which the survey design is in fact an incomplete experimental design).

Survey designs are characterised by a sufficiently large number (usually 1000 or more) of people being interviewed. Thus, based on a question (Have you ever been in a situation in which you came to the aid of another person and defended him or her against an attacker?), the division into the two groups could be made quasi during the survey with the help of a retrospective question. In addition, the survey asks about a number of other issues, thus indirectly inferring the effect of the experimental change.

Table 7.3 The survey design procedure

	Time	
	t_0	t_1
Experimental group	Experimental change	Measurement
Comparison group	–	Measurement

Table 7.4 The classical experimental design

		Time		
		t_0	t_1	t_2
R	Experimental group	Measurement	Experimental change	Measurement
R	Comparison group	Measurement	–	Measurement

Of course, this is not a classical experiment either, since the differences between the two groups that may already exist at the time t_0 have not been determined. Thus, it is also not clear whether the possibly determined differences between the experimental and the comparison group did not already exist *before* the experimental change (in this case, the assistance provided to another person). Survey designs of this kind are not suitable for uncovering causalities.

The classical experimental design is presented in Table 7.4. It is characterized by, first, the formation of two groups, second, the control of the boundary conditions (for example, by randomization [R], elimination of confounding variables, keeping boundary conditions constant), third, the manipulation of an independent variable, and fourth, the measurement of the initial conditions before the experimental change and at least one measurement after the experimental change.

Two special forms of the classical experiment are the blind test and the double-blind test. These forms are used in medicine, for example. In the blind experiment, the test subjects do not know which group they belong to within the experiment. In the double-blind experiment, this also applies to the other personnel involved in the experiment. If the aim is to test the effect of a new pharmaceutical preparation, the avoidance of artefacts is an essential task of the experimental design. Placebos are therefore administered to one group of patients and the drug of interest to the other. Provided that both groups are appropriately structured (have the same severity of illness, are similar in age and sex structure, harmonize in terms of other constitution, and so on), one can now observe the effect of the drug without having to fear that it is an artifact that has occurred, for example, because of the reactivity of the measurement.

Now it may be that the experimenters – quite unconsciously – take special care of those persons who have not been treated with the placebo. Thus, there would then again be unequal treatment in the two groups. In order to prevent this as well, the double-blind experiment can be resorted to, in which, for example, the doctors and nursing staff are not informed about the group allocation before the experiment begins.

In the following, typical *sources of error that* can occur in social experiments and corresponding counterstrategies are discussed:

- In the experimental setup, it is assumed that there is a relationship between the independent variable that is changed and the dependent variable. However, there may be other unknown relationships operating, that is, the dependent variable is not affected, or not only affected, by the manipulated independent variable. For example, an experiment has shown that the sale of chocolate Santa Clauses is promoted if they are wrapped in particularly beautiful colourful Christmas paper. However, it should not be forgotten that a third variable, the season, is at work here. Thus, in the summer months, the sale of Santas is not likely to be promoted by means of particularly attractive packaging.
- Effects of measurement. In the classical design of a social experiment, at least two measurements are provided in both groups. These can – unintentionally – themselves change the object being measured and thus make it difficult to transfer the results of the experiment to other external units of investigation. On the reactivity of measurements, see Sect. 4.7.
- Randomization can have a negative effect on the participants in the experiment. For example, if one draws lots for participation in an experiment, a loser problem may arise in the comparison group, which, according to the draw, was not administered the supposedly attractive experimental stimulus. This may be particularly the case when some form of reward, such as participation in training, is involved, the effect of which is to be tested in the experiment. Conversely, members of the experimental group may also feel like winners and be particularly receptive to the experimental stimulus. Thus, an unintended influence can be exerted on both groups.
- As already emphasized, it is possible that randomization does not succeed immediately. The random division into the two groups of an experiment, carried out according to a lottery drum model, can nevertheless lead to systematic differences between the two groups in the result.
- Maturation processes that may occur in the participants during the duration of the experiment must be taken into account. In the above example, where lollipops were administered to people with a cold (= experimental stimulus), such maturation processes may occur. A cure may take place here anyway. The more long-term an experiment is designed, the more likely maturation processes are.
- Especially in field experiments, the intervening events in the environment of the experiment can hardly be influenced. Competing products can appear on the market, new laws can be enacted, political crises can break out or be settled peacefully, and much more. All this can hardly be influenced in field experiments, but could have an effect on the dependent variable.
- Unintended influences can also have (already small) changes in the measuring instrument, for example in the wording of questions, a change of interviewer or experimenter and so on.
- Finally, incorrect selection of experimental participants as well as failures during the experiment, for example due to dropouts, can negatively influence the value of the result.

Various strategies are available as a countermeasure. One variant is the systematic elimination of unintended disturbance variables. This is particularly well possible in laboratory
experiments. In the event that laboratory experiments are out of the question, care should
be taken to keep the experimental conditions as constant as possible. The aim is to ensure
that the experimental conditions in both groups are always as similar as possible.

To validate the findings, the experiments should be replicated as often as possible.
The method of the double-blind experiment can be used. By including further groups in
the experiment, it becomes possible to control the influences emanating from the measurement at time t_0. This is Solomon's four-group experimental design (compare
Table 7.5).

The idea of the design in Table 7.5 is that possible influences of the measurement at
time t_0 could be revealed on the basis of comparisons between the corresponding groups at
time t_2. The price for this additional information is, of course, a correspondingly more
elaborate design.

Further aspects to be considered in social experiments are internal and external validity.
We speak of internal validity when no confounding variables occurred during the execution of the experiment and thus the experiment actually determines the intended facts.
External validity, on the other hand, refers to the generalizability of the results found in an
individual experiment. While a relatively high internal validity can be assumed for laboratory experiments due to the good control possibilities, external validity is a particular
problem here due to the artificiality of the situation. An advantage of laboratory experiments is also the relatively easy production or standardization of the experimental stimulus. In the case of field experiments, the situation is reversed. Here, the original conditions
allow a relatively easy transfer to external conditions. The lack of control or insufficient
monitoring of the environment of the experiment, on the other hand, endangers internal
validity.

Only rarely are all the prerequisites fulfilled for proceeding according to the classical
experimental design. Quasi-experiments are therefore used more frequently (see Table 7.6).
Such experiments do not fully meet the strict requirements of a classical design; for example, no randomization is used in the assignment to groups, so that the effect of any third-
party variables cannot be excluded. Thus, the ideal of an experiment can only be aspired
to here.

With the help of a social experiment, for example, the effect of a new financial performance incentive is to be tested in a company. The experimental stimulus is thus the new

Table 7.5 Solomon's four-group experimental design

		Time		
		t_0	t_1	t_2
R	Experimental group 1	Measurement	Experimental change	Measurement
R	Experimental group 2	–	Experimental change	Measurement
R	Comparison group 1	Measurement	–	Measurement
R	Comparison group 2	–	–	Measurement

Table 7.6 Experimental design without randomization

	Time		
	t_0	t_1	t_2
Experimental group	Measurement	Experimental change	Measurement
Comparison group	Measurement	–	Measurement

form of incentive. Now it must be ensured that, in addition to the experimental group, an equivalent comparison group is also available for the analysis. In the latter, the experimental stimulus is not used.

One possibility is to selectively compose the groups according to relevant aspects (= randomisation), for example according to the criteria of qualification, age and gender. The groups would then first be structured according to these criteria and then the test subjects randomly assigned to one or the other group. This would result in two groups with equivalent characteristics. It remains questionable, however, whether the characteristics included are actually those relevant to the purpose of the study. This targeted allocation according to certain characteristics is an example of the procedure also known as matching.

Now, in this experiment, it is necessary to observe further which environmental conditions have an effect during the duration of the experiment. For example, it could be that the experimental change also leads to changes in the composition of the experimental and comparison groups; certain employees leave these groups, others may join them. Here it must be ensured that such influences – which are unlikely to be prevented in practice – are at least observed, recorded and taken into account in the evaluation.

Time series experiments represent another variant. In this relatively complex form, several measurements are taken before and several measurements after the experimental change. In this way, the trend before the experimental change and the trend after the experimental change can be compared. This makes it easier to identify maturation processes in particular, and it is also possible, for example, to identify short-term effects of the experimental change and seasonal fluctuations (cf. Table 7.7).

An example where such a time-series experiment would be appropriate is the study of the effect of a new divorce law. Here, a wait-and-see attitude could arise in the run-up to the change in the law and thus lead to a postponement of the intention to divorce. It would then also be conceivable that immediately after the new law has been passed, its effect should first be awaited and this would in turn lead to a postponement of divorces. It is also possible, however, that the backlog created by the wait-and-see attitude leads to a particularly high divorce frequency. It would then be wrong to attribute this situation solely to the new law. Here, a time series experiment would offer the possibility of controlling for such influences. In this way, time series experiments could be used to prevent erroneous conclusions from being drawn in the case of such processes that have been conditioned over a longer period of time.

Table 7.7 Design of a time series experiment

		Timings	
	$t_0\ t_1\ t_2\ t_3$	t_4	$t_5\ t_6\ t_7\ t_8$
Experimental group	Measurements	Experimental change	Measurements
Comparison group	Measurements	–	Measurements

7.1.3   The GfK BehaviorScan: An Example of a Social Science Experiment from Market Research

In empirical social research, there are relatively few examples of the successful use of social experiments (cf. Petersen 2002, pp. 31, 67, 69). One experiment worth mentioning, however, is the one reported by Noelle-Neumann and Köcher (1997). In 1983 and 1984, they investigated the question of whether more television channels also lead to more family quarrels. To this end, a series of corresponding surveys were carried out in 1983, before numerous private broadcasting stations in Germany began broadcasting, which took place in 1984, and subsequently in 1985, and thus the original assumption was rejected.

In an experiment with released prisoners, Zeisel (1982) proved that there is no positive effect when they receive unemployment benefits for one year after their release from prison, regardless of their efforts to find a job. The aim of the experiment was to facilitate better reintegration of these individuals. However, the outcome of the experiment was negative after 12 months. In the control group and in the comparison group, the number of persons who reoffended was the same (compare Petersen 2002, p. 68 f.).

GfK-BehaviorScan by GfK Nuremberg can be seen as an interesting example of an extensive market research experimental study. Not infrequently, surprise is expressed at the immense knowledge of the market researchers, such as here:

"The bean counters of the market economy collect more information about Germans than the Federal Criminal Police Office, corporations and political parties. The computer data of the GfK computer center would fill 250 million A4 pages – and the trend is constantly rising. […] And in the test market Hassloch, a Palatinate community of 20,000 inhabitants, GfK, thanks to electronic recording, knows down to the last gum what households buy, how much they spend on what, which supermarkets they prefer and which commercials they most like to fall for".[1]

In this Hassloch test market operated by GfK since 1986 (compare Graf and Litzenroth 1991), the following experimental test conditions, among others, are systematically produced:[2]

---

[1] Compare: http://www.radiobremen.de/tv/daecher/archiv/086.html, last accessed 2004-06-15, unfortunately the source is no longer available.

[2] Compare: http://www.sueddeutsche.de/wirtschaft/922/341.765/text/ last accessed 05 May 2009, sorry source is no longer available, for more information see: http://www.stern.de/wirtschaft/unternehmen/meldungen/:Verbraucher-H%E4rtetest-Alltag:-Tests-Produkte-Ha%DFloch/517.312.html, and http://www.focus.de/finanzen/news/tid-12.695/gesellschaft-klein-deutschland-mitten-in--der-pfalz_aid_351.714.html both last accessed 07 March 2019.

- In a number of experimental households (n = 2000), targeted test commercials are broadcast on television (by superimposing other commercials); in the control group (n = 1000), these broadcasts are not broadcast.
- In all households, television viewing behaviour is registered with the help of electronic instruments (whether the television set is actually switched on during the advertising broadcast). The motivation to participate in this social experiment is supported by a partial reimbursement of the TV cable fees and by a free delivery of a TV magazine (HÖRZU).
- The allocation of households to the comparison or experimental group is carried out with the help of a matching procedure. In addition, certain structural characteristics of the households are determined. These range from sociodemographic structure to pet ownership. The size of the experimental groups also makes it possible, for example, to use a media mix design. In this case, certain households are only confronted with advertising in the supermarket, others additionally receive advertisements in the television newspaper and a further group can also be provided with corresponding advertising programmes via television.
- The supply of such advertising experiments can – supported by the structural data known about the households – be made goods group specific. In this case, comparison and experimental groups are formed (uniformly) with otherwise identical purchasing behaviour (= a criterion of randomisation).
- In television newspapers, which are specifically distributed to certain households only in Hassloch in the Palatinate, advertising has again been specifically placed in one experimental group; accordingly, the comparison group was not confronted with this advertising. This adds another interesting facet to the experiment.
- The purchasing behavior of the experimental and comparison households is also monitored throughout the region. According to GfK Nuremberg's own information, 90 to 95% of the food purchases of its test households are recorded in six regional test stores.
- Finally, the corresponding data is collected over a longer period of time. This historical data makes it possible, for example, to draw conclusions about the sustainability of the advertising[3] (cf. Table 7.7).

7.2 Case Studies

Case studies (sometimes the terms individual case studies or casuistry are also used) involve the complex, holistic analysis of a specific unit of inquiry. Such a unit of inquiry can be quite different objects (compare Alemann and Ortlieb 1975, p. 159 ff. and Hering and Jungmann 2019, p. 619 ff.). As an example, mention should be made of:

---

[3] More, information about GfK BehaviorScan can be found at http://ernaehrungsdenkwerkstatt.de/fileadmin/user_upload/EDWText/TextElemente/Soziologie/Empirisch_Umfragen/GfK_Hassloch_BEHAVIORSCAN_Infos_OLT.pdf, last accessed on March 07, 2019.

- One can study a single person. For example, clinical psychology and medicine are interested in the study of a certain disease that has occurred in a specific person and which is first examined in a single patient (= the case). Of interest to sociology are often also selected periods of the life of a specific person, which are then made the object of observation in the context of a biographical study.
- In case studies, a group of people can also be considered. This can be a community, a sect or, as in the study "The Unemployed of Marienthal" (cf. Jahoda et al. 1960), a group of people selected according to very specific criteria. The case study "The Polish Peasant in Europe and America" by Thomas and Znaniecki from the years 1918 to 1920 has become particularly well-known. Both small, real cooperating social groups, such as associations, as well as territorial units or specific temporal periods (compare Babbie 2002, p. 285) can become objects of analysis in case studies.
- A social organisation, such as the German Bundestag or a particular political party, can be examined in more detail with the help of a case study.
- Finally, the analysis of a society or a culture in the context of political or ethnological studies can also be carried out with the help of the case study instrument.

Case studies are not a survey technique in their own right, such as interviews and observations; rather, like social experiments, they are a special strategy that makes use of a variety of techniques. Goode and Hatt (1962, p. 300) describe the nature of this approach: "The single case study is […] not a special technique. Rather, it is a particular way of arranging the research material in such a way as to preserve the unitary character of the social object under study. In other words, the single case study is an approach in which each social unit is considered as a whole".

If you decide to work with a case study, you should consider several *methodological aspects:*

The respective units of analysis – regardless of whether it is a single person or a complete culture – are to be considered as a whole. In contrast to standardized survey research, no a priori dissection of the case into (the presumably relevant) dimensions is undertaken. It is precisely by considering the case as a whole as comprehensively as possible that this approach makes it possible to include all the determinants of a problem that come into question.

In case studies, the unit of analysis is not only considered as a complex whole, but also in the context of the respective environment. For example, the surprising electoral success of a political party (= the case) would not only be considered on the basis of the internal structure of this party, but the entire political system, for example other, competing parties, would also be included in the analysis. This means that the case to be analysed must be defined in such a complex way that all relevant determinants can be included in the analysis. On the other hand, the case must also be defined narrowly enough so that it is still possible to address the questions with the help of numerous methodological approaches.

Case studies are also characterized by the use of various methodological strategies for the acquisition of information. Numerous forms of questioning, observation and content analysis can be used. The list of possible "experimental plans for case studies is by no means exhaustive, because a plan determined by *content-related* aspects can be created for each specific question" (Roth and Holling 1999, p. 270; emphasis as in the original; compare Tack 1980). Particularly suitable for case studies are, for example, qualitative interviews, various forms of content analysis and group discussions.

In case studies, the degree of standardisation of the methods to be used will remain relatively low. At this point, the inventiveness of the respective researcher is in demand. This means that only relatively low demands can be made on the objectivity, reliability and, to some extent, validity of case studies. However, this shortcoming can be compensated for by other aspects, such as the greater attention to detail in case studies. Such attention to detail is used to reveal the specifics of the case in question particularly intensively.

It hardly needs to be mentioned that findings based on case studies cannot be used to make broader generalizations. Clarifying the question of what conclusions can be drawn from the study of one case for other cases remains problematic.

Case studies are also the method of choice when it comes to researching rare events, such as the solidarity experienced and the sense of togetherness during disasters. Since such events cannot be processed according to statistical selection rules, case studies offer themselves as alternatives.

Case studies offer the possibility to gain interesting information about the object of investigation in case of a lack of money and material resources. While quantifying studies are usually costly and time-consuming, case studies can be implemented with less effort. It should not be forgotten that the mix of methods to be used in case studies requires special methodological skills.

Last but not least, ethical reasons also speak in part for the use of case studies. For example, if a drug treatment is to be evaluated (see Sect. 7.4), it is usually not possible to include a random selection of subjects in the intervention programme to be evaluated. Instead, case studies can be used to qualitatively assess the impact of the programme on patients.

Within the framework of empirical social research, case studies are able to assume several important *functions:*

- The exploration of (still) unclear facts with the aim of arriving at a hypothesis. On the basis of a concrete case, suggestions for the generation of a scientific assumption can be gained. In this way, case studies prepare standardised surveys.
- Case studies can be used as the main method, for example in the analysis of typical or particularly rare cases.
- Case studies are well suited for the vivid illustration of certain events. Concrete descriptions can be used to vividly illustrate interrelationships.

The use of case studies – as a strongly qualitative methodology – can be compared to the work of a detective (see Sect. 3.8). A case is investigated in a number of different directions. Some of these may turn out to be unsuccessful, while other inquiries may lead to the case being solved. With this finding, only the case in question is then clarified for the time being. In principle, however, it is conceivable that in this way a contribution can also be made to the development of theory and the knowledge gained in the clarification of the case can be processed for generalizations. At the beginning of his activity, the detective does not have a specific strategy to solve the case, but is familiar with the use of different techniques. The design to be used arises successively, on the basis of the preceding findings.

However, following another model (compare Burawoy et al. 1991, p. 9 ff.; Babbie 2002, p. 285), case studies have "the purpose of discovering flaws in, and modifying, existing social theories". Case studies should be used to search for data that contradict existing theoretical approaches.

Finally, the advantages of case studies should be emphasized once again: It is an equal procedure in the social sciences, which enables a differentiated data collection and evaluation. In doing so, very different objects of investigation or problems can be placed in the focus of observation. Case studies can be applied to individuals as well as to entire cultures. This technique is also well suited when explorative intentions are pursued with the investigation, or when it is a question of organising studies with a pilot character. The approaches to be used in case studies are as varied as the objectives can be. For example, one may analyse letters, ask about life histories in narrative interviews, conduct in-depth interviews, examine newspaper notes, statistics, court records and photographs, or even conduct group discussions.

Following Petermann (1996); Bortz and Döring (2002, p. 579) name pedagogical, special education and clinical research as typical areas of application for case studies. Kromrey (1998, p. 507) cites ethnomethodology as a particularly frequent user of case studies (compare Whyte 1967). Baacke (1995, p. 45) highlights pedagogy, which is said to have an exceptionally long tradition of using case studies. Finally, Berger and Wolf (1989, p. 331) point to small group research and personality analysis as the most common uses of case studies.

7.3   Delphi Surveys

7.3.1   Concerns of the Delphi Approach

According to legend, an oracle was housed in a temple in the town of Delphi in the eighth century BC. Questions had to be submitted to the oracle in written form, whereby the answers were mostly ambiguous. The oracle's greatest heyday came to an end in 480 B.C. (compare Grupp 1995, p. 26 ff.).

The first reference to the use of a research approach named after the oracle in more recent times is dated 1948. In the late 1940s, the RAND Corporation (Santa Monica, USA) used Delphi in 14 experiments for military purposes (compare Linstone and Turoff 1975, p. 10; Dalkey 1969; Dalkey and Helmer 1963).

In 1964 Delphi surveys also became known to the public. The "Report on a Long Range Forecasting Study" was published (Gordon and Helmer 1964). Since the 1970s, the Delphi method has also spread to Western Europe, including Germany (cf. Albach 1970; Cuhls et al. 1995; Aichholzer 2000). In this context, it is also worth mentioning the first survey on the development of science and technology, which was started in 1971 by the National Institute of Science and Technology Policy (NISTEP) in Japan. This has been surveyed up to the present.

The Delphi method can be characterized as a highly structured group communication process in the course of which issues about which there is naturally uncertain and incomplete knowledge are assessed by experts (compare Häder and Häder 1995b, p. 12). The basic idea of Delphi is to use expert opinions in several waves to solve problems and to use anonymous feedback. The classic Delphi design consists of the following steps:

1. The operationalization of the research question. This involves determining the dependent and independent variables and preparing them for an empirical survey.
2. This is followed by the elaboration of a standardised questionnaire for a survey, which is usually conducted by post or via the Internet.
3. This instrument is then used for the anonymous survey of a group of experts (first wave).
4. After the survey results have been processed in the form of a statistical group response, anonymous feedback of these results is given to the participants of the survey.
5. Finally, against this background, the survey is repeated with a largely identical instrument.

In addition to this standard approach, a number of variants exist that provide a certain diversity in the use of the approach. For example, there are different views on the number of experts required and the way in which they are selected. There are different interpretations of the required number of waves, also the feedback is designed differently. In some cases, experts are asked self-ratings about their own competence and there is a relatively arbitrary use of different types of tasks in the survey. Finally, there are different views on the stopping criterion of Delphi surveys. This can be a consensus among participants or the stability of the obtained answers.

Four types of Delphi surveys are now distinguished (cf. Häder 2014, p. 24 ff.):

Type 1: Delphi Surveys for Idea Aggregation
This is an exclusively qualitative approach. It is characterized methodologically by a renunciation of the quantifying rounds. The participants are only confronted with a problem and then asked to comment on it in the form of an essay. As is usual in qualitative surveys, no pre-formulated answer categories are given. For the feedback to the participants, the received arguments are roughly processed. Afterwards, the qualitative survey is repeated and further arguments are collected. This approach is all the more successful the more ideas for problem solving are generated and qualified. This usually requires (compare Hasse 1999) only a relatively small number of participants. In certain cases, the

anonymity of the participants must be waived due to the individual chains of argumentation put forward by the participants.

Type 2: Delphi Surveys for the Most Precise Possible Determination of an Uncertain Issue

This type is used, similar to a weather forecast, to get clarity on a certain matter that is still unclear. This was the original aim of the Delphi approach. Underlying this is a certain idea of forecasting: the future is to be determined and partially planned with such studies. Here, reference must be made to the problem of self-fulfilling prophecies and self-destroying prophecies. Thus, the effect of an (initially correct) prediction can contribute to the fact that reality is specifically changed and the prediction is thus refuted. Even predictions that were initially wrong – for example, about the shortage of a commodity on the market – can be confirmed by triggering corresponding action, such as increased purchases.

This approach was used, for example, in well over 300 meetings to produce price forecasts in fruit-growing areas on Lake Constance and the Lower Elbe (compare Janssen 1976). After all, the quotation error that could be determined on the basis of the Delphi surveys was only −0.9% at the end of the season.

Type 3: Delphi Surveys to Identify and Qualify the Views of Experts on a Diffuse Issue

Delphi surveys can be used to collect the opinions of a very specific group of people, for example to draw specific conclusions about necessary interventions, to react to a foreseeable problem and to raise awareness about undesirable developments. The most important thing here is that the views of all participants are represented in a methodologically sound manner.

This new concept differs from the 'forecasting' typical of the post-war period in that it does not attempt to determine how the future will turn out, or even to plan it in detail, but rather emphasises communication about the future and its active shaping (cf. Cuhls 2000).

One example here is the identification of research needs in vocational education and training research. The aim was to determine the degree of urgency of various research directions with the help of an expert survey (compare Brosi et al. 1999).

Type 4: Delphi Surveys to Build Consensus Among Participants

Delphi surveys can also be used to prepare democratic decisions. Here, a specific group of participants (for example, certain interest groups) communicates about an issue that is as precisely pre-structured as possible. In this way, a continuous discussion process is to be initiated, which ultimately leads to consensus. To this end, the participants' views on certain issues are repeatedly asked. They trigger further thought processes and, not least due to a group norm, a certain consensus is worked out.

An example of this is a project which developed recommendations for consensus-oriented policies in the field of microelectronics and the labour market (cf. Mettler and Baumgartner 1997). The aim was to allow as many citizens as possible from the various social strata in North Rhine-Westphalia to develop a wide variety of visions of society in the form of normative scenarios. Reasoned proposals for measures were to be addressed to the political system; consensus on the desirable basic features of the future society was thus to be generated.

Table 7.8 shows again the most important characteristics of the four types.

In summary, it can be stated: Delphi surveys have so far been used to deal with very different substantive questions. Prognoses for developments in various fields are the most frequent application (compare Henry-Huthmacher and Wilamowitz-Moellendorff 2005). The clarification of retrospective facts, the determination of the state of the art, evaluation approaches as well as the identification of research needs are further concerns of Delphi surveys (compare Häder 2014). The Delphi technique is always used to clarify uncertain issues that are to be clarified with the help of this method.

Table 7.9 shows the respective procedures for recruiting experts and participants.

Table 7.8 Overview of the most important characteristics of the four types of Delphi surveys

Type of Delphi survey			
1	2	3	4
Qualitative	Qualitative and quantitative	Qualitatively and (above all) quantitatively	Quantitative
Hardly any operationalisation of the problem	The facts to be evaluated are defined as precisely as possible		Strongly driven operationalization
Goal: Aggregation of ideas	Objective: Determination of a factual situation	Objective: To identify the views of experts	Goal: Consensus
Example: Hasse (1999)	Janssen (1976)	Brosi et al. (1999)	Mettler and Baumgartner (1997)

Table 7.9 Principles for the recruitment of experts or participants for Delphi surveys

Type of study	
Idea aggregation	All conceivable points of view should be represented by at least one participant; a few experts, each with a different professional background, are sufficient
Determination of an uncertain situation	Hypotheses about the expertise required to solve the problem have to be developed, from this a specific strategy for recruiting the appropriate experts has to be found, general statements about the required number of participants are not possible, more participants do not necessarily improve the result
Quantification and qualification of expert opinions	Total survey of the group of persons of interest or random or deliberate selection; the more respondents, the more meaningful the result
Consensus	The structure of a population must be reflected in the sample; the selection error decreases as the number of participants increases

7.3.2   The Future of the Dresden Frauenkirche, Example of a Delphi Survey

The procedure for the conception, implementation and presentation of results of a Delphi survey will again be shown by means of an example. The aim of the study was to conduct a communication among experts on the future of the rebuilt Frauenkirche in Dresden (Häder and Kretzschmar 2005a, b).

Operationalization of the Research Question
Relevant dimensions of the discourse on the Frauenkirche were identified as:

- The historical dimension: Is the Frauenkirche still perceived as a symbol of the Protestant bourgeoisie?
- The architectural dimension: What consequences will the reconstruction have for the cityscape of Dresden and its design?
- The memory dimension: What is primarily remembered in connection with the Dresden Frauenkirche?
- The cultural dimension: What forms of cultural expression are associated with the Frauenkirche?
- The dimension of criticism: What is the criticism of reconstruction aimed at?
- The religious dimension: What religious function could the Frauenkirche fulfil?
- The political dimension: What function does the Frauenkirche fulfil in the context of political debates?
- The media dimension: How does the media stage the image of the Frauenkirche?
- The economic dimension: What economic expectations are associated with reconstruction?

Elaboration of the Question Programme
The survey aimed to make a forecast for the year 2013. In addition to a global assessment (How do you think people's affection for the Dresden Frauenkirche will develop over the next ten years, i.e. by about 2013? Will it tend to increase further, remain as it is at present, or tend to decrease?), the probability and degree of desirability of various scenarios grouped around the Frauenkirche were also to be assessed.

The scenarios were derived from the shown dimensions of the discourse about the Frauenkirche. Figure 7.1 shows a page of the questionnaire.

Consultation of a Group of Experts
100 well-known representatives from politics, business, culture, the church and the media were asked to take part. They were mainly sponsors of the reconstruction of the Dresden Frauenkirche.

With the following questions, we would like to try to predict what role the Frauenkirche will one day play after its completion. To what **extent do you think** the following statements about the Dresden Frauenkirche will be true in about ten years' time, and how desirable do you think it would be if such a development were to occur? Please try to imagine yourself in the year 2013.

First of all, we are interested in **religious and cultural** aspects.

Please make two crosses **in each line!**

I would consider such a development …

	Most likely	Pro-bably	Partly Partly	Un-likely	Very un-likely	Very desir-able	Desir-able	Partly Partly	Und-esir-able	At all Undesir-able
34. The Frauenkirche will be used by a congregation for religious services.	☐	☐	☐	☐	☐	☐	☐	☐	☐	☐
35. The Frauenkirche will have become an important venue for cultural events …	☐	☐	☐	☐	☐	☐	☐	☐	☐	☐
36. The income from tourism around the Frauenkirche will have laid the foundation for a comprehensive historical reconstruction of Dresden	☐	☐	☐	☐	☐	☐	☐	☐	☐	☐
37. The enthusiasm for the Frauenkirche will have led to a revival of religious life in Dresden	☐	☐	☐	☐	☐	☐	☐	☐	☐	☐
38. In 2013, the memory of the night of the destruction of Dresden will still be alive and there will be commemorations	☐	☐	☐	☐	☐	☐	☐	☐	☐	☐
39. The attention of the media towards the Frauenkirche will have decreased significantly after its inauguration	☐	☐	☐	☐	☐	☐	☐	☐	☐	☐
40. The Frauenkirche will have become a well-known landmark of western culture	☐	☐	☐	☐	☐	☐	☐	☐	☐	☐
41. The Frauenkirche will have to be constantly secured by the police	☐	☐	☐	☐	☐	☐	☐	☐	☐	☐

Fig. 7.1   Page 6 of the Delphi Questionnaire on the Future of the Frauenkirche

Although some of the participants were world-famous artists, representatives of the business world at the level of the CEOs of major German companies, as well as leading politicians and representatives of the Protestant regional church, a considerable willingness to respond was recorded – probably due to the attractive subject matter of the survey. After all, 76 experts took part in the discourse on the Frauenkirche in the first wave and 60 in the concluding second wave. The respondents included 27 experts from the field of religion, 38 people from business and politics, and 34 representatives from the field of culture.

Processing of the Results and Anonymous Feedback

The answers were statistically evaluated and – as usual in Delphi surveys – sent to the experts in anonymized form as percentages. At the same time, they were asked to give their opinion on the above-mentioned questions again in a second wave against the background of the group opinion conveyed to them.

Repetition of the Questioning

The repeated delivery of the corresponding judgments leads, as has been shown in the meantime (compare Häder 2014; Bardecki 1984; Becker 1974), to a renewed cognitive examination of the subject of the question and to a qualification of the original answers. The second wave followed the first at an interval of about four weeks.

Presentation of Results

For illustrative purposes, only one aspect of the study is shown here, the evaluation of some scenarios according to their desirability and the evaluation of the probability of their occurrence. Figure 7.2[4] shows both facets.

At the top right, for example, particularly likely and at the same time particularly desirable scenarios are grouped together, while at the bottom left are those that are considered both undesirable and unlikely. The arrows indicate the change in ratings in the second wave compared to the first. According to the experts, the memory of the night of Dresden's destruction (38) in particular will remain with people in 2013. At the same time, this is interpreted as a tradition that should be particularly preserved. Vehemently rejected, i.e. considered very unlikely and at the same time undesirable, are the two negatively formulated (41 and 46) assumptions: "the Frauenkirche will have to be secured by police" and "the surrounding area will lose tourists to the city of Dresden".

---

[4] The labelling of the graph follows the counting used in Fig. 7.1. The scenarios not mentioned there have the following meaning: (43) The reconstruction of the Frauenkirche provides impetus for numerous other building projects in the city. (44) Improved transport links increase the attractiveness of the city as an economic factor. (45) The Frauenkirche is one of the ten most visited sights in Germany. (46) The surrounding area loses tourists to the city of Dresden. and (47) Inner city development is accelerated and the attractiveness of inner city shopping areas is increased.

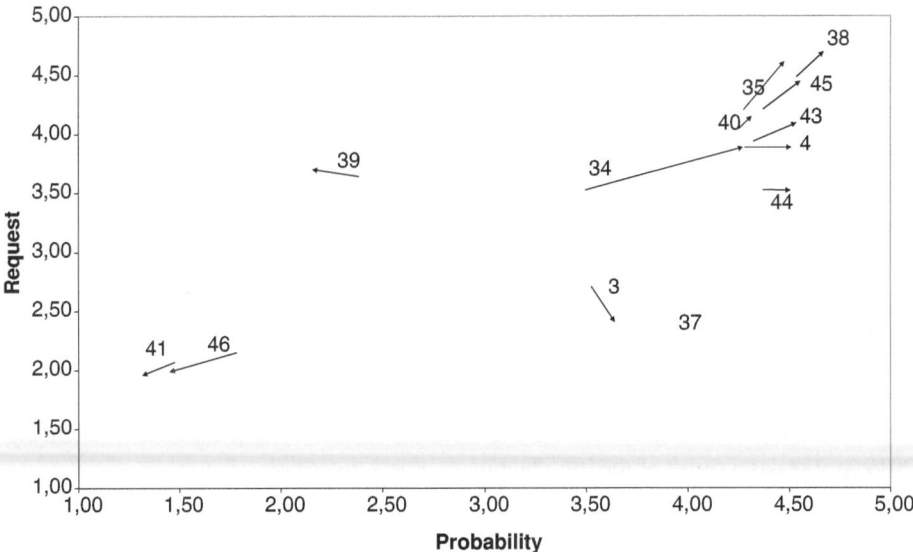

Fig. 7.2 Probability and desirability of different scenarios of the future use of the Frauenkirche, change in mean values from the first to the second wave

7.4   Evaluation Studies

7.4.1   The Concern of Evaluation Studies

Evaluation research makes an assessment (= evaluation) of measures or interventions. Evaluation studies are also not a method in their own right, but make use of a complex design that applies various, different techniques of data collection. Thus, evaluation studies are also not characterised by a specific common approach, but by their objective.

All those methodological rules apply to the evaluation design that were also presented for the other approaches to the objects of study. Above all, there are a number of analogies with social experiments.

The terms success monitoring, efficiency research, accompanying research, programme evaluation and impact monitoring are used synonymously with the term evaluation studies (cf. Thierau and Wottawa 1998, p. 13).

According to Rossi and Freeman (1993, p. 5), evaluation research involves "the systematic application of empirical research methods to assess the concept, study design, implementation, and effectiveness of social intervention processes". Bortz and Döring (cf. 2002, p. 101 f.) also define evaluation research in the same way.

The origins of scientific evaluation research lie in the 1930s in the evaluation of social policy measures in the USA in education and health care. Especially in the mid-1960s, the approach underwent a significant development. This led to evaluation research even becoming the strongest growth factor in the American social sciences at the end of the

1980s. One sign of the high status of evaluation research is, for example, special training programs for evaluators and the development of genuine standards that evaluation projects must meet (cf. Thierau and Wottawa 1998, p. 67).

The concept of social intervention processes whose impact is to be captured by evaluation studies must be particularly broad. Thus evaluations take place not only for anti-drug programmes or for anti-AIDS campaigns, but also, for example, to assess the success of marketing strategies and of new legal laws. Wottawa and Thierau (1998, p. 61) name a whole series of other objectives with which evaluation studies have been designed to date. These include the evaluation of therapeutic successes in the penal system (compare Waxweiler 1980) as well as the determination of the acceptance of forms of waste disposal and noise (compare Scharnberg et al. 1982) or a comparison of the suitability of different methods for the promotion of children's creativity (compare Levin et al. 1986). Döring refers to a study assessing the "living and working conditions of prostitutes in Germany" (2019, p. 174) as a result of the introduction of a new prostitution law.

Evaluation studies are also characterised by the fact that – in contrast to basic research, for example – they result in concrete recommendations for the client. It is thus clear that the results of such studies can be considerably explosive for those involved and must therefore be particularly well equipped to deal with criticism. At the latest when the results of an evaluation study claim that a measure or intervention has been ineffective (and should therefore better be terminated), the seriousness of the evaluation procedure must be demonstrated in detail. To ensure this, evaluation research must follow strict scientific and, above all, methodological criteria. Personal preferences and expectations of the client or the evaluator must not have any influence on the findings of evaluation projects. It is thus also obvious that self-evaluations are hardly capable of meeting such standards. For this reason, evaluation research is usually commissioned research, i.e. projects that are taken on by external parties.

For example, the management of a company may ask whether the funds it has spent for a particular purpose, for example for an image campaign, have also been sensibly budgeted. In this context, a number of different interests and wishes are then affected, which should be dealt with by the evaluators in as value-free a manner as possible. In addition to the client of an evaluation, the project executing agency that implemented the measure to be evaluated, such as the image campaign in this case, is likely to have interests in the (positive) outcome of the evaluation process. Thus, ideally, three actors are involved in an evaluation: firstly, the client, who is interested in information about the benefits of a measure or campaign financed by him; as here, this can be a company, for example. Secondly, the institution responsible for the implementation of the measure in question, for example an advertising agency, which is often interested in continuing the work it has started. Thirdly, of course, the evaluator is involved, who determines the impact of the measure. The evaluator has a particular obligation to be neutral, not to represent the interests of one actor or the other. At the same time, he is obliged to present the evaluation process and its results in such a way that the actors involved can understand them. It would also be conceivable to set up an additional scientific advisory board to monitor the evaluation or advise the evaluators.

A prerequisite for the acceptance of the evaluation result is the fully comprehensible documentation of the procedure as well as the use of undisputed facts as a basis for the evaluation. Thus, it is advantageous for the acceptance of the evaluation results if the three actors mentioned above cooperate in the design of the evaluation.

Here is an example of how *not to* handle the results of an evaluation: The result of a study commissioned by U.S. President Richard Nixon to determine the effects of pornography was: "After a diligent, multifaceted evaluation, the commission reported that pornografy did not appear to have any of the negative social consequences often attributed to it. Exposure to pornographic materials, for example, did not increase the likelihood of sex crimes" (Babbie 2002, p. 347). Now the intention was to make a corresponding change in the law on the basis of this finding. However, since Nixon declared the result of the evaluation to be wrong, this did not happen. Apparently, the result of the evaluation contradicted the deeply internalized views of the American president.

As a rule, the findings brought to light in an evaluation are not – as in basic research – made available to a broad public. Ultimately, as in the example of the American president, the client will decide on the use of the material collected during the evaluation.

The aim of an evaluation project is to assess the impact of an intervention. When using the evaluation results, it must be clear that such an assessment always remains relative. It is practically impossible to control or keep constant all the influences acting on a particular system. However, the aim must be to obtain the most accurate statements possible for the evaluation with the help of the latest research methodology. A further problem arises in particular when (despite all the effort) such an evaluation study has produced only an uncertain, risky result, which is then supposed to provide the basis for further decisions.

In this context, it should be pointed out once again that an evaluation presupposes that all the actors involved also accept the result of the study. Without the willingness in principle of all those concerned to make changes on the basis of the evaluation result, the whole process would be nonsensical. The highest priority would then be to confirm something that cannot be changed anyway. This also appears to be one reason why evaluation research has been or is without chance in totalitarian social systems, whereas in the USA, the country of origin of evaluation studies, there is a particular readiness to innovate. Reference has already been made to an interesting exception.

Theories form the basis for evaluation. Bortz and Döring (2002, p. 106 – italics as in the original, M.H.) distinguish between scientific and technological theories: "*Scientific theories* serve to describe, explain and predict facts; they are developed in basic research. *Technological theories* provide concrete instructions for the practical implementation of scientific theories; they fall within the scope of applied or evaluation research".

7.4.2   The Approach to Evaluation Studies

An essential prerequisite for evaluation research is that a concept of the intervention to be evaluated is available. Above all, information must be available on the overall objective pursued by the intervention to be evaluated. If such a goal has not been defined or if there are only vague ideas about the purpose of a measure, this makes evaluation difficult or impossible.

The target population affected by the measure must also be clearly identified. Only in this way can it be clarified during an evaluation whether the funds spent were actually justifiably invested. The more concretely the objective and the target population can be specified, the more meaningful (clearer) the result of an evaluation can be.

The fact that it is not necessarily easy to fix such an overall objective or the sub-objectives required to achieve it may be shown by the evaluation efforts for universities in Germany. Evaluation can look at the sum of third-party funds raised, at a fictitious recommendation by the university teachers for a place of study to be chosen, at the ability of graduates, at the average duration of studies, at the quota of graduates placed after a certain period of time, at the number of failed students, at the size of seminars, at the entertainment value of lectures, at satisfaction with supervision or/and at the number of students per professor. All individual facets are likely to be interesting and important for the evaluation of a university. However, it is not necessarily the case that all the criteria mentioned are on the same dimension.

Another aspect to be considered in evaluation research is the methodological feasibility of the research question. If, for example, an evaluator wanted to know whether the financial resources spent on improving the infrastructure in eastern Germany during the unification of the two German states were justified or not, it would probably have to be stated – at least in retrospect – that the resources and methods available to social researchers were not sufficient to be able to make a robust judgement on this. From a methodological point of view, it would have been desirable to use a comparison (group) concept. Thus, a randomly selected part of East Germany should have been supported financially in a different way, and it would then have been possible to observe how the remaining part developed in comparison.

This brings up another relevant dimension of evaluation research: the ethical responsibility of all actors. Thus, certain design elements that are considered desirable from a methodological point of view in terms of a reliable result may have to be omitted due to ethical concerns. This is especially true whenever interventions are involved that entail interference with the lives of individuals. Suppose a new medical treatment has been developed. Thus, it stands to reason that it should be applied primarily to patients who are particularly ill and who need this therapy most of all. From an evaluator's point of view, however, the new therapy might have had to be administered randomly and might even have been completely omitted in the comparison group.

Another ethical aspect is the renunciation of false information. In the meantime, a study conducted in Alabama (USA) in 1932 is unthinkable. In this, poor black men were recruited, ostensibly to test the effect of various remedies against syphilis free of charge. A few hundred men voluntarily participated in this project. In reality, however, these individuals received no medication at all; the aim was rather to study the normal course of the disease (compare Jones 1981 and Babbie 2002, p. 346).

Logistical and administrative problems may also arise in implementing an evaluation design.

Attention should be drawn to the following problems in evaluation studies, on which agreement should be reached between all those involved (cf. Babbie 2002, p. 331 ff.):

1. What is the actual objective of the measure to be evaluated. This must be determined precisely. Behavioural changes are particularly demanding goals here, for example in the context of AIDS.
2. This is followed by the operationalisation of the objective. What behaviour is to be changed by the measure, is the use of clean needles, the use of condoms, the completion of an HIV test or the acquisition of knowledge about AIDS to be achieved. This is the dependent variable.
3. The measurement instruments suitable for the evaluation are to be developed. They should measure the dependent variable. It would be conceivable to observe the sales figures of condoms, to register the frequency of HIV tests or to determine the number of new infections. A decision must be made as to whether self-reported behaviour or behaviour measured by statistics should form the basis of the evaluation.
4. The context of the measure must be determined. If, for example, a training course for the unemployed is to be evaluated, the aim of which is to improve the job opportunities of these persons, the situation on the labour market (= system environment) must also be included in the evaluation. An ideal solution here would be to form a control group that is not affected by the measure but is in the same system environment.
5. The actual effect of the intervention needs to be monitored. To continue with the example of the evaluation of a training course for the unemployed, it would be necessary to examine the number of days absent from work of the individual persons who were to take part in the training course in question. Sometimes, therefore, it is not sufficient to simply randomly assign the test subjects to certain groups.
6. The allocation to the control and the comparison group must be controlled. In order to be able to assess the effect of an intervention, both groups must have a comparable structure at the beginning of the intervention.
7. The target group of the intervention must be precisely defined. For example, it must be specified whether the measure is intended to reach all unemployed persons, only young unemployed persons or only unskilled unemployed persons.
8. The question is whether the evaluation requires the development of new measurement methods or whether it is possible to fall back on the use of proven approaches.
9. The success and failure of a measure must be defined. Such a determination is usually made on the basis of a cost-benefit analysis. If a measure has caused costs in a certain amount and the benefit is greater than these costs, then it was successful. However, if,

for example, a new learning approach is to be evaluated-which has incurred costs of a certain amount-and the result is that student achievement improves by a certain score on a standardized achievement test, such a cost-benefit evaluation is more difficult. One possibility here would be to use the required costs and benefits of another test to make the assessment.

10. It is also necessary to ask about the unintended effects of a measure. Wind turbines, for example, can certainly prove their worth in terms of environmentally friendly electricity generation, but it must also be asked how such wind turbines influence the aesthetic perception of the landscape.

7.4.3   Types of Evaluation Studies

Evaluation studies can be divided into accompanying evaluations – in which the intervention is assessed on an ongoing basis and, if necessary, modified at short notice – or into evaluation projects, which only assess the respective measure after it has been completed. Bortz and Döring (2002, p. 112) and Döring (2019, p. 180) also use the terms formative and summative evaluation for this. Thus, in the first form mentioned, the evaluation takes place at certain regular intervals already during the term of the intervention. In the second form, the evaluation is summative and takes place after the intervention has ended.

In the context of an evaluation study – as already emphasised – it must be demonstrated that the results in question *only* occurred as a result of the intervention and would otherwise not have occurred. The aim is therefore to exclude other influencing variables – such as seasonal fluctuations or ageing processes among the participants. To achieve this, an experimental and a comparison group must ideally be included in the design of the evaluation. Similar to an experimental design, only in the experimental group is the change made. This is obvious if one imagines that one wanted to evaluate the effect of a cold remedy. Here it might be important to check whether the cold could not have been fought in the same time completely without such additions – as in the control group.

The same applies to the random assignment of the target subjects to the experimental and comparison groups (randomization). To stay with our example, one could expect implementation problems if it is not the persons worst affected by a disease who are also to benefit from the new, supposedly modern and improved treatment method – but rather a random selection decides on the allocation to the new treatment method. In such a case, however, differences found between the experimental and comparison groups could no longer be attributed (only) to the effect of the measure being evaluated.

The requirements for an evaluation mentioned here describe an ideal-typical procedure. In practice, it will hardly be possible to implement this. Thus, evaluation studies always presuppose optimisations. In order to describe this in more detail, a corresponding example will be discussed.

The following three types of evaluation studies can be distinguished from each other:

1. *Experimental designs:* These are characterized by before and after measurements and randomization in group assignment. They allow a differentiated evaluation of the effect of the stimulus. For example, it might turn out that a measure taken only has an effect on a certain subpopulation.
2. Mainly because of the ethical problems, experimental designs can often not be used in evaluation. Instead, *quasi-experimental designs are* used. These can be further differentiated into:
   2.1. *Time-series designs:* They are repeated several times in the experimental group only, without the use of a comparison group. However, the absence of the comparison group means that no reliable control of external influences can take place.
   2.2. *Designs with non-equivalent comparison groups:* In this case, a comparison group that already exists is used as a control. If, for example, a new form of teaching is to be evaluated, the students would actually have to be divided into the two groups by randomisation. For practical reasons, this is hardly possible in most cases. Instead, a comparison of already existing classes that are similarly structured at the beginning of the measure is suitable.

   Another example would be the use of deliberate refusers of a medical therapy to be evaluated as a non-equivalent comparison group in the evaluation.
   2.3. The third distinguishing form of quasi-experimental designs are *multiple time series designs.*
3. *Qualitative evaluations* can also be used. Babbie (2002) reports on an anti-drug campaign among pregnant women. Here, participants were told that babies have a lower birth weight if women smoke during pregnancy. The qualitative interviews conducted on this revealed that there was, however, a belief among the participants that if babies were lighter, they would also have an easier birth. This information could not have been obtained even with a complete experimental design. Here it makes sense to complement the quantitative evaluation with a qualitative approach.

7.4.4 Evaluation of the Three Strikes Law in the USA, an Example

In some US states, a new law came into force in the 1990s, the so-called Three Strikes Law. The aim of this law is to reduce crime with the help of more drastic measures. To this end, criminal careers are to be prevented in particular by punishing repeat offenders more severely. In the case of a repeat offence, the punishment will be doubled. Should a person become a criminal for the third time, a 25-year prison sentence would be imposed. So at least the aim of the new legislation is clearly defined.

Apparently, the result of the law change now indicates a complete success: Over the course of just five years, there was a 51% drop in homicides in California. This was said to have saved $21.7 billion in costs and prevented one million crimes (compare BayInsider

1999).[5] At least in part, this result was interpreted as a consequence of the new legislation.

However, this finding did not stand up to scrutiny. In 1994, for example, the new law was found to cost about $5.5 billion a year, for example, in the prisons where criminals are housed (see Greenwood et al. 1994). At the same time, it is obvious that persons who are housed in prison are generally unable to commit new crimes on the streets. Thus, the new law may not necessarily have a preventive effect. Thus, the decline in the murder rate can only be attributed to the new law in certain ways.

Another study in 1996 (see Greenwood et al. 1996) found that the new legislation would require about $one million to prevent 60 crimes. However, if the same amount of money were invested in promoting retention in schools or reducing the rate of students leaving school without a diploma, 258 crimes could be prevented. According to such a comparison, it was a very ineffective law.

Apparently, however, the importance of prison sentences is generally overestimated in the USA (compare Irwing and Austin 1997; Babbie 2002, p. 349), while other variants of crime control are overlooked. At the same time, this example provides a very good illustration of various aspects of the problems discussed in the context of evaluation projects.

7.5   Intercultural Studies: The Example of the European Social Survey

Intercultural empirical studies have become very important. Not least against the background of the unifying and thus enlarging Europe, in view of internationally operating business enterprises and in connection with globalisation, it is plausible to assume that the importance of such studies will continue to increase. In addition to commercial intercultural studies, there are various academic research series that address this claim (compare O'Shea et al. 2003). Examples include the International Social Survey Programme (ISSP), the European Value Survey (cf. Sect. 6.1.4 and van Deth 2004), the World Value Survey, the Eurobarometer and, above all, the European Social Survey (ESS). The latter in particular is characterized by high methodological standards. It serves here as a model for survey studies of this kind.

Intercultural studies face the same methodological challenges as all other social science studies. A suitable instrument has to be developed for the surveys, a pretest has to be collected, the fieldwork has to be organised and the data obtained has to be processed, documented and evaluated. In addition, intercultural studies are confronted with various other problems. Reference should be made to the translation of the survey instruments into the respective national languages, to the development of a sample design that can be implemented in the respective countries and ultimately leads to comparable results, to the

---

[5] Compare http://www.bayinsider.com/news/1999/03/01/three_strikes.html, last accessed 07/25/2006 Source is unfortunately no longer retrievable. More information available at https://www.legalmatch. com/law-library/article/three-strikes-laws-in-different-states.html, last accessed on March 07, 2019.

organization of fieldwork that is comparable in all participating countries and, last but not least, to the creation of such an interculturally operating organizational structure that is able to guarantee the enforcement of the developed methodological standards among all participants. Braun (2019, p. 912 ff.) also refers above all to the problem of developing equivalent measuring instruments for the surveys in the individual countries.

7.5.1 Organisational Structure

It has already been established in the previous sections that a number of (preferably correct) decisions have to be made in empirical studies. These concern, for example, the wording of the questions to be asked, their order, the procedure for the pretest and much more. This also and especially applies to interculturally oriented research. At the same time, it is evident that the implementation of the high methodological standards set in the ESS itself must be monitored accordingly. For these purposes, it is necessary that there be a central study management body that bears responsibility for the overall project and that ultimately makes all necessary decisions. Such a body should also have access to the necessary methodological expertise. The organisational structure shown in Fig. 7.3 was

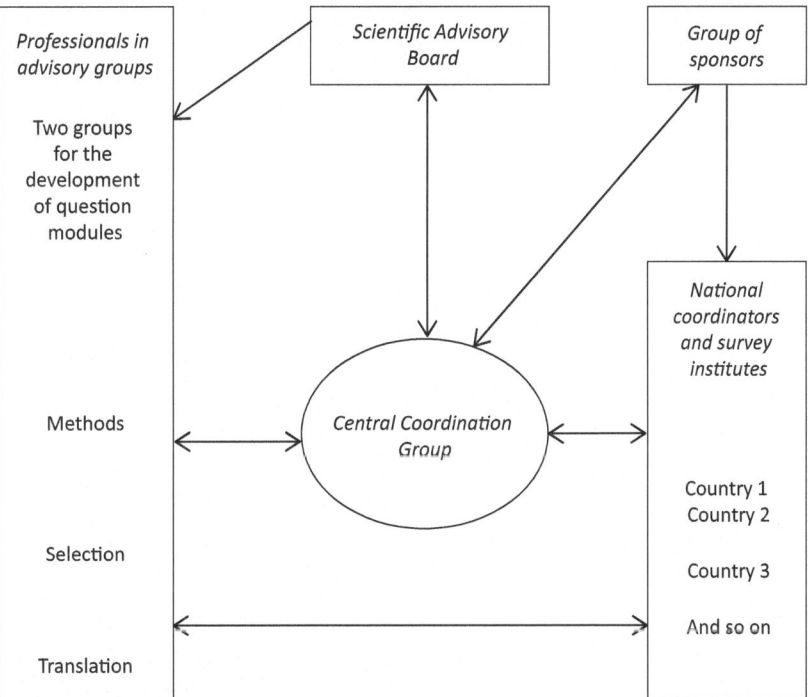

Fig. 7.3 The organisational structure of the ESS. (Compare ESS Round 1, Technical Report 2003, p. 5)

created for the ESS. In addition to a central coordination group, this also provides for project officers in each of the participating countries.

Never-ending questionnaire discussions, conflicts of interest with regard to the focal points to be set for the content of a study, differences of opinion regarding the methodological alternative to be chosen and, as a consequence, the overrunning of time schedules and the making of unbalanced compromises are experiences that can be gathered very quickly by those involved in empirical studies. The larger and the more homogeneous (after all, all participants want to have as equal a say as possible) a project team is, the more pronounced such disputes can become. It is therefore necessary to find organisational structures in order to avoid such conflicts, especially in intercultural comparative studies. Equipped with the relevant experience, the organisers of the ESS have successfully taken a different path.

Thus, organisational principles have been established which allow for a streamlined coordination of issues between the Central Coordinating Body and the individual country officers. In addition, various teams for methodological advice (sampling, translation and so on) have been created, acting independently of the national officers. In this way, the organisation of the ESS tries to keep a careful balance between the top-down and the bottom-up elements in the organisation.

In the following, selected methodological aspects of the ESS are presented as an example of the procedure for an intercultural study. In this context, it should not go unmentioned that the implementation of such a concept requires extensive financial and human resources as well as an immense time frame.[6] Given the immense resources required, it seems unrealistic to argue too strongly for a copy of the strategy practiced at the ESS. It is important at this point to refer to the problems to be solved in cross-cultural comparative studies.

7.5.2 Translation

Assuming an agreement in principle on the general content of an intercultural comparative survey, there is the problem of translating the survey instruments into the national languages of the countries involved. In most cases, the participants draw up a source questionnaire in English which, once completed, has to be translated into the respective national languages.

The term functional equivalence is used to describe the goal of such translations. It has proven relatively unpromising to use an external translation agency to accomplish this task. As shown above, seemingly minor changes in the wording of a question can lead to

---

[6] "The central coordination and design has been funded through the European Commission's Fifth and Sixth Framework Programmes and the European Science Foundation. The national scientific funding bodies in each country cover the costs of fieldwork." (http://www.europeansocialsurvey.org/ last accessed 08 March 2019).

different response behaviour. It is therefore important to use social science expertise to translate the original questionnaire into the national languages in a way that is equivalent in terms of content. These translations must be carried out by each individual country, and a central guideline for this has been drawn up by a group of experts within the framework of the ESS (cf: An Outline of ESS Translation Strategies and Procedures 2002).[7]

This policy contains a sequential procedure expressed using the acronym TRAPD. TRAPD stands for Translation, Review, Adjudication, *Pretesting and* finally Documentation. This is an approach in which the translation of the questionnaire is carried out by an interdisciplinary group, which distinguishes between five different steps.

Thus, the ESS deliberately refrains from using a translation-back-translation concept. This would provide for a translation of the original questionnaire into the respective national language and then a – relatively mechanical – back-translation of this questionnaire into the English original in order to check whether changes have been made in the formulations as a result of the translation.

The TRAPD model takes a more differentiated approach. Here, three different roles are first defined, firstly that of the translator, secondly that of the reporter and thirdly that of the decision-maker. Each role is then to be assumed by persons who possess certain skills. As a rule, two persons per questionnaire should act as translators. They prepare an initial translation of the questionnaire. The subsequent rapporteur should also have good translation skills. However, he should also be familiar with the design of the study and with the methodological principles of questionnaire design. For this activity, one person is sufficient in the envisaged model. Finally, the decision-maker – the third instance – works with the translators and the rapporteur. He finally determines which version of the translation is to be used.

The ESS also recommends that a parallel translation of the same questionnaire is carried out by different, independent translators, and that the entire translation is documented. Furthermore, the original questionnaire is annotated by the authors for translation in order to prevent misunderstandings among subsequent users of the data and to facilitate translation. This also includes background information on the actual objective of the question.

For example, the question text of the first question in the original reads: "On an average weekday, how much time, in total, do you spend watching television? Please use this card to answer." The following comments were made in the translation into German:

"We use 'working day' because this indicates that the time span in demand is Monday to Friday. Using list 1 we repeat the number of the card to use in every question – this is done throughout the whole questionnaire".

Translation instructions commenting on the initial questionnaire have proven helpful (compare Behr and Scholz 2011, p. 157 ff.).

---

[7] The extensive materials on the ESS are made available on the Internet. The homepage, where the links to the individual texts – also for translation – are located, can be found at the URL: http://www.europeansocialsurvey.org/ last accessed on 08/03/2019. The ESS was surveyed for the eighth time in 2016.

7.5.3 Sampling Strategy

For intercultural studies, manageable and also equivalent sampling strategies must be developed. Here, the problems result primarily from the very different national conditions for sampling. This applies above all to the possibility of accessing the most up-to-date and complete sampling frames possible, such as population registers, to the availability of experience in dealing with such registers, and to the associated costs of the sampling relationship.[8]

The ESS should proceed as recommended by Kish: "Sample designs may be chosen flexibly and there is no need for similarity of sample designs. Flexibility of choice is particularly advisable for multinational comparisons, because the sampling resources differ greatly between countries. All this flexibility assumes probability selection methods: known probabilities of selection for all population elements" (1994, p. 173 compare also Gabler 1999; Häder et al. 2003; Häder and Gabler 2003).

It can thus be stated that identical strategies should not be used in the selection process in the countries involved, but rather that the most suitable method should be used in each case. Now the problem is to find those strategies that lead to comparable results. In solving this question, the ESS starts from the confidence intervals.

Every estimate of a parameter of a population based on the distribution determined in a (random) sample is surrounded by a certain confidence interval. With a specifiable probability, it is possible to say within which limits the real value in the population lies. The width of this interval depends largely on the sampling strategy used. Identically large sample sizes do not necessarily imply identically large confidence intervals. For example, in the case of a simple random selection, the confidence interval is smaller (i.e. the statements can be made with a greater degree of certainty) than in the case of a multistage, stratified selection, where, for example, clumping can occur.

On the other hand, an identical confidence interval can be obtained with different sample designs. For example, if the sample size is increased in a stratified selection, the same effect can be achieved in this way with a smaller sample size as with a simple random selection. This consideration was put to use in the ESS.

For this purpose, a net sample size was defined for all participating countries. Now it was up to the national survey institutes to implement such a sample design that comes as close as possible to a simple random selection. This met with very different implementation conditions in the individual countries. Nevertheless, samples with a comparable confidence interval could be collected with the help of different sample sizes. Thus, equivalence was achieved using different sample sizes.

---

[8] Sampling for the European Social Survey – Round 4: Principles and Requirements, compare http://www.europeansocialsurvey.org/docs/round4/methods/ESS4_sampling_guidelines.pdf (last accessed 08 March 2019).

7.5.4 Intercultural Fieldwork, Organisation and Control

As a rule, a corresponding number of nationally operating field institutes is needed to carry out the fieldwork in intercultural studies. In this respect, too, there are naturally substantial differences between the individual countries or institutes. Thus, in this context, too, it is important to make efforts to ensure the desired equivalence during the fieldwork.

Again, the ESS is intended to serve as a model to demonstrate a possible model. In this study, emphasis was placed on compliance with various criteria.

Initially, it is suggested, the conclusion of contracts with each national field institute should be in English, if possible, to allow the Central Coordinating Group to control the field work. All contracts entered into should be concluded in accordance with the guidelines "Declaration on Ethics of the International Statistical Institute"[9] of August 1985. Strongly recommended here:

- the sending of announcement letters,
- visiting all intended addresses, households or persons (this prevents that only or especially easily accessible persons are interviewed) as well as appropriate documentation by means of special forms,
- random selection of all households and random selection of the target person in the household,
- an optional notification of the local police about the fieldwork taking place, in order to enable old or unsafe persons to make enquiries if necessary,
- finally – in the case of hard refusers – drawing up a description of the residential area in question.

Further regulations concern the personal training of all interviewers (often this is only done in writing), the use of certain contact protocols and finally a 14-day reporting on the respective progress of the survey phase including a forecast of the expected response rate to the central coordination team of the ESS.

Furthermore, all participating countries were called upon to achieve a certain response rate, in the case of the ESS 70%, or to aim for this. In addition, the proportion of persons who could not be contacted during the field period was not to exceed 3%. A further aim was that at least 90% of those who answered the main questionnaire should also complete the supplementary questionnaire. The fieldwork itself was to take place during only one month in a given period (in the first wave this concerned the period between 1 September and 31 December 2002). For Round 4, fieldwork was scheduled to begin on September 1 and to be completed by the end of December 2008.[10]

---

[9] Compare: https://www.isi-web.org/index.php/activities/professional-ethics/isi-declaration last accessed 03-08-2019.

[10] Compare: https://www.europeansocialsurvey.org/data/deviations_4.html last accessed 03-08-2019.

A quality control of the interviewer's work was to be carried out for at least 5% of the participants, for at least 10% of the refusers and for at least 10% of the persons not contacted. No more than 48 interviews were to be realized by one interviewer within the ESS survey. In addition, in all countries, important national events such as strikes, elections and so on that took place immediately before and during the field period in the respective country were to be recorded.

7.6   Time Budget Studies

7.6.1   Methodological Concepts for Time Use Studies

It is always exciting to ask how much time has been spent on which activities in life. Time budget studies serve to empirically determine the time, the duration and the sequence in which certain activities are carried out. In addition, other aspects of time use, such as secondary activities carried out at the same time, social contacts at certain times and the places where time is spent, can also be surveyed. Empirical social research has various methodological approaches at its disposal for researching time budgets:

- The activity-oriented survey (compare Blass 1990, p. 54 ff.; Haugg 1990, p. 76 ff.), in which an attempt is made to determine the frequency of certain activities in a specific time interval. For this purpose, a series of activities is presented to the respective target persons on lists. This is followed by a query about the typical average duration of these activities. From a methodological point of view, however, such a procedure is considered unsuitable for various reasons (cf. Gershuny 1990, p. 23 ff.). Above all, the duration of the individual activities can only be recorded imprecisely in this way. Furthermore, strong distortions due to socially desired responses must be assumed. In addition, it is not possible to distinguish between primary and secondary (the secondary occupations) activities. Also, no information can be provided on the placement of the individual activities in the course of the day. Finally, in order not to overtax the target persons, the questioning of time use must be limited to a relatively small number of activities, some of which can only be imprecisely distinguished from one another.
- The yesterday interview, in which the target persons are asked the following day about the time they spent on the previous day (compare Juster 1985; Holz 2000). In addition to the immense effort involved in interviewing, this methodological approach is characterized above all by the tendency of respondents to report a supposed average day rather than actually describing the preceding course of a day.
- The diary method currently offers a very good way of empirically recording time use. It is a form of standardised self-observation. In this method, the target persons record all activities of a certain minimum length over a defined period of time. The place where the activity is carried out, the persons present and other activities carried out at the same time can also be recorded. Finally, it is also possible to collect subjective

evaluations of these activities; for example, activities can be described in terms of how much the target persons like performing them.

- Finally, reference should be made to the mobile momentary inquiries. Here, the target persons are asked at certain times, for example, to state where they are and what activities they are engaged in at that moment. Such requests are made via mobile devices such as smartphones (cf. Trübner 2019, p. 1237).

In the Federal Republic of Germany (similar to Austria),[11] the diary method is the approach preferred by official statistics to determine the time budget of the population. It is characterized by several other advantages: The description of the activities carried out can be made by the target person in the diary in his own words, i.e. in terms he is familiar with. Such a form of data collection is also called open logging. The later coding of the activities is then done by the evaluator. In this way – as will be shown – up to 200 activities have proved to be surveyable. This contrasts with closed logging, which is more complicated to handle. In this strategy, the target persons are asked to describe their activities with the help of predefined categories.

With the help of the diary method, it is possible to keep to a 24-hour framework, i.e. entire days are logged by the target persons.[12] Both primary and secondary activities can be determined. Biases due to social desirability are also relatively low here, not least because of the absence of an interviewer. Advantages over other approaches also result from the more precise recording of completed activities and from the possibility of recording times of day and activities in chronological order.

Participants in time-use studies must be highly motivated, or special efforts must be made to generate such motivation in them. In order not to place too great a strain on the participants' motivation, or to keep the costs incurred by the target person low, such a time use protocol should be limited to a relatively short period of time. Here it could be about three days. Last but not least, it is also important here to preserve the anonymity of all persons involved.

A problematic aspect of this technique, and possibly a disadvantage, is the lack of control over the information provided by the target persons. Thus, the researchers have practically no possibility of checking the information provided by the recorders. Of course, this also applies to the above-mentioned other approaches to the time budget.

At this point, the time budget studies of the Official Statistics in Germany – the Federal Statistical Office – will be presented.

---

[11] For example, within the framework of the microcensus conducted in Austria in 1992, the time budget of more than 25,000 persons was also determined with the help of time use protocols. Other instruments, such as surveys of time spent the previous day, however, proved to be less practicable (cf. Beham et al. 1998, p. 2).

[12] In Finland – a country that famously did particularly well on the PISA test – children as young as ten filled out such time diaries with good results (compare Niemi 1983; Holz 2000, p. 184).

The study conducted in 1991/92 surveyed 7200 households. According to Ehling et al. (cf. 2001, p. 222), a stratified selection of households was used for this purpose. A total of 32,000 time budget questionnaires were distributed to the household members included in the population. These were all persons aged 12 and over. Their task was then to log all activities on two consecutive days, keeping to a five-minute rhythm. In addition, introductory and final interviews were conducted, for example, on the characteristics of the social and spatial context of the activities. The data from this study have been available since 2000 as a Scientific Use File or as a Public Use File containing 80% of the cases.

The study that followed in 2001/02 involved a survey of a total of 5000 households (net). Again, a stratified selection of households was used. The population in this study included all household members aged ten years and over. They were asked to record their time use in a diary on a total of three days (two of which were weekdays). In contrast to the study conducted in 1991/92, a ten-minute rhythm was specified. This survey took place over a whole year. It was prepared by a two-phase pretest.

In terms of content, the surveys focused on the following aspects:

- the actual main activity,
- the most important concurrent activity,
- on those people with whom these activities were spent together,
- the means of transport by which the journeys were made,
- the location at the beginning and at the end of the respective diary day,
- a general assessment of the course of the day described (in order to be able to identify any exceptional events),
- Traveling on diary days and finally,
- when the diary was kept.

Both data sets make it possible to address interesting questions. Reference is made to the following:

- Problems concerning working time: What changes, differentiated by gender, for example, can be detected between the two survey dates? What proportions, for example, do part-time and full-time jobs have, do people also have several jobs at the same time, how large is the proportion of marginal employment?
- Will there be a shift away from household production towards market production or, conversely, will goods and services increasingly be produced in households?
- What is the time use of teenagers?
- What is the division of labour within the family, to what extent does the man participate in the housework?
- Can you also observe changes in shopping times due to the changed shop opening times?
- Is the increase in women's employment at the expense of time spent together with children?

- Is the right to a nursery place a help for women to better combine work and family life?
- How do the new electronic media affect social contacts in leisure time?
- Based on the comparison of the two studies, can we speak of an increase in loneliness among older people?
- Is there a change in the total amount of time worked by certain social groups, for example single parents and working women?
- What aspects of time budgeting have seen the biggest changes over the last ten years?
- Based on the findings, is there a need for political or socio-political action?

7.6.2   Time Concepts

Theoretical time concepts come into play in the evaluation of time budget studies. These provide a structuring of the time spent on certain activities. While it may still have been sufficient for the consideration of the time fund in industrial society to differentiate only between working time and leisure time, more finely structured concepts are now used:

Geißler (1993) sees three elements of time use: first, *personal time;* second, *social time* spent in family or voluntary activities; and third, *paid work time.*

Opaschowski (1976) distinguishes the use of time on the basis of the different degree of disponibility with which time use can be decided individually. There are also three types of time. First is *determination time,* this can also be referred to as dependent time. It is spent on activities such as sleep, eating, personal hygiene, but also on gainful employment and school attendance. Traditional patterns of action in the family also constitute determination time. Secondly, we speak of *obligation time,* during which certain obligations such as schoolwork, household chores, activities for the church or for clubs are completed. Finally, thirdly, there is *disposition time,* which is the actual leisure time. This can now be used for hobbies, doing nothing and so on.

Another concept has been developed by von Schweitzer (1990). The likewise tripartite system focuses on the space in which time is spent as a differentiating characteristic. Thus, there is (first) *public time,* which is the time used for primary activities such as work and education. It is used to acquire money, goods, power and prestige. Alongside this is (secondly) *familial time, this is that* time which is used for domestic and craft activities. Time used for family members also falls under this category. It is time that has to be spent on the creation, maintenance and upkeep of family life. Finally, (thirdly) *personal time is characterized* by the fact that it is spent on personal regeneration and extra-occupational qualification. Personal time is time spent on one's own interests and needs. For example, lunch with colleagues is part of public time, lunch with family would be attributed to family time, and finally, the same activity would be part of personal time if it were done alone.

Table 7.10 shows the breakdown into these last three types of time presented for different groups.

Table 7.10 Time structure for different population groups

Behavioural groups of persons	Public time [min]	Family time [min]	Personal time [min]
Fully employed	570	280	590
Part-time employees	270	530	640
Pupils, students, trainees	430	240	770
Housewives, unemployed, pensioners without children	50	690	700

7.6.3 Examples of Results of Time Use Studies

Before describing the steps involved in a time budget study, two examples of the use of time budget data from the Federal Statistical Office will be presented. The first example concerns the choice of means of transport (compare Kramer 2001; Hertkorn et al. 2001).

First of all, the relevance of this issue seems to be obvious in the context of environmental preservation: road traffic causes not insignificant CO_2 emissions. However, the individual choice of transport mode is also an important aspect for regional planning, e.g. for the expansion of local public transport or for the construction of infrastructure. Appropriately designed time budget studies are now able to determine the part of the day or the part of the journeys that are made using public transport.

In determining time use, the interaction of elements at the micro and macro levels can also be empirically observed. Thus, correlations can be established between the objective characteristics of a place of residence at the macro level (the assessment of the residential location can, for example, be made via the price per square metre achieved in the sale of housing) and at the micro level (the furnishings of the households located there and the socio-demographic characteristics of the people living in this area). Finally, the concrete time use of individuals can be classified in a model.

This makes it possible to empirically process hypotheses that assert a relationship between the types of residence (at the macro level) and the time consumption for certain types of mobility among certain individuals (at the micro level). For example, the dependent variable is determined by the choice of the particular mode of transport. The independent variables in such a study may be:

- the respective types of residence, their centrality and their transport links,
- the purpose of the journey, is it, for example, journeys to work, to school or journeys to meet friends,
- the destination, i.e. a specific place,
- the time of day at which the journey takes place,
- the distance to be covered,
- the time required to overcome the distance,

- the duration of the journey as perceived by the target,
- the degree of reasonableness that the choice of alternative means of transport would entail,
- the extent to which people feel free to choose their mode of transport,
- individual characteristics of the persons concerned, such as their family situation, age and occupation.

From Meyer and Weggemann (2001) and Clauphein et al. (2001) comes the second example of a time budget study. This is about the time fund spent on eating. The relevance of the question can be relatively easily justified in this problem as well. Thus, as a purely quantitative argument, about 10% of daily activities are spent around eating. According to this, an average of 80 min is spent on eating and 50 min on preparing and following up meals (in the language of time budget research, Beköstigung). Another argument for the importance of the topic is of a content-related nature. In today's society, for example, there are fewer children and, at the same time, there seem to be increasing professional obligations, such as scheduling difficulties and the like. The question is whether life in such a flexible society – following the example of the USA – will also manifest itself in different eating habits. The following questions can be answered with the help of time budget studies:

- Is there a tendency towards situational eating, is there a noticeable detachment from fixed meal times, or do traditional eating habits continue to apply?
- Are less sumptuous meals being consumed and more fine, expensive and plain food being consumed instead?
- Is the trend towards eating out or do home meals dominate?
- Which is the main meal, lunch or dinner?
- Does it come to the partnership behavior with the Preparation and follow-up of the meals?
- Do people eat more alone or more in community?

On the basis of the time budget studies of the Federal Statistical Office, answers can be presented to the questions posed. This shows that no evidence of situational eating was found. Fixed meals continue to be eaten in Germany. On average, only 9 min are spent on meals outside the home and 70 min on meals at home. There are still clear peak times at which breakfast, lunch and dinner take place. Traditional patterns have thus also been preserved here. The main meal, the one that lasts the longest, takes place in the evening for men and for people over 35. For women, there is no difference between lunch and dinner. A decrease in traditional family meals cannot (yet) be observed in Germany. The distribution of roles between the sexes with regard to meals continues to be constant. Only men over the age of 65 participate significantly more. Finally, on average 2.9 meals are eaten per day, 2.7 of which are shared with other people.

7.6.4   Design of a Time Use Survey

Now we will demonstrate how the design for a time budget study can look. The focus will be on the question of the extent to which a paid part-time job (hereafter referred to as a job) influences students' study behaviour. Thus, correlations are to be expected between the pursuit of a job and study performance, study design and study duration. However, leisure time, subjective attitudes towards studying, the amount of time available for a partner and for social contacts are also likely to be influenced by whether or not a job is pursued alongside studies. The amount of free time available for friends, watching TV, listening to music, reading, hobbies and sports as well as financial constraints also play a role in this context. The empirical study is guided by the following problems:

- A job tightens the freely available time fund and causes students to manage their time more precisely in their studies and leisure time.
- Time for the job does not come at the expense of time for studies. However, this only applies as long as the time fund required for the job does not exceed a certain limit.
- Jobs that are closer in content to one's studies are held by older students and by students with better academic records.
- There are no gender differences in jobbing.
- Any time conflicts are resolved in favor of the job.
- Time spent on jobs comes at the expense of free time, not time spent studying.

When developing the design, the first thing to decide is for which population statements are to be made. For practical reasons, it usually makes sense to impose certain restrictions at this point. This can be done, for example, by making statements only for students in the main study period or only for students of one faculty.

In order to deal with the aforementioned problem, it is still necessary to adopt a comparative approach. Thus, both students with and without a job should be included in the study.

Furthermore, it must be ensured that a detailed recording of the entire daily routine is made. However, the time fund for the job can be logged as block time. It is not of interest what a person does during his secondary activity, but only the duration of the activity and its location in the daily routine (is night work done, for example?) play a role. This makes it much easier to determine the time budget.

Three instruments are needed: firstly, the actual time log, secondly, a sheet with supplementary questions, and thirdly, a cover letter explaining the purpose of the study to the target persons. Instructions for completion must also be provided.

The time log should – following the model of the surveys of official statistics – also record secondary activities in addition to the main activities. Furthermore, it must be asked how much an activity is enjoyed and with whom this activity is carried out.

In order to implement the comparative approach, distinctions must be made in the days to be logged. These are either days on which the students have a job or days on which they do not have a job.

The target persons are to be either those who have a job – in which case various working time models must be taken into account – or those who do not.

The individual instruments have the following appearance as shown in the figures. The time log (see Fig. 7.4) is structured in tabular form. The target persons are asked to enter all activities that take longer than 15 min. For this purpose, they are given a corresponding manual. In addition to a sample page, it contains a reference to the fact that a block time can be specified for the job, as well as the request to also log their own completion time.

Finally, an additional questionnaire must be developed. In order to be able to deal with the problems mentioned, this must contain questions on the success of the study, on satisfaction with the job and on the reasons why a job is being pursued. Furthermore, demographic data as well as special features during the time of the survey must be stated. Finally, the supplementary sheet has the appearance shown in Fig. 7.5.

The survey described was implemented as part of a course at the Institute of Sociology at the Technical University of Dresden. The data are also available on the internet.

Eighteen people participated in the pretest. These students each carried out structured self-observation over three days. This means that data is available for a total of $(18 \cdot 3 \cdot 24=)$

Zeit		Tatigkeit		Mit wem						Satisfaction with time spent				
Stunde	Minuten	Haupttatigkeit (bitte moglichst exakt beschreiben)	Gleichzeitige Tatigkeit (bitte moglichst exakt beschreiben)	Alterine	Partner	Andere Familienmitglied	Kollegen Kommenlionen	Freunde	Andere	sehr zutrieden	zutrieden	teils teils	unzutrieden	sehr unzutrieden
05	00-15			×							×			
05	15-30			×							×			
05	30-45			×							×			
05	45-60			×							×			
06	00-15			×							×			
06	15-30			×							×			
06	30-45					×						×		
06	45-60			×						×				
07	00-15			×							×			
07	15-30							×	×					×
07	30-45						×	×			×			
07	45-60						×	×			×			
08	00-15						×	×			×			
08	15-30						×	×			×			
08	30-45						×	×			×			
08	45-60						×	×			×			
09	00-15							×			×			
09	15-30						×	×	×			×		
09	30-45						×	×	×			×		
09	45-60						×	×	×					

Example for filling in the time log diary
If you have any queries, please contact
Thank you and have fun filling in

Fig. 7.4 Instructions for filling in the time protocol and time protocol

1296 hours. These data consist of a total of 5184 reported individual activities (15-min intervals).

The pretest should answer the following questions:

- How accurately or how reliably was the daily routine *presumably* recorded?
- Is the chosen 15 min rhythm in the time protocol appropriate, is it too rough or already too detailed?
- How many times a day should time use be logged?

Study with and without a job

Please mark the appropriate answer with a dark pencil in the corresponding answer field. No pencil!

For corrections, please color in and circle the valid box!

First, a few questions about studying:

	Very dissatisfied				Very satisfied
01) All in all, how satisfied or dissatisfied are you with your studies?	1 □	2 □	3 □	4 □	5 □

02) Will you complete your studies in the standard period of study specified for this purpose?
Probably yes □
Probably no □

03) What grade did you achieve in your intermediate examination? ....

04) In which of the following categories would you place yourself in terms of your achievements in your studies?
In the first third □
In the second (middle) third □
Last third □

The following questions refer to the job that is recorded in the time log

05) Do you have a paid job/work alongside your studies?
Yes □ (please continue with question 06)
No □ (please continue with question 14)

06) What is true about your job? How do you do it (several answers are possible)?
Regularly □
On call □
Days a week □
On weekends □
All day □
Half-day □
By the hour □
During the day □
Evening/night □

07) How many hours do you work on average per month?
Up to 20 hours □
20 to under 40 hours □
40 to under 60 hours □
60 to under 80 hours □
More than 80 hours □

08) For how many months have you been working alongside your studies?
For .... Months

09) What kind of job do you do? (please be as specific as possible)

Fig. 7.5 The additional questionnaire

10) Why do you work?	Does not apply				Applies
	1	2	3	4	5
I work mainly for financial reasons	☐	☐	☐	☐	☐
I work a job for a change.	☐	☐	☐	☐	☐
I'm working a job to do someone a favor	☐	☐	☐	☐	☐
I have a job for subject-specific and study-related reasons Interest.	☐	☐	☐	☐	☐

11) What do you use your income for?	Does not apply				Applies
	1	2	3	4	5
I need my income for living expenses (food, rent, etc.).	☐	☐	☐	☐	☐
I need my income to cover tuition costs (semester fee, books, etc.).	☐	☐	☐	☐	☐
I use the income to finance my leisure activities, personal preferences and hobbies.	☐	☐	☐	☐	☐

12) Is there a particular goal for which you save your income?

Yes   ☐, which one? ....

No   ☐

13) Were the days logged typical of your other daily life?

　　Yes, the logged days were typical　　　　　☐

　　No, something special happened　　　　　☐

　　Please justify briefly what!

Personal details:

14) Your sex　　　　　　　　　　　　Female　　☐

　　　　　　　　　　　　　　　　　　Male　　　☐

15) What year were you born?　　　　　..................

16) What is your marital status?　　　　☐ Single

　　　　　　　　　　　　　　　　　　☐ Married

　　　　　　　　　　　　　　　　　　☐ Departed

　　　　　　　　　　　　　　　　　　☐ Living separately

　　　　　　　　　　　　　　　　　　☐ Widowed

17) Do you have children?　　　　　　Yes　☐, how many? ....

　　　　　　　　　　　　　　　　　　No　☐

18) What subject(s) are you studying?

　　☐ In the main compartment ....

　　☐ In the minor ....

　　☐ In the diploma program ....

119) In which semester are you studying? ....

Thank you!

Fig. 7.5 (continued)

- How much time does it take to complete the protocol?
- How exactly does the memory of concrete periods of time and concrete activities work?
- What marginal distributions were obtained on the supplementary sheet?
- Is it only reasonable for the target to log three days or are logs also possible on four days?

For the evaluation of such an open time budget study, a classification scheme for the elements of the time budget is required. The elaboration of such classifications represents one of the most complicated tasks in the context of time budget studies. At this point, reference should be made to the approaches of Geißler (1993); Opaschowski (1976) and von Schweitzer (1990). Gross (1996) reports that in one study over 500 individual activities were recorded, which then had to be combined into main activities for the evaluation.

The pretest survey was followed by a coding of the reported activities into ten main groups. For this purpose, an existing model was used.[13] According to this, the coding scheme has the following appearance:

1. Personal area/physiological regeneration (01 Sleeping, 02 Eating and drinking, 03 Other personal area/physiological regeneration activities),
2. Gainful employment (12 secondary employment, 13 qualification, 14 time associated with gainful employment, 15 time associated with own job search, 16 lunch breaks associated with gainful employment),
3. Qualification/education (21 School/college [courses], 22 Homework/preparation and follow-up of courses [school/college], 23 Qualification/continuing education for personal reasons),
4. Household and family care (31 preparing meals, 32 maintaining house and home, 33 making, mending and caring for textiles, 34 gardening and caring for animals, 35 building and repairing, 36 shopping and using outside services, 37 household planning and organising, 38 looking after children, 39 caring for and looking after adults),
5. Voluntary work, informal help (41 Holding office or voluntary function, 42 Informal help for other households),
6. Social life and entertainment (51 Social contacts, 52 Entertainment and culture, 53 Participation in meetings),
7. Participation in sports activities (61 Physical activity, 64 Hunting, fishing, and gathering, 65 Set-up time for sports activities),
8. Hobbies and games (71 Arts, 72 Technical and other hobbies, 73 Games),
9. Mass media (81 reading, 82 watching TV and videos, 83 radio/music, 84 computer),
10. Travel time and indefinite use of time.

With the help of this scheme, the calculation of the amount of time spent, the determination of an average day of the pretest participants as well as an initial evaluation in the

---

[13] Federal Statistical Office: Concept and procedure of the time use survey of official statistics 2001/2002.

Table 7.11 Average time spent on different activities (in minutes) by different people in the pretest

Activity	All	With job (n = 6)	Without a job (n = 12)
Sleeping	505.0	517.5	498.8
Wining and dining	99.7	90.0	104.6
Other activities for personal regeneration	70.5	74.0	68.8
Paid secondary activity (job)	53.0	159.2	–
Preparation and follow-up of teaching	11.4	13.4	10.4
Meal preparation	13.9	10.0	15.9
Shopping	20.3	15.9	22.5
Entertainment and culture	66.0	51.5	72.9
Television	53.6	78.4	41.0
Pursue hobbies	7.0	14.2	3.4

One day (24 h) corresponds to 1140 min

subgroups (with and without a job) could be carried out. Table 7.11 shows an overview of this.

Thus, it could be established for the participants in the pretest, for example, that their not inconsiderable sideline activity is not at the expense of the preparation and follow-up of teaching.

Pretests

8

Abstract

Above (compare the examples in Sect. 6.1. "Forming judgements") two dialogues between an interviewer and a target person were reproduced. The interviewer asked the target person a question about his television consumption in the past seven days. He received a seemingly valid answer to this without further ado. He would then have moved on to the next question in a normal interview. But based on an inquiry ("How did you come up with that?"), it turned out that the target's answers were each strongly inaccurate. If the interviews had been stopped after the target person's answer or continued with the next question, there would obviously have been an immense discrepancy between the answer given and the actual time spent watching television. It is particularly interesting that there was no evidence of such a discrepancy up to that point. Neither did the subject ask a follow-up question, nor was the answer in any way illogical (such as "twelve tigers"). This must sensitize the designer of a survey instrument. Thus, the idea that a non-functioning instrument will stand out in some way in the pre-survey obviously proves to be so untenable. Further findings are presented below to illustrate this in detail. Thus the maxim "Even after years of experience, no expert can write a perfect questionnaire" continues to apply (Sudman and Bradburn 1982). The known rules for questionnaire design can only help to avoid gross errors. So far, however, all rules have always remained incomplete. They left room for questionnaire designers and did not exclude numerous exceptions. Social research is therefore dependent on special methods which are able to provide more reliable information about the quality of the instruments in the context of preliminary studies. Even the otherwise very helpful questionnaire rating system (FBS) does not spare the questionnaire developer from subjecting the questionnaire to a pretest (compare Faulbaum et al. 2009, p. 126). Thus, "If you don't have the resources to pilot test your questionnaire,

M. Häder, *Empirical Social Research*, https://doi.org/10.1007/978-3-658-37907-0_8

359

don't do the study" still applies (Sudman and Bradburn 1982). In the literature, the objectives to be pursued with the help of a pretest have so far been presented in a relatively colourful way. For example, pretesting is supposed to be about:

- To check the comprehensibility of the questions,
- To determine the variance that occurred in the responses,
- To test the clarity of the questionnaire,
- Identify any difficulties that targets may have in answering questions,
- To test the theoretical validity of the questionnaire and
- Anticipate field conditions to determine the functioning of the intended design, which means, for example, not using student surveys if possible.

8.1   Overview of the Procedures

The strategies used in pretesting can be divided into three groups:[1] *First, the* pretesting procedures used in the field: Here, the target subjects come from the same population and are examined under conditions as similar as possible to those later envisaged for the actual survey. *Secondly, the* cognitive pretest techniques or the laboratory procedures and *thirdly,* procedures based on expert judgements. The cognitive procedures attempt – as in the above example – to elucidate in the laboratory, i.e. under artificial conditions, the processes and steps that a target person goes through when answering questionnaire questions. For all three groups, social scientists again have various methods at their disposal for this purpose. Table 8.1 gives an overview. The individual procedures are discussed in more detail below. Afterwards, the procedure for a pretest will be demonstrated in detail using an example.

Table 8.1 Overview of various pretest procedures

Pretest procedure in the field	Cognitive methods	Other methods
Standard pretest	Think-aloud	Focus groups
Behaviour coding	Probing	Experts
Problem coding	Confidence rating	
Random sample	Paraphrasing	
Intensive interview	Sorting procedure	
Qualitative interviews	Response latency	
Analysis of the response distributions		
Split-ballot		

Compare Prüfer and Rexroth (1996a, p. 96)

---

[1] At this point, special attention is given to pretests for surveys. However, pretests must also be carried out for observations and for content analyses. The rules to be followed are discussed in the corresponding Sects. 6.2 and 6.3.

Various authors regard split-ballot or randomised experiments as an independent group of procedures for determining the quality of questionnaire questions. In some cases, the defining technique is also included in the cognitive procedures (Groves et al. 2004, p. 249 f.).

8.2 Pretesting in the Field

8.2.1 Standard Pretest/Observation Pretest

The standard pretest, also known as the observation pretest, is probably the most frequently used strategy. In this case, the present form of the questionnaire is presented to a number of target persons and their reactions are observed when answering or completing the questionnaire. Unfortunately, hardly any further binding rules can be given from the literature for the standard pretest. Different recommendations indicate that there is a wide scope for the design of this instrument:

- For example, the sample size should be between ten and 200 persons (Schnell et al. 2005, p. 327), but in most cases it is 20 to 50 persons.
- Either quota selections, mostly based on gender and age, or random selections can be used.
- It is recommended to use as interviewers either specially trained persons, or those interviewers who will also be active in the actual survey, or the members of the research group.

The standard pretest is usually a one-time interview under conditions that are as realistic as possible, whereby the focus is on observing conspicuous features and problems. Active questioning by the interviewers, for example to understand the indicators, is not planned.

The observational pretests bring several *advantages*. They are relatively cheap, can be carried out without a great expenditure of time and they allow the duration of a survey to be determined realistically. This is particularly important if a multi-topic survey is planned and the necessary costs have to be determined. However, even in the case of exclusive studies, the survey institutes calculate their offers not least on the basis of the time required for the survey.[2]

Limitations of the described procedure are to be mentioned: The idea that insufficiently functioning questions in the standard pretest stand out is – as shown above – at least not true in every case. The target persons also give the interviewer a seemingly logical answer to quite astonishing questions while maintaining the rules of conversation. Even in the case of not knowing or not understanding the question, one cannot safely assume that this will be admitted, that there will be queries or that in such a case an answer will be refused.

---

[2] Various survey institutes offer insertions in so-called multi-topic surveys. With such bus insertions, the fees to be paid are determined either/min or per question.

The procedure for the observation pretest is also not very systematic. Which reactions the interviewers report as conspicuous features is determined subjectively by them. Such anomalies may be a frown on the part of the target person, a conspicuously long pause between the question and the answer, or a direct comment on the question asked. Thus, observation tests are only a relatively rough method for determining the quality of a survey instrument.

8.2.2 Behaviour Coding

Behaviour coding is a more refined procedure than observation tests. This is a structured observation of the behaviour of interviewers and interviewees. In Behaviour Coding, too, an interviewer presents the questions of a questionnaire to a target person and requests the corresponding answers. The behaviour of both persons is systematically coded by an external observer. Tape or video recordings of the respective pretest interviews can also be used as aids for this purpose.

Such a coding system can capture the behaviour of the interviewer and the interviewee per question and per answer, respectively. For example (compare Prüfer and Rexroth 1985; Oksenberg et al. 1991) the coding scheme shown in Table 8.2 can be used.

The obvious advantage of behaviour coding over observation tests is the more comparable results of this method due to the coding scheme used. The disadvantage, however, is that the causes of registered abnormalities remain unexplained. In addition, Behaviour Coding is based on the same idea as the standard pretest, namely that unclear questions would lead to conspicuous response behaviour. The procedure is also relatively costly and the influence of the required recording of the interview remains unclear.

Table 8.2 Coding system for behaviour coding

Code	Description	
Interviewer … (select only one category)		
E	Exact	Reads exactly the wording
S	Slight change	Makes slight changes to the questions
M	Major change	Makes major changes so that the meaning changes
Respondent … (note all that apply)		
1	Interruption	Answers prematurely (interrupts)
2	Clarification	Wants a repetition of the question or an explanation of the question, or makes a comment indicating problems of understanding
3	Adequate answer	Answers adequately
4	Qualified answer	Answers adequately and makes an additional comment suggesting uncertainty
5	Inadequate answer	Answers inadequately
6	Don't know	Answer does not know
7	Refusal to answer	Refuses to answer

Compare Oksenberg et al. (1991)

Faulbaum et al. (2003) present a special form of behaviour coding. They modify the procedure in order to be able to apply it to CATI questionnaires in telephone surveys. The basic idea of the approach is to have interviewers systematically code abnormalities in respondent behavior in a somewhat larger sample (the authors describe its use in 100 interviews), using the computer technology of telephone interviewing. Interviewers assess respondent behavior according to whether a spontaneous response occurred or whether the response was delayed. Both behaviors are further distinguished. Thus, a spontaneous "response can either be completely adequate in the sense that the response can be assigned by the interviewer to one of the permissible response categories, or it is inadequate in the sense that the interviewer cannot make an assignment without asking" (Faulbaum et al. 2003, p. 23).

All in all, the following types of spontaneous answers are distinguished: first, the answer corresponds exactly to an answer specification. Secondly, the answer does not correspond exactly to an answer specification. Thirdly, the answer can only be assigned by asking. Fourthly, the answer was given prematurely, i.e. before the question was read out in full, and fifthly, the answer was refused or the answer was "don't know".

In the case of non-spontaneous responses, a distinction is made between problems, firstly, that have arisen because of acoustics. Second, which occur because of an unclear meaning of a term. Thirdly, which arise from the understanding of the question and fourthly, which occur because of unclear answer categories.

During the telephone interview, the interviewers not only code the substantive answers of the respondents, but also make a characterization of the answers according to the pattern described here (spontaneous versus non-spontaneous and so on). They have the computer's function keys (F) at their disposal for this purpose. Since coding has to be done additionally during a normal interview, the scheme to be used for this can only be relatively simple.

The advantages of the described method are mainly the rapid availability of the obtained results and the relatively easy handling.

8.2.3 Problem Coding

Problem coding is a procedure developed and practiced at ZUMA[3] in Mannheim. In this case (similar to behavior coding) a systematic evaluation of the behavior of only the target person takes place during the interview. However, in order not to overtax the interviewers,

---

[3]"Founded in 1986 as the "Gesellschaft Sozialwissenschaftlicher Infrastruktureinrichtungen" (Society of Social Science Infrastructure Institutions), GESIS initially consisted of three legally independent institutes, the "InformationsZentrum Sozialwissenschaften" (IZ) in Bonn, the "Zentralarchiv für Empirische Sozialforschung" (ZA) in Cologne, and the "Zentrum für Umfragen, Methoden und Analysen" (ZUMA) in Mannheim." (https://www.gesis.org/institut/ last visited Mar. 11, 2019).

a reduced code system is used for the observation. This is limited to the values zero for adequate and one for not adequate. The problem coding is done by the interviewers already during the interview and not afterwards. It thus requires interviewers who are particularly well acquainted with and trained in this technique.

If inadequate answers are given during an interview, the interviewers are asked in a second stage after the end of the interview to comment on this more specifically or to explain their experiences.

8.2.4   Random Sample

The random probe technique (compare Schuman 1966) can only be used for closed questions. Random rehearsal involves asking the target person additional questions on randomly selected indicators in order to check the understanding of the question. For example, ten out of 200 questions of an interview could be checked more closely in this way per questionnaire. For this purpose, following Schuman (Schuman 1966, p. 241), such questions are asked as:

- Please give an example of what you mean!
- I understand you and why did you answer that?
- Can you please tell me a little more about this?

8.2.5   Intensive Interview

The aim of the Intensive Interview (cf. Belson 1981, 1986) is also to identify a possibly incorrect understanding of the question in a formally correct answer. In a standardized interview, a formally correct answer exists as soon as the target person names a permissible answer category. However, in order to detect a wrong understanding of the question, the target person must be asked specific follow-up questions. For this purpose, certain comprehension questions are presented to the interviewee after the end of the interview. In the first step, the question to be examined and the answer already given are read out again by the interviewer and the target person is then asked to justify his or her original answer. As is usual in an in-depth interview, the interviewer asks further questions if necessary. In a second step, standardised follow-up questions are also asked on the understanding of the question.

8.2.6   Qualitative Interviews

Qualitative interviews can also be used at an early stage of questionnaire development. These are non-standardised in-depth interviews in which the interviewer asks the target person follow-up questions. He questions the answers previously obtained in the standardised survey, he inquires about possible alternatives and so on. It is important to note that here – in contrast to the quantitative survey – the personal views of the target person largely determine the course of the interview. For example, they have the opportunity to report on the extent to which they found themselves and their problems in the standardised questionnaire, whether problem areas were fully covered, where the particular relevance of the topic of the survey lies from their point of view, which given answers were unsatisfactory or incomplete for them and so on.

Qualitative interviews thus also require interviewers who are particularly well trained and familiar with the subject of the survey.

8.2.7   Analysis of Response Distributions

An initial analysis of the response distributions determined in the pretest can also provide information about the quality of the instruments used. The prerequisite for this is that a relatively high number of persons to be surveyed have been confronted with the instrument to be evaluated. Then it will be possible to pay particular attention to categories that are only minimally populated or not populated at all. Extreme, i.e. not expected, implausible frequency distributions can also provide indications for a necessary revision of the questionnaire. Finally, the frequent choice of alternative categories such as "don't know", "refused" and the like can provide at least rough indications of remaining deficiencies in the development of the instrument.

8.2.8   Split-Ballot Technique

The split-ballot technique involves the use of several variants of a question in the (pre-) survey. This technique would be useful if the questionnaire developers do not agree on how best to ask about the use of illegal drugs in the context of a survey, for example, face-to-face or with the help of a questionnaire to be completed by the target person himself. To resolve this issue, the population surveyed at pretest must be randomly divided. Each sub-population is then given one of the two possible options. The results of the survey are used to look for differences in frequency distributions between the two split groups. These differences are then attributed to the methodological variants used. Unfortunately, this does not clarify which of the tested versions would be the better one. In part, a solution can be arrived at on the basis of theoretical assumptions. Thus, in this example, the variant that determines a higher reported drug consumption is probably the better one.

A prerequisite for the split-ballot technique is that relatively large samples are used. The split-ballot technique is well suited to be used in combination with other methods.

In some cases, this strategy is also used during the actual survey. It then provides information for the design of follow-up surveys, for example.

8.3   Cognitive Procedures

These procedures pursue the goal of uncovering the respondents' understanding of the question as well as their procedures for obtaining information, finding answers and editing. In doing so, they are based on the cognitive psychological model of answering questionnaire questions already discussed above. The cognitive interviews are based on the technique of protocol analysis developed by Simon (compare Ericsson and Simon 1980, 1984) and his colleagues. The original interest here is the question of how mathematical tasks or even chess problems can be solved.

Here again, various approaches have become established. A corresponding overview can be found, for example, in Prüfer and Rexroth (1996a, b, 1999, 2005) and in Groves et al. (2004, p. 246 ff.).

8.3.1   Think Aloud Method

The Think Aloud method involves asking the target to think aloud while answering a question. A corresponding instruction could read like this:

> I'm going to ask you a question right now. After that, I'm going to ask you to think out loud. You are to please speak aloud any thoughts that are going through your mind. Please begin as soon as you hear the question and continue until you have arrived at an answer. These do not have to be complete sentences, of course. If you ever get stuck, I will remind you to keep talking.

On the basis of the tape recordings made in the process, the answers are transcribed and then qualitatively evaluated.

Two strategies can be distinguished in the Think Aloud technique: First, thinking aloud directly while answering the question (Concurrent Think Aloud) and second, recapitulating one's own thought processes only after answering the question (Retrospective Think Aloud). The first approach in particular places high demands on the target. It has been shown that by far not all persons are able to cope with this task.

A particular problem with Restrospective Think Aloud is that it cannot be ruled out that the target person tries too hard in his report to justify the answer given and in doing so reports too little or not at all about the thought processes he has completed.

The Think Aloud technique is particularly good at uncovering problems from the understanding of the question to the formulation of the particular answer.

8.3.2   Probing/Demand Technique

In the probing technique, additional questions are asked about specific indicators in order to get to know the response behaviour of the target persons better. This can be done in many different ways. The variety relates on the one hand to the time at which the follow-up questions are asked and on the other hand also to the objective pursued in probing.

Follow-up probings take place after the question has been answered and post-interview probings are asked after the interview. Comprehension probings are used for question comprehension. For this purpose, the target person can be asked to explain the meaning of individual terms from the question text or to explain aspects of the answer he/she gave. Information-Retieval Probings are used especially for retrospective questions. They pursue the goal of making the step of information acquisition transparent. Finally, Confidence Ratings focus on the subjectively assumed reliability of the answers given.

A few examples will be used to clarify the probing technique in more detail. First, an example of follow-up probing to justify the chosen response level will be cited, followed by an example of post-interview probing to understand the term.

First Example of Probing in the Context of a Cognitive Pretest
Question text: "The goal of the Nazis was to eliminate Jewish Bolshevism by waging war against the Soviet Union."

Text of the probing immediately after answering the question using a seven-point scale: "And why did you choose this scale value?"

Answers to this probing question.

Scale value of the spontaneously given answer to the question	
4	I don't know. I can't do anything with the statement
2	I don't know, but I don't believe it. I gave the value "two" based on my gut feeling.
4	Am relatively unsure, thought I'd take a middle ground. The four therefore, because I do not know
1	Because I just can't imagine that the Nazis had any other goal than to gain land in that case. But suspect it, do not have a clue
6	I don't like to give one and seven. With the six I want to relativize and I believe that it was the goal of the Nazis
5	It was always the goal of the National Socialists to act against the Jews

Second Example of Probing in the Context of a Cognitive Pretest
Question text: "Behind the official criticism of communism, many Nazis and fellow travelers hid the old National Socialist anti-Bolshevism after 1945."

Text of the probing after the interview was completed: "What do you understand by National Socialist anti-Bolshevism?"

Answers to the probing question.

Scale value of the spontaneously given answer to the question	
5	These are people who were against bolshevism, that is, against leftist ideals. And national socialist is then one level more, who were extremely against it
Don't know	It was obvious. National Socialism went totally against communism
6	That was in the third Reich everything that was connected with communism, that you rejected all that
3	I don't know about that. Bolshevism equals communism. The three should be the middle, because I do not know for sure
3	Oh dear, there's not much I can say about that, I don't really know …

It is quite obvious that almost all target persons have problems in understanding the two questions and also in finding answers. This technique also makes it very clear that the scale value four, as the midpoint of the seven-point scale, is also used by the target persons as a fallback category and is given the meaning "don't know" by them.

8.3.3 Paraphrasing

Paraphrasing involves asking the target person to repeat the question in their own words after the answer they have given. In this way, problems in understanding the question in particular can be uncovered. The following is an example of this.

Third Example of Paraphrasing in the Context of a Cognitive Pretest
Question text: "Compared to how others live here in Germany, do you think you get your fair share, more than your fair share, a little less, or a lot less?"

"Please repeat the question I just asked you in your own words."
Reply

1. "In your current job, compared to others living in Germany, do you think you get your fair share, less fair, somewhat fair, or quite unfair?"
2. "That I should say that I benefit excessively from the welfare state compared to other segments of the population."
3. "Whether I'm actually satisfied with what I own, what I have, what I can do."

The result of paraphrasing also provides the questionnaire developer with concrete information on question comprehension. In this example, the target persons do not succeed in recapitulating the question and also the intended answer levels in a meaningful way.

8.3.4 Sorting

Sorting is also about understanding terms. Here, the target persons are asked to sort a stack of cards. On the individual cards, for example, different leisure activities are printed, namely those that can be done with relatively high and those that can only be done with little effort. In the free sort, these cards are grouped according to the user's own criteria. In the dimensional sort, on the other hand, the corresponding criteria according to which the cards are to be sorted are predefined.

8.3.5   Response Latency

The idea behind the response latency method – i.e. measuring the time it takes respondents to find an answer – is that long response times indicate problems in question comprehension. Particularly in CATI interviews, it is technically relatively easy to determine this time period very accurately. But it is also possible to use this technique in interviews in which the target person processes the questionnaire independently on the computer (Computer Assisted Self Interviewing – CASI). However, there is not yet much experience with the technique of response time measurement.

Stocké used this procedure to determine the availability of information among the target persons. The response time serves as an indicator to show "the intensity with which the attitude object and the reported evaluation are linked in the respondent's memory" (Stocké 2002, p. 31). If a target person has not yet formed an attitude towards an issue asked about, this would entail a time-intensive search.

For more information on the use of this technique in telephone surveys, compare Häder and Neumann (2019, pp. 279–287).

In summary, some major advantages of cognitive methods can be identified. These include a relatively quick implementation of the various tests and, due to the small number of cases, only relatively low costs. It should be noted, however, that the target subjects, some of whom are asked to perform complicated tasks in the pretest, such as thinking aloud, should be given credit for their sometimes not inconsiderable efforts. While observational pretests require a relatively complete questionnaire, cognitive methods can be used at various stages of question development, i.e. also with questionnaires that are still relatively unfinished, as well as only with selected questions.

A disadvantage of the cognitive methods is that they do not allow any statements to be made about the functioning of the questionnaire as a whole. Due to the mostly small number of cases, it is also difficult to arrive at generalizations of the findings.

8.4   Expert Evaluations

In many cases it has proved useful to involve experts in the development of survey instruments. These can and should also be persons who are able to assess the developed instruments with the distanced view of an external expert. They should therefore not themselves have been involved in the questionnaire development in question.

In focus groups, which can be called together in the early stages of the development of the instrument, indications of the acceptance and understanding of the topic of the survey or of individual terms and questions can be determined. A possible procedure for the evaluation of a written questionnaire could be that the participants of the focus group session first fill in the questionnaire themselves. Afterwards, this questionnaire is repeated in the group and combined with a discussion of the impressions. However, such focus groups can also be formed by members of the population to be surveyed.

Consultation with individual experts on questionnaire design can also yield evidence. For example, the wording of the questions, the structure of the questions, the answer options, the succession of questions and the interviewer instructions can be assessed by the experts. However, this low-cost strategy is considered to have low reliability. In other words, different experts might identify different problems.

Finally, reference should be made to a comparative study. Presser and Blair (1994) report on an investigation into the success of different pretest strategies. For this purpose, four groups were formed, each using different instruments to review the same question-naire consisting of 140 individual questions. In group one, eight interviewers conducted telephone interviews in two waves ($n = 35$ and $n = 43$, respectively) and then discussed their experiences as a group. In group two, the researchers created behavioral codings of the interviewers, while in group three, cognitive interviews (Follow-Up Probes and Think Aloud) were conducted with 30 individuals. Finally, in group four, two panels of experts discussed the instrument.

While the expert team identified 160 problems, only about 90 were located in the others. While cognitive tests were particularly useful for detecting comprehension problems, behavior coding and, in group one, mainly problems encountered by inter-viewers could be located. Finally, the expert panel turned out to be the most cost-effective approach.

In conclusion, it is often not sufficient to use only one method in pretesting in order to identify all possible problems in question construction. Prüfer and Rexroth (1999) there-fore advocate a multi-method or two-phase pretest.

8.5   The Pretest Design of the Dresden Self-Defence Survey 2001

Section 4.2 presented the design of the Dresden Self-Defence Project 2001 and described the purpose and procedure of this study in more detail. The pretest strategy will now be presented as an example.

8.5.1   Issues to Be Addressed in the Pre-Test

The starting point for the creation of the instrument was formed by case descriptions in which people invoked the right to self-defence. One task in the development of the ques-tionnaire was to construct an instrument from these case studies, with the help of which the attitudes of the general population to self-defence could be surveyed (as validly and reliably as possible) by telephone. The source was mostly court records from which these cases were collected. Various problems arose in the process:

- The case histories presented here were too long to be reproduced verbatim over the telephone. Such questions would not only have overtaxed the cognitive receptivity and patience of the interviewee, but the interviewer would also have been overly burdened by the abundance of text to be read out. In addition, the targeted time frame for the average duration of an interview of 25 min would not have been adhered to. Therefore, in a first step, all cases were shortened. The pretest now had to show whether this was successful.
- Case 1 (see Table 3.1 in Sect. 3.4) read after the shortening: "A woman is standing in the last free parking space in order to reserve it. A car driver asks her to make room for him, otherwise he will drive towards her. The woman stops because she believes she has a right to the parking space. The driver then pushes her out of the parking space with his car. The woman suffers abrasions as a result" (in questionnaire case 1).
- The cases were not originally available in a colloquial form necessary for a population survey. They had to be translated accordingly.
- The pretest had to show – with the help of the facet theory (compare Borg 1992) – whether the cases described in the questionnaire actually (still) contained the three desired facets after the shortening (compare Table 8.3).

Case 1 cited above, for example, should contain the facts a1 (the attack is on an immaterial good, namely the parking space), b1 (the injury is slight) and c2 (the attacker is not superior). The pretest should now show whether this was actually achieved with the formulation used.

8.5.2   The Pretest Design

There are several variants for conducting pretests. For the self-defence project, a specific cognitive procedure was chosen. The techniques of paraphrasing and probing were to be used, primarily to check conceptual understanding and to observe whether the target subjects are able to find an adequate answer to these questions.

For this purpose, in a first step – as recommended by Prüfer and Rexroth (1996b, 1999) – the case was read out by an interviewer and the respondent was asked for an evaluation. This was still the same procedure as was intended for the actual interview. In addition, however, all spontaneous utterances of the interviewee were then to be registered by

Table 8.3   Overview of the contents targeted by the questionnaire questions

a: Attack is perpetrated on	b: Consequences of the act of self-defence	c: Status of the attacker
a1: Intangible asset	b1: Slight injury	c1: Attacker is superior
a2: Asset(s) in kind	b2: Serious injury	c2: Attacker is not superior
a3: Life and limb	b3: Death	

the interviewer. This was the *first* cognitive element of the pretest. In a *second* step, the respondents were then given the task of repeating the question text in their own words.

It was now interesting to see to what extent the interviewee was actually able to correctly represent all dimensions of a case (cf. Table 8.3). *Thirdly, in the case* of unclear descriptions or an obviously incorrect representation, follow-up questions (probes) were used. The aim was to establish whether, in the case of non-mentioning, the corresponding dimensions were at least still established in the respondent's consciousness. In the case of incorrect mentioning, further inquiries were to identify possible causes.

Fourthly, each individual case was concluded with a General Probe, which is understood to be a general request for further problems (here: with the respective case) addressed to all respondents.

The other indicators planned for the survey differed in their structure and construction from the cases just discussed. For this reason, a different approach had to be adopted here for the pretest. In a first step, 19 questions were presented in mixed form to 13 randomly selected persons with the request to identify those ten that were most difficult to understand. This served as a pre-selection, as the inclusion of all questions in the pretest seemed too extensive. A total of eleven indicators emerged as possibly problematic. These were each printed on a separate sheet together with a scale (compare the example below). With the help of the scale, it was then asked to what extent there was clarity about the item listed above.

The respondent should first go through all the indicators and make their ratings. In all cases where respondents did not choose a value of zero, a prompt followed. This required the respondent to justify the value they chose. This was to shed light on possible ambiguities.

Pre-Testing Template Used for the Pre-Test
"How clear is this sentence to you? Please rate this on a scale. The value 0 means that the sentence is completely clear to you, the value 10 means that the sentence is not clear to you at all. You can use the values in between to grade your opinion. Please tick the appropriate value."

8.5.3 Implementation

The pretest was conducted on only nine individuals. These were recruited based on the quota characteristics of age and education. To keep the size of the pretest from becoming too large, not every respondent was presented with all the cases. Instead, the cases were divided so that each case was assessed and recounted by six people in total. Respondents received an honorarium for their participation.

8.5.4   Results

Overall, the pretest confirmed the methodological implementation of the content concept. However, changes had to be made with regard to individual cases. The pretest results for case 1 are shown here as an example. Due to the findings, this case had to be slightly modified.

Pretest Findings on Case 1 of the Cognitive Test (N = 6)
First step, spontaneous remarks

ID		Reply
1	–	Not
2	–	Not
5	–	Not
6	This is not justified, though I can muster some sympathy for it	Not
7	Coercion	Not
9	–	Not

Second step, *paraphrasing*

ID	
1	A woman is standing in the parking space to keep the parking space clear. A motorist comes, wants to occupy the parking space, but the woman does not allow it and the motorist then injures her
2	A *female driver*[a] is pushed by a driver for a parking space, the driver threatens her – I think – for parking. The woman says that she has a right to do so and gets abrasions from the driver
5	A woman wanted to park car, the other driver also wanted to park and he wanted the woman out
6	A *young* woman stands in the last free parking space to reserve the parking space: A motorist then pulls up and asks her to make room. The *young* woman stops anyway. The man, who has previously told her that he will push her aside with the car, does so, causing her abrasions in the process

[a]Comments from the subjects that are inconsistent with the research design are italicized

ID	
7	A woman keeps a parking space free by standing there. A car driver comes along and wants to push into this parking space and he asks the woman to make room, but this woman says: "No, I'll stay here", and when the car driver then gently pushes her away, she suffers skin abrasions. I think it was a mistake on the part of the driver, he shouldn't have been driving, it was clearly coercion, although the woman was also partly to blame
9	A woman is standing in a parking space, whereby I try to imagine this, I don't quite succeed, a man asks her to clear the parking space because he thinks it belongs to him, she doesn't, whereupon he pushes her out, so spatially factually I can't imagine this, but if I assume it like this, in the process, she suffers abrasions, I don't understand, yes

Third step, targeted *probings for* unclear representations in the Paraphrasing

ID	N1: Why did an altercation occur here?
5	That both wanted to park in the same parking space, but the woman was first. And the man had no right to park there

ID		N2: Was the woman injured? If so, how seriously?
1		She has suffered abrasions
5		*I don't think*
7		Yeah, skin abrasions, oh that wasn't hard. Minor skin abrasions

Fourth step, general demand for problems

ID	
9	ZP.: Yes, so I do have problems imagining how someone can be in the parking space and someone else can think it's his, yes, so I don't really know what the parking space should look like, so how someone, well, a car can push another car out of the parking space, that's perhaps possible, although it's not actually possible Int.: The woman stood there ZP: She was standing in the parking space, but I immediately imagined her sitting in her car

8.5.5   The Pretest at the Survey Institute

Finally, a standard pretest was agreed with the survey institute. This was primarily intended to check the technical functioning of the CATI instrument and to supplement the findings obtained during the cognitive pretest.

Preparation and Evaluation of the Data

9

Abstract

The processing of data in a qualitative study differs from that in primarily quantitative research. First, we will take a closer look at the procedure for evaluating qualitative data. Here, it is mainly a matter of transcribing the text material in order to make both a comprehensible processing and a reconstruction as complete as possible, for example of the qualitative interview, possible. There are different systems for transcription. The concrete decision on how exactly an interview should be transcribed depends on the goal of the analysis. For example, the researcher must ask him/herself whether it is relevant to his/her research goal to also reproduce a speaker's dialect. Transcribing requires a special playback device and a lot of perseverance. It is important not only to fix the spoken text, but also to capture all other features that have shaped a conversation. These include laughter, pauses, different people speaking at the same time, the raising or lowering of the voice, special intonations and so on. In addition, incomplete sentences, filler words, special rules for upper and lower case letters and the like make for what at first glance is not exactly a clear representation of the conversation.

9.1 Preparation of Data and Troubleshooting

9.1.1 Preparation of Qualitative Data

Various methodological approaches to data collection generate qualitative, textual data, for example when answering open-ended questions in the context of an otherwise standardised survey, in the various qualitative surveys such as narrative interviews, in group discussions, in pretests and so on. Various software programs such as MAXQDA can be used to transcribe such texts (compare Kuckartz and Rädicker 2019, p. 444 ff.). In addition, multimedia data such as images, photos and videos can also be collected.

The conversation analytic transcription system (GAT) according to Selting et al. (1998; other approaches are presented by Ehlich and Switalla 1976 and Ramge 1978) will be presented here (compare Table 9.1) as an example of similar transcription arrangements for textual data.

A three-column table is generally used for the transcription. In the first column, the lines are numbered consecutively. This makes it possible to refer to specific parts of the conversation during later analysis. In the second column the speaker is named. This should usually be either the target person (ZP) or the interviewer (I). Finally, in the third column, the content of the conversation is written down. Figure 9.1 shows an example of the transcription of a narrative interview about the experience of the Elbe flood in 2002.

The original material, i.e. the tape, should then be archived in compliance with data protection rules, which require above all that the text be made anonymous. The transcription should also be carried out in such a way that names of persons and, if necessary, also place names are designated with the help of suitable abbreviations. This makes it possible to publish the transcript without de-anonymising the target person.

9.1.2   Preparation of Quantitative Data

Preparing the results of a quantitative study involves creating a computer-readable data file for the statistical analysis that follows. This requires several steps:

The Coding and the Data Transmission
Apart from CATI and CAPI surveys, the field phase ends with a number of completed questionnaires, observation protocols or – as in the case of a content analysis – with the coding sheets created in the process. In the best case, these can be recorded in a machine-readable form using scanners. If this possibility does not exist, the results must be entered manually into the mask of an EDP program such as SPSS, Stata or Excel.

Error Control and Correction
Particular attention must be paid to subsequent error checking, especially in the case of manual data entry. The error checks should be carried out in different directions:

- The search for wild codes. These are values that lie outside the permissible range. For example, if the gender is coded with the values 1 for male, 2 for female and 9 for no specification, all other values appearing at this point are wild codes. Wild codes can be detected with the help of simple frequency analyses. They should – if this is still possible – be clarified by a comparison with the survey documents. If this is not possible, such values should be removed or reported as special levels such as "not specified".

Table 9.1 Conversation-analytical transcription system according to Selting et al. (1998)

Sequential structure/sequential structure	
[ ]	Overlaps and simultaneous speech
=	Fast, immediate connection of new contributions or units
Breaks	
(.)	Micropause
(–), (– –), (– .– . –)	Short, medium, longer pause of about 0.25 to 0.75 s duration
(2,0)	Pause of more than 1 s duration
Other segmental conventions	
Un-uh	Wear within units
: , : : , : : :	Elongation, elongation, depending on duration
uh, uh, etc.	Delay signals, so-called filled pauses
'	Abort through glottal occlusion
Laughter	
Sun (h) o	Laughing particles during speech
Haha hehe hihi	Silvery laughter
(laughs)	Description of laughter
Reception signals	
Hm, yeah, no, nope	Monosyllabic signals
H = hm, ja = a, nei = ein	Two-syllable signals
, hm 'hm	With glottal closure, mostly negative
Accentuation	
acceNt	Primary or main accent
akzEnt	Secondary or secondary accent
AK! TENT!	Extremely strong accent
Unit-end tone	
?	High rising
,	Means increasing
–	Consistent
;	Mean falling
.	Low falling
Other conventions	
(coughs)	Para/extra-linguistic actions/events
<<cough>>	Speech-accompanying para- and extra-linguistic actions and events with reach
<<amazed>>	Interpretive comments with scope
( )	Incomprehensible passage depending on length
(such)	Presumed wording
al (s) o	Suspected sound or syllable
(such/which)	Alternative presumption
((…))	Transcript omission
→	Reference to a transcript line discussed in the text

(continued)

Table 9.1 (continued)

Sequential structure/sequential structure	
Inhalation and exhalation	
. h, . hh, . hhh	Inhalation depending on duration
H, hh, hhh	Exhale according to duration
Volume and speech rate changes	
<<F>>	Forte, loud
<<fortissiom>>	Fortissimo, very loud
<<p>>	Piano, soft
<<pp.>>	Pianissimo, very soft
<<all>>	Allegro, fast
<<len>>	Lento, slow
<<cresc>>	Crescendo, getting louder
<<dim>>	Diminuendo, becoming quieter
<<acc>>	Accelertando, speeding up
<<rall>>	Rallentando, slowing down

- The search for implausible values. In CATI and PAPI surveys, this step can also be carried out during data collection. However, when manually transferring the data, special attention must be paid to the search for implausible values. For example, it is possible that a target person is 18 years old and at the same time a widow, and it is also possible that a professor does not have a high school diploma, but such constellations are not particularly likely. It is possible that the widow is not 18 but already 81 years old. As far as the original documents from the survey are still accessible, an attempt should be made to clarify the origin of implausible values. In the event that such clarification is not (or no longer) possible, these values should remain in the data set. However, it is expected that such values are handled carefully during the evaluation.
- The search for inconsistent values. These are values that cannot occur for logical reasons. For example, it is not possible that a 21-year-old target person has already lived in the same apartment for 30 years. Furthermore, in ranking questions, it is not possible for two declarations to be in the same place. Sometimes a whole series of such logical checks can be carried out in a data set. They ultimately serve to improve data quality and should therefore be taken particularly seriously. Such inconsistent values should also be checked against the original documents and corrected or eliminated from the data set.
- The treatment of missing values. Missing values can occur for a number of reasons. For example, the interviewer may have forgotten to ask the relevant question, the indicator may have been skipped due to a filter, the target person may refuse to provide information on the question or may signal their ignorance, and so on. For such cases, the so-called special levels are to be provided in the analysis program.

Finally, it is important that the error control procedure is documented. The corrections or deletions of cases made during data entry can be important for secondary analytical or methodological studies. It also becomes clear at this point once again what advantages the

01	I	i would like to know how the tide (.)
02		gone on
03		= what happened how you
04		and what you perceived
05		thought and felt
06	ZP	yes (.) , so =, that began on
07		monday afternoon yes with pirna AN (--)
08		where the flood <<acc>> yes already came out
09		<<rall>>
10		respectively the godlEUba
11	I	yes the gottleuba
12	ZP	hm: and <<acc>> in pirna lives yes
13		My <<rall>> SISTER
14		with her two children
15		= and it's complete
16		been sloshed
17		on the bahnhofstrasse
18	I	Hm
19	ZP	= and she called me in the afternoon AN, that
20		just already the water in her
21		apartment door flows by (– -)
22		thought that can not be (– -)
23		anyway: she took us down
24		ordered us to come down
25		pick up the CHILDREN
26		my husband drove down and (.)
27		there was already everything beforeBEI = so there
28		<<all>> he already stood <<all>>
29		up to the chest in the water (– -) na:
30		the children were allowed to stay there
31		and on Tuesday morning:
32		Then it started here
33		(4,0)
34		so and there I was at work early
35		gone (.) there then came
36		Suddenly in the morning my big one
37	I	home from school =

Fig. 9.1 Beginning of a transcript of a narrative interview

CATI and CAPI technique brings with it, in which these steps can be omitted – provided that they are properly programmed.

The Transformation of Variables
In preparation for the statistical data analysis, it is also necessary to transform a number of variables or to create new variables. For example, it is advisable to form a corresponding

index from the indicators that surveyed the materialistic or post-materialistic values according to Inglehart, i.e. to construct a new variable.

Recoding is also necessary in the event that – as described and recommended in Sect. 6.1.3 – the income question is first posed openly to the target persons and then, if not answered, as a list question. In this case, therefore, the data set initially contains two variables which, for practical reasons, should now be merged for the following analysis.

A reshaping of variables also becomes necessary, for example, when the Unfolding Technique was used in telephone surveys. This was the case in the survey on the image of surveys.. As already shown, a polarity profile was created in the process. In order not to overtax them cognitively, the target persons were asked in a first step whether they considered surveys to be popular or unpopular, or whether neither of the two terms applied to surveys. Depending on the answer given, the second step asked to what extent surveys were (dis)liked: only somewhat or very much. For the evaluation, these two questions could now be synthesized into one.

Depending on the intention of the analysis, it may be helpful to combine other variables that were collected separately in the original questionnaire. This step should also be documented in a comprehensible manner for reanalyses.

The Recoding of Variables
For the evaluation, it may be useful to recode variables. The question about the highest school-leaving qualification in the Dresden self-defence study provided for the following possible answers: firstly, "still a pupil", secondly, "without a secondary school leaving certificate", thirdly, "secondary school leaving certificate", fourthly, "secondary modern school leaving certificate", fifthly, "A-levels", sixthly, "entrance qualification for a university of applied sciences", seventhly, "completed technical college studies", eighthly, "completed university studies" and ninthly, "other school leaving certificate". In some cases, such differentiated information is not necessary for further analysis, so that it is possible, for example, to recode the nine groups into only three groups. It would be conceivable, for example, to divide them into: First, low, second, medium, and third, high educational attainment.

For certain evaluations, it can also be useful to condense the variables into dichotomous characteristics. For example, the Sunday question is used to find out which party a target person intends to vote for with their second vote (if elections were actually held next Sunday). If we were only interested in potential CDU voters, we could create a corresponding dichotomous variable from the Sunday question. This would have the values 0 (the respondent does not intend to vote for the CDU) and 1 (=she intends to vote for the CDU). The term dummy variable is also used for such recoding.

The Treatment of Open and Semi-Open Questions
Only rarely are standardised surveys collected exclusively with the help of fully standardised indicators. Instead, it is often the case that the target persons are also asked a few (usually few) open-ended questions. In such cases, the interviewer usually writes down the

answers given verbatim or records the answers on a laptop. These notes must also be transcribed. It is also possible to create categories ex post on the basis of these open answers. In this sometimes very elaborate step, numbers are assigned to certain answers.

This also applies to semi-open or hybrid questions. In these, a series of standardised answers is provided, but in addition, open questions are also asked about the relevant facts.

In the Dresden self-defence study, for example, it was not only of interest how the respondents thought about and evaluated individual cases of self-defence, but also what experiences they themselves had already had with self-defence. To this end, an introductory question was asked as to whether the target person had ever come to the aid of another person (self-defence assistance). After this indicator, which also functioned as a filter question, it was of interest who the person attacked was. For this purpose, it was first asked whether it was a known person or a stranger. If the former was the case, the following question, shown in Fig. 9.2, was asked.

The CATI interviewers recorded the answers given at q24c_04 verbatim. The evaluation[1] produced the picture shown in Table 9.2.

Several things are interesting about this table: *First,* the individual interviewers each wrote down the same answers differently. For example, with: work colleague, Work colleague, workcolleague, Work Colleague, Colleague, colleague and colleagues.. The eight different answers thus become only one when summarized accordingly.

Secondly, various target persons have not understood the upstream filter correctly. According to this, the question shown here should actually only be asked if it is about an "other" known person. Thus, at least the following answers should not have appeared here: Stranger, Stranger Person, Unknown, unknown, and Completely Unknown People.

Thirdly, it can be assumed that the answer specification "a distant acquaintance", which was read out to the target persons on the telephone, was not perceived by them in the way originally intended by the questionnaire constructors. Thus, answers such as: "people you know by sight" and "by sight" should actually be placed in category q24c_03 (compare Fig. 9.2).

q24c Was the known person attacked …

q24c_02 … a boyfriend/girlfriend   1
q24c_04 … someone else, and that is: 2
q24c_01 … a family member(s)   3
q24c_03 … a distant acquaintance/acquaintance

Interviewer: Please note: _____4
q24c_ka: no answer

Fig. 9.2 Example of a hybrid question from the Dresden self-defence study

---

[1] The data set of this survey is available under the study number ZA4253 at the Data Archive for Social Sciences Department of GESIS in Cologne. https://doi.org/10.4232/1.4253/.

Table 9.2 Responses to the open-ended question about the known person for whom self-defence assistance was provided

	Number
TNZ (filter)	3420
Colleague	2
Colleague	3
Colleagues	1
Colleague	1
Colleague	1
A stranger	1
A customer	1
A teenager to be cared for	1
Foreign	2
Stranger	1
Friends and acquaintances or even strangers	1
Newlywed couple	1
Child	3
Child on schoolyard	1
Client	1
Colleague	1
Colleague	1
Colleague	1
People you know by sight	1
Roommate apartment building	1
Neighbor	1
Neighbors	1
Person in the train	1
Students	2
Students	4
Pupil of her	1
Schoolgirl	1
Schoolchild	1
Unknown	1
Unknown	1
Unknown	1
Completely unknown people	1
From see	1

This means that a decision must now be made as to which changes are to be made to the original data set.

Due to its great importance, reference should be made here once again to the documentation of all changes made to the original data set during data processing. These should be included in the corresponding field and method reports (cf. Chap. 10).

Treatment of Missing Values

When coding standardized questions, a specific numerical code must be provided for each answer choice. However, a specific coding must also be assigned in the case of item non-response or questions that do not apply. In the Dresden self-defence study, for example, questions were asked about the number of people living in the household. The corresponding information was then included in the data file. However, in the event that a target person refused to provide information on this, the value −1 was noted in the appropriate place in the data record. Before an evaluation of the average household size of the respondents could be made, the value −1 had to be agreed as a missing value in the data set and thus excluded from the calculation of the mean value. If this had not been done, a mistake would of course have been made. In the actual survey, however, this error – due to only a very small number of missing values – did not turn out to be particularly large. The mean household size changed from 2.65 (without the agreement of missing values) to 2.67 (with the exclusion of the missing values).

9.2   Basic Principles of Statistical Analysis

After completion of the data file and after documentation of the changes made to the original data set, statistical data analysis can begin. For this purpose, a number of guidebooks are available that contain corresponding recommendations. Reference should be made to Blasius and Baur (2019); Kühnel and Krebs (2001); Backhaus et al. (2000); Bortz (1999); Andreß et al. (1997); Benninghaus (1998); Wittenberg and Cramer (2000); Wittenberg (1998) and to Zöfel (2003).

The statistical analysis of the data is highly dependent on the investigator's particular interest in knowledge and the design he or she has chosen. For this reason, it is difficult to standardise the steps involved in statistical data analysis. However, some fundamental issues in statistical data analysis will be discussed. In addition, a separate section shows examples of the steps in which such an analysis can be carried out.

In the evaluation, it is important to note the difference between descriptive statistics and inferential statistics. Descriptive statistics are concerned with the description of the target persons studied by statistical means. In principle, it does not matter whether it is a sample or a total survey. The findings obtained with descriptive statistics refer in each case (only) to the units investigated or to the persons. As a rule, data analysis begins with such evaluations.

In an inferential statistical approach, on the other hand, certain characteristics (also referred to as parameters) of the population are estimated on the basis of the characteristic values determined in the sample. In contrast to descriptive statistics, these estimates also include the specification of error ranges (confidence intervals) and significance tests. Both strategies are briefly discussed below.

9.2.1 Descriptive Evaluations

In order to obtain an overview of the findings, descriptive analyses should be prepared in a first evaluation step. In univariate analyses, the aim is to describe the empirically determined manifestations of *a* characteristic. *Bivariate* analyses include two and *multivariate* evaluations include several facts in the analysis.

For the univariate representations, two measures in particular can be used to describe the units under investigation: the mean and the mean variation. An important prerequisite for data analysis is an understanding of the scale level in question, as this determines the methods that can be used for an analysis. The following techniques can be used to describe the population under investigation – depending on the scale level:

Classificatory Nominal Scale

The nominal scale is used to assign numbers to certain objects in order to be able to distinguish between them. An indicator taken from the Dresden self-defence study that yields results at this scale level is the question about the target's occupation (compare Fig. 9.3).

Figure 9.4[2] shows the response distribution obtained.

Only the modal value can be used to represent a mean value. It indicates which object or, in this case, which occupation is most frequently represented in the sample. This involves more than 2000 persons who stated that they worked as employees (63%).

In order to assess the quality of the sample, this distribution can be compared with a reference sample, for example with that of the Microcensus[3] from the year 2000. At the same time, this reveals a typical problem of survey research: the sample of the Dresden self-defence survey apparently contains too few blue-collar workers and too many white-collar workers.

Ordinal Scale

With an ordinal scale, the individual objects can be placed in an order so that the next object is larger or more pronounced than the previous one. With the ordinal scale, no statement is made about the distances between the individual objects. In the Dresden self-defence study, after the activity of a target person had been roughly determined (compare Fig. 9.3), more specific questions were asked. Thus, all 511 workers were asked the following question, shown in Table 9.3. It can be assumed that the qualification of the target persons increases with ascending scale.

Table 9.4 shows the distribution of responses found in the sample.

In order to be able to make a statement about the mean value, the median can be used. This value indicates where the cumulative 50% of the answers lie on the scale. This can be

---

[2] Individuals still in education at general education schools were not presented with this question.

[3] Comparisons to the 2000 microcensus: http://www.gesis.org/Methodenberatung/Untersuchungs-planung/Standarddemografie/dem_standards/demsta2004.pdf last accessed 09/03/2006, now unfortunately no longer available.

Are you/were you …

Worker 1	⇨S16b
Employee 2	⇨S16c
Civil servant (professional soldier/judge) 3	⇨S16e
Farmer 4	⇨S16f
Freelancer 5	⇨S16f
Other self-employed (previously according to the old spelling) or	⇨S16f
Entrepreneur 6	
Assisting family member 7	⇨S17
Trainee 8	⇨S16d
Member of a PGH 9	⇨S16f
Not specified .... −1	⇨S17

The information contained in the last column concerns a filter

Fig. 9.3 Question about the profession of the respondent. (Source: Dresden self-defence study)

		Frequency	Valid percentages	Microcensus
Valid	Not specified	37	1.1	3.1
	Workers	511	15.5	29.3
	Employee	2.085	63.3	46.4
	Civil servant (professional soldier/judge)	239	7.3	6.8
	Farmer	21	.6	–
	Freelancer	71	2.1	
	Other self-employed/entrepreneur	228	6.9	10.1
	Assisting family member	8	.2	1.1
	Trainee	93	2.8	4.4
	Total	3.292	100	
Missing	Filter	171		
Total		3.463		

Fig. 9.4 Occupations of the target persons interviewed in the Dresden Self-Defence Study and occupations of those interviewed in the Microcensus

seen in the column "Cumulated percentages". According to this, the median lies in the group of skilled workers.

The table once again clearly shows how missing values and special levels are dealt with. For example, four people did not answer the question about their occupation and 2952 people were not even asked the question in the first place (since they previously stated that they were not workers, the filter caused the question to be skipped). Of course, these respondents were then not included in the formation of the median, but were classified as missing.

Table 9.3 Question to workers about their qualifications

Are you/were you …?	
… Unskilled	1
… Semi-skilled	2
… Skilled workers	3
… Foreman, column leader	4
… Foreman, foreman, brigadier	5

Source: Dresden self-defence study

Table 9.4 Structure of the workers interviewed in the Dresden self-defence study

		Frequency	Percent	Valid percentages	Cumulated percentages
Valid	Unskilled	57	1.7	11.3	11.3
	Learned	114	3.3	22.6	33.9
	Skilled workers	300	8.7	59.1	93.0
	Foreman, column leader	21	0.6	4.2	97.1
	Foreman, Brigadier	15	0.4	2.9	100
	Total	507	14.6	100	
Missing	Filter	2952	85.3		
	Not specified	4	0.1		
	Total	2956	85.4		
Total		3463	100		

Interval Scale

With an interval scale, the distances between the individual objects can now also be determined to be the same. The Dresden self-defence study also contains a corresponding indicator for this.

One question first asked whether the person in question had ever been in a situation in which he or she had come to the aid of another person and defended him or her against an attacker. If the answer was yes, the question then went on to ask how often this had happened to them. The interviewers were encouraged to note the corresponding number. The result was as shown in Table 9.5 and Fig. 9.5.

In the case of an interval scale, the arithmetic mean can be calculated as the average value, as is often done in schools with grading scales. In this case it is 0.8.

Further information about the responses obtained is provided by the *measures of dispersion*. These are something like the quality criteria of the mean values. The lower the dispersion, the more accurate the estimator for the population parameter sought.

A further look at Table 9.5 or at the mean value may make this clear. On the basis of the arithmetic mean of just under 1, it could be assumed that the persons interviewed have each had cause to provide self-defence assistance about once. However, this interpretation would not be particularly successful. If one also looks at the distribution of the answers or

Table 9.5 Frequency of self-defence assistance provided to date

		Frequency	Percent	Valid percentages	Cumulated percentages
Valid	0	2369	68.4	68.7	68.7
	1	477	13.8	13.8	82.5
	2	294	8.5	8.5	91.0
	3	139	4.0	4.0	95.0
	4	52	1.5	1.5	96.5
	5	36	1.0	1.0	97.6
	6	8	0.2	0.2	97.8
	7	9	0.3	0.3	98.1
	8	5	0.2	0.2	98.2
	10	32	0.9	0.9	99.1
	12	3	0.1	0.1	99.2
	15	9	0.3	0.3	99.5
	20	17	0.5	0.5	100
	Total	3450	99.6	100	
Missing	System	13	0.4		
Total		3463	100		

Source: Dresden self-defence study

the dispersion, one finds that most people have not yet provided self-defence assistance –
and that others have done so all the more frequently. A concrete statement on this is provided by a measure of dispersion. There are, in turn, different variants for its calculation:

- The Range. This is the range or the difference between the maximum and minimum
  value. In the above example (cf. Table 9.5), the range is (20–0 =) 20. The wide range
  already indicates that an arithmetic mean of 0.8 self-defence aids per respondent is not
  very meaningful.
- For interval-scaled data, the standard deviation can be determined. This is the square
  root of the average squared deviation of the individual values from the mean. The value
  in question here is 2.15 self-defense aids. The standard deviation has the dimension of
  the respective measured values.
- The variance, which is nothing other than the square of the standard deviation, can still
  be calculated. However, it is given without a dimension.
- For ordinal and interval scaled data, it is possible to designate the quartile distance. It
  is also suitable when response specifications are used that have open categories at the
  edges. These are often used, for example, when asking about income. For this, one uses
  specifications such as "up to 500 euros" at the lower end and "10,000 euros and more"
  at the upper end. The following values result for the frequency of self-defence
  experience:
  25% of respondents have no such experience,
  50% of the respondents also have no such experience,
  75% of respondents have a unique experience with self-defense assistance.

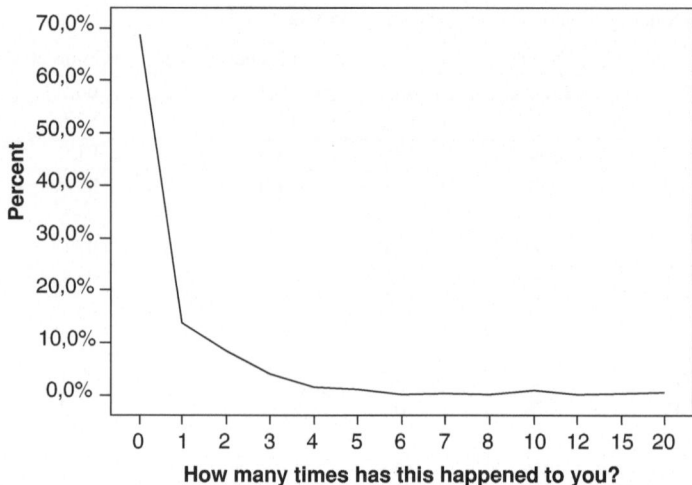

Fig. 9.5 Frequency of self-defence assistance provided to date, shown as a graph. (Source: Dresden self-defence survey)

These values are also a good aid to interpreting the arithmetic mean.
- Graphs such as charts and boxplots can also be created to illustrate the distribution. Diagrams are the simplest form of representation (compare Fig. 9.6).

Boxplots show the median, interquartile range, any outliers and extreme cases of individual variables. Figure 9.6 shows the income of those with self-defence experience compared to the income of those respondents for whom this is not the case. It can be seen that respondents with experience of self-defence reported a higher income than those without. This finding contradicts assumptions according to which assisting in self-defence is to be linked with a thug mentality and suggests, on the contrary, that the responsible protection of other people's property should be assumed as a motive for assisting in self-defence.

In the boxplots, the thick line symbolizes the position of the median. The upper and lower lines – particularly visible in the group of persons *without* experience of self-defence – indicate the position of the upper and lower extreme values respectively. The dark area, approximately in the middle of the diagrams, refers to the middle 50%.

In addition to univariate representations, bivariate evaluations are also frequently carried out. These allow statements to be made as to whether or not there is a connection between two facts in the persons studied. The findings presented in Fig. 9.6 already allow such a statement. Various other procedures are available:

- Four-field tables can be created for dichotomous situations. Here again an example from the Dresden self-defence study (cf. Table 9.6): The following event was to be assessed: "A woman was violently attacked by a strange man. She was only able to defend herself by stabbing the man with a knife. The man was fatally injured." Gendered

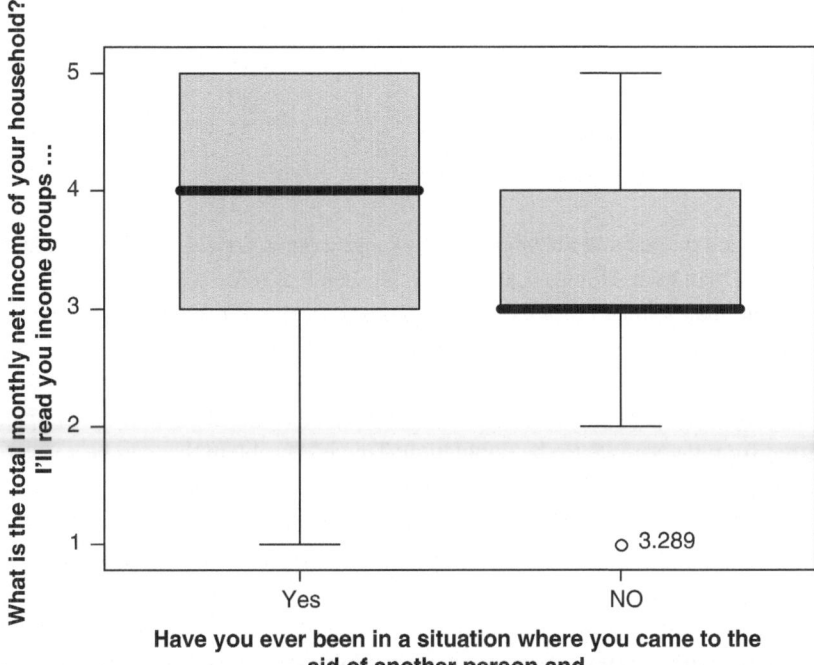

Fig. 9.6 Income of individuals with and without self-defense assistance experience, shown as boxplots. (Source: Dresden self-defence study)

Table 9.6 Evaluation of a self-defence case by men and women (observed frequencies)

		Men	Women	Total
	Justified	1052	1213	2265
	Not justified	437	623	1060
Total		1489	1836	3325

Source: Dresden self-defence study

responses were expected here. The hypothesis was that women, due to their gender-specific socialization, would be more likely than men to consider the described actions of their gender counterpart to be justified.

By convention, the independent variable is included in the columns of the table and the dependent variable in the rows. Only valid cases are reported here. Individuals who declined to rate the case in question were excluded from the analysis. Also shown are the marginal distributions. These result from the summation of the respective column and row frequencies. Finally, the number of valid cases is noted at the bottom right. This information is important because the number of cases was somewhat reduced due to item non-response.

Table 9.7 Evaluation of a self-defence case by men and women (column percentages)

	Men	Women	Total
Justified	71	66	68
Not justified	29	34	32
Total	100	100	100

Source: Dresden self-defence study

It can be seen in the table that significantly more women considered the case described to be justified than men. However, it can also be seen that more women were represented in the sample than men. In order to obtain a more meaningful picture here, the column percentages are now given instead of the absolute frequencies (cf. Table 9.7).

The finding is surprising: the behaviour described is rejected more frequently by women (34%) than by men (29%). The question now arises as to whether such a finding can be transferred from the sample to the population. This is a problem that can be answered by inferential statistics.

To look more closely at such frequency differences and to determine whether this difference is due to chance, one can use χ^2-(chi-square) methods. Such methods can be used for nominally scaled data, but also for interval scaled data.

One way of graphically displaying multivariate distributions are iconplots, for example in the version of the Chernoff faces. Such evaluations can be made with the Statistica program package, among others. "Iconplots are multidimensional symbols that represent cases or units of observation" (Statistica Manuel 2005, p. 1). The values of different variables are assigned to different elements of a face in the Chernoff faces.

In addition to Chernoff faces, circles, stars, rays, polygons, columns, lines and profiles can also be used. This is based on the idea that graphical representations and also human faces in particular are well suited for quickly identifying conspicuous features in the data. For this purpose, variables are assigned to certain features of a face, such as the length of the nose or the height of the eyebrows. As an example, the simultaneous evaluation of 18 cases (compare also Table 3.1) as well as the gender and age of the interviewees are shown. In Table 9.8 it is first determined which facial elements correspond to which variables. Then (compare Fig. 9.7) the response profiles of twelve target persons are presented using the characteristics defined in this way.

Correlations

In the following example, we will now be interested in the extent to which the answers to two questions about control beliefs are related to each other. The two questions are:

- (x) Unexpected situations, such as a robbery, completely overwhelm you.
- (y) There are always ways to fight back, even if you are physically attacked.

It is assumed here that an opposite trend can be observed when answering the questions. People who think they can control a self-defence situation would have to agree more

Table 9.8 Iconplot definition of the Chernoff faces

Variable	Face element	Situation
q1	Face width	A woman is standing in the last free parking space in order to reserve it. A driver asks her to make room for him, otherwise he will drive towards her. The woman stops because she believes she has a right to the parking space. The driver then pushes her out of the parking space with his car. The woman suffers abrasions as a result
q2	Ear level	A man watches as another man steals his bicycle and rides off with it. He chases the thief and brings him down. The bike thief is slightly injured in the process
q3	Half face height	A boxer is threatened by four young men in front of a discotheque. He beats – by targeted blows to the face – the attackers into flight and injures them slightly in the process
q4	Eccentricity upper half	A farmer observes from a distance that a man is about to set fire to his barn, in which the harvest and valuable machines are stored. Since the man does not react to the farmer's shouts and the farmer cannot run to the barn fast enough, he shoots at the man and injures him seriously
q5	Eccentricity lower half	Because his car parked on the street was scratched recently, someone is eyeing four young men in front of his house. They feel provoked by this. Therefore the young men start throwing stones. As the stone-throwers approach, the attacked man fires a few untargeted shots. One shot hits a stone thrower and causes paraplegia
q6	By a nose	A stranger intrudes into an apartment. The owner of the apartment first defends his premises in vain with a walking stick, then he fends off the intruder with a fatal knife thrust
q7	Position center of the mouth	A man was beaten up for no reason by another, *stronger one*. After a few hours he is again massively attacked by the same person. This time he is able to defend himself by stabbing the attacker with a knife he had taken as a precaution
q8	Curvature of the mouth	A motorist harasses another driver on a country road due to his driving style. At a traffic stop, the harassed driver gets out of the car and approaches the other driver threateningly. The latter then pulls out a pistol and forces the other to retreat
q9	Mouth length	Someone watches a couple making love in a public park at night. When the lover asks the spectator to leave, the spectator thinks he has the same right to be in the park as the couple. When the lover then attacks the bystander, the bystander proves to be *physically inferior*. He can only defend himself with a knife and in doing so injures the lover fatally
q10	Eye center level	A holiday home has already been broken into 13 times. The owner then sets up a radio with detonating powder and signs warning of bombs. During another burglary the radio explodes and the burglar loses his hand

(continued)

Table 9.8   (continued)

Variable	Face element	Situation
q11	Centre-to-centre distance	A landlord has had a dispute with a group of youths for some time. When they invade his pub and attack him, he falls backwards and finally fights back with the help of an iron bar, severely injuring two attackers
q12	Eye tilt	One man insults another by accusing him of being to blame for his own imprisonment in the Soviet Union. A brawl ensues, in which the insulter proves *clearly superior.* The insulted person, however, then accidentally brings the other down, and the insulter suffers such serious injuries that he dies
q2a	Eyes eccentricity	A man sees another man stealing his bicycle. He confronts the man. The physically superior thief then attacks the bicycle owner. In the ensuing confrontation, however, the thief is slightly injured by the bicycle owner
q3a	Eyes half length	How would you rate the boxer's behavior if it had just been *an attacker* threatening the boxer?
q6a	Position of the pupil of the eye	Suppose the intruder had been armed. How would you then evaluate the behavior of the apartment owner
q7a	Eyebrow level	How would you evaluate the behaviour of the attacked person if his tormentor had *not* been physically superior to him
q9a	Corner of the eyebrow	Suppose the person watching the couple make love at night in a public park had been *physically superior to them*
q12a	Eyebrow length	Suppose the man who uttered the insult had been physically *inferior*
Gender of the target person		Ear radius
Age of the target person		Nose width

strongly with the second question and disagree correspondingly more strongly with the first question. To examine this, a correlation can be calculated. For a simplified demonstration, only the responses of those 14 individuals who are foremen, or brigadiers and who answered these two questions are shown here (see above). Further, it is assumed that the two questions produced interval scale level data. Table 9.9 shows the responses to both questions and the steps taken to calculate the correlation coefficient.

The correlation coefficient (r_{xy}) is calculated according to the following formula (compare Kühnel and Krebs 2001, p. 404):

$$r_{xy} = \frac{n \cdot \sum x_i \cdot y_i - \left(\sum x_i\right) \cdot \left(\sum y_i\right)}{\sqrt{n \cdot \sum x_i^2 - \left(\sum x_i\right)^2} \cdot \sqrt{n \cdot \sum y_i^2 - \left(\sum y_i\right)^2}}$$

The individual calculation steps required for this can be traced in Table 9.9. These then enable the formula to be applied, as demonstrated further below.

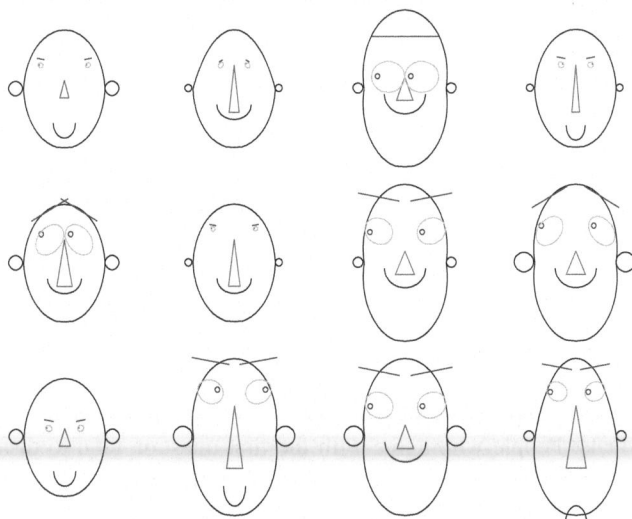

Fig. 9.7 The representation of twelve respondents from the Dresden self-defence study using Chernoff faces

Table 9.9 Responses to the above questions (x_i) and (y_i) by foremen and brigadiers

ID	x_i	y_i	$(x_i)^2$	$(y_i)^2$	$x_i \cdot y_i$
1	3	3	9	9	9
2	3	2	9	4	6
3	4	4	16	16	16
4	2	4	4	16	8
5	4	2	16	2	8
6	1	3	1	9	3
7	4	2	16	4	8
8	3	1	9	1	3
9	2	4	4	16	8
10	3	4	9	16	12
11	4	3	16	9	12
12	4	4	16	16	16
13	4	4.	16	16	16
14	2	4	4	16	8
Σ	43	44	145	150	133

n = 14
Source: Dresden self-defence survey

$$r_{xy} = \frac{14 \cdot 133 - 43 \cdot 44}{\sqrt{14 \cdot 145 - (43)^2} \cdot \sqrt{14 \cdot 150 - (44)^2}} \quad r_{xy} = -0.16$$

The correlation coefficient can take values between -1 and $+1$. A value of -1 indicates a perfectly negative correlation. A value of 0 indicates no correlation between the two variables. As a rule of thumb (compare Kühnel and Krebs 2001, p. 404 f.), the following verbal interpretations of the correlation coefficient apply:

Amount of r_{xy}	
0 to 0.005	To be neglected
Above 0.005 up to 0.20	Low
Over 0.20 to 0.50	Medium
Over 0.50 to 0.70	High
Over 0.70	Very high

Thus, looking at the sign alone, the assumption has been confirmed for the 14 randomly selected individuals: People who agree with the first question tend to disagree with the second. However, this correlation exists only to a small extent.

9.2.2 Inferential Statistical Analyses

Whereas the analysis steps shown so far have been concerned with merely describing the cases studied, procedures are now discussed which allow conclusions to be drawn from the elements studied to a population. With such conclusions, however, errors cannot be excluded by their very nature.

It was shown above that the women who were interviewed in the study consider a certain behaviour to be unjustified more often than men. Inferential statistical analyses are now to provide information on whether this finding can be transferred to the population. To this end, hypotheses are generated. One hypothesis is: There is no difference between men and women in the evaluation of the self-defence case in question. This hypothesis is also referred to as H_0 and the null hypothesis, respectively. The other hypothesis looks like this: There is a difference between men and women with respect to the evaluation of the self-defense case in question. This is then the H_1 hypothesis, which is the alternative to the null hypothesis. It is therefore also called the alternative hypothesis.

The next step is to see which hypothesis can be accepted on the basis of the data. Since this is a sample survey, such a statement cannot be made with certainty. Therefore, a probability must now be determined that can be used as a benchmark for such a decision.

The α- and the β-Error
The α-error, also called error of the first kind, indicates the probability that an H_0-hypothesis is rejected, although this null hypothesis is actually true. This is also referred to as a false positive. The β-error, or error of the second kind, is the probability that the H_0-hypothesis is retained in error, although the null hypothesis is incorrect. In such a case, therefore, a

correlation that actually exists would then be overlooked. This would result in a false-negative result (cf. Dubben and Beck-Bornholdt 2004, p. 61 ff.).

Transferred to a court hearing, in the course of which arguments for and against the guilt of a defendant are compiled, this would mean in the case of the α-error that an actually innocent person is erroneously found guilty. Thus, to avoid the α-error as much as possible, no defendants would be allowed to be convicted. However, then the β-error occurs. By analogy, this indicates the probability that a guilty person will be acquitted by mistake or, in the social sciences: that a connection that actually exists will be overlooked. In order to reliably avoid the β-error, therefore, all accused persons would also have to be convicted or all suspected connections would also have to be confirmed. At the same time, this makes clear the relationship between the α-error and the β-error; they are complementary to each other.

In the social sciences – similar to jurisprudence – it is common to work with the α-error. This indicates the probability that a correlation is negated in an investigation, even though it actually exists. The aim (in the social sciences as well as in jurisprudence) is to keep this error as low as possible and thus consciously accepts that connections that actually exist are overlooked or that a guilty person is *not* convicted (for example, because the evidence is insufficient).

Two further issues need to be considered in inferential statistical data analysis:

First, if an H_0-hypothesis has been accepted after an empirical investigation, this is not definitive proof that a suspected relationship does not in fact exist. There always remains a certain uncertainty that necessitates further investigation.

Second, imagine that an H_0 hypothesis is true. That is, there is no correlation between two variables under consideration. If now 100 researchers work on the respective H_0-or analogously on the corresponding H_1-hypothesis, then with a probability of error of 0.05 as many as 95 researchers will accept the H_0-but also five of them will accept the H_1-hypothesis. The latter will then claim that there is (with 95% probability) a relationship between the characteristics. However, since it is now usually far more interesting to publish about a determined relationship than to publicize the failure of a conjecture, there is a certain danger that this will also and especially be done by those researchers who have established the H_1 result. This would then lead to a whole series of errors in the literature. To prevent this, one should attach importance to increasingly tackle replication studies, i.e. to empirically test a conjecture several times.

Significance Tests and the Strength of a Relationship

Significance tests are often used in data analysis. At this point, we would like to point out the difference between the result of a significance test and the strength of a correlation. These are different issues, which often leads to misunderstandings in data analysis.

Significance represents a test statistic at a certain level, for example 5%. If such a significance level is set, this means that every 20th test detects an effect, although there is no such effect in the population. Here there is a strong dependence on the number of cases examined. The more cases there are in the sample, the more reliable statements can be made (the prerequisite here is that a random sample was collected).

Now, for example, a certain significance level (the probability of error is, for example, $\alpha = 0.05$) can be used to determine which of the two hypotheses (H_0 or H_1) is true. It would be meaningless to determine such a level using data from a total survey. Since all elements of the population have already been examined, the findings cannot be transferred to other persons, and thus there is no possibility of an error, or of a probability of error, or of a significance level.

The situation is different when it comes to the *strength of a relationship*. The strength of the relationship between two variables should be independent of the number of cases examined. The corresponding statement initially refers only to the units studied. It is a statement of measures of association among the cases studied. The strength of a correlation can also be determined within the framework of a descriptive evaluation (cf. section on correlation). A significance level, on the other hand, cannot.

The χ^2 (Chi-Square) Test

Further above, the evaluations of a self-defence case by men and women were shown (compare Tables 9.6 and 9.7). Here it turned out that the women interviewed apparently answered differently than expected. This may have happened by chance due to sampling error. However, it is also conceivable that the difference found in the sample is indicative of a difference in the population. This can now be clarified by the χ^2-test.

The χ^2-test makes a comparison between the observed distribution and a theoretical distribution. The first cell (men justifying the behavior) contains 1052 individuals. In determining the theoretical distribution, the procedure is as follows: It is calculated from the product of the marginal distributions of the column and row in question (in the example above: 2265 multiplied by 1489 gives 3,372,585) divided by the sample size of 3325 This leads to the value 1014 This value is therefore lower than that of the actual observed cases. Accordingly, the expected distribution can also be determined for the remaining cells. The rest of the calculation is shown in Table 9.10.

The χ^2-value is the sum of the column $(a - b)^2/b$. Whether the value of 8.08 just calculated here still represents a random deviation, a comparison with the known χ^2-distribution reveals.

Before this, however, the number of degrees of freedom must be determined. If the marginal distributions of the columns and rows are given, only one cell can be freely

Table 9.10 Calculation of a χ^2 value

	a	B	b	a − b	$(a - b)^2$	$(a - b)^2/b$
1	1052	(2265 • 1489)/3325	1014	38	1444	1.42
2	1213	(1836 • 2265)/3325	1251	−38	1444	1.15
3	437	(1060 • 1489)/3325	475	−38	1444	3.04
4	623	(1836 • 1060)/3325	585	38	1444	2.47
Σ						*8.08*

a: Contents of the cells from Table 9.6.
b: row or column marginal totals and sample size from Table 9.6.

selected. The other cells can then be determined by subtracting this value from the marginal distributions.

On the basis of tables published in the literature (cf. Bortz and Döring 2002) or by means of an EDP statistics program, it can now be determined that the difference determined in the sample is also to be found in the population with 95% probability.

χ^2-calculations can also be used for facts with a larger number of characteristics. However, it should be noted that the individual cells must each be occupied by at least five cases in order to be able to perform such calculations. If this is not possible, the values of a variable, such as age or income, must be combined into larger groups. Furthermore, the calculation rule only applies to unrestricted random selections, as otherwise the design effect must be taken into account.

Comparison of Mean Values (T-Test)
Mean comparisons can be made for interval-scaled data. The t-test can be used for this purpose. Continuing the above example, a simple problem is to be worked on. It should now be of interest whether a higher level of agreement with the case described is not only gender-specific, but also dependent on the age of the target person.

It could be argued, for example, that younger people have a higher degree of control in a self-defence situation. They feel more able than older people to stand their ground in such a situation. From this, in turn, the assumption could be derived that they also advocate a more sustainable form of counter-defence.

To investigate this, one can determine the arithmetic mean of the age of the target subjects, separately for the group that considers the behaviour to be justified and for those that interpret this as not being justified. Such a descriptive description reveals that the individuals in the sample who rate the described behavior as justified are on average 43.8 years old and those who do not are 46.7 years old. Thus, the expected difference in the sample is indeed present. With the aid of boxplots (cf. Fig. 9.8), this impression can be further substantiated graphically.

Now it would have to be examined again whether the determined difference of almost three years could have come about due to the sampling error, or whether it is significant, i.e. with a certain probability it also indicates a difference in the population. For this purpose, a t-test would have to be applied.

The corresponding t-value is calculated according to the formula:

$$t = \frac{m_1 - m_2}{\sqrt{\left[\frac{n_1 + n_2}{n_1 \cdot n_2}\right]\left[\frac{(n_1 - 1)\cdot s_1^2 + (n_2 - 1)\cdot s_2^2}{n_1 + n_2 - 2}\right]}}$$

Here, too, the value determined must be compared with the corresponding value either from a table or with the aid of a computer statistics program. Table 9.11 shows the output of the calculations made with the help of SPSS (compare also Wittenberg and Cramer 2000, p. 201 ff.). Finally, it is shown that there is a 95% probability that the age differences

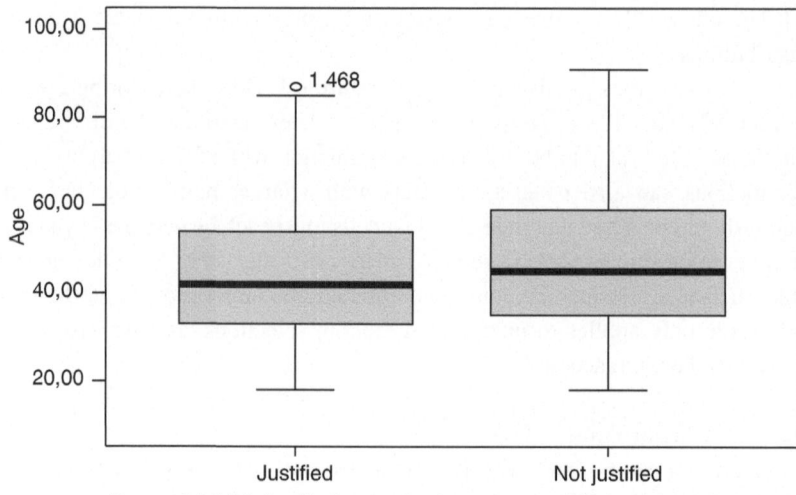

Fig. 9.8 The age of targets who perceive a case as justified verus as not justified. (Source: Dresden self-defence survey)

determined between the two groups do not occur by chance. It should again be noted that this applies to unrestricted random samples, i.e. no design effect needs to be taken into account.

The null hypothesis is that the age of persons who evaluate a case in a certain way does not differ significantly from the age of persons who evaluate this case differently. First, the result of the variance homogeneity test (Levene's F-test, cf. Dayton 1970, p. 34 f.) must be asked. SPSS calculates two t-tests. One for the case that the variances of the dependent variable for the two subgroups are equal (in the SPSS output above) and the other for the case that they are unequal. The significance given here is 0.017. It follows that the t-test for non-equal variances is competent (i.e. the lower one in the above Table 9.11). Now here it can be seen that the age differences found between the two groups are not random. Thus, it can be argued that age has a significant impact on the evaluation of the case.

9.2.3   The CHAID Analysis

Another – admittedly not yet widely used – method for data analysis is the Chi-Square-Automatic-Interaction-Detector (CHAID).

CHAID is a procedure for examining categorical data. It was developed by Kass in 1980 (compare Magidson 1994, for another example of the application of the procedure compare Sievers 1994). This procedure, which has been further developed in the meantime, is part of a module supplementing the SPSS program package and is called Answer Tree. The CHAID algorithm contained in it is described as follows:

Table 9.11 SPSS output for the calculation of a t-test

		Levene test of variance equality		T-test for equality of means					95% confidence interval of the difference	
		F	Significance	T	df	Sig. (2-sided)	Medium Difference	Standard error of the difference	At	Upper
Age	Variances are equal	5.693	0.017	−5.000	3322	0.000	−2.86120	0.57227	−3.98324	−1.73916
	Variances are not equal			−4.912	1981.145	0.000	−2.86120	0.58245	−4.00348	−1.71892

It is a

"Sequence of segmentations and summaries driven by the results of various association analyses (e.g., cross-tabulation analyses), each of which assesses the association of the criterion
with a predictor.

- In the first step, the sample is divided into the categories of the best predictor, previously combining categories without significant differences. A predictor is *good if* it
  demonstrates a significant association with the criterion (…). Among the good predictors, the one with the smallest p-level in the association test is chosen as the best.
- In the second step, another decomposition is attempted for each of the obtained groups,
  using only the remaining predictors, of course. ….

The algorithm runs in each branch of the resulting tree until no significant predictor is
found."[4]

An attempt will now be made to demonstrate this procedure using a meaningful example
from the Dresden self-defence survey. So far, various empirical indications have been
found that self-defence assistance is practised primarily by certain personality types. This
idea is now to be pursued more concretely with the help of the CHAID analysis. A corresponding question could thus be: How can persons be described in more detail who have
experience in self-defence assistance? Various socio-demographic criteria will be used to
answer this question. It will be examined to what extent each criterion can help to explain
the experiences mentioned. To do this, the first step is to determine which variables can be
considered at all. The following variables are included as predictors for this example:

- Income grouped into three income groups: below €511 (=1), between €511 and below
  €2300 (=2) and above €2300 (=3),
- Gender of the target person,
- German citizenship: yes (=1) or no (=0),
- City: Residents of a city with 100,000 inhabitants or more (=large) compared to resi-
  dents of a smaller city or municipality (=small),
- Formal education in three groups: Hauptschule (=1), Realschule (=2) and Abitur (=3),
- Origin: from West or East Germany,
- Election: an election decision is reported (=1) or not reported (=0) in the Sunday survey,
- the existence of own self-defence experiences: yes (=1) or no (=0),
- Trainees: yes (=1) or no (=0),
- Churchgoing frequency: never and not regularly (=1) as well as regularly (=0).

Figure 9.9 shows the result of the CHAID analysis as a tree.

---

[4] Cf: http://www.uni-trier.de/urt/user/baltes/docs/chaid6/chaid6.pdf,   last   accessed   13.04.2006,
unfortunately now no longer available.

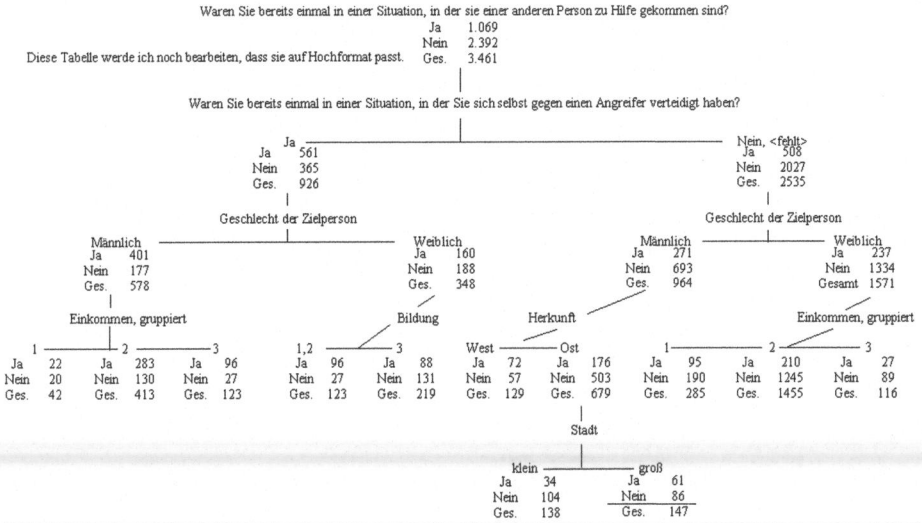

Fig. 9.9 Chaid analysis

This figure will now be evaluated from top to bottom. The respondent population (n = 3461) is considered in terms of how they answered the first question ("Have you ever been in a situation where you came to the aid of another person?"). Of all the predictor variables (see above), the question about whether you yourself have ever been in such a situation (second question) most strongly explains the responses to the first question. Now, on the left side of the tree, we see that people with their own self-defense experience (n = 926) were significantly more likely to have also helped others in such a situation (n = 561). On the right-hand side, on the other hand, we find all persons without their own self-defence experience (n = 2535). Proportionally, significantly fewer of them (n = 508) have provided self-defence assistance.

From now on we have to deal with two sub-populations: on the left those with and on the right those without self-defence experiences. In both populations, we now examine which of the remaining variables has the strongest explanatory power. In the case of those with their own self-defence experience (left), the variable gender now explains the response behaviour most strongly. However, this also applies to the right-hand side. Now the original population has already been divided into four subpopulations. The response structures can be seen in Figure.

Now the procedure goes into the next round. For all four populations, the tests are again calculated with the remaining predictor variables. This shows (left) that men who have their own self-defence experiences (n = 578) now differ most on the basis of their income. For women with their own self-defense experiences (n = 348, the branch to the right), education level is the strongest explanatory variable.

It should be noted that the program automatically merges categories (as in the example, for the variable "Income, grouped", groups 1 and 2, and for the variable "Education", also

Table 9.12   SPSS output for the explained total variance (section)

Component	Initial eigenvalues			Rotated sum of squared charges		
	Total	% of variance	Cumulated %	Total	% of variance	Cumulated %
1	9.56	31.85	31.85	4.35	14.50	14.50
2	2.56	8.54	40.39	3.67	12.22	26.73
3	2.30	7.67	48.06	2.78	9.27	36.00
4	1.97	6.57	54.63	2.74	9.13	45.13
5	1.49	4.96	59.59	2.48	8.27	53.40
6	1.34	4.47	64.06	2.09	6.97	60.37
7	1.19	3.98	68.04	1.68	5.60	65.96
8	1.03	3.43	71.47	1.65	5.51	71.47
9	0.98	3.26	74.73			
10	0.91	3.03	77.76			

groups 1 and 2) if the two categories do not differ significantly in terms of criterion distribution.

The calculation of the tree yielded ten terminal nodes in this analysis. These each represent groups that differ from one another on the basis of a particular combination of characteristic values. The groups found thus differ as much as possible with regard to their response rates, i.e. with regard to the self-defence assistance experience present.

All predictors that no longer appear in the presentation did not reach the required significance level of $p = 0.05$. This is the case for the four variables "German citizenship" and "church attendance". This applies to the four variables "German citizenship", "voting decision", "trainee (apprentice)" and "churchgoing frequency".

The persons summarised in node three have the largest proportion of people experienced in self-defence, at 78%. These are men who themselves already report experiences of self-defence and at the same time have a particularly high income. This contrasts with the people in node nine. These are women with no self-defence experience of their own, who have only a medium or low income. In the survey, only 14% of them stated that they had experience of self-defence. The difference between these two groups is obviously immense.

The strongest predictors of self-defence assistance behaviour have been shown to be one's own self-defence experience and gender. In addition, it is important to point out some further special features:

• If we look at those people who have their own experience of self-defence (these are summarised in nodes one to five), we see that in the case of men with a higher income and women with a high level of education (A-levels), the experience of self-defence assistance also increases in each case. It can be assumed that we are dealing here in each case with particularly responsible people who, through their behaviour, help to support the legal system (compare Schroeder 1972).

- People who do not have any experience of self-defence of their own have significantly lower experiences of self-defence assistance. Here, an income dependency can be discerned among women: The lowest experiences of self-defence assistance are found among those with medium and low incomes.
- The situation is more complicated for males who have not experienced self-defence assistance. In total, only 28% of them report such experiences. However, if one takes a closer look at this group of people, it becomes apparent that at this point (and only at this point) there is a West-East difference: men without experience of self-defence who live in the western federal states have significantly more experiences of self-defence than their counterparts in the east. If the former live in the west of Germany, the proportion is still almost 42% (compare node eight).
- The CHAID analysis allows an assessment of the influence of the individual predictors. A distinction is made primarily (a) on the basis of the existence of self-defence experiences, (b) gender and (c) qualification or income. The influence of the two characteristics "city" and "origin from West or East Germany" remains relatively low. These are not able to contribute to a general clarification of the facts examined and have only partial explanatory power in the case of a specific group of people.
- A particularly strong explanatory power emanates from the fact that self-defence has already been practised once. Learning theories (compare Holland and Skinner 1971; Opp 1972) could be used here as an explanation. People who have experienced that self-defence leads to positive action results are also more willing to provide self-defence assistance. It is assumed that previous self-defence experiences have actually led to positive action results.

9.3 Multivariate Methods for Data Analysis

Multivariate methods for data evaluation provide information about the relationships between more than two variables. Various methods are used for this purpose in empirical social research. At this point, only a general overview can be given. The user is referred to the special literature at the appropriate places.

9.3.1 Factor Analysis

This procedure assumes that latent facts can be made responsible for the determined expressions of different variables. This idea will be illustrated by means of a school report; likewise, the results of an intelligence test[5] could also serve as an illustration. It contains a set of scores that express a person's performance. The idea now is that the individual

---

[5]The procedure was originally developed in the early twentieth century to evaluate such intelligence tests.

marks do not come about completely independently of each other, but that there are, for example, special talents behind each of them. This explains why the grades in mathematics, physics and chemistry are particularly similar. The same may be true for German, Latin and English. We are therefore dealing with a certain structure in the data. Factor analysis now reveals this structure. Two approaches can be distinguished: the exploratory and the confirmatory.

Exploratory factor analysis involves searching for correlations (factors) in the data. The correlations found in this way must then be interpreted in a way that makes sense in terms of content. An example of this is given in Table 4.5 in Chap. 4, which shows a rotated component matrix with factor loadings of 0.100 and greater. The analysis of the 30 variables initially resulted in eight factors. Table 9.12 shows an excerpt from the overview of the explained variance produced by SPSS. This shows that the eight factors explain over 70% of the variance.

At the same time, another possible application of factor analysis is shown here. The original intention was to form a one-dimensional index expressing attitudes towards social science surveys. This index now only includes those variables that load strongly on the first factor. A repetition of the factor analysis then yields a one-dimensional solution. Factor analyses are thus also suitable for testing the one-dimensionality of indices.

Confirmatory factor analysis starts from a certain presumed structure. Conceivable – albeit somewhat far-fetched – would be a theory according to which it is not the intelligence of the students but the person of the teacher that is responsible for school performance. Thus, a very specific, theoretically hypothesized structure would also have to be present in the data. The confirmatory factor analysis now examines to what extent such an assumption can actually be held on the basis of the empirical data.

The SPSS program package enables a relatively simple calculation of the exploratory factor analysis (compare Wittenberg and Cramer 2000, p. 148 ff.). The programs EQS and LISREL, but also SPSS, are suitable for the preparation of a confirmatory factor analysis (compare Bentler 2001; Jöreskog et al. 2001).

9.3.2   Cluster Analyses

Another multivariate procedure that is also relatively frequently used in the social sciences is cluster analysis. This procedure can also be used to uncover structures in the data. In doing so, similarities are searched for. Both similar, homogeneous units of investigation (target persons) and similar variables (answers to questionnaire questions) can be grouped into clusters.

After determining which variables are to be used for the similarity check, a measure must be found that is capable of expressing this similarity numerically. A wealth of variants is now available for this in program packages such as SPSS or ClustanGraphics.

The user of cluster analysis will find specific guidance in the following sources: Wiedenbeck and Züll (2001); Backhaus et al. (2000) and Bortz (1999).

Table 9.13 Results of a binary logistic regression analysis – Exp(B) – on the granting of self-defence assistance (yes = 1, no = 2)

Constant	2.12***
Number of own self-defence experiences	0.55***
Gender: Reference category male	1.98***
Income grouped: Reference category: Low	0.85***
Education grouped: Reference category: Low	0.85***
Urban dwellers: Reference category: No	0.87
Origin: Reference category: West	1.13
Nagelkerkes R^2	0.23
$n = 3339$	

9.3.3   Regression Analyses

In Sect. 3.8, regression models were already presented as an approach characteristic of quantitative research. The aim here was to uncover the causes of the human feeling of happiness. Happiness was the dependent variable. It was to be estimated using the level of income, living with a partner (as a binary variable), health status and similar variables. Since the dependent variable was considered to be interval scaled, linear regression was used for this purpose. However, if the dependent variable is a dichotomous variable, a binary logistic regression must be used for the calculation.

If the values shown in Table 9.13 are less than 1, the probability that the person in question has already performed self-defence assistance once increases accordingly. Thus, for example, it can be seen that the more often a corresponding situation has already been experienced, the more frequently self-defence assistance is provided.

The user of this procedure can find further information in Andreß et al. (1997) and in Backhaus et al. (2000).

9.4   The Use of Facet Theory for Data Evaluation

The data evaluation is always guided by the knowledge interests of the organizer of the study. To this end, it makes use of various methods and strategies, such as the χ^2 test or multivariate approaches. In the following, the Dresden self-defence study is used as an example to demonstrate the use of facet theory for the empirical treatment of certain questions.

9.4.1   Basic Principles

The basic assumption of the study can be summarized in the form of a mapping theorem (compare Fig. 9.10). According to this, the justification of a more or less violent defensive

A target pi evaluates a case in which a <u>given good</u> (a) is being used by a
(a1 = intangible asset)
(a2 = asset value[e])
(a3 = life & limb)

<u>Specific person</u> (b) is attacked and in the defence of which the attacker is
(b1 = the person is superior)
(b2 = the person is not superior)

Takes a <u>certain damage</u> (c) _ whether this conduct of the person attacked
(c1 = slight injury)
(c2 = serious injury)
(c3 = death)

Is justified or nott
(1 = yes)
(2 = no)

Fig. 9.10 Image sentence for the justification of a violent defence

Table 9.14 Correlation coefficients of variables q1 to q12a

	q2	q2a	q3	q3a	q4	q5	q6	q6a	q7	q7A	q8	q9	q9A	q10	q11	q12	q12a
q1	.00	-.01	.00	.00	.05	.02	.04	.02	.02	.03	.02	.01	.01	.03	.00	.00	.03
q2	-	.60	.23	.19	.11	.07	.17	.19	.11	.05	.00	.05	.04	.09	.25	.11	.07
q2a		-	.19	.13	.08	.06	.12	.19	.09	.06	.00	.04	.03	.08	.22	.07	.05
q3			-	.57	.12	.09	.19	.19	,14	.07	.01	.06	.03	.10	.21	.08	.08
q3a				-	.11	.08	.18	.12	.16	.14	.02	.09	.05	.08	.18	.08	.12
q4					-	.18	.26	.13	.19	.11	.06	.06	.03	.23	.14	.10	.07
q5						-	.14	.08	.15	.11	.04	.11	.07	.08	.09	.03	.02
q6							-	.39	.31	.19	.02	.15	.09	.19	.26	.13	.12
q6a								-	.20	.09	.01	.10	.05	.12	.27	.10	.07
q7									-	.50	.06	.13	.07	.22	.19	.08	.10
q7a										-	.05	.09	.13	.14	.10	.06	.13
q8											-	.07	.06	.07	.00	.02	.01
q9												-	.51	.06	.09	.01	.02
q9a													-	.07	.05	.00	.03
q10														-	.17	.10	.09
q11															-	.16	.11
q12																-	.71

action is based on three criteria. The decisive factors are, firstly, the respective good attacked, secondly, the superiority of the attacker and, thirdly, the damage caused during the defence.

As already shown, each case to be assessed in the questionnaire contains all three facets (a, b and c). In the empirical testing of this assumption, the first step is to correlate all indicators with each other and thus create a distance matrix (compare Table 9.14).

In evaluations that make use of the facet theory, it is common to use the procedures of Multidimensional Scaling (MDS). The corresponding calculations were made with the procedure PROXSCAL of SPSS. This allows the distance matrix (Table 9.14) to be graphically represented in two-dimensional space in a descriptive manner. Figure 9.11 shows

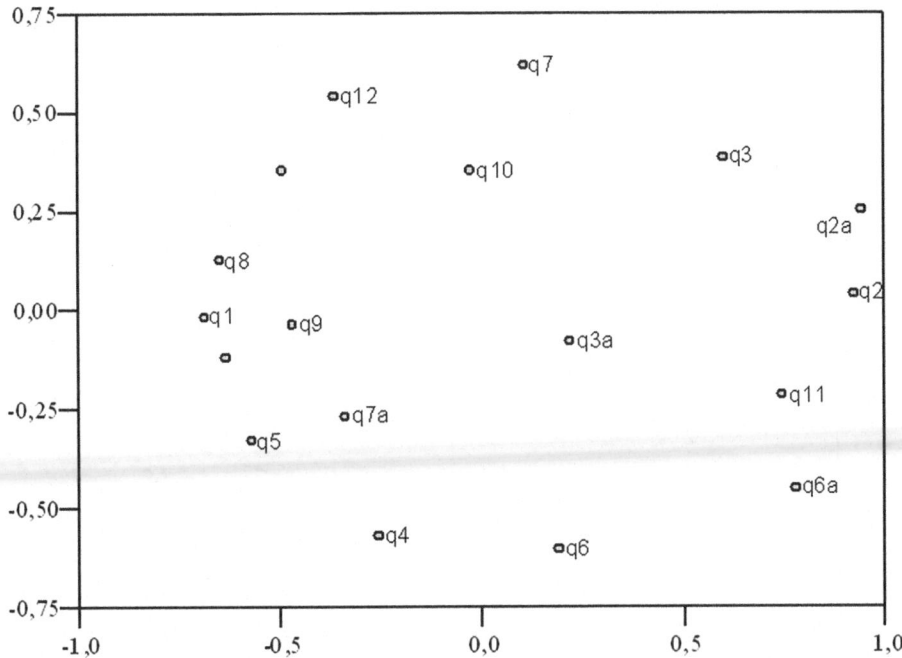

Fig. 9.11 MDS solution for all respondents

the result. The designation of the points in the configuration plots created in this way is analogous to the designations used in the questionnaire (q1 to q12a).

According to Borg (2000, p. 3), one could say that this picture tells more than the 143 figures from Table 9.14. For example, while the correlation analyses yielded a value for p of 0.00 between the variables q1 and q2, they yielded a value of 0.71 for q12 and q12a. Thus, q1 and q2 were represented with a correspondingly large distance, while q12 and q12a appeared with a relatively small distance in the configuration plot.

Now an attempt is made to partition the individual facets (compare Fig. 9.11) in the MDS solution found. A successful partitioning of the two-dimensional space structured by the answers of the target persons would indicate a confirmation of the initial assumptions. To this end, the first step is to enter the individual facet labels into the MDS solution. Then, an attempt is made to identify the facets in the MDS solutions that, according to the mapping set, cause the cases to be rated as justified or not justified, respectively. Figures 9.12, 9.13 and 9.14 present these findings.

Figure 9.12 turns to facet (a) Property damaged in defense. Here the partitioning of a modular pattern succeeds, with two errors. (These are each illustrated with the aid of an arrow pointing to the region actually expected). A modular pattern refers to ordered facets and was to be expected given the scale level of the independent variable. While at the center are those cases in which minor injuries occurred (a1), at the periphery are those in which defense of an asset resulted in the death of the intervener (a3). In between are (a2) – according to logic – the serious injuries.

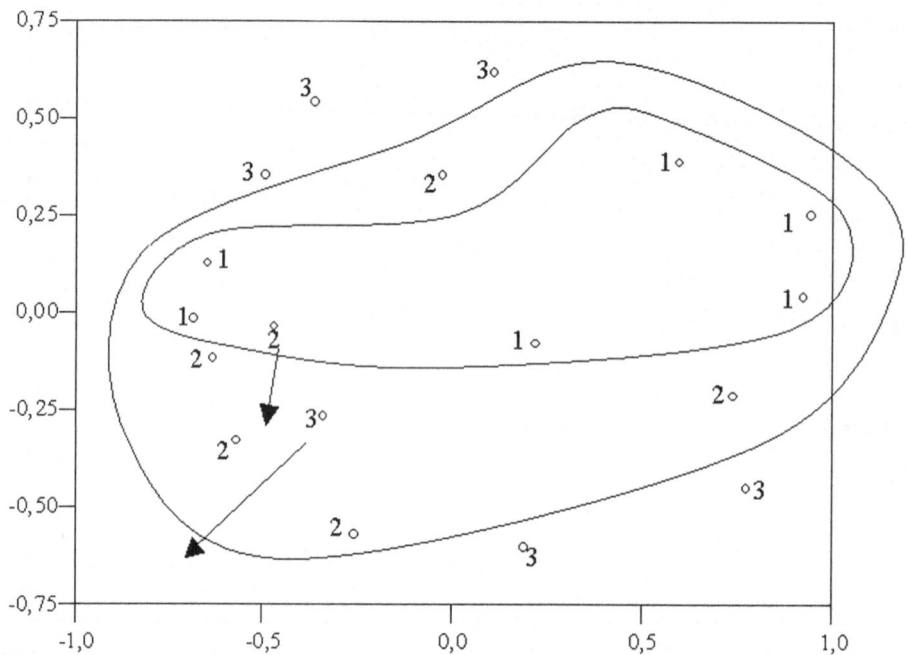

Fig. 9.12 MDS solution for all respondents, showing facet (a) Property damaged in defence, with labels: *1*: minor injury; *2*: serious injury; and *3*: death

Figure 9.13 illustrates the partitioning of facet (b) attacker superiority. Here the two-dimensional space has been partitioned with the help of a polar structure. This resulted in four incorrect groupings. Due to the dichotomous expression of the independent variables, no particular pattern was to be expected. If one combines the polar representation of the facet superiority of the attacker with the modular pattern of the facet (a) damaged goods in defence, one obtains the well-known Radex partitioning (compare Borg 2000).

Finally, Fig. 9.14 shows facet (c) attacked material. The partitioning succeeds here again relatively well with the help of an axial pattern. The space is thus decomposed by a simple partitioning. There are only two errors in this process.

To summarize: Facet (b) attacker superiority causes the most problems in partitioning. Four variables could not be classified as suspected. The reasons for this are either a flawed theoretical concept or its lack of empirical implementation. Both assumptions will be further investigated.

9.4.2   Search for Causes of Mispartitioning

Cases q2 and q10 have been classified in the superior attacker region. However, the concept assumed that it was *not the* attack of a superior person. The question was:

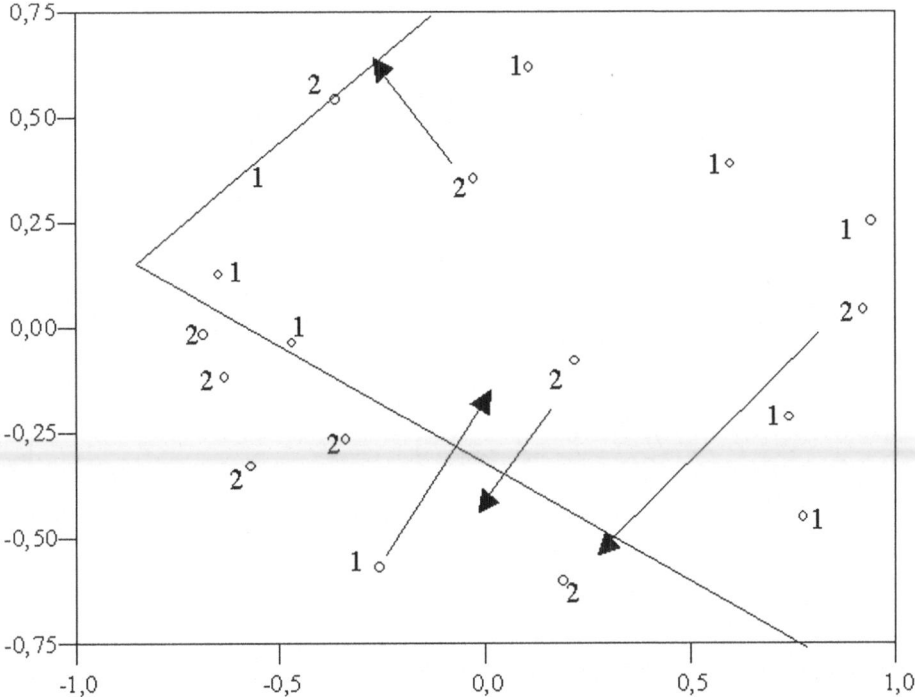

Fig. 9.13 MDS solution for all respondents, showing facet (b) superiority of the attacker with labels *1*: the attacker is superior and *2*: the attacker is not superior

q2
A man watches as another man steals his bicycle and rides off with it. He chases the thief and brings him down.

The wording he "rides off with it" could have given the target the impression that the bicycle thief was superior to the owner because of the distance he had gained in the meantime. This could justify the deviating partitioning determined.

In the following case, too, it cannot be ruled out that, contrary to the original expectations, the classification came about because of a different understanding of the situation by the target persons. The case was described thus:

q10

A holiday home has already been broken into 13 times. The owner then sets up a radio with detonating powder and signs warning of bombs. During another break-in, the radio explodes and the burglar loses his hand.

It was expected that this was *not a* superior attacker. However, the impression of the attacker's superiority could have been created in the target persons due to the 13 preceding (and presumably successful) burglaries. If one follows these considerations, the theoretical concept can be maintained, but its operationalisation would have to be examined again.

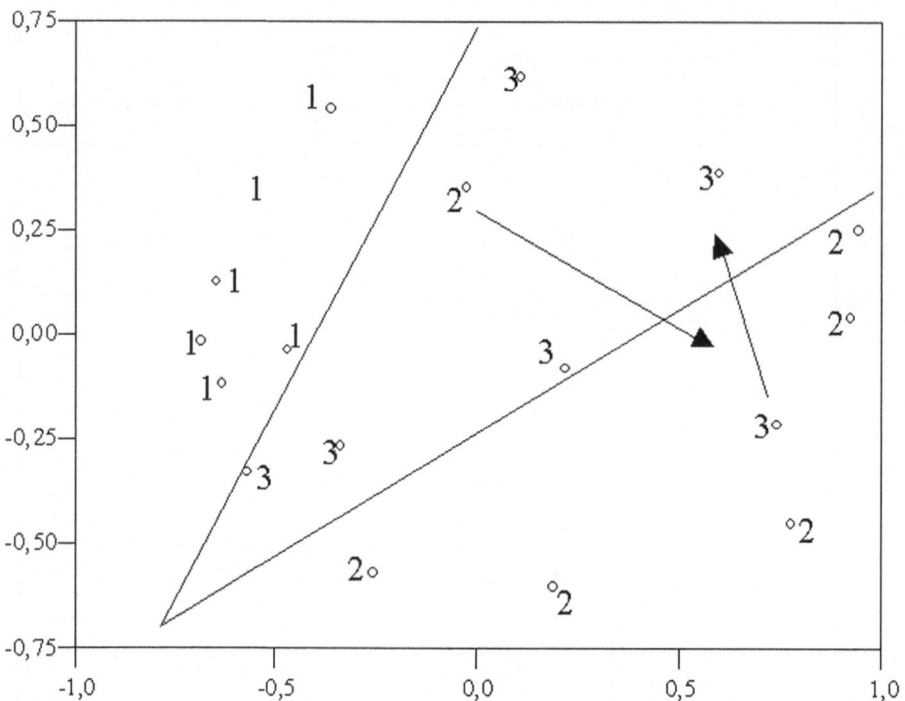

Fig. 9.14 MDS solution for all respondents, illustration of facet (c) attacked good with the designations *1*: intangible good, *2*: tangible assets and *3*: life and limb/health

The following case is also classified contrary to expectations – in the region of non-superior attackers:

q4
A farmer observes from a distance that a man is about to set fire to his barn, in which the harvest and valuable machines are stored. Since the man does not react to the farmer's shouts and the farmer cannot run to the barn fast enough, he shoots at the man and injures him seriously.

The presumption was that it was a superior attacker because of the distance and because the attack was already relatively advanced. Here too, however, the wording appears to have been interpreted differently. Thus, it may not be obvious to the respondents why the unarmed attacker should be superior to the armed farmer.

Three of the four partitioning errors encountered are thus likely to have arisen from a divergent understanding of the situation of the attack. Accordingly, the theoretical starting assumption is upheld.

For the other two facets, it is relatively easy to partition the cases in analogy to the theoretical concept developed. As shown, an allocation was only not possible for two cases in each case.

It is also obvious here that the target persons had a different understanding of the case presented to them. Thus, instead of – as originally assumed – an attack on physical integrity, the following act was understood as an attack on material assets – intrusion into another's premises.

q11
A landlord has been arguing with a group of youths for some time. When they invade his pub and attack him, he falls backwards and finally fights back with the help of an iron bar, seriously injuring two attackers.

This means that the explanation for the misclassifications can be found above all in deficiencies in the empirical implementation of the concept. The theoretical concept should be retained, but the operationalisation should be reviewed. Table 9.15 provides a further overview of the classifications made.

With Borg (2000, p. 10), it should be noted that these regionalizations are not trivial. To this end, he poses the following thought experiment. Take 18 ping-pong balls. Each one is labelled with the characteristics of the respective facets. Each ball then bears three labels. In our example, for example, "intangible good", "the attacker is superior" and "slight injury". Then throw the balls in the air and finally try to partition the recovered balls into regions, as was done here. This is not generally possible, certainly not with such simple boundary lines as the straight lines and circles shown in Figs. 9.12, 9.13 and 9.14.

Table 9.15 Target/actual comparison of the partitionings, empty cells indicate correct grouping

Case	(a) Property damaged in the course of the defence	(b) Superiority of the attacker	(c) Damaged goods
q1			
q2		Instead of 2 in 1	
q2a			
q3			
q3a		Instead of 2 in 1	
q4		Instead of 1 in 2	
q5			
q6			
q6a			
q7			
q7a	Instead of 3 in 2		
q8			
q9	Instead of 2 in 1		
q9a			
q10		Instead of 2 in 1	Instead of 2 in 3
q11			Instead of 3 in 2
q12			
q12a			

Basis: all respondents

9.5   Qualitative Analyses

One concern of qualitatively oriented social research is to do justice to the uniqueness of each person studied or each case, as far as possible, and in so doing to work out their typology and ultimately also to recognise structural regularities (Friebertshäuser and Prengel 1996, p. 148, cf. Sect. 3.8). These premises determine the evaluation of qualitative studies. As shown above, the results of qualitative interviews are usually transcribed and are thus available for further evaluation. The qualitative approach also mostly follows certain rules in the evaluation. The principle of openness and that of sequentiality are to be mentioned here above all.

Practicing this, the first step is to break down the qualitative interview or the transcript into temporal sequences. This procedure can be well illustrated by a biographical narrative in which a target person has described his or her life. This is followed, for example, by a search for aspects of the narrative that have not been addressed. It is important that neither prior knowledge nor theories are used in this context.

Reference has already been made elsewhere to ethnomethodological conversation analysis (cf. Sect. 6.1) and objective hermeneutics (cf. Sect. 6.3). Bortz and Döring (2002, p. 331 ff.) highlight the following four evaluation techniques: global evaluation, qualitative content analysis, grounded theory and linguistic evaluation methods.

- The global evaluation is based on Legewie (1994). It comprises the following ten steps: First, orientation (skimming the text), second, activating contextual knowledge (the genesis and antecedents of the text are to be explored), third, working through the text (ideas and questions are noted and relevant passages are marked), fourth, elaboration of ideas (interesting ideas are noted on index cards), fifth, creating an index (three to five notes per page are recommended), sixth, the summary (the most important contents are arranged), seventh, the evaluation of the text (opinions are drawn up on credibility, comprehensibility, distribution of roles, gaps, ambiguities in the text and so on), eighth, the collection of evaluation keywords (the relevance to the research question is to be assessed), ninth, the evaluation of the consequences for further work (which analyses still need to follow?) and finally, tenthly, a presentation of the results (summary of the text, evaluative statement, index, further evaluation plans). One advantage of the global evaluation is that it can be mastered in a relatively short time.
- Qualitative content analysis according to Mayring (1989, 1993), on the other hand, is more elaborate. Here, a distinction is made between three forms of content analysis: firstly, summary content analysis (a reduction of the source text to an abridged version is carried out), secondly, explicative analysis (additional material is used for explanations, additional information on ambiguities is obtained in this way) and thirdly, structuring analysis (this involves the creation of a category scheme and its refinement).
- Grounded theory according to Glaser and Strauss (1967) and Strauss (1987, 1994), respectively, is a technique for developing and testing theories that is anchored in the data. It provides for an unbiased, open coding of the text. This involves going through

the text line by line several times. This results in a coding protocol. In such a protocol it is noted which text features or indicators refer to which latent facts or constructs. This is followed by axial coding, in which the constructs are further elaborated and linked together. This in turn leads to memos, the basic building blocks for a theory. For grounded theory, what is known as theoretical sampling is used. This means that particularly interesting cases – these can also be contrasting cases – are selected for the research on the basis of individual considerations.

- Linguistic evaluation methods include text analysis (Brinker 1997) and conversation analysis (Brinker and Sager 1996). While the former looks at the linguistic design of a text (for example, e-mail communication in everyday life, the wording of an interview or the entries in a diary), conversation analysis is devoted to dialogue. The latter thus go beyond the design of the text.

In addition to these approaches, there are other efforts to establish evaluation techniques. First of all, *photography and film analysis* should be mentioned here.

9.5.1 Photography and Film Analysis

According to Petermann (2001, p. 230), this qualitative type of analysis builds on the still unadulterated positivist views. According to this, it is not the finished photograph that is to be evaluated, but the "process of taking the photograph, which is a subject/object relationship, as a *method of social interaction*" (italics as in the original). Extremely helpful guidance for this, as well as numerous references to further reading, can be found in Tuma and Schnettler (2019); in Akremi (2019b) and in Georg Peez (2003). They are reproduced here – very abbreviated.

Rules for photo-analysis (compare Haupert 1994; Haupert and Schäfer 1991) are first understood as questions guiding interpretation, the aim of which is to direct the attention of the interpreter. They are:

- What is the actual theme of the interpretation?
- Photographs have more content that is suitable for analysis in principle than can be realized in a single examination. Thus, the focus of the interpretation has to be determined precisely in a first step.
- What are the first impressions? (pictorial experience)
- This concerns the context of the image (the place, the time, the respective weather and so on) and the subjective associations of the viewer. What stands out first points to certain relevancies. It is also important whether there is a consensus among different interpreters on these impressions.
- What do the performers feel when they recreate the people in the picture themselves? (Physicality)

- The body language of the persons depicted is not only to be deciphered by looking at them, but the interpreters are to feel them through their own bodies and then interpret them. Again, it is of interest whether different interpreters agree in these sensations.
- Determination of the image genre
- Is it a posed picture or a snapshot? Wedding or communion photos, holiday and friendship pictures are often posed. Can strategies of self-portrayal of the depicted persons be used in the interpretation of the pictures?
- How can the image be described verbally?
- The picture elements and contents are to be verbalized in this question. After all, every photo interpretation makes use of language.
- Is there something that was expected in the picture but is missing?
- What reasons could there be for such an absence? What specific meaning, if any, is expressed in this?
- How is the picture structured? What is central? (optical weighting)
- Can it be assumed from the photo that the photographer has worked with certain stylistic and emphasis devices? This requires photographers who are trained accordingly. What effect was the photographer trying to achieve?
- What do the people depicted look like and what is their relationship to each other?
- How do the characters want to portray themselves, how do they portray the interpreters, and what is the relationship between these levels of meaning?
- According to Peez (2003), these are the core questions of a qualitative photo analysis that follows the methodology of objective hermeneutics. They are also directed at the factual reality depicted.
- What contextual information can interpreters use in their analysis?
- Knowledge about the context of a photograph should not be used to exclude possible readings. However, contextual information should be used if it can generate new readings.
- What is the intention of the photographer? Does it express the personal relationship to the object?
- Does the photographer follow certain clichés in the composition of the picture as well as in the chosen perspective and what is then expressed in it?
- What is the purpose and who are the addressees of the image? (Interaction)
- If the purpose and addressees are unknown, the interpreters can attempt to (re)construct plausible contexts by means of thought experiments.
- Can the interpretations obtained be combined into a global characteristic? (Case structure)
- This heuristic question is about the relation of subjective-intentional and objective-latent structure of meaning.
- In the case of competing interpretations, it should be decided which is the most probable
- Calculations should be made as to which of the possible readings of the interpreters seem more or less plausible.
- What technique was used to take the picture?

- The technical side (for example telephoto or wide-angle lens) of the respective shot must also be found out and possibly understood as a stylistic device.
- In order not to miss anything, one should finally analyze the recording systematically-schematically: from top to bottom, from left to right.
- What else then strikes the viewer?

Further references to this method can be found in: Beck (2003); Mietzner and Pilarczyk (2003); Peez (2003); Sontag (1978) and Tenorth and Lüders (1994).

9.5.2 Body Language Analysis

Since the 1950s, the analysis of body language has been understood as a sub-field of social psychology and defined here as a "study of speaking movement" (Reutler 1988, 1993 compare Wallbott 2001). The aim is to scientifically interpret the indicator function of such movements as those of the head, the eyes of the hands and the feet, and finally also of the entire body.

The analysis of body language is currently mostly carried out in the context of recommendations for speech performances or job interviews as well as for the examination of the effect of salespersons on their customers. But also the body-language handling of complaints, the role of this non-verbal communication in disputes or in conflicts have been addressed so far. Nevertheless, the following remains to be said: "In empirical social science, word and writing are recognised as objects of investigation. Body language, on the other hand, has to date been in a special scientific position" (Dieball 2004, p. 2).[6]

The roots of this approach go back to Darwin (1872), who based his evolutionary theory on "patterns of expression and physiological changes" (Wallbott 2001, p. 233). However, Darwin was mainly interested in movements of the mind in humans and animals.

This also makes it necessary to transcribe body language – for example in the course of a group discussion – with the help of an appropriate set of rules.

As an example, an interpretation guide[7] will now be shown again, to be applied especially when considering referents:

- The smile
- People who laugh have a positive effect on their environment. However, a distinction must be made between real and fake smiles: The real smile is usually accompanied by raised cheeks, small skin thickening and wrinkles under the eyes (crow's feet) and

---

[6] Compare: http://miami.uni-muenster.de/servlets/DerivateServlet/Derivate-2011/03_einleitung_diss_dieball.pdf (last accessed Feb. 17, 2014; but now unfortunately no longer available).

[7] Source: http://www.gilthserano.de/personality/body/0001.html, last accessed 04/21/2009, sorry, no longer available.

lowering of the eyebrows. In a fake smile, the muscles around the eyes are not active. Often fake smile abruptly breaks or gradually disappears from the face.

- The mouth
- The pout is designed to make the other person feel guilty. A struggle for dominance within a relationship is taking place here. Just like the crooked smile, a crooked mouth does not seem very credible. If even only one corner of the mouth is raised, this facial expression always signals cynicism, arrogance or a feeling of superiority.
- The eyes
- The most important signals are sent by the eyes. A friendly open look puts the listener in a positive mood. If you don't give the other person a look, you are seen as arrogant and arrogant. But who fixes too long and too intensively his counterpart, is quickly seen as threatening and belligerent.
- The height of the hands
- For the effect of gestures, the height at which the hands are located is decisive. All gestures that take place below the waist are evaluated as negative statements; gestures at the height of the waist are evaluated as neutral and above as positive.
- The visibility of the hands
- In addition, the visibility of the hands is an important criterion. Hidden hands – in trouser pockets or behind the back – are perceived as negative.
- Hand-neck gestures
- In principle, gestures above the waist are considered positive. However, the so-called hand-neck gestures are excluded from this. The hand often reaches for the neck when it actually wants to touch the face, the nose or the mouth. Both are extremely negative gestures.

However, numerous misunderstandings should also be pointed out. For example, it is mostly a matter of probabilistic coding, i.e. that a certain element of body language only indicates a certain meaning with a certain probability. The assumption that character traits are expressed in expressive behaviour is also no longer tenable in this oversimplified form. Thus, there is obviously still a great deal of research to be done at this point as well.

Documentation of Empirical Projects

10

Abstract

If, in the empirical treatment of a problem, it turns out that the H_1-hypothesis can be maintained, this is usually done with a certain probability of error. Finally, (repeated) replication of this initially one-time study will reveal whether the finding there occurs again, thus further strengthening the H_1-hypothesis. Laws and theories can then be developed in this way. Replication of empirical findings is thus a prerequisite for being able to work scientifically. A prerequisite for replication is now the documentation of the empirical project. The following section is devoted to this important step.

10.1 The Quality Criteria of Survey Research and the Handling of Empirical Data

Friedrichs (2019, p. 291) suggests the following standard structure for the presentation of results of quantitative empirical studies, which he demonstrates with an example: (1) problem, (2) theory, (3) hypotheses, (4) methods and samples, and (5) summary/conclusions. In the following, the main focus will be on the presentation of the quality criteria of empirical studies.

One of the principles of empirical research is that the results of such a study are also determined to a large extent by the method used. It has already been shown in various places how seemingly insignificant changes in the design of a survey can lead to very significant shifts in the marginal distributions of the responses. For example, the experiment by Schwarz and Bless (1992) revealed sequence effects in the answers to the question about the general attitude towards the CDU depending on the respective preliminary question about Richard von Weizsäcker. An experiment by Schwarz et al. (1985), which asked about television viewing time, also yielded interesting findings. Here, the answer to the question was clearly influenced by the scales presented to the target subjects. The same methodological effect was found in a study by the Allensbach Institute for Public Opinion Research (cf. Petersen 2002, p. 205), which involved differently designed list templates for a question about success in one's own life. It is also worth mentioning the study from the USA in which the word "forbid" was replaced by the expression "not allowed", which also resulted in different answer distributions. Finally, it is worth recalling Noelle-Neumann (1996), who used a balanced version ("Should all employees be members of the union or should this be left up to each individual?") and a non-balanced version ("Should all employees be members of the union?") for a question on union membership and in this way arrived at quite different results.

Since such effects cannot be prevented, it is necessary to precisely document the respective methodological procedure, e.g. the design of the questionnaire. The maxim here must be that a project is presented in such a way that it can be replicated on the basis of the information on the procedure. Since it is impossible to prevent the instruments from influencing the results, the procedure adopted must be documented in order to make such influences comprehensible. The following aspects, which can influence the results, must be disclosed in such documentation:

1. the client of an investigation,
2. the aim of the investigation,
3. the questionnaire or the survey standard used in each case,
4. the population to which the study was directed,
5. the sampling design, insofar as no total survey took place,
6. the procedure for field work,
7. the steps taken during data cleansing,
8. the way in which the findings are presented.

The client or the institution financing the survey can modify the answers to questions in a certain sense. This phenomenon, also referred to as sponsorship, assumes that the target persons are consciously- or unconsciously – influenced in their answers by the organiser of the survey. Thus, it could lead to different answers from the target persons depending on whether the interviewers ask their questions on behalf of the government or the opposition, whether they ask on behalf of scientific social research or on behalf of a well-known tabloid newspaper.

Studies (cf. Koch 2002) evaluating non-response have shown that the institute commissioned with the fieldwork can also have an influence on the results of the survey. If it is an institution organised in the ADM or an institute that is still relatively new on the market, if the survey was carried out under the auspices of a university department, if it was an exclusive study or a multi-topic survey. All aspects can influence the results and should therefore be documented. But also the understanding of what nonresponse is in the first place and which failures are to be understood as systematic nonresponse can vary from institute to institute.

The same applies to the goals pursued with the investigation. It is conceivable that science, an interest in market research, journalism or simply curiosity are the triggers for the empirical investigation.

A particularly sensitive element of the experimental design is the survey standard, usually the questionnaire. This should be made accessible to the consumer of the results of a study. Only in this way can sequence effects, the halo effect (the position of the questions in the questionnaire), a possible radiation of questions on the following ones, possibly a preference for the middle category (was there such a category at all?) and so on be detected. The possible use of aids such as envelopes, template sheets and the like – for example, to counter the tendency towards socially desirable answers or to better visualise the answer scale – should also be described.

It is also important to ask whether it is possible to name sources from which specific questions have been borrowed for a survey. Are these indicators based on the operationalisation of certain theories? Are they new developments and if so, what was the concrete approach to this instrument development?

The population covered by the survey shall be defined. The sampling frame used for a selection, such as the telephone directory, the registers of the population registration office, a personnel register of a company, shall also be indicated.

The sample design must be presented. Is it a probability selection (random selection), a deliberate selection (for example, using the quota method), or an arbitrary selection in which sampling was not controlled. What type of probability selection was used? Has stratification of the sample been undertaken. How frequently have the target persons been contacted.

The fieldwork itself also needs to be described. As an example, reference should again be made to the field report of the Dresden self-defence study (cf. Table 5.8). Here, information on the failures, for example, can provide important indications of the quality of the data.

The data cleaning procedure should be described. What changes were made to the raw data matrix, how were missing values handled, how were invalid codes dealt with, were plausibility checks used and what was the result of these checks?

Finally, when presenting results, care must be taken to ensure that it is always possible to replicate the calculations. If possible, the user of data must be able to obtain clarity on the following questions: Was weighting used and if so, how were the weighting factors determined. Was indexing performed and according to which criteria? How were missing values dealt with? If the calculation of a sampling error is possible, the confidence interval can be determined. Finally, it should be stated whether and where the data file is accessible for reanalysis.

The following list shows the high standard of the American Association for Public Opinion Research (AAPOR) for the documentation of surveys. Adherence to the standards of AAPOR should also be the goal for social science studies in Germany.

Codes of Ethics of the American Association for Public Opinion Research: AAPOR from 1991 (Kaase 1999, p. 137 f)

It must be documented:

- the person commissioning an investigation and those carrying out the investigation,
- the purpose of the study, including specific objectives,
- the questionnaire and/or the exact wording of all questions, including all visual templates, as well as the wording of all instructions and explanations given to interviewers and respondents that could be expected to influence the answers,
- the definition of the population for which the survey is intended to be representative and a description of the sampling frame used to identify this population (including its sources and possible biases),
- a description of the sampling design, including the lump size, the number of contact trials, information on the selection criteria and screening procedures, the method of determining the sample elements, the mode of data collection and any other essential information,
- a description of the target selection, making clear how respondents were identified by the researchers or whether respondents were recruited entirely through self-selection, and other details of the sampling, in sufficient detail to allow adequate replication,
- The sample size and results of sample realization including a complete enumeration of all cases: Number of people contacted, addresses not used, people not reached, refusers, dropouts, neutral dropouts, and complete interviews/questionnaires,
- where applicable, documentation and full description of any response rates quoted in reports (for quota samples, the number of refusals) and (whenever available) information on differences between participants and non-participants,
- a description of the formation of all special values, data cleaning, data weighting or index formation,
- a presentation of the precision of the results, including, where appropriate, estimates of sampling error and references to other sources of error, so as not to mislead as to accuracy and precision, and a description of any weighting and estimation procedures used,
- a description of all percentage values that serve as a basis for conclusions,
- a clear delineation of results based only on parts of the sample, not on the whole sample,
- Method(s), location(s) and date(s) of interviews, fieldwork or data collection,
- Interview characteristics,
- Copies of interviewer instructions or manuals, field inspection results, code books, and other important working papers,
- any other relevant information that a lay person would need to make a reasonable assessment of the reported results.

Binding rules for dealing with poll results exist in Germany above all for election forecasts. Driven by the fear of influencing voters on the basis of predictions, these may only be published on election day after voting has ended (cf. critically Donsbach 2001). It is to be hoped that appropriate regulations will also be established to protect the consumer of polling data – similar to the description of the contents of a fruit yoghurt in Germany.

At this point it should have become clear (once again) that social science surveys have a high scientific value. In addition, the figures obtained also have a certain appeal, which is used by the media for entertainment purposes, for example. Thus, the findings of survey research – between elections – not least reveal the democratic will of the population. The interest of the media in the USA in the results of polls is shown in Table 10.1.

In the following (compare above all Example I), some examples of how the media deal with the results of empirical surveys are shown.

Examples (I) for Dealing with Empirical Survey Results in the Media

Focus': Railways' Reputation at New Low According to Survey

MUNICH (dpa-AFX) – According to a report in the news magazine "Focus", the reputation of the railways has reached a new low. Citing a survey by Deutsche Bahn, "Focus" reported that more than half of 100,000 respondents in the fourth quarter of 2000 had given Deutsche Bahn grades between four and six. The average grade at the end of 2000 was 3.6. In 1997, it had been 2.8. According to "Focus", the opinion research institute Infas regularly carries out surveys for the Railway Board of Management for internal information purposes.

According to Infas, a record 17% of respondents in the latest survey criticised the deterioration of long-distance transport. More people than ever before criticised the unpunctuality of trains. Most of the respondents criticized fares that were too high. Railway spokesman Dirk Große-Leege explained that with the clearing of the tariff jungle, a renewed fleet of cars and the elimination of construction sites, everything would be better again./ys/DP/rh/js 01.04.2001, 13:55.

Survey: Germans' Confidence in the Euro Continues to Wane

ALLENSBACH (dpa-AFX) – The euro, which will replace the German mark next year, is increasingly losing prestige among Germans. Only 20% still have great or very great confidence in the European currency, the Institute for Public Opinion Research in Allensbach reported on Saturday. Two years ago the figure was 37%.

In the Allensbach survey completed in March, 26% of Germans expressed no or little confidence in the euro (January 1999: 17%). The number of those who have "less confidence" in the European currency rose from 33% to 47% in this period. The rest were undecided.

Table 10.1 Number of reports in the USA in election years in which the phrases "polls say" or "polls have shown" appeared

2004	2000	1996	1992
11,327	11,890	7984	4489

Evaluated were: U.S. Newpapers, Magazines, various TV as well as CNN on election day
Compare Frankovic (2005)

60% of those surveyed now see the introduction of the euro as not a good decision. At 69%, the number of sceptics in the east is significantly higher than in the west at 58%. Only 30% of East Germans and 40% of West Germans believe that the benefits of the euro will outweigh the disadvantages in the long term.

After surveying 2094 people, the Lake Constance demoscopists drew the following conclusion: "Regardless of which side of the coin the subject of the euro is now examined with demoscopic questions – the majority of the population only answers negatively."/DP/as/jb 25.03.2001, 13:30.

Germans Know Much Less About the Euro than Expected

BERLIN (dpa-AFX) – Germans clearly know less about the euro than they think – and the majority of them do not yet trust the new currency. These are two results of an EMNID study published in Berlin on Friday on behalf of the German Savings Banks and Giro Association (DSGV). Subjectively, the Germans are of the opinion that they are well aware of the upcoming cash introduction, explained the DSGV; objectively, however, only a small part of the population is well informed about it. According to the study, only 38% of the 1005 respondents expressed confidence in the single currency, which will finally replace the D-Mark at the turn of the year. 59% answered in the negative to the question whether they had confidence in the euro. Among those with a high school diploma, the majority (59%) had confidence in the single currency. According to the data provided by the public financial institutions, 53% of those questioned thought they knew the exact date from which the euro banknotes would be issued. Only just under a quarter, however, gave the correct date of 1 January 2002, from which notes are to be available at least at cash dispensers throughout the euro zone. That the issue of the coins in Germany already on 17 December this year with the sales of so-called household mixtures for 20 Marks (10,23 EUR) in the banks begins, knew accordingly even only 5%. Information and clearing-up must be clearly strengthened, demanded the industry federation. According to the survey, as many as 51% of those questioned know that they will initially still be able to issue coins and notes in D-marks in Germany after the turn of the year; the deadline for this is 28 February 2002. In the survey, 42% of those questioned also said that they wanted to take part in a special exchange campaign run by the banks this May. With this campaign, the banks want to spread out the reflux of billions of coins and notes over a longer period of time./pin/FP/js 23.03.2001, 16:07.

The three examples were randomly selected from the Internet. Nevertheless, some of the texts contain methodological references in addition to content-related findings, such as the sample size, the institution that organised the survey, the survey period, the commissioner of the study and – in one case – even a trend design used. It is quite obvious, however, that the authors are primarily interested in reporting interesting results of the respective survey. Readers are assumed – probably rightly – to be interested in such figures. After all, they also assume a certain methodological understanding on the part of the recipient and do not completely disregard corresponding references. The situation is somewhat different in the action described in Example II.

Example (II) of How to Deal with Survey Results in the Media

Saxon Newspaper Monday, 28 June 2004

Debt wall against Waldschlösschen Bridge, citizens' initiative protests/Regular response to SZ action, By Thilo Alexe.

The planned Waldschlösschen Bridge continues to heat up tempers. Opponents of construction from the citizens' initiative Waldschlösschen want to erect a symbolic wall of debt in front of the city hall during the city council meeting on Thursday to illustrate Dresden's drastic financial situation.

"The city's current debts amount to around 850 million euros," the initiative emphasizes. For the construction, the municipality will have to inject money despite state funding – probably in the double-digit millions. "The city would have to take on additional debt for this dubious project and would thus block the future," criticizes the initiative. Also the action of the SZ, which had asked the readers to vote on the bridge, triggers a lively response. On Saturday alone, more than **1000 voting coupons** reached the editorial office. This means that around **3000 Dresdeners** have taken part in **the campaign** so far.

The tenor is still clear: 2118 readers are for the construction of the bridge, 861 against. The vote runs until Wednesday. Only **the coupons from** the **SZ of** 23 June are valid.

Meanwhile, Dresden political scientist Dietrich Herrmann, once a Roßberg supporter during the election campaign, is calling for a debate on the city's priorities. Whoever wants a bridge must also know that there is less money available for other tasks.

From the point of view of empirical social research, criticism must be levelled at various aspects of this action (after the presentation of the procedure has been positively highlighted). Assuming that a certain claim to credibility and scientificity is associated with this survey, the procedure contradicts various methodological standards:

- Participation in the poll takes place via self-selection. It can be assumed that participation in the survey is more likely if the persons concerned are particularly affected.
- The population about which a statement is to be made is hardly definable. It could be the buyers or the subscribers of the newspaper on a certain day. (Does the real estate market also appear in this newspaper on this day of the week, so that certain reader groups could have a special interest in this issue?) However, only one coupon entitling to vote was apparently issued per newspaper. But it is possible that several people living in the same household read the same newspaper. If so, not all readers are eligible to vote. Should a generalization be made to the citizens of Dresden, the social profile of the readers of this newspaper would have to be scrutinized more closely.
- A figure of 3000 participants alone says nothing about the quality of the sample. Without knowing the exact number of copies sold on that day, it is also impossible to say anything about the response rate. If we assume that about 30,000 copies were sold, we would be dealing with a participation rate of only about 10%.
- The coupon system used appears suitable to ensure the casting of only one vote per person: It is probably fair to assume that a number of people did not purchase multiple

newspapers that day in order to manipulate the outcome. However, the prevention of multiple voting is only one methodological aspect of many.

In summary, the relatively good methodological documentation of the action has made it possible to assess its scientific seriousness and to identify any weaknesses. From the strict point of view of empirical social research, the design is not tenable. The alternative here would have been a random selection of the target persons, for example via the population register or an equally randomly drawn telephone sample.

10.2   Method Reports

Certain research programmes such as the ALLBUS or the ESS (cf. Sect. 4.8) serve the purpose of collecting data for empirical social research and making them generally accessible to users as promptly as possible. The use of such data sets in secondary analyses and, above all, for teaching purposes requires that each phase of the research process be made as transparent as possible. Therefore, in order for users to be able to understand the process of data collection and critically engage with the data obtained, the design and conduct of the study must also be documented in detail in the relevant method reports. The following example quotes the table of contents of the method report of the ALLBUS study of 2014.

The Content of the Methodology Report for ALLBUS 2014
(compare Wasmer et al. 2017, p. 5 f.)

1. Introduction
2. The Basic Concept of the ALLBUS and ISSP Studies
    2.1. The Basic Concept of the ALLBUS
    2.2. The Basic Concept of the ISSP
    2.3. Overview of the Methodological-Technical Characteristics of the ALLBUS Studies
3. The Questionnaire of the ALLBUS 2014
    3.1. General Overview
    3.2. The Focus Topic "Social Inequality and the Welfare State"
        3.2.1. Social Mobility
        3.2.2. Position in the Inequality Structure
        3.2.3. Attitudes Towards Social Inequality
    3.3. Additional Focus on Health
    3.4. Additional Focus on Lifestyles and Cultural Capital
    3.5. Other Content Replication Issues
    3.6. Demographic Information in ALLBUS 2014
    3.7. Other Variables of the ALLBUS 2014
        3.7.1. Derived Variables

Certainly, not every empirical study will be able to provide such detailed documentation. At this point, however, it should be shown which information is in principle relevant for a replication. For the design of method and field reports for telephone surveys, see Gramlich and Häder (2019).

References

ADM. Arbeitskreis Deutscher Markt- und Sozialforschungsinstitute e. V. (Hrsg.). 1999. *Stichproben-Verfahren in der Umfrageforschung. Eine Darstellung für die Praxis.* Opladen: Leske + Budrich.

ADM. Arbeitskreis Deutscher Markt- und Sozialforschungsinstitute e. V. 2002. *Jahresbericht 2001.* http://www.adm-ev.de/.

Aichholzer, Georg. 2000. Innovative Elemente des österreichischen Technologie-Delphi. In *Die Delphi-Technik in den Sozialwissenschaften – Methodische Forschungen und innovative Anwendungen*, Hrsg. Michael Häder, Sabine Häder, 67–94. Opladen: Westdeutscher Verlag.

Ajzen, Icek. 1985. From intension to action: A theory of planned behavior. In *Action-control: From cognition to behavior*, Hrsg. Julius Kuhl, Jürgen Beckmann, Berlin, Heidelberg: Springer.

Ajzen, Icek. 1988. *Attitudes, Personality and Behaviour.* Milton Keynes: Open University Press.

Ajzen, Icek. 1991. The theory of planned behavior. *Organizational Behavior and Human Decision Processes* 50(2): 179–211.

Ajzen, Icek, und Martin Fishbein. 1980. *Understanding attitudes and predicting social behavior.* Englewood-Cliffs, NJ: Prentice-Hall.

Akremi, Leila. 2019a. Stichprobenziehung in der qualitativen Forschung. In *Handbuch der Methoden der empirischen Sozialforschung,* 2. Auflage, Bd. 1, Hrsg. Nina Baur, Jörg Blasius, 313–331, Wiesbaden: Springer VS.

Akremi, Leila. 2019b. Filme. In *Handbuch der Methoden der empirischen Sozialforschung,* 2. Auflage, Bd. 2, Hrsg. Nina Baur, Jörg Blasius, 1203–1214, Wiesbaden: Springer VS.

Albach, Horst. 1970. Informationsgewinnung durch strukturierte Gruppenbefragung. *Zeitschrift für Betriebswirtschaft* 40: 11–26.

Albers, Ines. 1997. Einwohnermelderegister-Stichproben in der Praxis. Ein Erfahrungsbericht. In *Stichproben in der Umfragepraxis*, Hrsg. Siegfried Gabler, Jürgen H.P. Hoffmeyer-Zlotnik, 117–126. Opladen: Westdeutscher Verlag.

Albert, Hans. 1956. Das Werturteilsproblem im Lichte der logischen Analyse. *Zeitschrift für die gesamte Staatswissenschaft* 112: 410–439.

Albert, Hans. 1960. Wissenschaft und Politik. Zum Problem der Anwendbarkeit einer wertfreien Wissenschaft. In *Probleme der Wissenschaftstheorie. Festschrift für Viktor Kraft*, Hrsg. Ernst Topitsch, 201–232. Wien: Springer.

Albert, Hans. 1964. Probleme der Theorienbildung. Entwicklung, Struktur und Anwendungen sozialwissenschaftlicher Theorien. In *Theorie und Realität. Ausgewählte Aufsätze zur Wissenschaftslehre der Sozialwissenschaften*, Hrsg. Hans Albert Tübingen: Mohr.

Albert, Hans. 1965. Wertfreiheit als methodisches Prinzip. Zur Frage der Notwendigkeit einer normativen Sozialwissenschaft. In *Probleme der normativen Ökonomik und der wirtschaftlichen
Beratung*, Hrsg. Erwin von Beckerath, Herbert Giersch, Berlin: Duncker & Humbolt. sowie in:
Topitsch, Ernst (Hrsg.): Logik der Sozialwissenschaften. Köln, Berlin: Kiepenheuer & Witsch,
183–210.

von Alemann, Heine, und Peter Ortlieb. 1975. Die Einzelfallstudie. In *Techniken der empirischen
Sozialforschung*, Bd. 2, Hrsg. Jürgen von Koolwijk, Maria Wieken-Mayser, 157–177. München:
Oldenbourg.

Althoff, Stefan. 1997. Quoten-Auswahlverfahren – Warum nicht?. In *Stichproben in der
Umfragepraxis*, Hrsg. Siegfried Gabler, Jürgen H.P. Hoffmeyer-Zlotnik, 19–32. Opladen:
Westdeutscher Verlag.

Amelung, Knut. 2002. Der frühe Luhmann und das Gesellschaftsbild bundesrepublikanischer
Juristen. Ein Beitrag zur deutschen Rechtsgeschichte im 20. Jahrhundert. In *Festschrift für Klaus
Lüderssen. Zum 70. Geburtstag am 2. Mai 2002*, Hrsg. Cornelius Prittwitz, Michael Baurmann,
Klaus Günther, Lothar Kuhlen, Reinhard Merkel, Cornelius Nestler, Lorenz Schulz, 7–16.
Baden-Baden: Nomos Verlagsgesellschaft.

Amelung, Knut. 2003. Sein und Schein bei der Notwehr gegen die Drohung mit einer Scheinwaffe.
Jura 1(2): 91–97.

Amelung, Knut, und Michael Häder. 1999. *Vorstellungen über und potentielle Verhaltensintentionen
bei Notwehr in der Allgemeinbevölkerung der Bundesrepublik Deutschland*. Antrag an die
VW-Stiftung im Rahmen des Förderschwerpunkts Recht und Verhalten: Entstehung, Wirkung
und Fortentwicklung von Recht im Kontext menschlichen Verhaltens. Dresden, Mannheim,
unveröffentlichtes Manuskript.

Amelung, Knut, und Ines Kilian. 2003. Zur Akzeptanz des deutschen Notwehrrechts in der
Bevölkerung. Erste Ergebnisse der Dresdner Notwehrerhebung. In *Strafrecht, Biorecht,
Rechtsphilosophie: Festschrift für Hans-Ludwig Schreiber zum 70. Geburtstag am 10. Mai 2003*,
3–11. Heidelberg: Müller.

Andreß, Hans-Jürgen, Jacques A. Hagenaars, und Steffen Kühnel. 1997. *Analyse von Tabellen und
kategorialen Daten*. Heidelberg: Springer.

Aquilino, William S. 1993. Effects of spouse presence during the interview on survey responses
concerning marriage. *Public Opinion Quarterly* 57(3): 358–376.

Ardelt, Elisabeth. 1989. Soziogramm. In *Sozialwissenschaftliche Methoden. Lehr- und Handbuch
für Forschung und Praxis*, Hrsg. Erwin Roth, 184–195. München: R. Oldenbourg.

Aries, Elizabeth F. 1973. *Interaction patterns and themes of male, female, and mixed groups*.
PhD. Dissertation. Harvard University.

Aries, Elizabeth F. 1977. Male-female interpersonal styles in all male, female, and mixed groups. In
Beyond Sex Roles, Hrsg. Alice G. Sargent, 292–299. St. Paul, MN: West.

Atteslander, Peter unter Mitarbeit von Roland Buchheit. 1984/2003. *Methoden der empirischen
Sozialforschung*. 10. Auflage. Berlin: de Gruyter, Sammlung Göschen.

Baacke, Dieter. 1995. Pädagogik. In *Handbuch Qualitative Sozialforschung. Grundlagen, Konzepte,
Methoden und Anwendungen*, Hrsg. Uwe Flick, Ernst von Kardorff, Heiner Keupp, Lutz von
Rosenstiel, Stephan Wolff, 44–47. Weinheim: Beltz Psychologie Verlags Union.

Babbie, Earl. 2002. *The practice of social research*, 9. Auflage, Blemont, CA: Wadsworth Thomson
Learning.

Backhaus, Klaus, Bernd Erickson, Wulff Plinke, und Rolf Weiber. 2000. *Multivariate
Analysemethoden*. Berlin Heidelberg: Springer.

Bär, Mathias. 2013. *Sicherheits- und verteidigungspolitische Meinungsbildung*. Eine empirische
Anwendung des Modells der Frame-Selektion. Diplomarbeit am Institut für Soziologie der TU
Dresden.

Bales, Robert K. 1950. *Interaction process analysis. A method for the study of small groups.* Cambidge: Addison-Wesley Press.

Bandilla, Wolfgang. 2003. Die Internet-Gemeinde als Grundgesamtheit. In *Online-Erhebungen 5. Wissenschaftliche Tagung* Sozialwissenschaftlicher Tagungsband, Bd. 7, Hrsg. Statistisches Bundesamt Wiesbaden, Arbeitsgemeinschaft Sozialwissenschaftlicher Institute e. V. (ASI), 71–82. Bonn: GESIS Leibniz Institut für Sozialwissenschaften.

Bandilla, Wolfgang, und Michael Bosnjak. 2000. Perspektiven der Online-Forschung. In *Neue Erhebungsinstrumente und Methodeneffekte* Schriftenreihe Spektrum Bundesstatistik, Bd. 15, Hrsg. Statistisches Bundesamt, 166–174. Wiesbaden: Metzler Pöschel.

Bardecki, Michal J. 1984. Participants' response to the Delphi method: An attitudinal perspective. *Technological Forecasting and Social Change* 25(3): 281–292.

Bartsch, Simone. 2012. *... würden Sie mir dazu Ihre E-Mail-Adresse verraten? Internetnutzung und Nonresponse beim Aufbau eines Online Access Panels.* Baden Baden: Nomos.

Bathelt, Harald, und Katrin Griebel. 2001. *Die Struktur und Reorganisation der Zulieferer- und Dienstleisterbeziehungen des Industriepark Höchst (IPH).* Forschungsbericht. Institut für Wirtschafts- und Sozialgeographie, Johann Wolfgang Goethe-Universität Frankfurt, Working Papers 02-2001.

Baur, Nina. 2019. Quantitative Netzwerkanalysen. In *Handbuch der Methoden der empirischen Sozialforschung,* 2. Auflage, Bd. 2, Hrsg. Nina Baur, Jörg Blasius, 1281–1299, Wiesbaden: Springer VS.

Baumert, Jürgen, Eckhard Klieme, Michael Neubrand, Manfred Prenzel, Ulrich Schiefele, Wolfgang Schneider, Petra Stanat, Klaus-Jürgen Tillmann, und Manfred Weiß. 2001a. *PISA 2000. Basiskompetenzen von Schülerinnen und Schülern im internationalen Vergleich.* Opladen: Leske + Budrich.

Baumert, Jürgen, Petra Stanat, und Anke Demmrich. 2001b. PISA 2000: Untersuchungsgegenstand, theoretische Grundlagen und Durchführung der Studie. In *PISA 2000. Basiskompetenzen von Schülerinnen und Schülern im internationalen Vergleich,* Hrsg. Jürgen Baumert, Eckhard Klieme, Michael Neubrand, Manfred Prenzel, Ulrich Schiefele, Wolfgang Schneider, Petra Stanat, Klaus Jürgen Tillmann, Manfred Weiß, 15–68. Opladen: Leske + Budrich.

Bayer, Klaus. 1999. *Wiesbaden* Argument und Argumentation. Logische Grundlagen der Argumentationsanalyse, Studienbücher zur Linguistik, Bd. 1. Opladen: Westdeutscher Verlag.

Beck, Christian. 2003. Fotos wie Texte lesen. Anleitung zur sozialwissenschaftlichen Fotoanalyse. In *Film- und Fotoanalyse in der Erziehungswissenschaft. Ein Handbuch,* Hrsg. Yvonne Ehrenspeck, Burkhard Schäffer, 55–71. Opladen: Leske + Budrich.

Becker, Dirk. 1974. *Analyse der Delphi-Methode und Ansätze zu ihrer optimalen Gestaltung.* Inaugural-Dissertation zur Erlangung der Würde eines Doktors der Wirtschaftswissenschaften der Universität Mannheim.

Beham, Martina, Daniela Huter, und Vera Nowak. 1998. *Was machen Mütter, Väter und Kinder mit ihrer Zeit? Familienbezogene Auswertung der Zeitbudgeterhebung 1992, Forschungsbericht.* Johannes Keppler Universität Linz.

Behr, Dorotée, und Evi Scholz. 2011. Questionnaire Translation in Cross-National Survey Research. On the Types and Value of Annotations. *methoden – daten – analysen* 5(2): 157–179.

Belson, William A. 1981. *The design and understanding of survey questions.* Aldershot, England: Gower.

Belson, William A. 1986. *Validity in survey research.* Aldershot, England: Gower.

Benninghaus, Hans. 1998. *Einführung in die sozialwissenschaftliche Datenanalyse,* 5. Auflage, München Wien: Oldenbourg.

Bentler, Peter M. 2001. *EQS 6: Structural equations program manual.* Encino: Multivariate Software.

Berelson, Bernard. 1952. *Content analysis in communications research*. Glencoe, Ill.: Free Press.

Beretta, Lorenzo und Santaniello, Alessandro. 2016. *Nearest neighbor imputation algorithms: a critical evaluation*. In BMC Medical Informatics & Decision Making. 16 (Suppl 3): 74.

Berger, Fred. 2006. Zur Wirkung unterschiedlicher materieller Incentives in postalischen Befragungen. Ein Literaturbericht. *ZUMA-Nachrichten* 1(58): 81–100.

Berger, Horst, und Herbert F. Wolf. 1989. *Handbuch der soziologischen Forschung. Methodologie Methoden Techniken*. Berlin: Akademie Verlag.

Bergmann, Gustav. 1966. Sinn und Unsinn des Operationalismus. In *Logik der Sozialwissenschaften*, 3. Auflage, Hrsg. Ernst Topitsch, 104–112. Köln Berlin: Kiepenheuer & Witsch.

Bergmann, Jörg R. 1981. Ethnomethodologische Konversationsanalyse. In *Dialogforschung. Jahrbuch 1980 des Instituts für deutsche Sprache*, Hrsg. Peter Schröder, Hugo Steger, 9–51. Düsseldorf: Schwann.

Bergmann, Jörg R. 1991. Deskriptive Praktiken als Gegenstand und Methode der Ethnomethodologie. In *Sinn und Erfahrung: Phänomenologische Methoden in den Humanwissenschaften*, Hrsg. Max Herzog, Carl F. Graumann, 87–101. Heidelberg: Asanger.

Bergmann, Jörg R. 1993. Alarmierendes Verstehen: Kommunikation in Feuerwehrnotrufen. In *Wirklichkeit im Deutungsprozess. Verstehen und Methoden in den Kultur- und Sozialwissenschaften*, Hrsg. Thomas Jung, Stefan Müller-Dohm, 283–328. Frankfurt am Main: Surkamp.

Bergmann, Jörg R. 1995. Konversationsanalyse. In *Handbuch Qualitative Sozialforschung. Grundlagen, Konzepte, Methoden und Anwendungen*, 2. Auflage, Hrsg. Uwe Flick, Ernst von Kardorff, Heiner Keupp, Lutz von Rosenstiel, Stephan Wolff, 213–218. Weinheim: Beltz.

Bieber, Ina Elisabeth, und Evelyn Bytzek. 2012. Umfragen: Eine geeignete Erhebungsmethode für die Wahlforschung? Ein Vergleich unterschiedlicher Befragungsmodi am Beispiel der Bundestagswahl 2009. *Methoden Daten Analysen* 6(2): 185–211.

Biemer, Paul P., Deborah Herget, Jeremy Morton, und Gordon Willis. 2003. *The feasibility of monitoring field interview performance using computer audio recorded interviewing (CARI)*. In *Proceedings of the Survey Research Methods Section*, 1068–1073. Washington, DC: American Statistical Association.

Biemer, Paul P., und Lars Lyberg. 2003. *Introduction to survey quality*. New York: Wiley.

Biernacki, Patrick, und Dan Waldorf. 1981. Snowball sampling: Problems and techniques of chain referral sampling. *Sociological Methods and Research* 10(2): 141–163.

Bishop, George F. 1987. Experiments with the middle response alternative in survey questions. *Public Opinion Quarterly* 51(2): 220–232.

Bittner, Jochen. 2002. Deutschland: Wo jeder sich vor jedem fürchtet. *DIE ZEIT*, 7. November.

Bjerstedt, Ake. 1956. *Interpretations of sociometric choice status*. Gleerup & Copenhagen: Munksgaard.

Blanchflower, David G., und Andrew J. Oswald. 2002. *Well-being over time in Britain and the USA*. www.nber.org/papers/w9395.pdf.

Blasius, Jörg, und Karl-Heinz Reuband. 1996. Postalische Befragungen in der empirischen Sozialforschung. Ausschöpfungsquoten und Antwortqualität. *planung & analyse* 23(1): 35–41.

Blasius, Jörg. 2019. Skalierungsverfahren. In *Handbuch der Methoden der empirischen Sozialforschung*, 2. Auflage, Bd. 2, Hrsg. Nina Baur, Jörg Blasius, 1437–1449, Wiesbaden: Springer VS.

Blasius, Jörg, und Nina Baur. 2019. Multivariate Datenstrukturen. In *Handbuch der Methoden der empirischen Sozialforschung*, 2. Auflage, Bd. 2, Hrsg. Nina Baur, Jörg Blasius, 1379–1400, Wiesbaden: Springer VS.

Blass, Wolf. 1990. Theoretische und methodische Grundlagen der Zeitbudgetforschung. In *Zeitbudgeterhebungen*, Hrsg. Rosemarie von Schweitzer, Manfred Ehling, Dieter Schäfer, 54–75. Stuttgart: Metzler-Poeschel.

Bless, Herbert, und Norbert Schwarz. 1999. Suffcient and necessary conditions in dualprocess models. In *Dual-process theories in social psychology*, Hrsg. Shelly Chaiken, Trope Yaacov, 423–440. New York: Guilford Press.

Böcker, Franz. 1988. Scale forms and their impact on rating's reliability and validity. *Journal of Business Research* 17(1): 15–26.

Böhme, Rainer. 2003. *Fragebogeneffekte bei Online-Befragungen*. Wissenschaftliche Arbeit zur Erlangung des akademischen Grades Magister Artium im Fach Kommunikationswissenschaft an der Philosophischen Fakultät der Technischen Universität Dresden.

Bohnsack, Ralf. 2003. Dokumentarische Methode. In *Hauptbegriffe qualitativer Sozialforschung*, Hrsg. Ralf Bohnsack, Winfried Marotzki, Michael Meuser, 40–44. Opladen: Leske + Budrich.

Borg, Ingwer. 1992. *Grundlagen und Ergebnisse der Facettentheorie. Methoden der Psychologi* Bd. 13. Göttingen: Huber.

Borg, Ingwer. 2000. *Explorative multidimensionale Skalierung, ZUMA How-to- Reihe Nr.1* https://www.gesis.org/fileadmin/upload/forschung/publikationen/gesis_reihen/howto/how-to1ib.pdf.

Borg, Ingwer. 2003. *Führungsinstrument Mitarbeiterbefragung. Theorien, Tools und Praxiserfahrungen*. Göttingen Bern: Hogrefe.

Borg, Ingwer, und Thomas Staufenbiel. 1997. *Theorien und Methoden der Skalierung. Eine Einführung*. Bern Göttingen: Huber.

Borg, Ingwer, und Christian Treder. 2003. Item-Nonresponse in Mitarbeiterbefragungen. *ZUMA-Nachrichten* 2(53): 58–76.

Borg, Ingwer, Michael Braun, und Michael Häder. 1993. Arbeitswerte in Ost- und Westdeutschland: Unterschiedliche Gewichte, aber gleiche Struktur. *ZUMA-Nachrichten* 2(33): 64–82.

Bortz, Jürgen. 1999. *Statistik für Sozialwissenschaftler*. Berlin Heidelberg New York: Springer.

Bortz, Jürgen, und Nicola Döring. 2002. *Forschungsmethoden und Evaluation für Human- und Sozialwissenschaftler*, 3. Auflage, Berlin Heidelberg New York: Springer.

Bosnjak, Michael. 2001. Teilnahmeverhalten bei Web-Befragungen. Nonresponse und Selbstselektion. In *Online-Marktforschung*, Hrsg. Axel Theobald, Marcus Dreyer, Thomas Starsetzki, 79–95. Wiesbaden: Gabler.

Bosnjak, Michael. 2002. *(Non)Response bei Web-Befragungen*. Aachen: Shaker.

Braun; Michael. 2019. Interkulturell vergleichenden Umfragen. In *Handbuch der Methoden der empirischen Sozialforschung*, 2. Auflage, Bd. 2, Hrsg. Nina Baur, Jörg Blasius, 907–918, Wiesbaden: Springer VS.

Braun, Michael, und Peter Ph. Mohler. 1991. Die Allgemeine Bevölkerungsumfrage der Sozialwissenschaften (ALLBUS): Rückblick und Ausblick in die neunziger Jahre. *ZUMA-Nachrichten* 2(29): 7–25.

Bridgman, Percy W. 1927. *The logic of modern physics*. New York: Macmillan.

Brinker, Klaus. 1996. Zur Analyse der narrativen Themenentfaltung am Beispiel einer Alltagserzählung. In *Varietäten der deutschen Sprache. Festschrift für Dieter Möhn*, Hrsg. Jörg Hennig, Jürgen Meier, 279–289. Frankfurt am Main: Peter Lang.

Brinker, Klaus. 1997. *Linguistische Textanalyse. Eine Einführung in Grundbegriffe und Methoden*. Berlin: Erich Schmidt Verlag.

Brinker, Klaus, und Sven Frederik Sager. 1996. *Linguistische Gesprächsanalyse: eine Einführung*. Berlin: Erich Schmidt.

Brosi, Walter, Elisabeth M. Krekel, und Joachim Gerd Ulrich. 1999. Delphi als ein Planungsinstrument der Berufsbildungsforschung? Erste Ergebnisse einer BIBB-Studie. *Berufsbildung in Wissenschaft und Praxis* 28(6): 11–16.

Brosius, Felix. 2002. *SPSS 11*. Bonn: mitp Verlag.

Brückner, Erika. 1985. Telefonische Befragungen – Methodischer Fortschritt oder erhebungsökonomische Ersatzstrategie? In *Herausforderungen der Empirischen Sozialforschung*, Hrsg. Max Kaase, Manfred Küchler, 66–70. Mannheim: ZUMA.

Brüsemeister, Thomas. 2000. *Qualitative Forschung. Ein Überblick.* Wiesbaden: Westdeutscher Verlag.

Burawoy, Michael, Alice Burton, Ferguson Ann Arnet, Kathryn J. Fox, Joshua Gamson, Nadine Gartrell, Leslie Hurst, Charles Kurzman, Leslie Salzinger, Josepha Schiffman, und Shiori Ui. 1991. *Ethnography unbound: Power and resistance in the modern metropolis.* Berkeley: University of California Press.

Campbell, Donald Thomas, und Donald W. Fiske. 1959. Convergent und discriminant valididation by the multitrait-multimethod-matrix. *Psychological Bulletin* 56(2): 81–105.

Campbell, Donald Thomas, und Julian C. Stanley. 1963. *Experimental and quasi-experimental designs for research.* Chicago: Rand MacNally.

Cantril, Hadley. 1944. *Gauging public opinion.* Princeton: University Press.

Carr, Leslie G. 1971. The srole items and acquiescence. *American Sociological Review* 36(2): 287–293.

Cartwright, Dorwin, und Alwin Zander. 1953. *Group dynamics.* New York: Harper & Row.

Cary, Charles D. 1976. Patterns of emphasis upon Marxist-Leninist ideology. A computer content analysis of Soviet school history, geography and social science textbooks. *Comparative Education Review* 20: 11–29.

Chaiken, Shelly, und Trope Yaacov (Hrsg.). 1999. *Dual-process theories in social psychology.* New York: Guilford Press.

Chalmers, Alan F. 2001. *Wege der Wissenschaft. Einführung in die Wissenschaftstheorie.* Berlin: Springer.

Chesler, David J., Neil J. van Steenberg, und Joyce E. Brueckel. 1956. Effect on morale of infantry team replacement and individual replacement systems. In *Sociometry and the science of man*, Hrsg. Jakob L. Moreno, 331–341. New York: Beacon House.

Church, Allan H. 1993. Estimating the effect of incentives on mail survey response rates: A meta-analysis. *Public Opinion Quarterly* 57(1): 62–79.

Clark, Andrew E., und Andrew J. Oswald. 2002. A simple statistical method for measuring how life events affect happiness. *International Journal of Epidemiology* 31(6): 1139–1144.

Clauphein, Erika, Ulrich Oltersdorf, und Georg Walker. 2001. Zeit fürs Essen – Deskriptive Auswertung der Zeitbudgeterhebung. In *Zeitbudget in Deutschland – Erfahrungsberichte der Wissenschaft* Schriftenreihe Spektrum Bundesstatistik, Bd. 17, Hrsg. Manfred Ehling, Joachim Merz, u. a. 202–213. Wiesbaden: Metzler Poeschel.

Coates, Joseph F. 1975. In defense of Delphi: A review of Delphi assessment, expert opinion, forecasting and group process by H. Sackman. *Technological Forecasting and Social Change* 7(2): 193–194.

Cobben, Fannie, Barry Schouten, und Jelke Bethlehem. 2012. Weighting to Adjust for Non-observation Errors in Telephone Surveys. In *Thelephone Surveys in Europe. Research and Practice*, Hrsg. Sabine Häder, Michael Häder, Mike Kühne, 169–185. Berlin, Heidelberg: Springer.

Cranach, Mario v., und Hans-Georg Frenz. 1969. Systematische Beobachtung. In *Sozialpsychologie*, Bd. 7/1, Hrsg. Carl Friedrich Graumann, 269–331. Göttingen: Hografe.

Cuhls, Kerstin. 2000. *Wie kann ein Foresight-Prozess in Deutschland organisiert werden?* Düsseldorf: Friedrich-Ebert-Stiftung.

Cuhls, Kerstin, Sybille Breiner, und Hariolf Grupp. 1995. *Delphi-Bericht 1995 zur Entwicklung von Wissenschaft und Technik – Mini-Delphi.* Karlsruhe: Fraunhofer-Institut für Systemtechnik und Innovationsforschung.

Dahrendorf, Ralf. 1971. *Homo Sociologicus*. Opladen: Westdeutscher Verlag.

Dalkey, Norman C. 1969. *The Delphi method: An experimental study of group opinion. RAND RM 5888-PR, June.*

Dalkey, Norman C., und Olaf Helmer. 1963. An experimental application of the Delphi method to the use of experts. *Management Science* 9(3): 458–467.

Datenreport, Zahlen und Fakten über die Bundesrepublik Deutschland. Hrsg. vom Statistischen Bundesamt in Zusammenarbeit mit WZB und ZUMA und der Bundeszentrale für Politische Bildung (2002/2004), Bonn.

Dayton, und C. Mitchell. 1970. *The design of educational experiments*. New York: McGraw-Hill.

Delbecq, André L., Andrew Van de Ven, und David H. Gustafson. 1975. *Group techniques for program planning: A guide to nominal group and Delphi processes*. Glenview, Ill.: Scott, Foresman and Company.

Deppermann, Arnulf. 1999. *Gespräche analysieren. Eine Einführung in konversationsanalytische Methoden*. Opladen: Leske + Budrich.

Denzin, Norman K., und Yvonna S. Lincoln (Hrsg.). 2000. *Handbook of Qualitative Research*. Sage: Thousand Oaks.

Der Spiegel. 1994. Ohrfeige an der Haustür. 41(26): 41–46.

van Deth, Jan. 2004. *Deutschland in Europa: Ergebnisse des European Social Survey 2002–2003*. Wiesbaden: VS Verlag für Sozialwissenschaften.

Deutschmann, Marc, und Sabine Häder. 2002. Nicht-Eingetragene in CATI-Surveys. In *Telefonstichproben*, Hrsg. Siegfried Gabler, Sabine Häder, 68–84. Münster New York: Waxmann.

Dias de Rada, V. 2005. The effect of follow-up mailings on the response rate and response quality in mail surveys. *Quality & Quantity* 1(39): 1–18.

Dieball, Werner. 2004. *Körpersprache und Kommunikation im Bundestagswahlkampf: Gerhard Schröder versus Edmund Stoiber*. Berlin, München: Wissenschaftliche Schriften.

Diekmann, Andreas. 2001. *Empirische Sozialforschung: Grundlagen, Methoden, Anwendungen*. Reinbek bei Hamburg: Rowohlt. weitere Auflagen 2004, 2012.

Diekmann, Andreas. 2004. *Empirische Sozialforschung: Grundlagen, Methoden, Anwendungen*. Reinbech bei Hamburg: Rowohlt.

Diekmann, A., und D. Wyden. 2001. *Empirische Sozialforschung: Grundlagen, Methoden, Anwendungen*. Reinbek bei Hamburg: Rowohlt. Weitere Auflagen 2004, 2012.

Diekmann, Andreas, und Ben Jann. 2001. Anreizformen und Ausschöpfungsquote bei postalischen Befragungen. Eine Prüfung der Reziprozitätshypothese. *ZUMA-Nachrichten* 25(48): 18–27.

Diekmann, Andreas, und David Wyder. 2002. Vertrauen und Reputationseffekte bei Internet-Auktionen. *Kölner Zeitschrift für Soziologie und Sozialpsychologie* 54(4): 674–693.

Dillman, Don A. 1978. *Mail and telephone surveys: The total design method*. New York: Wiley.

Dillman, Don A. 1983. Mail and other self-administered questionnaires. In *Handbook of survey research*, Hrsg. Peter H. Rossi, James D. Wright, Andy B. Anderson, 359–376. New York: Academic Press.

Dillman, Don A. 2000. *Mail and internet surveys. The tailored design method*. New York: Wiley.

Dillman, Don A., T.L. Brown, J. Carlson, E.H. Carpenter, F.O. Lorenz, M. Robert, S. John, und R.L. Sangster. 1995. Effects of category order on answers to mail and telephone surveys. *Rural Sociology* 60(4): 674–687.

Döring, Nicola. 2019. Evaluationsforschung. In *Handbuch der Methoden der empirischen Sozialforschung*, 2. Auflage, Bd. 1, Hrsg. Nina Baur, Jörg Blasius, 173–189, Wiesbaden: Springer VS.

Donsbach, Wolfgang 2001. *Who's afraid of election polls? Normative and empirical arguments for freedom of pre-election surveys*. Amsterdam: ESOMAR.

Dubben, Hans-Hermann, und Hans-Peter Beck-Bornholdt. 2004. Die Bedeutung der statistischen Signifikanz. In *Methoden der Sozialforschung* Kölner Zeitschrift für Soziologie und Sozialpsychologie, Sonderheft 44, Hrsg. Andreas Diekmann, 61–74.

Duffield, Christine. 1993. The Delphi technique: a comparision of results obtaining using two expert panels. *International Journal of Nursing Studies* 30(3): 227–237.

Edwards, Allan L. 1957. *The social desirability variable in personality assessment and research.* New York: Holt Rinehart & Winston.

Ehlich, Konrad, und Bernd Switalla. 1976. Transkriptionssysteme. Eine exemplarische Übersicht. *Studium Linguistik* 2: 78–105.

Ehling, Manfred et al. 2001. Zeitverwendung 2001/2002 – Konzeption und Ablauf der Zeitbudgeterhebung der amtlichen Statistik. In *Zeitbudget in Deutschland – Erfahrungsberichte der Wissenschaft* Schriftenreihe Spektrum Bundesstatistik, Bd. 17, Hrsg. Manfred Ehling, Joachim Merz u. a., 214–239. Wiesbaden: Metzler Poeschel.

Eifler, Stefanie, und Heinz Leitgöb. 2019. Experiment. In *Handbuch der Methoden der empirischen Sozialforschung,* 2. Auflage, Bd. 2, Hrsg. Nina Baur, Jörg Blasius, 203–218, Wiesbaden: Springer VS.

Endruweit, Günter, Gisela Trommsdorff, und Nicole Burzan (Hrsg.). 2014. *Wörterbuch der Soziologie,* 3. Auflage, Konstanz: UVK Verlagsgesellschaft.

Engel, Uwe, und Björn Oliver Schmidt. 2019. Unit- und Item-Nonresponse. In *Handbuch der Methoden der empirischen Sozialforschung,* 2. Auflage, Bd. 1, Hrsg. Nina Baur, Jörg Blasius, 385–404, Wiesbaden: Springer VS.

Erffmeyer, Robert C., Elizabeth S. Erffmeyer, und Irving M. Lane. 1986. The Delphi technique: An empirical evaluation of the optimal number of rounds. *Group and Organization Studies* 11(1–2): 120–128.

Ericsson, Anders K., und Herbert A. Simon. 1980. Verbal reports as data. *Psychological Review* 87: 215–251.

Ericsson, Anders K., und Herbert A. Simon. 1984. *Protocol analysis: Verbal report as data.* Cambridge, MA: The MIT Press.

Esser, Hartmut. 1986a. Können Befragte Lügen? Zum Konzept des „wahren Wertes" im Rahmen der handlungstheoretischen Erklärung von Situationseinflüssen bei der Befragung. *Kölner Zeitschrift für Soziologie und Sozialpsychologie* 38(2): 314–336.

Esser, Hartmut. 1986b. Über die Teilnahme an Befragungen. *ZUMA-Nachrichten* 1(18): 38–47.

Faulbaum, Frank. 2019. Total Survey Error. In *Handbuch der Methoden der empirischen Sozialforschung,* 2. Auflage, Bd. 1, Hrsg. Nina Baur, Jörg Blasius, 505–521, Wiesbaden: Springer VS.

Faulbaum, Frank, Marc Deutschmann, und Martin Kleudgen. 2003. Computerunterstütztes Pretesting von CATI-Fragebögen: Das CAPTIQ-Verfahren. *ZUMA-Nachrichten* 1(52): 20–34.

Faulbaum, Frank, Peter Prüfer, und Margit Rexroth. 2009. *Was ist eine gute Frage? Die systematische Evaluation der Fragenqualität.* Wiesbaden: VS Verlag für Sozialwissenschaften.

Feger, Hubert. 1983. Planung und Bewertung von wissenschaftlichen Beobachtungen. In *Datenerhebung. Enzyklopädie der Psychologie* Forschungsmethoden, Bd. 1, Hrsg. Hubert Feger, Jürgen Bredenkamp, 1–75. Göttingen: Hogrefe.

Fest, Joachim. 2003. Was wir aus der Geschichte nicht lernen. Die Verkettung von Vernunft und Verhängnis: Warum Historiker gut daran tun, die biografische Methode stärker zu achten. *DIE ZEIT,* 20. März, S. 38.

Fiedler, Fred E. 1954. Assumed similarity measures as predictors of team effectiveness. *Journal of Abnormal and Social Psychology* 49: 381–388.

Fink, Alexander, Oliver Schlake, und Andreas Siebe. 2001. *Erfolg durch Szenario-Management: Prinzip und Werkzeuge der strategischen Vorausschau.* Frankfurt am Main, New York: Campus-Verlag.

Finkel, Steven E., Thomas M. Guterbock, und Marian J. Borg. 1991. Race- of- interviewer effects in a pre-election poll, Virginia 1989. *Public Opinion Quarterly* 55(2): 313–327.

Fischer-Rosenthal, Wolfram. 1996. Strukturale Analyse biographischer Texte. In *Quantitative Einzelfallanalysen und qualitative Verfahren*, Hrsg. Elmar Brähler, Corinne Adler, 147–208. Gießen: Psychosozial-Verlag.

Florian, Michael. 2000. Das Ladenburger „TeleDelphi": Nutzung des Internets für eine Expertenbefragung. In *Die Delphi-Technik in den Sozialwissenschaften – Methodische Forschungen und innovative Anwendungen*, Hrsg. Michael Häder, Sabine Häder, 195–216. Wiesbaden: Westdeutscher Verlag.

Fowler, Floyd Jackson Jr. 2001. Why it is easy to write bad questions. *ZUMA-Nachrichten* 25(48): 49–66.

Frankovic, Kathleen A. 2005. Reporting, 'The Polls' in 2004. *Public Opinion Quarterly* 69(5): 682–697.

Freedman, Davis, Robert Pisani, und Roger Purves. 1978. *Statistics*. New York: Norton.

French, Robert L., und Ivan N. Mensh. 1948. Some relations between interpersonal judgment and sociometric status in a college group. *Sociometry* 11(4): 335–345.

Frey, James H., Gerhard Kunz, und Günther Lüschen. 1990. *Telefonumfragen in der Sozialforschung. Methoden Techniken Befragungspraxis*. Opladen: Westdeutscher Verlag.

Freyer, Hans. 1928. *Theorie des objektiven Geistes. Eine Einleitung in die Kulturphilosophie*. Leipzig Berlin: Teubner.

Friebertshäuser, Barbara, und Annedore Prengel. 1996. *Handbuch Qualitative Forschungsmethoden in der Erziehungswissenschaft*. Weinheim München: Juventa Verlag.

Friedrichs, Jürgen. 1990. *Methoden empirischer Sozialforschung*, 14. Auflage, Opladen: Westdeutscher Verlag.

Friedrichs, Jürgen. 2000. Effekte des Versands des Fragebogens auf die Antwortqualität bei einer telefonischen Befragung. In *Methoden der Telefonumfragen*, Hrsg. Volker Hüfken, 171–182. Wiesbaden: Westdeutscher Verlag.

Friedrichs, Jürgen. 2019. Forschungsethik. In *Handbuch der Methoden der empirischen Sozialforschung*, 2. Auflage, Bd. 1, Hrsg. Nina Baur, Jörg Blasius, 67–76, Wiesbaden: Springer VS.

Früh, Werner. 2001. *Inhaltsanalysen. Theorie und Praxis*, 5. Auflage, München: Ölschläger.

Fürst, Dietrich, Herbert Schubert, Ansgar Rudolph, und Holger Spieckermann. 1999. *Das Regionale Netzwerk Hannover. Erste Ergebnisse*. Opladen: Leske + Budrich.

Gabler, Siegfried. 1992. Schneeballverfahren und verwandte Stichprobendesigns. *ZUMA-Nachrichten* 2(31): 47–69.

Gabler, Siegfried. 1994. Die Ost-West-Gewichtung der Daten der ALLBUS-Baseline-Studie 1991 und des ALLBUS 1992. *ZUMA-Nachrichten* 2(35): 77–81.

Gabler, Siegfried. 2004. Gewichtungsprobleme in der Datenanalyse. In *Methoden der Sozialforschung* Kölner Zeitschrift für Soziologie und Sozialpsychologie, Hrsg. Siegfried Gabler, 128–147.

Gabler, Siegfried, und Sabine Häder. 1997. Überlegungen zu einem Stichprobendesign für Telefonumfragen in Deutschland. *ZUMA-Nachrichten* 2(41): 7–18.

Gabler, Siegfried, und Sabine Häder. 2000. Über Design-Effekte. In *Querschnitt. Festschrift für Max Kaase*, Hrsg. Peter Ph Mohler, Paul Lüttinger, 73–97. Mannheim: ZUMA.

Gabler, Siegfried, und Sabine Häder (Hrsg.). 2002. *Telefonstichproben. Methodische Innovationen und Anwendungen in Deutschland*. Münster, New York, München: Waxmann.

Gabler, Siegfried, und Sabine Häder. 2006. Bericht über das zweite Treffen der Arbeitsgruppe Mobilsample. *ZUMA-Nachrichten* 1(58): 120.

Gabler, Siegfried, und Jürgen H.P. Hoffmeyer-Zlotnik. 1997. *Stichproben in der Umfragepraxis*. Opladen: Westdeutscher Verlag.

Gabler, Siegfried. 1999. A model based justification of Kish's formula for design effects for weighting and clustering. *Survey Methodology* 25(1): 105–106.

Gabler, Siegfried, Jürgen H.P. Hoffmeyer-Zlotnik, und Dagmar Krebs (Hrsg.). 1994. *Gewichtung in der Umfragepraxis*. Opladen: Westdeutscher Verlag.

Geißler, Karlheinz A. 1993. *Zeit leben. Vom Hasten und Rasten, Arbeiten und Lernen, Leben und Sterben*, 5. Auflage, Weinheim, Berlin: Beltz.

Gershuny, Jonathan. 1990. International comparisons of time budget surveys. Methods and opportunities. In *Zeitbudgeterhebungen – Ziele, Methoden und neue Konzepte. Forum der Bundesstatistik*, Bd. 13, Hrsg. Rosemarie von Schweitzer, 23–53. Stuttgart: Metzler Poeschel.

Geschka, Horst. 1977. Delphi. In *Langfristige Prognosen. Möglichkeiten und Methoden der Langfristprognostik komplexer Systeme*, Hrsg. Gerhart Bruckmann, 27–44. Würzburg, Wien: Physica-Verlag.

Ghanbari, S. Azizi. 2002. *Einführung in die Statistik für Sozial- und Erziehungswissenschaftler*. Berlin Heidelberg New York: Springer.

Gigerenzer, Gerd, Ullrich Hoffrage, und Hans Kleinbölting. 1991. Probabilistic mental models: A Brunswikian theory of confidence. *Psychological Review* 98(4): 506–528.

Glaser, Barney G., und Anselm L. Strauss. 1967. *The discovery of grounded theory*. Chicago, Illinois: Aldine Publ.

Glendall, Philip, und Janet Hoek. 1990. A question of wording. *Marketing Bulletin* 1: 25–36.

Glück, Gerhard. 1971. Methoden der Beobachtung. In *Forschungstechnik für die Hochschuldidaktik*, Hrsg. Günther Dohmen, 57–66. München: Beck.

Goode, William, und Paul K. Hatt. 1962. Die Einzelfallstudie. In *Beobachtung und Experiment in der Sozialforschung*, Hrsg. René König, 299–313. Köln: Kiepenheuer & Witsch.

Gordon, Theodore J., und Olaf Helmer. 1964. *Report on a long range forecasting study*. Rand Paper P-2982, Santa Monica, Cal.

Götze, Hartmut. 1992. Das Stichprobendesign der Empirisch-Methodischen Arbeitsgruppe (EMMAG): Darstellung und Bewertung. *ZUMA-Nachrichten* 1(30): 95–108.

Gouldner, Alvin W. 1960. The norm of reciprocity: A preliminary statement. *American Sociological Review* 25(2): 161–178.

Goyder, John. 1985. Face-to-face interviews and mail questionnaires: The net difference in response rate. *Public Opinion Quarterly* 49(2): 243–252.

Goyder, John. 1987. *The silent minority: Nonrespondents on sample surveys*. Cambridge: Polity Press.

Graf, Christine, und Heinrich Litzenroth. 1991. Bringt mehr Werbung mehr Umsatz? Ökonomische Werbewirkungsmessung mit GfK-BehaviorScan. *Media Perspektiven* 11–12/1993 Information Resources Inc. (IRI) (Hrsg.): Brochure describing IRI BehaviorScan. Chicago, Illinois autumn.

Grau, Ina, Ullrich Mueller, und Andreas Ziegler. 2000. Die Verzerrung von Erinnerungen durch das Vorwissen der Befragten: Die Rolle impliziter Theorien. *ZUMA-Nachrichten* 2(47): 20–35.

Graumann, Carl F. 1966. Grundzüge der Verhaltensbeobachtung. In *Fernsehen in der Lehrerbildung*, Hrsg. Ernst Meyer, 86–107. München: Manz.

Gramlich, Tobias, und Sabine Häder. 2019. Methoden- und Feldberichte. In *Telefonumfragen in Deutschland*. Hrsg. Sabine Häder, Michael Häder, und Patrick Schmich. 393–404. Wiesbaden: Springer VS.

Granovetter, Mark. 1973. The Strenght of Weak Ties. *American Journal of Sociology* 78(6): 1360–1380.

Granovetter, Mark. 1995. *Getting a job. A study of contacts and careers*. Chicago: University of Chicago Press.

Greenwood, Peter W., C. Peter Rydell, und Karyn Model. 1996. *Diverting children from a life of crime: Measuring costs and benefits*. Santa Monica, CA: Rand Corporation.

Greenwood, Peter W., C. Peter. Rydell, A.F. Abrahamse, J.P. Caulkins, J. Chiesa, Karyn E. Model, und S.P. Klein. 1994. *Three strikes and you're out: Estimated benefits and costs of California's new mandatory-sentencing law*. Santa Monica, CA: Rand Corporation.

Greve, Werner, und Dirk Wentura. 1997. *Wissenschaftliche Beobachtung. Eine Einführung.* Weinheim. Psychologie Verlags Union: Beltz.

Gronlund, Norman E. 1955. Sociometric status and sociometric perception. *Sociometry* 18(2): 122–128.

Gross, Irene. 1996. Tagesablauf 1981 und 1992. Ergebnisse des Mikrozensus. *Statistische Nachrichten* 51(2): 93–103.

Grossman, Beverly, und Joyce Wrighter. 1948. The relationship between selection-rejection and intelligence, social status, and personality among 6th grade children. *Sociometry* 11(4): 346–355.

Groves, Robert M., und Mick P. Couper. 1998. *Nonresponse in household interview surveys.* New York Chichester Weinheim: Wiley.

Groves, Robert M., und Nancy H. Fultz. 1985. Gender effects among telephone interviewers in a survey of economic attitudes. *Sociological Methods & Research* 14(1): 31–52.

Groves, Robert M., Robert B. Cialdini, und Mick P. Couper. 1992. Understanding the decision to participate in a survey. *Public Opinion Quarterly* 56(4): 475–495.

Groves, Robert M., Floyd J. Fowler Jr., Mick P. Couper, James M. Lepkowski, Eleanor Singer, und Roger Tourangeau. 2004. *Survey Methodology. New Jersey.* Hoboken: Wiley.

Grunert, K.G. 1981. *Informationseffizienz und Möglichkeiten ihrer Verbesserung auf dem Automobilmarkt.* unveröffentlichtes Arbeitspapier, zitiert nach Mohler/Frehsen (1989).

Grupp, Hariolf. 1995. *Der Delphi-Report: Innovationen für unsere Zukunft.* Stuttgart: Deutsche Verlagsanstalt.

Gühne, Julia. 2013. *Soziale Einbettung und Kriminalitätsfurcht. Diplomarbeit.* TU Dresden: Institut für Soziologie.

Guttman, Louis. 1944. A basic for scaling qualitative data. *American Sociological Review* 9: 139–150.

Guttman, Louis. 1947. The Cornell technique for scale and intensity analysis. *Educational and Psychological Measurement* 7(2): 247–279.

Häder, Michael. 1998. Wird die DDR im Rückblick immer attraktiver? Zur retrospektiven Bewertung der Zufriedenheit. In *Sozialer Wandel in Ostdeutschland. Theoretische und methodische Beiträge zur Analyse der Situation seit 1990*, Hrsg. Michael Häder, Sabine Häder, 7–38. Opladen: Westdeutscher Verlag.

Häder, Michael. 2000. *Mobilfunk verdrängt Festnetz. Übersicht zu den Ergebnissen einer Delphi-Studie zur Zukunft des Mobilfunks. ZUMA-Arbeitsbericht 2000/05.* Mannheim: ZUMA.

Häder, Michael. 2012. Qualität empirischer Daten. *Weiterbildung, Zeitschrift für Grundlagen, Praxis und Trends* 4: 9–12.

Häder, Michael. 2014. *Delphi-Befragungen. Ein Arbeitsbuch.* 3. Auflage. Wiesbaden: Westdeutscher Verlag.

Häder, Michael. 2008a. Die Delphi-Methode. In *Politikberatung. Stuttgart:* Lucius & Lucius, Hrsg. Stephan Bröchler, Rainer Schützeichel, 33–46.

Häder, Michael. 2008b. The use of scales. In *Handbook of public opinion research*, Hrsg. Wolfgang Donsbach, Michael W. Traugott London, New Delhi, Singapore: Sage Publishing: Los Angeles.

Häder, Michael. 2009. *Der Datenschutz in den Sozialwissenschaften.* Anmerkungen zur Praxis sozialwissenschaftlicher Erhebungen und Datenverarbeitung in Deutschland. Working Paper No 90. DIW Berlin.

Häder, Michael, und Sabine Häder. 1995a. *Turbulenzen im Transformationsprozess. Die individuelle Bewältigung des sozialen Wandels in Ostdeutschland 1990–1992.* Opladen: Westdeutscher Verlag.

Häder, Michael, und Sabine Häder. 1995b. Delphi und Kognitionspsychologie. Ein Zugang zur theoretischen Fundierung der Delphi-Methode. *ZUMA-Nachrichten* 2(37): 8–34.

Häder, Michael, und Sabine Häder. 1997. Adressvorlaufverfahren: Möglichkeiten und Grenzen. Eine Untersuchung am Beispiel der Erhebung Leben Ostdeutschland. In *Stichproben in der*

Umfragepraxis, Hrsg. Siegfried Gabler, Jürgen H.P. Hoffmeyer-Zlotnik, 33–42. Opladen: Westdeutscher Verlag.

Häder, Michael, und Sabine Häder. 1998. *Sozialer Wandel in Ostdeutschland. Theoretische und methodische Beiträge zur Analyse der Situation seit 1990.* Opladen: Westdeutscher Verlag.

Häder, Michael, und Sabine Häder. 2009. *Telefonbefragungen über das Mobilfunknetz. Konzept, Design und Umsetzung einer Strategie zur Datenerhebung.* Wiesbaden: VS Verlag für Sozialwissenschaften.

Häder, Michael, und Sabine Klein. 2002. Wie wenig das Recht unser Verhalten regelt. *ZUMA-Nachrichten* 1(50): 86–113.

Häder, Michael, und Gerald Kretzschmar. 2005a. Mit welcher Zukunft darf die Dresdner Frauenkirche rechnen?. *Pastoral-Theologie Monatszeitschrift für Wissenschaft und Praxis in Kirche und Gesellschaft* 94(9): 330–340.

Häder, Michael, und Gerald Kretzschmar. 2005b. Die ,Religion' der Dresdner Frauenkirche. Empirische Befunde zur Bindung an ein schillerndes Phänomen. *International Journal of Practical Theology* 9: 4–24.

Häder, Michael, und Mike Kühne. 2009. Die Prägung des Antwortverhaltens durch die soziale Erwünschtheit. In *Telefonbefragungen über das Mobilfunknetz. Konzept, Design und Umsetzung einer Strategie zur Datenerhebung*, Hrsg. Michael Häder, Sabine Häder, 175–186. Wiesbaden: VS Verlag für Sozialwissenschaften.

Häder, Michael, Sabine Häder, und Kerstin Hollerbach. 1996. Methodenbericht zur Untersuchung „Leben Ostdeutschland 1996", *ZUMA-Arbeitsbericht 1996/06*. Mannheim: ZUMA.

Häder, Michael, und Jan Hermelink. 2017. Wie konfliktträchtig ist das ephorale Leitungsamt? Befunde einer sozialwissenschaftlichen Untersuchung. *Pastoral-Theologie Monatszeitschrift für Wissenschaft und Praxis in Kirche und Gesellschaft* 312–322.

Häder, Michael, und Robert Neumann. 2019. Datenqualität. In Telefonumfragen in Deutschland, Hrsg. Michael Häder, Sabine Häder, und Patrick Schmich, 241–292. Wiesbaden: Springer VS.

Häder, Sabine, und Siegfried Gabler. 2003. Sampling and estimation. In *Cross cultural survey methods*, Hrsg. Janet Harkness, Fonds van de Vijver, Peter Ph. Mohler, 117–134. New Jersey: Wiley.

Häder, Sabine, und Axel Glemser. 2004. Stichprobenziehung für Telefonumfragen. In *Methoden der Sozialforschung* Kölner Zeitschrift für Soziologie und Sozialpsychologie, Sonderheft 44, Hrsg. Andreas Diekmann, 148–171. Deutschland: Springer VS.

Häder, Sabine, Siegfried Gabler, und Christiane Heckel. 2009. Stichprobenziehung, Gewichtung und Realisierung. In *Telefonbefragungen über das Mobilfunknetz. Konzept, Design und Umsetzung einer Strategie zur Datenerhebung*, Hrsg. Michael Häder, Sabine Häder, 21–49. Wiesbaden: VS Verlag für Sozialwissenschaften.

Häder, Sabine, Michael Häder, und Mike Kühne (Hrsg.). 2012. *Telephone Surveys in Europe. Research and Practice.* Heidelberg: Springer Berlin.

Häder, Sabine, Gabler, Siegfried, Laaksonen, Seppo, und Peter Lynn. 2003. The sample. *ESS* 2002/2003: Technical Report, Kapitel 2, unter: http://www.europeansocialsurvey.com.

Häder, Sabine, und Matthias Sand. 2019. Telefonstichproben. In *Telefonumfragen in Deuschland*, Hrsg. Sabine Häder, Michael Häder und Patrick Schmich, 45–80, Wiesbaden: Springer VS.

Häder, Sabine Häder, Michael Häder and Patrick Schmich. (Hrsg.). 2019. *Telefonumfragen in Deuschland*, Wiesbaden: VS Verlag.

Halfmann, Jost. 2001. Bad dreams. Technische Dyspotien von Ingenieuren. *Wissenschaftliche Zeitschrift der Technischen Universität Dresden* 5–6: 78–82.

Hallworth, H.J. 1953. Sociometric relationships among grammar school boys and girls between the age of eleven and sixteen. *Sociometry* 16(1): 39–70.

Hartmann, Petra. 1995. Response behavior in interview settings of limited privacy. *International Journal of Public Opinion Research* 8(4): 383–390.

Hartmann, Peter, und Bernhard Schimpl-Neimanns. 1992. Sind Sozialstrukturanalysen mit Umfragedaten möglich? Analysen zur Repräsentativität einer Sozialforschungsumfrage. *Kölner Zeitschrift für Soziologie und Sozialpsychologie* 44(2): 315–340.

Hasse, Jürgen. 1999. Bildstörung: Windenergien und Landschaftsästhetik. In *Wahrnehmungsgeographische Studien zur Regionalentwicklung*, Hrsg. Rainer Krüger. Oldenburg: Bibliotheks- und Informationssystem der Universität Oldenburg.

Hastie, Reid. 1981. Schematic principles in human memory. In *Social cognition: The Ontario symposium*, Hrsg. Edward Tory Higgins, C. Peter. Herman, Mark P. Zanna, 39–88. Hillsdale, NJ: Erlbaum.

Hastie, Reid, und Bernadette Park. 1986. The relationship between memory and judgement depends on whether the judgement task is memory-based or online. *Psychological Review* 93(3): 258–268.

Haugg, Kornelia. 1990. Die bisherige Erfassung des Zeitbudgets von Personen und Familien – Zielsetzungen und ausgewählte Forschungsergebnisse. In *Zeitbudgeterhebungen – Ziele, Methoden und neue Konzepte* Schriftenreihe der Bundesstatistik, Bd. 13, Hrsg. Rosemarie von Schweitzer, Manfred Ehling, Dieter Schäfer, 76–87. Stuttgart: Metzler-Poeschel.

Haupert, Bernhard. 1994. Objektiv-hermeneutische Fotoanalyse am Beispiel von Soldatenfotos aus dem Zweiten Weltkrieg. In *Die Welt als Text*, Hrsg. Detlef Garz, Klaus Kraimer, 281–314. Frankfurt a. M.: Suhrkamp.

Haupert, Bernhard, und Franz J. Schäfer. 1991. *Jugend zwischen Kreuz und Hakenkreuz. Biografische Rekonstruktionen als Allgemeingeschichte des Faschismus*. Frankfurt a. M.: Suhrkamp.

Heberlein, Thomas A., und Robert Baumgartner. 1978. Factors affecting response rates to mailed questionnaires: A quantitative analysis of the published literature. *American Sociological Review* 43(4): 447–462.

Heckel, Christiane. 2002. Erstellung der ADM-Telefonauswahlgrundlage. In *Telefonstichproben. Methodische Innovationen und Anwendungen in Deutschland*, Hrsg. Siegfried Gabler, Sabine Häder, 11–31. Münster New York München: Waxmann.

Heckel, Christiane. 2003. Online gewonnene Stichproben – Möglichkeiten und Grenzen. In *Online-Erhebungen 5. Wissenschaftliche Tagung* Sozialwissenschaftlicher Tagungsband, Bd. 7, Hrsg. Arbeitskreis Deutscher Markt- und Sozialforschungsinstitute e. V., Arbeitsgemeinschaft Sozialwissenschaftlicher Institute e. V. (ASI), Statistisches BundesamtWiesbaden, 83–94. Bonn: GESIS Leibniz-Institut für Sozialwissenschaften.

Heise, Nele. 2015. Big Data – small problems? Ethische Dimensionen der Forschung mit Online-Kommunikationsspuren. In: *Digitale Methoden in der Kommunikationswissenschaft*, Hrsg. Axel Maireder, Julian Ausserhofer, Christina Schumann, Monika Taddicken, Berlin (Digital Communication Research 2).

Helfferich, Cornelia. 2019. Leitfaden und Experteninterviews. In *Handbuch der Methoden der empirischen Sozialforschung*, 2. Auflage, Bd. 2, Hrsg. Nina Baur, Jörg Blasius, 669–685, Wiesbaden: Springer VS.

Hempel, Carl Gustav. 1952. *Fundamentals of concept formation in empirical science* International Encyclopaedia of United Science II, Bd. 7. Chicago: Chicago Press.

Hempel, Carl Gustav, und Paul Oppenheim. 1948. Studies in the logic of explanation. *Philosophy of Science* 15(2): 135–175.

Henry-Huthmacher, Christine, und Ulrich von Wilamowitz-Moellendorff. 2005. *Deutschland im Umbruch. Delphi-Studie 2004/2005. Befragung ausgewählter Expertinnen und Experten über die Zukunft Deutschlands*. Bonn: Konrad Adenauer Stiftung.

Henken, V.J. 1976. Banality reinvestigated – Computer content analysis of suicidal and forced death documents. *Suicide And life-threatening Behavior* 6(1): 36–43.

Hering, Linda, und Robert Jungmann. 2019. Einzelfallanalyse. In *Handbuch der Methoden der empirischen Sozialforschung*, 2. Auflage, Bd. 1, Hrsg. Nina Baur, Jörg Blasius, 619–632, Wiesbaden: Springer VS.

von Hermanni, Hagen. 2019. Das Total Survey Error Modell in telefonischen Befragungen. In *Telefonumfragen in Deutschland*, Hrsg. Sabine Häder, Michael Häder, Patrick Schmich. 17–34, Wiesbaden: Springer VS.

Hertkorn, Georg, Claudia Hertfelder, und Peter Wagner. 2001. Klassifikation von Zeitverwendungstagebüchern. In *Zeitbudget in Deutschland – Erfahrungsberichte der Wissenschaft* Schriftenreihe Spektrum Bundesstatistik, Bd. 17, Hrsg. Manfred Ehling, Joachim Merz, 78–90. Wiesbaden: Metzler Poeschel.

von der Heyde, Christian. 2002. Das ADM-Telefonstichproben-Modell. In *Telefonstichproben. Methodische Innovationen und Anwendungen in Deutschland*, Hrsg. Siegfried Gabler, Sabine Häder, 32–45. Münster New York München: Waxmann.

Higgins, E.T., W.S. Rholes, und C.R. Jones. 1977. Category accessibility and impression formation. *Journal of Experimental Social Psychology* 13(2): 141–154.

Hippler, Hans-Jürgen. 1985. Schriftliche Umfragen bei repräsentativen Bevölkerungsstichproben oder: Wie erreicht man 78 %?. In *Herausforderungen der empirischen Sozialforschung*, Hrsg. Max Kaase, Manfred Küchler, 71–74. Mannheim: ZUMA.

Hippler, Hans-Jürgen. 1988. Methodische Aspekte schriftlicher Befragungen: Probleme und Forschungsperspektiven. *planung & analyse* 6(6): 244–248.

Hippler, Hans Jürgen, und Kristiane Seidel. 1985. Schriftliche Befragung bei allgemeinen Bevölkerungsstichproben – Untersuchungen zur Dillmanschen „Total Design Method". *ZUMA-Nachrichten* 1(16): 39–56.

Hippler, Hans-Jürgen, und Norbert Schwarz. 1986. Not forbidding isn't allowing: the cognitive basis of the forbid – allow asymmetry. *Public Opinion Quarterly* 50(1): 87–96.

Hoffmann-Riem, Christa. 1980. Die Sozialforschung einer interpretativen Soziologie. Der Datengewinn. *Kölner Zeitschrift für Soziologie und Sozialpsychologie* 32(2): 339–372.

Hoffmeyer-Zlotnik, Jürgen H.P. 1997. Random Route-Stichproben nach ADM. In *Stichproben in der Umfragepraxis*, Hrsg. Siegfried Gabler, Jürgen H.P. Hoffmeyer-Zlotnik, 33–42. Opladen: Westdeutscher Verlag.

Hoffmeyer-Zlotnik, Jürgen H.P., Doris Hess, und Alfons J. Geis. 2006. Computergestützte Vercodung der International Standard Classification of Occupation (ISCO-88). Vorstellen eines Instrumentes. *ZUMA-Nachrichten* 1(58): 101–113.

Holland, James G., und Burrhus F. Skinner. 1971. *Analyse des Verhaltens*. München Berlin: Urban & Schwarzenberg.

Holleman, Bregje. 2000. *The forbid/allow asymmetrie. On the cognitive mechanism underlying wording effects in surveys*. Amsterdam Atlanta: Radopi.

Hollerbach, Kerstin. 1998. Ranking oder Rating? Die Wahl der Skala in der Werteforschung. In *Sozialer Wandel in Ostdeutschland. Theoretische und methodische Beiträge zur Analyse der Situation seit 1990*, Hrsg. Michael Häder, Sabine Häder, 221–255. Opladen: Westdeutscher Verlag.

Holm, Kurt. 1974. Theorie der Frage. *Kölner Zeitschrift für Soziologie und Sozialpsychologie* 26(1): 91–114.

Holz, Erlend. 2000. *Zeitverwendung in Deutschland – Beruf, Familie, Freizeit* Schriftenreihe Spektrum Bundesstatistk, Bd. 13. Wiesbaden: Metzler-Poeschel.

Hopf, Christel. 1993. Soziologie und qualitative Forschung. In *Qualitative Sozialforschung*, Hrsg. Christel Hopf, Elmar Weingarten, 11–37. Stuttgart: Klett-Cotta.

Hoppe, Michael. 2000. *Aufbau und Organisation eines Access-Panels*. In: *Neue Erhebungsinstrumente und Methodeneffekte* Schriftenreihe Spektrum Bundesstatistik, Hrsg. vom Statistischen Bundesamt, Bd. 15, 145–165. Wiesbaden: Metzler Pöschel.

Hox, Joop J., Edith D. de Leeuw, und H. Vorst. 1996. A reasoned action explanation for survey nonresponse. In *International perspectives on nonresponse*, Hrsg. Laaksonen, Seppo, 101–110. Helsinki: Statistics Finland.

Huber, Günter L. 2001. Computergestützte Auswertung qualitativer Daten. In *Handbuch Qualitative Forschung. Grundlagen, Konzepte, Methoden und Anwendungen*, Hrsg. Uwe Flick, Ernst v. Kardorff, Heiner Keupp, Lutz v. Rosenstiel, Stephan Wolff, 243–248. Weinheim: Beltz.

Hüfken, Volker. 2000. Kontaktierung bei Telefonumfragen. Auswirkungen auf das Kooperations- und Antwortverhalten. In *Methoden in Telefonumfragen*, Hrsg. Volker Hüfken, 11–31. Wiesbaden: Westdeutscher Verlag.

Hüfken, Volker, und A. Schäfer. 2003. Zum Einfluss stimmlicher Merkmale und Überzeugungsstrategien des Interviewers auf die Teilnahme in Telefonumfragen. *Kölner Zeitschrift für Soziologie und Sozialpsychologie* 55(2): 321–339.

Information Resources Inc. (IRI). 1991. *(Hrsg.): How advertising works: analyses of 400 Behaviorscan cases*. Chicago, Ill: University of Chicago Press.

Inglehart, Ronald. 1977. *The silent revolution: Changing values and political styles among western publics*. Princeton. N.Y.: University Press.

Irwing, John, und James Austin. 1997. *It's about time: America's imprisonment binge*. Belmont, CA: Wadsworth.

Jacob, Rüdiger, und Willy H. Eirmbter. 2000. *Allgemeine Bevölkerungsumfragen*. München: Oldenbourg.

Jahoda, Marie, Paul Lazarsfeld, und Hans Zeisel. 1960. *Die Arbeitslosen von Marienthal*. Allensbach Bonn: Verlag für Demoskopie.

Jansen, Dorothea. 2003. *Einführung in die Netzwerkanalyse. Grundlagen, Methoden, Forschungsbeispiele*. Opladen: Leske + Budrich.

Janssen, H. 1976. *Die Notierung von geschätzten Gleichgewichtspreisen – Ein Beitrag zur Preisstabilisierung auf dem Obstmarkt* Schr. Ges. Wi. So. Landbau, Bd. 13, 379–392. München, Bern, Wien: BLV-Verlag.

Jenkel, Dorett, und Susanne Lippert. 1998. Politische Proteste in Leipzig von 1990 bis 1996. Befragungen und Dokumentenanalysen im Vergleich. In *Sozialer Wandel in Ostdeutschland. Theoretische und methodische Beiträge zur Analyse der Situation seit 1990*, Hrsg. Michael Häder, Sabine Häder, 256–285. Opladen: Westdeutscher Verlag.

Jobe, J.B., und D.J. Herrmann. 1996. Implications of models of survey cognition for memory theory. In *Basic and applied memory research* Practical applications, Bd. 2, Hrsg. Herrmann J. Douglas, M. Johnson, Christopher Herzog, Paula Hertel, 193–205. Hillsdale, NJ: Erlbaum.

Jones, James H. 1981. *Bad blood: The Tuskegee syphilis experiments*. New York: Free Press.

Jöreskog, Karl G., Dag Sorbom, Stephen du Toit, und Mathilda du Toit. 2001. *LISREL 8*. Lincolnwood, Ill: SSI Scientific Software Internat.

Juster, Thomas F. 1985. Conceptual and methodological issues involved in the measurement of time use. In *Time, goods and well being*, Hrsg. Thomas F. Juster, Frank P. Stafford, 19–31. Ann Arbor Michigan: Institute for Social Research.

Jütte, Wolfgng. 2002. *Soziales Netzwerk Weiterbildung: Analyse lokaler Institutionslandschaften*. Bielefeld: Bertelsmann.

Kaase, Max. 1986. unter Mitwirkung von Robert Schweizer und Erwin K. Scheuch: Stellungnahme zum Entwurf eines Gesetzes zur Änderung des Bundesdatenschutzgesetzes. *ZUMA-Nachrichten* 1(18): 3–20.

Kaase, Max. 1999. (Hrsg.): *Qualitätskriterien der Umfrageforschung: Denkschrift*. Berlin: Akademie-Verlag.

Kalton, Graham. 1983. *Compensating for missing survey data. Research Report Series*. Ann Arbor Michigan: University of Michigan.

Kane, Emily W., und Laura J. Macaulay. 1993. Interviewer gender and gender attitudes. *Public Opinion Quarterly* 57(1): 1–28.

Kaynak, Erdener, Jonathan Bloom, und Marius Leibold. 1994. Using the Delphi technique to predict future tourism potential. *Marketing Intelligence & Planning* 12(7): 18–29.

Kelle, Udo, und Susann Kluge. 1999. *Vom Einzelfall zum Typus. Fallvergleich und Fallrekonstruktion in der qualitativen Sozialforschung.* Opladen: Leske + Budrich.

Kiesl, Hans. 2019. Gewichtung. In *Handbuch der Methoden der empirischen Sozialforschung,* 2. Auflage, Bd. 1, Hrsg. Nina Baur, Jörg Blasius, 405–412, Wiesbaden: Springer VS.

Kiesler, Sara, und Lee S. Sproull. 1986. Response effects in the electronic survey. *Public Opinion Quarterly* 50(3): 402–413.

Kilian, Ines. 2011. *Die Dresdner Notwehrstudie: Zur Akzeptanz des deutschen Notwehrrechts in der Bevölkerung.* Baden-Baden, Zürich; St. Gallen: Nomos, Dike.

Kinsey, Alfred C., Wardell B. Pomeroy, und Martin E. Clyde. 1948. *Sexual behavior in the human male.* Philadelphia: Saunders.

Kirsch, Anke. 2000. Delphi via Internet: Eine Expertenbefragung zu Trauma und Trauma(re) konstruktion. In *Die Delphi-Technik in den Sozialwissenschaften – Methodische Forschungen und innovative Anwendungen,* Hrsg. Michael Häder, Sabine Häder, 217–234. Opladen: Westdeutscher Verlag.

Kish, Leslie. 1965. *Survey Sampling.* New York: Wiley.

Kish, Leslie. 1988. A taxonomy of elusive populations. In *American Statistical Association. Proceedings of the section on Survey Research Methods,* 44–46.

Kish, Leslie. 1994. Multipopulation survey designs: Five types with seven shared aspects. *International Statistical Review* 62(2): 167–186.

Klein, Sabine, und Rolf Porst. 2000. *Mail-Survey. Ein Literaturbericht. ZUMA-Technischer Bericht 10/2000.* Mannheim: GESIS Leibniz Insitut für Sozialwissenschaften.

Kleine-Brockhoff, Thomas. 2003. Der verwundete Krieger. *DIE ZEIT,* 11. September, S. 14 ff.

Klemperer, Viktor. 1968. *LTI Notizbuch eines Philologen.* Leipzig: Reclam.

Koch, Achim. 1995. Gefälschte Interviews: Ergebnisse der Interviewerkontrolle beim ALLBUS 1994. *ZUMA-Nachrichten* 1(36): 89–105.

Koch, Achim. 1997. ADM-Design und Einwohnermelderegister-Stichprobe. Stichprobenverfahren bei mündlichen Bevölkerungsumfragen. In *Stichproben in der Umfragepraxis,* Hrsg. Siegfried Gabler, Jürgen H.P. Hoffmeyer-Zlotnik, 99–116. Opladen: Westdeutscher Verlag.

Koch, Achim. 2002. 20 Jahre Feldarbeit im ALLBUS: Ein Blick in die Blackbox. *ZUMA-Nachrichten* 2(51): 9–37.

Koch, Achim, und Rolf Porst (Hrsg.). 1998. *Nonresponse in Survey Research* ZUMA-Nachrichten Spezial, Bd. 4. Mannheim: GESIS Leibniz Institut für Sozialwissenschaften.

Koch, Achim, und Martina Wasmer. 2004. Der ALLBUS als Instrument zur Untersuchung sozialen Wandels: Eine Zwischenbilanz nach 20 Jahren. In *Sozialer und politischer Wandel in Deutschland. Analysen mit ALLBUS-Daten aus zwei Jahrzehnten* Blickpunkt Gesellschaft, Bd. 7, Hrsg. Rüdiger Schmitt-Beck, Martina Wasmer, Achim Koch, 13–41. Wiesbaden: VS Verlag für Sozialwissenschaften.

Koch, Achim, Martina Wasmer, Janet Harkness, und Evi Scholz. 2001. *Konzeption und Durchführung der „Allgemeinen Bevölkerungsumfrage der Sozialwissenschaft" (ALLBUS) 2000. ZUMA-Methodenbericht 2001/05.* Mannheim: ZUMA.

Köhler, Gabriele. 1992. Methodik und Problematik einer mehrstufigen Expertenbefragung. In *Analyse verbaler Daten. Über den Umgang mit qualitativen Daten,* Hrsg. Jürgen H.P. Hoffmeyer-Zlotnik, 318–332. Opladen: Westdeutscher Verlag.

König, René. 1967. *Handbuch der empirischen Sozialforschung* Bd. I. Stuttgart: Ferdinand Enke.

König, René. 1969. *Handbuch der empirischen Sozialforschung* Bd. II. Stuttgart: Ferdinand Enke.

König, René. 1973. Die Beobachtung. In *Grundlegende Methoden und Techniken der empirischen Sozialforschung. Erster Teil* Handbuch der empirischen Sozialforschung, Bd. 2, Hrsg. René König, 1–65. Stuttgart: Ferdinand Enke.

König, René. 1957. *Das Interview. Formen, Technik, Auswertung. 2., völlig umgearbeitete, verbesserte und erweiterte Auflage.* Köln: Verlag für Politik und Wirtschaft. unter Mitarbeit von Dietrich Rüschemeyer.

von Koolwijek, Jürgen 1974. Das Quotenverfahren. Paradigma sozialwissenschaftlicher Auswahlpraxis. In *Techniken der empirischen Sozialforschung*, Bd. 6, Hrsg. Jürgen von Koolwijk, Maria. Wieken-Mayser, 81–99. München: Oldenbourg.

Kops, Manfred. 1984. Eine inhaltsanalytische Bestimmung von Persönlichkeitsbildern in Heiratsanzeigen. In *Computerunterstütze Inhaltsanalyse in der empirischen Sozialforschung*, Hrsg. Hans-Dieter Klingemann, 54–97. Frankfurt am Main: Campus.

Kramer, Caroline (Hrsg.). 2001. *FREI-Räume und FREI-Zeiten: Raumnutzung und Zeit-Verwendung im Geschlechterverhältnis.* Baden-Baden: Nomos Verlagsgesellschaft.

Krausch, Stefanie. 2005. *Interviewerverzerrungen: Effekte auf die Antworten in Telefonbefragungen unter besonderer Berücksichtigung der Einstellung des Interviewers. Diplomarbeit.* Technische Universität Dresden: Institut für Soziologie.

Krebs, Dagmar, und Natalja Menold. 2019. Gütekriterien quantitativer Sozialforachung. In *Handbuch der Methoden der empirischen Sozialforschung*, 2. Auflage, Bd. 1, Hrsg. Nina Baur, Jörg Blasius, 489–504, Wiesbaden: Springer VS.

Krebs, Dagmar. 1991. *Was ist sozial erwünscht? Der Grad sozialer Erwünschtheit von Einstellungsitems. ZUMA-Arbeitsbericht Nr. 1991/18.* Mannheim: ZUMA.

Kretzschmar, Gerald. 2001. *Distanzierte Kirchlichkeit. Eine Analyse ihrer Wahrnehmung.* Neukirchen-Vluyn: Neukirchener.

Krieg, Sabine. 2004. *Wohnungslose in Dresden: Biographische Verläufe, Alltagsorganisation, Soziale Hilfssysteme. Diplomarbeit.* Technische Universität Dresden: Institut für Soziologie.

Kromrey, Helmut. 1998. *Empirische Sozialforschung. Modelle und Methoden der standardisierten Datenerhebung und Datenauswertung.* Opladen: Leske + Budrich. 2002.

Kroneberg, Clemens. 2005. Die Definition der Situation und die variable Rationalität der Akteure. Ein allgemeines Modell des Handelns. *Zeitschrift für Soziologie* 34: 344–363.

Kroneberg, Clemens. 2011. *Die Erklärung sozialen Handelns. Grundlagen und Anwendung einer integrativen Theorie.* Wiesbaden: VS Verlag für Sozialwissenschaften.

Krosnick, Jon A. 1991. Response strategies for coping with the cognitive demands of attitude measures in surveys. *Applied Cognitive Psychology* 5(3): 213–236.

Krosnick, Jon A. 1999. Survey research. *Annual Review of Psychology* 50: 537–567.

Krosnick, Jon A., und Alwin F. Duane. 1987. An evaluation of a cognitive theory of response-order effects in survey measurement. *Public Opinion Quarterly* 51(2): 201–219.

Krysan, Maria, Howard Schuman, Lesli Jo Scott, und Paul Beatty. 1994. Response rates and response content in mail versus face-to-face surveys. *Public Opinion Quarterly* 58(3): 381–399.

Kuckartz, Udo. 2005. *Einführung in die computergestützte Analyse qualitativer Daten.* Wiesbaden: VS Verlag für Sozialwissenschaften.

Kuckartz, Udo, und Stefan Rädicker. 2019. Datenaufbereitung und Datenbereinigung on der qualitativen Sozialforschung. In *Handbuch der Methoden der empirischen Sozialforschung*, 2. Auflage, Bd. 1, Hrsg. Nina Baur, Jörg Blasius, 441–456, Wiesbaden: Springer VS.

Kühne, Mike, und Michael Häder. 2009. Das Kriterium Antwortlatenzzeiten. In *Telefonbefragungen über das Mobilfunknetz. Konzept, Design und Umsetzung einer Strategie zur Datenerhebung*, Hrsg. Michael Häder, Sabine Häder, 185–190. Wiesbaden: VS Verlag für Sozialwissenschaften.

Kühnel, Steffen-M., und Dagmar Krebs. 2001. *Statistik für die Sozialwissenschaften. Grundlagen Methoden Anwendungen.* Reinbeck: rowolths enzyklopädie.

Kurt, Ronald, und Regine Herbrik. 2019. Sozialwissenschaftliche Hermeneutik und hermeneutische Wissenssoziologie. In *Handbuch der Methoden der empirischen Sozialforschung*, 2. Auflage, Bd. 1, Hrsg. Nina Baur, Jörg Blasius, 545–564, Wiesbaden: Springer VS.

Küsters, Ivonne. 2019. Narrative Interviews. In *Handbuch der Methoden der empirischen Sozialforschung,* 2. Auflage, Bd. 2, Hrsg. Nina Baur, Jörg Blasius, 687–693, Wiesbaden: Springer VS.

Kunz, Franziska. 2010. Mahnaktionen in postalischen Befragungen. Empirische Befunde zu Auswirkungen auf den Rücklauf, das Antwortverhalten und die Stichprobenzusammensetzung. *methoden daten analysen. Zeitschrift für empirische Sozialforschung* 4(2): 127–155.

Lamminger, Thomas, und Zander E. Frank. 2002. Access-Panel als Grundlage für Online-Erhebungen. In *Online-Erhebungen 5. Wissenschaftliche Tagung* Sozialwissenschaftlicher Tagungsband, Bd. 7, Hrsg. Arbeitskreis Deutscher Markt-, Arbeitsgemeinschaft Sozialwissenschaftlicher Institute e. V. (ASI), Statistisches Bundesamt, 95–108. Bonn: GESIS Leibniz Institut für Sozialwissenschaften.

Lamnek, Siegfried (Hrsg.). 1995. *Methoden und Techniken* Qualitative Sozialforschung, Bd. 2. Weinheim: Beltz Psychologie Verlags Union.

Lamnek, Siegfried (Hrsg.). 1998. *Gruppendiskussion. Theorie und Praxis.* Weinheim: Beltz Psychologie Verlags Union.

Lasswell, Harold D. 1952: The Structure and Function of Communication in Society. In *Lyman Bryson (Hrsg.): The Communication of Ideas.* A Series of Addresses. Harper & Brs., New York.

Latcheva, Rossalia, und Eldad Davidov. 2019 Skalen und Indizes. In *Handbuch der Methoden der empirischen Sozialforschung,* 2. Auflage, Bd. 2, Hrsg. Nina Baur, Jörg Blasius, 893–905, Wiesbaden: Springer VS.

de Leeuw, Edith D. 1992. *Data quality in mail, telephone and face to face surveys.* Amsterdam: Univ. Diss.

de Leeuw, Edith D. (Hrsg.). 1999. *Journal of Official Statistics.* Special Issue on Survey Nonresponse. Bd. 15. Statistics Sweden.

Legewie, Heiner. 1994. Globalauswetung von Dokumenten. In *Texte verstehen. Konzepte, Methoden, Werkzeuge. Konstanz: Universitätsverlag,* Hrsg. Andreas Boehm, Andreas Mengel, Thomas Muhr, 177–182.

Lemann, Thomas B., und Richard L. Solomon. 1952. Group characteristics as revealed in sociometric patterns and personality ratings. *Sociometry* 15(1): 7–90.

Levi-Strauss, Claude. 1967. *Strukturale Anthropologie.* Frankfurt am Main: Suhrkamp.

Levin, H.M., G.V. Glass, und G.R. Meister. 1986. Different approaches to improving performance at school: A cost-effectiveness comparisation. *Zeitschrift für internationale erziehungs- und sozialwissenschaftliche Forschung* 3(2): 155–176.

Lewin, Kurt, und Ronald Lippitt. 1938. An experimental approach to the study of autocracy and democracy. *Sociometry* 1(3): 292–300.

Lewis, Oscar et al. 1953. Controls and experiments in field work. In *Anthropology today: An encyclopedic inventory,* 452–475. Chicago: University of Chicago Press.

Lienert, Gustav. 1969. *Testaufbau und Testanalyse.* Weinheim: Beltz.

Likert, Rensis. 1932. A technique for the measurement of attitudes. *Archives of Psychology to the Study of Functional Groups* (140): 1–55.

Lindeman, Eduard C. 1924. *Social discovery: An approach.* New York: Republ. Publ. Co.

Linsky, Arnold S. 1975. Stimulating responses to mailed questionnaires: A review. *Public Opinion Quarterly* 39(1): 82–101.

Linstone, Harold A., und Murray Turoff. 1975. *The Delphi method: Techniques and applications.* London: Reading, Mass: Addison-Wesley Company.

Lockhart, Daniel C. 1986. *Mailed questionnaire returning behaviour: A comparison of Triandis' and Fishbein's theories of the predictors.* Dissertation Abstracts International (47) Heft 7, University Microfilms No. AAC 8622995.

Longmore, T. Wilson. 1948. A matrix approache to the analysis of rank and status in a community in Peru. *Sociometry* 11(3): 192–206.

Loosveldt, Geert. 1997. Interaction characteristics in some question wording experiments. *Bulletin de Méthodologie Sociologique BMS* 56: 20–31.

Lüdtke, Oliver et al. 2007. *Umgang mit fehlenden Werten in der psychologischen Forschung. Probleme und Lösungen.* Psycholog. Rundschau 58(2) 103–117.

Lüttinger, Paul (Hrsg.). 1999. *Sozialstrukturanalysen mit dem Mikrozensus* ZUMA-Nachrichten Spezial, Bd. 6.

Lundberg, George A., und Mary Steele. 1937. Social attraction patterns in a rural village. *Sociometry* 1(1): 77–80.

Magidson, Jay. 1994. *SPSS for Windows CHAID-Release 6.0.* Chicago: SPSS inc.

Mahrt, Merja. 2015. Mit Big Data gegen das "Ende der Theorie"?. In: *Digitale Methoden in der Kommunikationswissenschaft*, Hrsg.: Axel Maireder, Julian Ausserhofer, Christina Schumann, Monika Taddicken. Berlin (Digital Communication Research 2).

Maisonneuve, Jean. 1952. Selective choices and propinquity. *Sociometry* 15(2): 135–140.

Mangold, Werner. 1960. Gegenstand und Methode des Gruppendiskussionsverfahrens. In *Grundlegende Methoden und Techniken der empirischen Sozialforschung. Erster Teil*, 3. Auflage, Handbuch der empirischen Sozialforschung, Bd. 2, Hrsg. René König, 228–259. Stuttgart: Enke.

Manns, Marianne, Jona Schultze, Claudia Herrmann, und Hans Westmayer. 1987. *Beobachtungsverfahren in der Verhaltensdiagnostik. Eine systematische Darstellung ausgewählter Beobachtungsverfahren.* Salzburg: Müller.

Marlowe, Douglas P., und David Crowne. 1960. A new scale of social desirability independent of psychopathology. *Journal of Consulting Psychology* 24(4): 349–354.

Marscher, Konstantin. 2004. *Partnerpräferenzen in homosexuellen Beziehungen. Eine Inhaltsanalyse von Kontaktanzeigen. Diplomarbeit.* Technische Universität Dresden: Institut für Soziologie.

Marx, Karl. 1971. Zur Kritik der politischen Ökonomie. In Karl Marx, Friedrich Engels *Werke.*, Bd. 13, 7–160. Berlin: Dietz.

Mayer, Karl Ulrich, und Peter Schmidt. 1984. *Allgemeine Bevölkerungsumfrage der Sozialwissenschaften – Beiträge zu methodischen Problemen des ALLBUS 1980. ZUMA-Monografien Sozialwissenschaftliche Methoden.* Frankfurt/New York: Campus.

Mayring, Philipp. 1989. Qualitative Inhaltsanalyse. In *Qualitative Forschung in der Psychologie*, Hrsg. Gerd Jüttemann, 187–211. Heidelberg: Asanger.

Mayring, Philipp. 1993. *Qualitative Inhaltsanalyse. Grundlagen und Techniken*, 4. Auflage, Weinheim: Deutscher Studienverlag.

Mayring, Philipp. 2001. Qualitative Inhaltsanalyse. In *Handbuch Qualitative Forschung. Grundlagen, Konzepte, Methoden und Anwendungen*, Hrsg. Uwe Flick, Ernst v. Kardorff, Heiner Keupp, Lutz v. Rosenstiel, Stephan Wolff, 209–213. Weinheim: Beltz.

Mayring, Philipp, und Thomas Fenzl. 2019. Qualitative Inhaltsanalyse. In *Handbuch der Methoden der empirischen Sozialforschung*, 2. Auflage, Bd. 1, Hrsg. Nina Baur, Jörg Blasius, 633–648, Wiesbaden: Springer VS.

Mees, Ulrich, und Herbert Selg. 1977. *Verhaltensbeobachtung und Verhaltensmodifikation.* Stuttgart: Klett.

Meier, Gerd, Michael Schneid, Yvonne Stegemann, und Angelika Stiegler. 2005. Steigerung der Ausschöpfungsquote von Telefonumfragen durch geschickte Einleitungstexte. *ZUMA-Nachrichten* 2(57): 37–55.

Merten, Klaus. 1995. *Inhaltsanalyse. Einführung in die Theorie, Methode und Praxis.* Opladen: Westdeutscher Verlag.

Merton, Robert K. 1957. The self-fulfilling prophecy. In *Social theory and social structur*, Hrsg. Robert K. Merton Glencoe, Ill: Free Press.

Merton, Robert K. 1987. The focussed interview and focus groups. *Public Opinion Quarterly* 51(4): 550–566.

Merton, Robert K., und Patricia L. Kendall. 1979. Das fokussierte Interview. In *Qualitative Sozialforschung*, Hrsg. Christel Hopf, Elmar Weingarten, 171–204. Stuttgart: Klett-Cotta. 1993.

van Meter, Karl M. 1990. *Methodological and design issues: Techniques for assessing the representatives of snowball sampling. The collection and interpretation of data from hidden population. NIDA Research Monograph, Nummer 98.* Rockville: US Government Printing Office.

Metschke, Rainer, und Rita Wellbrock. 2002. *Datenschutz in Wissenschaft und Forschung.* www.datenschutz-berlin.de/attachments/47/Materialien28.pdf?1166527077.

Mettler, Peter H., und Thomas Baumgartner. 1997. *Partizipation als Entscheidungshilfe. PARDIZIPP – ein Verfahren der (Langfrist-)Planung und Zukunftsforschung.* Opladen: Westdeutscher Verlag.

Meyer, Simone, und Sigrid Weggemann. 2001. Mahlzeitmusteranalyse anhand der Daten der Zeitbudgeterhebung 1991/92. In *Zeitbudget in Deutschland – Erfahrungsberichte der Wissenschaft* Schriftenreihe Spektrum Bundesstatistik, Bd. 17, Hrsg. Manfred Ehling, Joachim Merz, u. a. 188–201. Wiesbaden: Metzler Poeschel.

Mietzner, Ulrike, und Ulrike Pilarczyk. 2003. Methoden der Fotografieanalyse. In *Film- und Fotoanalyse in der Erziehungswissenschaft. Ein Handbuch*, Hrsg. Yvonne Ehrenspeck, Burkhard Schäffer, 19–36. Opladen: Leske + Budrich.

Milgram, Stanley, Leon Mann, und Susan Harter. 1965. The lost-letter technique: A tool of social research. *Public Opinion Quarterly* 29(3): 437–438.

Mochmann, Ekkehard. 2019. Quantitative Daten für die Sekundäranalyse. In *Handbuch der Methoden der empirischen Sozialforschung*, 2. Auflage, Bd. 1, Hrsg. Nina Baur, Jörg Blasius, 259–270, Wiesbaden: Springer VS.

Mohler, Peter Ph. 1978. *Reflektionen des Verhältnisses zwischen Individuum und kollektiver Macht in Abituraufsätzen* Europäische Hochschulschriften, 1917–1971. Frankfurt a. Main: Lang.

Mohler, Peter Ph, Katja Mitarbeit Hauk, und Ute von Frehsen. 1989. *cui Computerunterstütze Inhaltsanalyse – Grundzüge und Auswahlbibliographie zu neueren Anwendungen, ZUMA-Arbeitsbericht 1989/09.* Mannheim: ZUMA.

Mohr, Hans-Michael. 1986. Dritte beim Interview. Ergebnisse zu Indikatoren aus dem Bereich Ehe und Partnerschaft mit Daten des Wohlfahrtssurvey 1984. *ZA-Information* 2(19): 52–71.

Moreno, Jakob L. 1934. *Who shall survive?* D.C.: NervousandMentalDiseasePubl.: Washington.

Müller, Walter. 1999. Der Mikrozensus als Datenquelle Sozialwissenschaftlicher Forschung. In *Sozialstrukturanalysen mit dem Mikrozensus* ZUMA-Nachrichten Spezial, Bd. 6, Hrsg. Paul Lüttinger, 7–27.

Muhr, Thomas. 1997. *Atlas-ti: the knowledge workbench; visual qualitative data; analysis, management, model building; short user's manual; version 4.1 for Windows 95 and Windows NT/ Scienitfic Software Development.*

Murry Jr., J.W., und J.O. Hammons. 1995. Delphi: A versatile methodology for conducting qualitative resarch. *The Review of Higher Education* 8(4): 424–436.

Namenwirth, J. Zvi. 1986. The wheels of time and the interdependence of cultural change in America. In *Dynamics of culture*, Hrsg. J.Zvi. Namenwirth, R.F. Weber, 57–88. Boston, Mass.: Allen & Unwin.

Narayan, Sowmya, und Jon A. Krosnick. 1996. Education moderates some response effects in attitude measurement. *Public Opinion Quarterly* 60(1): 58–88.

Nederhof, Anton J. 1985. A survey on suicide: Using a mail survey to study a highly threatening topic. *Quality and Quantity* 19(3): 293–302.

Nehnevajsa, Jiri. 1973. Soziometrie. In *Grundlegende Methoden und Techniken der empirischen Sozialforschung. Erster Teil.* Handbuch der empirischen Sozialforschung, Bd. 2, Hrsg. René König, 260–299. Stuttgart: Ferdinand Enke.

Niemi, Iiris. 1983. *The 1979 time use study method.* Helsinki: Central Statistical Office of Finland.

Nießen, Manfred. 1977. *Gruppendiskussion. Interpretative Methodologie – Methodenbegründung – Anwendung*. München: Fink.

Noelle-Neumann, Elisabeth. 1996. *Öffentliche Meinung. Die Entdeckung der Schweigespirale*. Frankfurt am Main Berlin: Ullstein.

Noelle-Neumann, Elisabeth, und Renate Köcher. 1997. *Computer machen nicht einsam und Privatfernsehen bringt keinen Familienstreit*. Allensbacher Jahrbuch der Demoskopie 1993–1997, Bd. 10, 463–467. München: Allenbach.

Noelle-Neumann, Elisabeth, und Thomas Petersen. 1998. *Alle, nicht jeder. Einführung in die Methoden der Demoskopie*. München: Deutscher Taschenbuch Verlag.

Noelle-Neumann, Elisabeth, und Thomas Petersen. 2000. Das halbe Instrument, die halbe Reaktion. Zum Vergleich von Telefon- und Face-to-Face-Umfragen. In *Methoden in Telefonumfragen*, Hrsg. Volker Hüfken, 183–200. Wiesbaden: Westdeutscher Verlag.

Northway, Mary L. 1940. A method for depicting social relations by sociometric testing. *Sociometry* 3(2): 144–150.

Obermann, Christof. 1992. *Assessment Center: Entwicklung*. Wiesbaden: Gabler.

Oevermann, Ulrich, Tilman Allert, und Elisabeth Konau. 1984. Zur Logik der Interpretation von Interviewtexten. In *Interpretation einer Bildungsgeschichte* Hermeneutisch lebensgeschichtliche Forschung, Bd. 2, Hrsg. Thomas Heinze, 7–61. Hagen: Studienbrief der FernUniversität Hagen.

Oevermann, Ulrich, Tilman Allert, Elisabeth Konau, und Jürgen Krambeck. 1979. Die Methodologie einer ‚objektiven Hermeneutik‘ und ihre allgemeine forschungslogische Bedeutung in den Sozialwissenschaften. In *Interpretative Verfahren in den Sozial- und Textwissenschaften*, Hrsg. Hans Georg Soeffner, 352–434. Stuttgart: Metzler.

Oksenberg, Lois, und Charles Cannell et al. 1988. Effects of interviewer vocal characteristics on nonresponse. In *Telephone survey methodology*, Hrsg. Robert M. Groves, 257–269. New York: Wiley.

Oksenberg, Lois, Charels Cannell, und Graham Kalton. 1991. New strategies of pretesting survey questions. *Journal of Official Statistics* 7(3): 349–366.

Ono, Ryota, und Dan J. Wedemeyer. 1994. Assessing the valididy of the Delphi technique. *Futures* 26(3): 289–304.

Opaschowski, Horst W. 1976. *Pädagogik der Freizeit: Grundlegung für Wissenschaft und Praxis*. Bad Heilbrunn: Klinkhardt.

Opp, Karl-Dieter. 1972. *Verhaltenstheoretische Soziologie*. Reinbeck: Rowolth.

Opp, Karl-Dieter. 1999. *Methodologie der Sozialwissenschaften. Einführung in Probleme ihrer Theorienbildung und praktischen Anwendung*. Opladen: Westdeutscher Verlag.

Opp, Karl-Dieter. 2004. *Arten von Problemen, der Beitrag der Politikwissenschaft zur Lösung praktischer Probleme*. http://www.politikon.org/ilias2/.

O'Shea, Ruth, Caroline Bryson, und Roger Jowell. 2003. *European Social Survey, Comparative Attitudinal Research in Europe, European Social Survey Directorate*. London: National Centre for Social Research. http://www.europeansocialsurvey.com/.

Otto, Birgit, und Thomas Siedler. 2003. *Armut in West- und Ostdeutschland – Ein differenzierter Vergleich. In: DIW-Wochenbericht 4/03*.

Patzelt, Werner J. 1986. *Sozialwissenschaftliche Forschungslogik. Einführung*. München Wien: Oldenbourg.

Payne, Stanley L. 1949. Case study in question complexity. *Public Opinion Quarterly* 13(4): 653–658.

Payne, Stanley L. 1951. *The art of asking questions*. Princeton: University Press.

Peez, Georg. 2003. *Kunstpädagogik und Biografie. Ein Seminarbericht und erste Ergebnisse zur Feldforschung nach Merkmalen einer Profession. Vortrag am 28.11.2003b zur Tagung „Wenn Kunstpädagogik Sinn macht". Frauen – Kunst – Pädagogik*. Paderborn: Universität Paderborn. http://www.georgpeez.de/texte/fkp03.htm.

Petermann, Franz. 1996. *Einzelfalldiagnostik in der klinischen Praxis*, 3. Auflage, Weinheim: Psychologie Verlags Union.

Petermann, Werner. 2001. Fotografie- und Filmanalyse. In *Handbuch Qualitative Forschung. Grundlagen, Konzepte, Methoden und Anwendungen*, Hrsg. Uwe Flick, Ernst v. Kardorff, Heiner Keupp, Lutz v. Rosenstiel, Stephan Wolff, 228–232. Weinheim: Beltz.

Petersen, Thomas. 2002. *Das Feldexperiment in der Umfrageforschung*. Frankfurt New York: Campus.

Petty, Richard E., und John T. Cacioppo. 1986. *Communication and persuasion: Central and peripheral routes to attitude change*. New York: Springer.

Pfleiderer, Rolf. 2000. Methodeneffekte beim Umstieg auf CATI-Techniken. In *Neue Erhebungsinstrumente und Methodeneffekte* Schriftenreihe Spektrum Bundesstatistik, Bd. 15, Hrsg. Statistisches Bundesamt, 57–70. Wiesbaden: Metzler Pöschel.

Pollock, Friedrich. 1955. *Gruppenexperiment. Ein Studienbericht*. Frankfurt am Main: Europäische Verlagsanstalt.

Polte, Henry. 2017. *Zwischen Kontaktieren und Ignorieren: Ein Ansatz zur multivariaten Analyse großer sozialer Netzwerke*. Diplomarbeit im Studiengang Diplomsoziologie an der TU Dresden, Philosophische Fakultät, Institut für Soziologie.

Popper, Karl R. 1971. *Logik der Forschung*. Tübingen: Mohr.

Porst, Rolf. 2000. *Praxis der Umfrageforschung*, 2. Auflage, Stuttgart Leipzig Wiesbaden: Teubner.

Porst, Rolf, und Christa von Briel. 1995. *Wären Sie vielleicht bereit, sich gegebenenfalls noch einmal befragen zu lassen? Oder: Die Gründe für die Teilnahme an Panel-Befragungen. ZUMA Arbeitsbericht 04/1995*. Mannheim: ZUMA.

Porst, Rolf, und Klaus Zeifang. 1987. Wie stabil sind Umfragedaten? Beschreibung und erste Ergebnisse der Test-Retest-Studie zum ALLBUS 1984. *ZUMA-Nachrichten* 1(20): 8–31.

Powell, M. Reed. 1951. The nature and extent of group organization in a girls' dormitory. *Sociometry* 14(3): 317–339.

Presser, Stanley, und J. Blair. 1994. Survey pretesting: Do different methods produce different results? In *Sociology Methodology (24)*, Hrsg. P. Marsden, 73–104. Washington DC: American Sociological Association.

Proctor, Charles H., und Charles P. Loomis. 1951. Sociometry. In *Research methods in social relations*, Bd. 2, Hrsg. Marie Jahoda, Morton Deutsch, Stuart W. Cook, 561–584. New York: Holt, Rinehart & Winston.

Prüfer, Peter, und Margit Rexroth. 1985. Zur Anwendung der Interaction-Coding-Technik. *ZUMA-Nachrichten* 2(17): 2–49.

Prüfer, Peter, und Margit Rexroth. 1996a. Verfahren zur Evaluation von Survey Fragen: Ein Überblick. *ZUMA-Nachrichten* 1(39): 95–116.

Prüfer, Peter, und Margit Rexroth. 1996b. *Verfahren zur Evaluation von Survey Fragen: Ein Überblick. ZUMA-Arbeitsbericht 1996/05*. Mannheim: ZUMA.

Prüfer, Peter, und Margit Rexroth. 1999. Zwei-Phasen-Pretesting. In *Querschnitt. Festschrift für Max Kaase*, Hrsg. Peter Ph. Mohler, Paul Lüttinger, 203–219. Mannheim: ZUMA.

Prüfer, Peter, und Margit Rexroth. 2005. *Kognitive Interviews. In: ZUMA How-to Reihe Nr. 15*. http:// www.gesis.org/Publikationen/Berichte/ZUMA_How_to/Dokumente/pdf/How_to15PP_MR.pdf.

Prüfer, Peter, Lisa Vazansky, und Darius Wystup. 2003. *Antwortskalen im ALLBUS und ISSP. Eine Sammlung. ZUMA-Methodenbericht 2003/11*. Mannheim: ZUMA.

Ramge, Hans. 1978. *Alltagsgespräche*. Frankfurt am Main: Diesterweg.

Rammert, Werner. 1994. *Die Technik in der Gesellschaft. Forschungsfelder und theoretische Leitdifferenzen im Deutschland der 90er Jahre. Verbund Sozialwissenschaftliche Technikforschung* Mitteilungen, Bd. 13. Köln: Max-Planck-Institut für Gesellschaftsforschung.

Redfield, Robert, und Alfonso Villa Rojas. 1934. *ChanKom: aMayavillage*. Washington: Carnegie Inst.

Reichertz, Jo. 2019. Empirische Sozialforschung und soziologische Theorie. In *Handbuch der Methoden der empirischen Sozialforschung*, 2. Auflage, Bd. 1, Hrsg. Nina Baur, Jörg Blasius, 31–47, Wiesbaden: Springer VS.

Reichertz, Jo. 2001. Objektive Hermeneutik. In *Handbuch Qualitative Forschung. Grundlagen, Konzepte, Methoden und Anwendungen*, Hrsg. Uwe Flick, Ernst v. Kardorff, Heiner Keupp, Lutz v. Rosenstiel, Stephan Wolff, 223–228. Weinheim: Beltz.

Reinecke, Jost. 1991. *Interviewer- und Befragtenverhalten – Theoretische Ansätze und methodische Konzepte*. Opladen: Westdeutscher Verlag.

Reuband, Karl-Heinz. 1990. Interviews, die keine sind. „Erfolge" und „Misserfolge" beim Fälschen von Interviews. *Kölner Zeitschrift für Soziologie und Sozialpsychologie* 42(4): 706–733.

Reuband, Karl-Heinz. 1992. On third persons in the interview situation and their impact on responses. *International Journal of Public Opinion Research* 4(3): 269–274.

Reuband, Karl-Heinz. 2003. Variationen der Permissivität: Wie Frageformulierungen unterschiedliche Antwortverteilungen erbringen, wenn von ‚Erlauben' oder ‚Verbieten' die Rede ist. *ZA-Information* 2(53): 86–96.

Reutler, Bernd H. 1988. *Körpersprache verstehen*. München: Humboldt-Taschenbuch.

Reutler, Bernd H. 1993. *Körpersprache im Bild. Die unbewussten Botschaften; ihre Merkmale und Deutung auf einen Blick*. Augsburg: Weltbildverlag.

Richey, J.S., B.W. Mar, und R.R. Horner. 1985. The Delphi technique in environmental assesssment. *Journal of Environmental Management* 21(2): 135–159.

Richter, Hans-Jürgen. 1970. *Die Strategie schriftlicher Massenbefragungen: ein verhaltenstheoretischer Beitrag zur Methodenforschung*. Bad Harzburg: Verlag für Wissenschaft, Wirtschaft und Technik.

Riggs, Walter E. 1983. The Delphi technique, an experimental evaluation. *Technological Forecasting and Social Change* 23(1): 89–94.

Ritsert, Jürgen. 1964. *Inhaltsanalyse und Ideologiekritik. Ein Versuch über kritische Sozialforschung*. Frankfurt: Athenäum.

Rogers, Theresa F. 1976. Interview by telephone and in person: Quality of responses and field performance. *Public Opinion Quarterly* 40(1): 51–65. nachgedruckt auch in: Singer/Presser 1989.

Rösch, Günter. 1994. Kriterien der Gewichtung einer nationalen Bevölkerungsstichprobe. In *Gewichtung in der Umfragepraxis*, Hrsg. Siegfried Gabler, Jürgen H.P. Hoffmeyer-Zlotnik, Dagmar Krebs, 7–26. Opladen: Westdeutscher Verlag.

Rosentahl, Gabriele. 1987. *„Wenn alles in Scherben fällt …" Von Leben und Sinnwelt der Kriegsgeneration. Typen biographischer Wandlungen*. Opladen: Leske + Budrich.

Rosenthal, Robert, und Leonore Jacobson. 1968. *Pygmalion in the classroom*. New York: Holt.

Rossi, Peter H., und Howard E. Freeman. 1993. *Evaluation*. Beverly Hills: Sage.

Rössler, Patrick. 2005. *Inhaltsanalyse*. Konstanz: UVK Verlagsgesellschaft.

Roth, Erwin, und Heinz Holling. 1999. *Sozialwissenschaftliche Methoden. Lehr- und Handbuch für Forschung und Praxis*. München Wien: Oldenbourg Verlag.

Rottmann, Verana, und Holger Strohm. 1987. *Was Sie gegen Mikrozensus und Volkszählung tun können. Ein Ratgeber*. Berlin: Zweitausendeins.

Rowe, Gene, George Wright, und Fergus Bolger. 1991. Delphi. A reevaluation of research and theory. *Technological Forecasting and Social Change* 39(3): 235–251.

Roxin, Claus. 1997. *Strafrecht. Allgemeiner Teil I*, 3. Auflage, München: Beck.

Rugg, Donald. 1941. Experiments in wording questions. *Public Opinion Quarterly* 5(1): 91–92.

Sacks, Harvey. 1971. Das Erzählen von Geschichten innerhalb von Unterhaltungen. In *Zur Soziologie der Sprache. Ausgewählte Beiträge zum 7. Weltkongreß der Soziologie* Kölner Zeitschrift für Soziologie und Sozialpsychologie, Sonderheft 15, Hrsg. Rolf Kjolseth, Fritz Sack, 307–317.

Sand, Matthias. 2018. *Gewichtungsverfahren in Dual-Frame-Telefonerhebungen bei Device-Specific Nonresponse*. GESIS Schriftenreihe 20. Köln: GESIS.

Sand, Matthias, und Siegfried Gabler. 2019. Gewichtung von (Dual-Frame-) Telefonstichproben. In *Telefonumfragen in Deutschland*, Hrsg. Sabine Häder, Michael Häder, und Patrick Schmich, 405–424. Wiesbaden: Springer VS.

Saris, E. Willem. 1998. Ten years of interviewing without interviewers: The telepanel. In *Computer-Assisted Survey Information Collection*, Hrsg. Mick P. Couper, R.P. Baker, J. Bethlehem, C.Z. Clark, J. Martin, W.L.I.I. Nicholls, J.M. O'Reilly, 409–431. New York: Wiley.

Schaller, S. 1980. Beobachtungsverfahren in der Verhaltensdiagnostik. In *Methoden der klinisch-psychologischen Diagnostik. Hamburg: Hoffmann & Campe* Handbuch der Klinischen Psychologie, Bd. 1, Hrsg. Werner Wittling, 130–157.

Scharnberg, Torsten, Klaus Wühler, H.-O. Finke, und Rainer Guski. 1982. *Beeinträchtigung des Nachtschlafs durch Lärm*. Berlin: Umweltbundesamt.

Scheffler, Hartmut. 2003. Online-Erhebungen in der Marktforschung. In *Online-Erhebungen 5. Wissenschaftliche Tagung*, Hrsg. ADM Arbeitskreis Deutscher Markt- und Sozialforschungsinstitute e. V., Arbeitsgemeinschaft Sozialwissenschaftlicher Institute e. V. (ASI) und Statistisches Bundesamt, Wiesbaden. Sozialwissenschaftlicher Tagungsband 7. Bonn, 31–41.

Schegloff, Emanuel. 1968. Sequencing in conversational openings. *American Anthropologist* 70(6): 1075–1095.

Scheuch, Erwin K. 1967a/1973. Das Interview in der Sozialforschung. In *Handbuch der empirischen Sozialforschung*, 3. Auflage, Bd. 1, Hrsg. René König, 136–196. Stuttgart: Enke.

Scheuch, Erwin K. 1967b. Auswahlverfahren in der Sozialforschung. In *Handbuch der empirischen Sozialforschung*, 3. Auflage, Bd. 1, Hrsg. René König, 309–347. Stuttgart: Enke.

Scheuch, Erwin K. 1974. Auswahlverfahren in der Sozialforschung. In *Handbuch der Empirischen Sozialforschung*. Bd. 1, Hrsg. René König. Stuttagrt: Enke.

Scheuch, Erwin K., und Hansjörgen Daheim. 1970. Sozialprestige und soziale Schichtung. In *Soziale Schichtung und soziale Mobilität*, 4. Auflage, Kölner Zeitschrift für Soziologie und Sozialpsychologie, Bd. Sonderheft 5, Hrsg. David W. Glass, René König, 65–103.

Schlinzig, Tino, und Götz Schneiderat. 2009. Teilnahmebereitschaft und Teilnahmeverhalten bei Telefonumfragen in der Allgemeinbevölkerung über das Mobilfunknetz. In *Telefonbefragungen über das Mobilfunknetz. Konzept, Design und Umsetzung einer Strategie zur Datenerhebung*, Hrsg. Michael Häder, Sabine Häder, 83–97. Wiesbaden: VS Verlag für Sozialwissenschaften.

Schneid, Michael. 1995. *Disk-By-Mail: Eine Alternative zur schriftlichen Befragung? ZUMA-Arbeitsbericht 1995/02*. Mannheim: ZUMA.

Schneid, Michael. 1997. *Einsatz computergestützter Befragungssysteme in Europa (Eine computerisierte Fax-Umfrage). ZUMA-Arbeitsbericht 1997/01*. Mannheim: ZUMA.

Schneider, Steffen. 1995. *Optimale Randomized-Response-Designs-Techniken mit abhängiger Randomisierung*. Hochschulschrift: Dortmund, Univ., Diss.

Schnell, Rainer. 1991. Der Einfluss gefälschter Interviews auf Survey-Ergebnisse. *Zeitschrift für Soziologie* 20(1): 25–35.

Schnell, Rainer. 1997. *Nonresponse in Bevölkerungsumfragen. Ausmaß, Entwicklung und Ursachen*. Opladen: Leske + Budrich.

Schnell, Rainer. 2001. *Notizen aus der Provinz. Eine Fallstudie zur Methodenforschung in der BRD anlässlich der Publikation „Möglichkeiten und Probleme des Einsatzes postalischer Befragungen" von Karl-Heinz Reuband in der KZfSS 2001, 2*, S. 307–333. www.uni-konstanz.de/FuF/Verwiss/Schnell/NotizenausderProvinz.pdf.

Schnell, Rainer. 2012. *Survey-Interviews. Methoden standardisierter Befragungen*. Wiesbaden: VS Verlag für Sozialwissenschaften.

Schnell, Rainer, Paul B. Hill, und Elke Esser. 2013. *Methoden der empirischen Sozialforschung*. München Wien: Oldenbourg. weitere Auflage 2013.

Schneller, Johannes. 1997. Stichprobenbildung nach dem repräsentativen Quoten-Verfahren. In *Stichproben in der Umfragepraxis*, Hrsg. Siegfried Gabler, Jürgen H.P. Hoffmeyer-Zlotnik, 5–18. Opladen: Westdeutscher Verlag.

Schnepper, Markus. 2004. *Robert K. Mertons Theorie der self-fulfilling prophecy. Adaption eines soziologischen Klassikers*. Europäische Hochschulschriften: Reihe 22, Soziologie, Bd. 392. Frankfurt am Main: Peter Lang Internationaler Verlag der Wissenschaften.

Schreiber, Dieter. 1975. Skalierungsprobleme. In *Der sozialwissenschaftliche Forschungsprozess*, Hrsg. Walter Friedrich, Werner Hennig, 277–334. Berlin: Deutscher Verlag der Wissenschaften.

Schröer, Norbert, und Jo. Reichertz. 1992. *Polizei vor Ort. Studien zur empirischen Polizeiforschung*. Stuttgart: Enke.

Schroeder, Friedrich-Christian. 1972. Die Notwehr als Indikator politischer Grundanschauungen. In *Festschrift für Reinhart Maurach zum 70. Geburtstag*, Hrsg. Friedrich-Christian Schroeder, Heinz Zipf, 127–142. Karlsruhe: Müller.

Schulz, Winfried (Hrsg.). 1970. *Kausalität und Experiment in den Sozialwissenschaften. Methodologie und Forschungstechnik*. Mainz: Hase & Koehler.

Schulz, Winfried. 2003. Inhaltsanalyse. In *Fischer Lexikon Publizistik/Massenkommunikation*, Hrsg. Elisabeth Noelle-Neumann, Winfried Schulz, Jürgen Wilke, 41–63. Frankfurt: Fischer.

Schulze, Ingo. 1998. Simple Storys. *Ein Roman aus der ostdeutschen Provinz*. Deutscher Taschenbuch Verlag.

Schuman, Howard. 1966. The random probe: A technique for evaluating the validity of closed questions. *American Sociological Review* 31(2): 218–222.

Schuman, Howard. 1992. Context effects: State of the past/State of the art. In *Context effects in social and psychological research*, Hrsg. Norbert Schwarz, Seymour Sudman, 5–20. New York: Springer.

Schuman, Howard, und Stenley Presser. 1981. *Questions and answers in attitude surveys*. New York: Academic Press.

Schütze, Fritz. 1977. *Die Technik des narrativen Interviews in Interaktionsfeldstudien – dargestellt an einem Projekt zur Erforschung von kommunalen Machtstrukturen*. Mimeo: Universität Bielefeld.

Schütze, Fritz. 1983. Biographieforschung und narratives Interview. *Neue Praxis* 13(3): 283–293.

Schütze, Fritz. 1987. *Das narrative Interview in Interaktionsfeldstudien*. Hagen: Fernuniversität Hagen.

Schupp, Jürgen. 2014. Paneldaten für die Sozialforschung. In *Handbuch der Methoden der empirischen Sozialforschung*, 2. Auflage, Bd. 2, Hrsg. Nina Baur, Jörg Blasius, 1265–1277, Wiesbaden: Springer VS.

Schwarz, Norbert, und Herbert Bless. 1992. Assimilation and contrast effects in attitude measurement. An inclusion/exclusion model. *Advances in Consumer Research* 19: 72–77.

Schwarz, Norbert, und Fritz Strack. 1991. Context effects in attitude surveys: Applying cognitive theory to social research. *European Review of Social Psychology* 2: 31–50.

Schwarz, Norbert, C.E. Gryson, und Bärbel Knäuper. 1998. Formal features of rating scales and the interpretation of question meaning. *International Journal of Public Opinion Research* 10(2): 177–183.

Schwarz, Norbert, Hans-Jürgen Hippler, Brigitte Deutsch, und Fritz Strack. 1985. Response scales: Effects of category range on reported behaviour and comparative judgements. *Public Opinion Quarterly* 49(3): 388–395.

von Schweitzer, Rosemarie. 1990. Einführung in die Themenstellung. In *Zeitbudgeterhebungen. Ziele, Methoden und neue Konzepte*, Hrsg. Statistisches Bundesamt, 9–22. Stuttgart: Metzler-Poeschel.

Schweizer, Thomas, und Michael Schnegg. 1998. *Die soziale Struktur der ‚Simple Storys': Eine Netzwerkanalyse*. http://www.uni-koeln.de/phil-fak/voelkerkunde/doc/simple.html. Accessed on 04.05.2009.

Seeger, Thomas. 1979. *Die Delphi-Methode. Expertenbefragungen zwischen Prognose und Gruppenmeinungsbildungsprozessen; überprüft am Beispiel von Delphi-Befragungen im Gegenstandsbereich Information und Dokumentation.* Freiburg: HochschulVerlag. Dissertation.

Selting, Margret, und Peter Auer et al, 1998. Gesprächsanalytisches Transkriptionssystem. *Linguistische Berichte* 173: 91–122.

Sibberns, Heiko, und Jürgen Baumert. 2001. Anhang A: Stichprobenziehung und Stichprobengewichtung. In *PISA 2000. Basiskompetenzen von Schülerinnen und Schülern im internationalen Vergleich,* Hrsg. Jürgen Baumert, Eckhard Klieme, Michael Neubrand, Manfred Prenzel, Ulrich Schiefele, Wolfgang Schneider, Petra Stanat, Klaus-Jürgen Tillmann, Manfred Weiß, 511–524. Opladen: Leske + Budrich.

Sievers, Helga. 1994. Zeitbewusstsein, Handlungsintensionen und Eigenverantwortung. Eine Analyse der Zusammenhänge zwischen handlungsbestimmenden Orientierungen. *ZUMA-Nachrichten* 1(42): 144–167.

Silbermann, Alphons. 1974. Systematische Inhaltsanalyse. In *Handbuch der empirischen Sozialforschung,* Bd. 4, Hrsg. René König, 253–293. Stuttgart: Enke.

Silver, Brian D., Paul R. Abramson, und Barbara A. Anderson. 1986. The presence of others and overreporting of voting in American national elections. *Public Opinion Quarterly* 50(2): 228–239.

Singer, Eleonore, und Stanley Presser (Hrsg.). 1998. *Survey research methods. A reader.* Chicago: University of Chicago Press.

Sirken, Monroe G., und S. Schechter. 1999. Interdisciplinary survey methods research. In *Cognition and survey research,* Hrsg. Monroe G. Sirken, Herrmann J. Douglas, Susan Schechter, Norbert Schwarz, Judith M. Tanur, Roger Tourangeau, 1–10. New York: Wiley.

Soeffner, Hans-Georg. 1979. *Interpretative Verfahren in der Sozial- und Textwissenschaften.* Stuttgart: Metzler.

Soeffner, Hans-Georg. 1989. *Auslagung des Alltags – Der Alltag der Auslegung.* Frankfurt am Main: Suhrkamp.

Sontag, Susan. 1978. *Über Fotografie.* München Wien: Carl Hanser Verlag.

Spöhring, Walter. 1989. *Qualitative Sozialforschung.* Stuttgart: Teubner.

Statistica Manuel. 2005. https://documentation.statsoft.com/STATISTICAHelp.aspx?path=Common/CommonMenus/HelpQuickReferenceGuide

Stadtler, Klaus. 1983. *Die Skalierung in der empirischen Forschung.* München: Infratest.

Steinhäuser, Martin. 2002. *Gemeindliche Arbeit mit Kindern begleiten. Empirische Studien zur Entwicklung der Aufgaben und Strukturengemeindepädagogischer Fachaufsicht.* Münster: Lit.

Stember, Herbert, und Herbert Hyman. 1949. Interviewer Effects in the Classification of Responses. *Public Opinion Quarterly* 12(4): 669–682.

Stevens, Stanley S. 1959. Measurement, psychophysics and utility. In *Measurement: definitions and theories,* Hrsg. Charles West Curchman, Philburn Ratoosh, 18–63. New York: Wiley.

Stocké, Volker. 2002. Die Vorhersage von Fragereihenfolgeeffekten durch Antwortlatenzen: Eine Validierungsstudie. *ZUMA-Nachrichten* 1(50): 26–53.

Stocké, Volker. 2004. Entstehungsbedingungen von Antwortverzerrungen durch soziale Erwünschtheit. Ein Vergleich der Prognosen der Rational-Choice Theorie und des Modells der Frame-Selektion. *Zeitschrift für Soziologie* 33: 303–320.

Stocké, Volker, und Birgit Becker. 2004. Determinanten und Konsequenzen der Umfrageeinstellung. Bewertungsdimensionen unterschiedlicher Umfragesponsoren und die Antwortbereitschaft der Befragten. *ZUMA-Nachrichten* 1(54): 89–116.

Stoop, Ineke A.L. 2005. *The hunt for the last respondent. Nonresponse in sample surveys.* Social and Cultural Planning Office of the Netherlands.

Strack, Fritz. 1992. Order effects in survey research: Activation and information functions of preceeding questions. In *Context effects in social and psychological research,* Hrsg. Norbert Schwarz, Seymour Sudman, 23–34. New York: Springer.

Strack, Fritz. 1994. *Zur Psychologie der standardisierten Befragung. Kognitive und kommunikative Prozesse.* Berlin: Springer.

Strack, Fritz, und Leonard L. Martin. 1988. Thinking, judging, and communicating: A process account of context effects in attitude surveys. In *Social information processing and survey methodology*, 2. Auflage, Hrsg. Hans-Jürgen Hippler, 123–148. New York: Springer.

Strauss, Anselm L. 1987. *Qualitative analysis for social scientists.* Cambridge: Cambridge University Press.

Strauss, Anselm L. 1994. *Grundlagen qualitativer Sozialforschung.* München: Fink.

Strauss, Anselm L., und Juliet Corbin. 1996. *Grounded Theory. Grundlagen Qualitativer Sozialforschung.* Weinheim: Psychologie Verlags Union.

Stroebe, Wolfgang, Klaus Jonas, und Miles Hewstone. 2002. *Sozialpsychologie: Eine Einführung*, 4. Auflage, Berlin: Springer.

Struck, Eckart, und Helmut Kromrey. 2002. *PC-Tutor Empirische Sozialforschung. Version 2.0.* Opladen: Leske + Budrich.

Strübin, Jörg. 2019. Grounded Theory und Theoretical Sampling. In *Handbuch der Methoden der empirischen Sozialforschung*, 2. Auflage, Bd. 1, Hrsg. Nina Baur, Jörg Blasius, 525–544, Wiesbaden: Springer VS.

Sudman, Seymour, und Norman M. Bradburn. 1982. *Asking questions: A practical guide to questionnaire design.* San Francisco: Jossey-Bass.

Sudman, Seymour, Norman M. Bradburn, und Norbert Schwarz. 1996. *Thinking about answers. The application of cognitive processes to survey methodology.* San Francisco: Jossey-Bass.

Suppes, Patrick 1957. *Introduction to logic.* Toronto Princton London: VanNostrand.

Tack, Werner H. 1980. Einzelfallstudien in der Psychotherapieforschung. In *Klinische Psychologie in Forschung und Praxis Handbuch der klinischen Psychologie*, Bd. 6, Hrsg. Werner Wittling, 42–71. Hamburg: Hoffmann & Campe.

Tenorth, Heinz-Elmar, und Christian Lüders. 1994. Methoden erziehungswissenschaftlicher Forschung 1: Hermeneutische Methoden. In *Erziehungswissenschaft. Ein Grundkurs*, Hrsg. Dieter Lenzen, 519–542. Reinbek: Rowohlt.

Terwey, Michael. 1998. Analysen zur Verbreitung von ALLBUS. *ZA-Information* 1(42): 44–52.

Thierau, Heike, und Heinrich Wottawa. 1998. *Lehrbuch Evaluation.* Bern: Huber.

Thimm, Caja, und Patrick Nehls. 2019. Digitale Methoden im Überblick. In *Handbuch der Methoden der empirischen Sozialforschung*, 2. Auflage, Bd. 2, Hrsg. Nina Baur, Jörg Blasius, 973–990, Wiesbaden: Springer VS.

Thoma, Michaela, und Matthias Zimmermann. 1996. Zum Einfluss der Befragungstechnik auf den Rücklauf bei schriftlichen Umfragen. Experimentelle Befunde zur ‚Total-Design-Methode'. *ZUMA-Nachrichten* 2(39): 141–157.

Thomas, William I., und Florian Znaniecki. 1918. *The Polish peasant in Europe and America.* Chicago: University of Chicago Press.

Thorpe, J.G. 1955. An investigation of some correlates of sociometric status within school classes. *Sociometry* 18(1): 49–61.

Thurston, Louis L., und Ernest J. Chave. 1929. *The measurement of attitude.* Chicago: University of Chicago Press.

Tourangeau, Roger, Lance J. Rips, und Kenneth A. Rasinski. 2000. *The psychology of survey response.* Cambridge: University Press.

Trappmann, Mark, Hans J. Hummell, und Wolfgang Sodeur. 2005. *Strukturanalyse sozialer Netzwerke: Konzepte, Modelle, Methoden.* Wiesbaden: VS Verlag für Sozialwissenschaften.

Trübner, Miriam. 2019. Zeitverwendungsdaten. In *Handbuch der Methoden der empirischen Sozialforschung*, 2. Auflage, Bd. 2, Hrsg. Nina Baur, Jörg Blasius, 1233–1240, Wiesbaden: Springer VS.

Trübner, Miriam, und Andreas Mühlichen. 2019. Big Data. In *Handbuch der Methoden der empirischen Sozialforschung*, 2. Auflage, Bd. 1, Hrsg. Nina Baur, Jörg Blasius, 143–158, Wiesbaden: Springer VS.

Tuma, René, und Bernt Schnettler. 2019. Videographie. In *Handbuch der Methoden der empirischen Sozialforschung*, 2. Auflage, Bd. 2, Hrsg. Nina Baur, Jörg Blasius, 1191–1202, Wiesbaden: Springer VS.

Urdan, James, und Gardner Lindzey. 1954. Personality and social choice. *Sociometry* 17(1): 47–63.

Vetter, Berndt. 1975. Das Forschungsproblem. In *Der sozialwissenschaftliche Forschungsprozess*, Hrsg. Walter Friedrich, Werner Hennig, 151–169. Berlin: Verlag der Wissenschaften.

Wagner-Schelewsky, Pia, und Linda Hering. 2019. Online-Befragung. In *Handbuch der Methoden der empirischen Sozialforschung*, 2. Auflage, Bd. 2, Hrsg. Nina Baur, Jörg Blasius, 787–800, Wiesbaden: Springer VS.

Walker, T. Jeffry. 1994. Fax machines and social surveys: Teaching an old dog new tricks. *Journal of Quantitative Criminology* 10(2): 181–188.

Wallbott, Harald G. 2001. Analyse der Körpersprache. In *Handbuch Qualitative Forschung. Grundlagen, Konzepte, Methoden und Anwendungen*, Hrsg. Uwe Flick, Ernst v. Kardorff, Heiner Keupp, Lutz v. Rosenstiel, Stephan Wolff, 232–236. Weinheim: Beltz.

Wardlow, Mary E., und James E. Greene. 1952. An exploratory sociometric study of peer status among adolescent girls. *Sociometry* 15(3/4): 311–318.

Warner, Stanley L. 1965. Randomized response: a survey technique for eliminating evasive answer bias. *Journal of the American Statistical Association* 60(309): 63–66.

Wasmer, Martina, Michael Blohm, Jessica Walter, Regina Jutz, und Evi Scholz. 2017. Konzeption und Durchführung der „Allgemeinen Bevölkerungsumfrage der Sozialwissenschaften" (ALLBUS) 2014. *GESIS Papers* 2017|20.

Waterplas, Lina, Jean Billet, und Geert Loosveldt. 1988. De verbieden versus niet toelaten asymmetrie. Een stabiele formuleringseffect in survey-onderzoek?. *Mens en Maatschappij* 63: 399–417.

Watzlawick, Paul. 1976. *Wie wirklich ist die Wirklichkeit?* München: Pieper.

Waxweiler, Richard. 1980. *Psychotherapie im Strafvollzug. Eine empirische Erfolgsuntersuchung am Beispiel der sozialtherapeutischen Abteilung in einer Justizvollzugsanstalt*. Weinheim: Beltz.

Webb, Eugene T., Donald T. Campbell, Richard D. Schwartz, Lee Sechrest, und Janet Belew Grove. 1966. *Nonreactive measures in the social sciences*. Boston: Houghton Mifflin.

Weber, Max. 1951. Gesammelte Aufsätze zur Wissenschaftslehre. In *Gesammelte Werke*. Hrsg. Johannes Winckelmann Tübingen: Mohr.

Weber, Robert Philip. 1986. The long-term dynamics of cultural problem solving. In *Dynamics of culture*, Hrsg. J. Zvi. Namenwirth, Robert Philip Weber, 57–88. Boston: Allen & Unwin.

Weiß, Bernd, Henning Silber, Bella Struminskaya, und Gabiele Durrant. 2019. Mobilde Befragungen. In *Handbuch der Methoden der empirischen Sozialforschung*, 2. Auflage, Bd. 2, Hrsg. Nina Baur, Jörg Blasius, 801–812, Wiesbaden: Springer VS.

Whyte, William Foote. 1951. Observational field work methods. In *Research methods in social relations*, Bd. 2, Hrsg. Marie Jahoda, Morton Deutsch, Stuart W. Cook, 492–513. New York: Dryden.

Whyte, William Foote. 1967. *Street corner society – The social structure of an Italian slum*. Chicago: University Chicago Press.

Wiedenbeck, Michael, und Cornelia Züll. 2001. *Klassifikation mit Clusteranalyse: grundlegende Techniken hierarchischer k-means-Verfahren*. ZUMA How-to-Reihe Nr. 10. http://www.gesis. org/fileadmin/upload/forschung/publikationen/gesis_reihen/howto/how-to10mwcz.pdf.

Wiegand, Erich. 1998. Telefonische Befragungen: Datenschutz und Ethik. In *Telefonstichproben in Deutschland*, Hrsg. Siegfried Gabler, Sabine Häder, Jürgen H.P. Hoffmeyer-Zlotnik, 19–29. Opladen: Westdeutscher Verlag.

Wiegand, Erich. 2000. Chancen und Risiken neuer Erhebungstechniken in der Umfrageforschung. In *Neue Erhebungsinstrumente und Methodeneffekte* Schriftenreihe Spektrum Bundesstatistik, Bd. 15, Hrsg. Statistisches Bundesamt, 12–21. Wiesbaden: Metzler.

Wiegand, Erich. 2003. Qualitätsstandards und Standesregeln web-basierter Datenerhebungen. In *Online-Erhebungen 5. Wissenschaftliche Tagung*, Hrsg. ADM Arbeitskreis Deutscher Markt- und Sozialforschungsinstitute e.V., Arbeitsgemeinschaft Sozialwissenschaftlicher Institute e.V. (ASI) und Statistisches Bundesamt, Wiesbaden. Sozialwissenschaftlicher Tagungsband 7. Bonn, 61–70.

Wieseman, Frederick. 1972. Methodological bias in public opinion surveys. *Public Opinion Quarterly* 36(1): 105–108.

Windelbrand, Wilhelm. 1894. Geschichte und Naturwissenschaft. In *Präludien, Aufsätze und Reden zur Philosophie und ihrer Geschichte*, Bd. 2, Hrsg. E. Windelbrand Tübingen: Mohr.

Wintzenberg, J.B. 2004. Wie geht es Deutschland? *Stern*, Nr. 18 vom 22. April, 46–62.

Wirth, Heike. 1992. Die faktische Anonymität von Mikrodaten: Ergebnisse und Konsequenzen eines Forschungsprojekts. *ZUMA-Nachrichten* 1(30): 7–65.

Wittemann, Frank. 1997. *Grundlinien und Grenzen der Notwehr in Europa. Europäische Hochschulschriften*. Frankfurt am Main: Lang.

Wittenberg, Reinhard. 1998. *Grundlagen computerunterstützter Datenanalyse*. Stuttgart: Lucius & Lucius.

Wittenberg, Reinhard, und Hans Cramer. 2000. *Datenanalyse mit SPSS für Windows*, 2. Auflage, Stuttgart: Lucius & Lucius.

Witzel, Andreas. 1992. Das problemzentrierte Interview. In *Qualitative Forschung in der Psychologie. Grundlagen, Verfahrensweisen, Anwendungsfelder*, Hrsg. Gerd Jüttemann, 227–255. Weinheim: Beltz.

Wolf, Christof. 2004. Egozentrierte Netzwerke. Erhebungsverfahren und Datenqualität. In *Methoden der Sozialforschung* Kölner Zeitschrift für Soziologie und Sozialpsychologie, Sonderheft 44, Hrsg. Andreas Diekmann, 244–273. Wiesbaden: VS Verlag für Sozialwissenschaften.

Wottawa, Heinrich, und Heike Thierau. 1998. *Lehrbuch Evaluation*, 2. Auflage, Göttingen: Huber.

Wüst, Andreas M. 1998. *Die Allgemeine Bevölkerungsumfrage der Sozialwissenschaften als Telefonumfrage. ZUMA-Arbeitsbericht 1998/04*. Mannheim: ZUMA.

Zapf, Dieter. 1989. *Selbst- und Fremdbeobachtung in der psychologischen Arbeitsanalyse: Methodische Probleme bei der Erfassung von Stress am Arbeitsplatz*. Göttingen: Hogrefe.

Zeifang, Klaus. 1987a. *Die Test-Retest-Studie zum ALLBUS 1984 – Tabellenband. ZUMA-Arbeitsbericht 1987/01*. Mannheim: ZUMA.

Zeifang, Klaus. 1987b. *Die Test-Retest-Studie zum ALLBUS 1984 – Abschlussbericht. ZUMA-Arbeitsbericht 1987/02*. Mannheim: ZUMA.

Zeisel, Hans. 1982. Disagreement over the evaluation of a controlled experiment. *American Journal of Sociology* 88(2): 378–389.

van Zelst, Raymond H. 1952. Validation of a sociometric regrouping procedure. *Journal of Abnormal and Social Psychology* 47(2): 299–301.

Zimmermann, Ekkart. 1972. *Das Experiment in den Sozialwissenschaften*. Stuttgart: Teubner.

Zins, Stefan. 2018. *Geht die Ära der Zufallsstichproben ihrem Ende entgegen?* Marktforschung Das Portal für Markt-, Medien- und Meinungsforschung. https://www.marktforschung.de/dossiers/themendossiers/repraesentativitaet-und-zufallsstichprobe/dossier/geht-die-aera-der-zufallsstichproben-ihrem-ende-entgegen/.

Zöfel, Peter. 2003. *Statistik für Psychologen*. München: Pearson Studium.

Name Index

Subject Index

A

Access panels, 14, 70, 105, 150–153, 159, 164, 221, 263, 265, 268, 270, 271
Address listing, 138–140
ADM design, 135–143, 146–148, 153, 168, 221
Advantages and limitations of group discussions, 251
ALLBUS, 71, 80, 81, 84, 85, 96, 97, 117, 142–144, 158, 163, 169, 195–198, 201, 208, 209, 211, 216, 217, 220, 232, 233, 259, 265, 272–274, 424, 425
All statements, 36, 39–41, 48, 91
Alpha error, 131, 394
Alternative hypothesis, 47, 56, 394
Anonymity, 118–124, 176, 178, 180, 195, 196, 221, 222, 224, 228, 230, 233, 239, 252, 304, 328, 347

B

Behavioural traces, 111, 113, 114, 284
Behaviour coding, 360, 362–363
Beta error, 394
Big Data, 117, 118
Body language, 414–416
Body language analysis, 415–416
Boxplots, 388, 389, 397

C

Case study, 245, 323–326, 370
CHAID analysis, 398–403
Chernoff faces, 390–393
Chi-square test, 396
Chi-Square-Automatic-Interaction-Detector (CHAID), 398
Classification, 31, 41, 64, 85, 86, 125–135, 173–179, 238, 242, 273, 277, 281, 288, 300–303, 306, 356, 409, 411
Coding schemes, 64, 70, 102, 290, 303, 306, 356, 362
Cohort effects, 108
Collective hypotheses, 40
Comparison of mean values, 397, 398
Conducting group discussions, 252
Confidence intervals, 130, 132–135, 148, 154, 157, 161, 344, 383, 399, 419
Confidentiality, 118–124, 224
Consistency analysis, 95, 99
Construct validity, 99, 100
Content analysis
    computer-assisted, 306, 308
    qualitative, 298, 300, 310, 325, 412
Content validity, 99, 100
Context effects, 184, 188, 189
Contextual hypotheses, 40
Contingency analyses, 302, 303
Contrast effects, 189
Conversational analysis, ethnomethodological, 246, 412
Correlations, 36, 43, 47, 58, 74, 78, 88–91, 95, 100–102, 108, 134, 148, 154, 164, 165, 170, 215, 258, 297, 350, 352, 390, 392, 394–396, 404, 406, 407
Correspondence rules, 26, 42, 43
Criterion of scientific observations, 279